BODILY FLUIDS IN ANTIQUITY

From ancient Egypt to Imperial Rome, from Greek medicine to early Christianity, this volume examines how human bodily fluids influenced ideas about gender, sexuality, politics, emotions, and morality, and how those ideas shaped later European thought.

Comprising 24 chapters across seven key themes—language, gender, eroticism, nutrition, dissolution, death, and afterlife—this volume investigates bodily fluids in the context of the current sensory turn. It asks fundamental questions about physicality and fluidity: how were bodily fluids categorised and differentiated? How were fluids trapped inside the body perceived, and how did this perception alter when those fluids were externalised? Do ancient approaches complement or challenge our modern sensibilities about bodily fluids? How were religious practices influenced by attitudes towards bodily fluids, and how did religious authorities attempt to regulate or restrict their appearance? Why were some fluids taboo and others cherished? In what ways were bodily fluids gendered? Offering a range of scholarly approaches and voices, this volume explores how ideas about the body and the fluids it contained and externalised are culturally conditioned and ideologically determined. The analysis encompasses the key geographic centres of the ancient Mediterranean basin, including Greece, Rome, Byzantium, and Egypt. By taking a *longue durée* perspective across a richly intertwined set of territories, this collection is the first to provide a comprehensive, wide-ranging study of bodily fluids in the ancient world.

Bodily Fluids in Antiquity will be of particular interest to academic readers working in the fields of classics and its reception, archaeology, anthropology, and ancient to Early Modern history. It will also appeal to more general readers with an interest in the history of the body and history of medicine.

Mark Bradley is Professor of Classics and Associate Pro-Vice-Chancellor at the University of Nottingham, UK. Together with Shane Butler (Johns Hopkins University, USA), he is editor of a series of volumes on 'The Senses in Antiquity' for Routledge, for which he has contributed a volume on *Smell and the Ancient Senses* (2015).

Victoria Leonard is a Research Fellow at the Centre for Arts, Memory and Communities at Coventry University, UK, and at the Institute of Classical Studies, University of London, UK. Her research focuses on the late antique and early medieval western Mediterranean. She has published on religious conflict, gender and violence, and ancient historiography.

Laurence Totelin is Reader in Ancient History at Cardiff University, UK. She has published widely on Greek and Roman botany, pharmacology, and gynaecology.

BODILY FLUIDS IN ANTIQUITY

Edited by Mark Bradley, Victoria Leonard, and Laurence Totelin

Routledge
Taylor & Francis Group

LONDON AND NEW YORK

First published 2021
by Routledge
2 Park Square, Milton Park, Abingdon, Oxon OX14 4RN

and by Routledge
52 Vanderbilt Avenue, New York, NY 10017

Routledge is an imprint of the Taylor & Francis Group, an informa business

British Library Cataloguing-in-Publication Data
A catalogue record for this book is available from the British Library

Library of Congress Cataloging-in-Publication Data
Names: Bradley, Mark, 1977– editor. | Leonard, Victoria, editor. | Totelin, Laurence M. V., editor.
Title: Bodily fluids in antiquity / edited by Mark Bradley, Victoria Leonard, and Laurence Totelin.
Description: Abingdon, Oxon ; New York, NY : Routledge, 2021. | Includes bibliographical references and index.
Identifiers: LCCN 2020051403 (print) | LCCN 2020051404 (ebook) | ISBN 9781138343726 (hardback) | ISBN 9780367764067 (paperback) | ISBN 9780429438974 (ebook)
Subjects: LCSH: Body fluids—History—To 1500. | Civilization, Classical. | Civilization, Western—Classical influences.
Classification: LCC QP90.5 .B34 2021 (print) | LCC QP90.5 (ebook) | DDC 612/.01522—dc23
LC record available at https://lccn.loc.gov/2020051403
LC ebook record available at https://lccn.loc.gov/2020051404

ISBN: 978-1-138-34372-6 (hbk)
ISBN: 978-0-367-76406-7 (pbk)
ISBN: 978-0-429-43897-4 (ebk)

DOI: 10.4324/9780429438974

This book is dedicated to our partners and children,
Richard, Philip, Thomas, Llewellyn, and Gwilym.

CONTENTS

CONTENTS

CONTENTS

FIGURES

TABLES

ACKNOWLEDGEMENTS

Laurence Totelin, Victoria Leonard, and Mark Bradley would like to thank all participants in the original conference 'Bodily Fluids/Fluid Bodies in Greek and Roman Antiquity', which was held at St Michael's College, Cardiff University, 11–13 July 2016. They helped to shape discussion and ensured that the conference was intellectually stimulating and thoroughly enjoyable. We would especially like to thank those who presented their research but did not contribute to the proceeding volume, including: Deborah Lyons, Rosie Jackson, Laura Mareri, Heather Hunter-Crawley, Colin Webster, Christiaan Caspers, Emilio Capettini, Blossom Stefaniw, Jan Stenger, Dawn LaValle Norman, Susanne Turner, Calloway Brewster-Scott, and Leyla Ozbek, in addition to all the other participants in the conference. We would also like to thank St Michael's College for hosting the conference and Rachel Roberts for photographing the event. Without the generous support of the Wellcome Trust, Cardiff University, the Institute of Classical Studies, the Classical Association, and the University of Bristol, the conference would not have been possible, and we would not have been able to provide bursaries to enable the participation of postgraduates and early career academics. The conference website is still available, and includes the original programme and photographs of the event. Please see: https://bodilyfluids.wordpress.com/ (Accessed February 2021). Thanks especially to Gary Fisher for his careful and expert copy-editing, and to Jane Draycott and Kristi Upson-Saia for their guidance in the final stages.

Figure 0.1 All participants of the conference 'Bodily Fluids/Fluid Bodies in Greek and Roman Antiquity', 11 July 2016, St Michael's College, Cardiff University.

Source: © The Editors.

CONTRIBUTORS

Mark Bradley is Professor of Classics and Associate Pro-Vice-Chancellor at the University of Nottingham, UK. He is author of *Colour and Meaning in Ancient Rome* (2009) and was editor of *Papers of the British School at Rome* (2011–17). Together with Shane Butler (Johns Hopkins University, USA), he is editor of a series of volumes on 'The Senses in Antiquity' for Routledge, for which he has contributed a volume on *Smell and the Ancient Senses* (2015). He has also edited a volume on classics and the British Empire (2010) and a volume on ideas about dirt, pollution, and corporeality in the ancient and modern city of Rome (2012).

Jane Burkowski, having earned a doctorate in classical languages and literature at Oriel College, Oxford, UK, in 2013, now works as a freelance proofreader, copyeditor, and translator. As an independent researcher, her interests include Latin poetry (especially love elegy), Roman social history, and the interaction between the two. Her past publications include a chapter on the figure of Apollo in Tibullus 2.3 and 2.5, in *Augustan Poetry and the Irrational* (2016), edited by Philip Hardie.

Claude-Emmanuelle Centlivres Challet is Scientific Collaborator at the Institut d'Archéologie et des Sciences de l'Antiquité at the University of Lausanne, Switzerland. Her current research interests focus on gender and couple relationships. She has published on these subjects as well as on Roman breastfeeding, breast pumps, infant feeding, family, Pliny, and Juvenal. She is the editor of *Married Life in Greco-Roman Antiquity* (forthcoming 2021, Routledge).

Amy Coker held teaching and research positions in classics and ancient history at various institutions in the United Kingdom between 2010 and 2017, including a Leverhulme Early Career Fellowship at the University of Manchester (2013–16) to work on the sexual and scatological vocabulary of Classical Greek. She is now Honorary Research Fellow at the University of Bristol, UK, and teaches classics at Cheltenham Ladies' College, UK. Her current work focuses on offensive language and the language of the human body in Greek. She has published in the fields of historical linguistics, pragmatics, and classics, and her most recent pieces are on the treatment of obscene language in the most well-known lexicon of Ancient Greek, Liddell and Scott (2019), and on a filthy joke told by Cleopatra involving a ladle (2019).

Tasha Dobbin-Bennett is Assistant Professor of Art History at Emory University: Oxford College, USA. Her primary research examines the ancient Egyptian conceptions of decomposition and putrefaction through an interdisciplinary study applying forensic anthropological research to the ancient Egyptian medical and religious texts. Having spent a number of years on archaeological excavations in Egypt, she is interested in how Egypt's unique climate contributed to the ancient Egyptian perception of the post-mortem human body.

Rebecca Fallas is an early career researcher and holds a Visiting Research Fellowship in Classical Studies at The Open University, UK, and previously in History and Philosophy of Science at the University of Leeds, UK. Her research focuses on health and reproduction in the ancient world. She is currently working on her first monograph, *Experiencing Infertility in Ancient Greece*, based on her PhD thesis.

Rebecca Flemming is Senior Lecturer in Ancient History in the Classics Faculty, and a Fellow of Jesus College, at the University of Cambridge, UK. Her research focuses on the society and culture of the Roman Empire, and she has published widely on classical medicine, gender and sexuality, both together and separately. Her book *Medicine and the Making of Roman Women: Gender, Nature and Authority from Celsus to Galen* came out in 2000, and the volume she co-edited with Nick Hopwood and Lauren Kassell, *Reproduction: Antiquity to the Present Day*, was published in 2018. She is currently writing a book on medicine and empire in the Roman world.

Andreas Gavrielatos is currently teaching at the University of Reading, UK, and he has previously taught at the University of Edinburgh, UK, and the Open University of Cyprus, Greece. His research revolves around Roman satire and in particular Persius, and he is currently preparing a commentary on his satires alongside a series of articles. He has also published on Roman onomastics and cultural and linguistic contacts in the Roman world. He is the editor of volumes on Roman identities, ancient Cyprus, and Roman multiculturalism.

Michael Goyette is Instructor of Classics at Eckerd College in St Petersburg, Florida, USA. His research and teaching interests include ancient medicine and medical humanities, ancient science, gender, tragedy, reception, and pedagogy. With each of these interests, he frequently engages with representations of bodies and their implications for lived experiences. A major goal of both his research and teaching is to demonstrate how the humanities and the sciences complement and speak to each other in multiple time periods, including the present. He is currently working on a book comparing discourses of illness in Senecan tragedy with Latin medical prose.

Rosalind Janssen is Honorary Lecturer in Education at UCL's Institute of Education, UK. She was previously a curator in UCL's Petrie Museum and then a lecturer in Egyptology at UCL's Institute of Archaeology. She currently teaches Egyptology classes at both Oxford University, UK, and London's City Lit. Rosalind has published a series of articles on the ancient Egyptian village of Deir el-Medina, covering aspects such as juvenile hooliganism and the position of older women within the community. She and her Egyptologist husband Jac. Janssen were the authors of *Growing Up and Getting Old in Ancient Egypt* (2007).

Emily Kearns is Lecturer and Senior Research Fellow at St Hilda's College, Oxford, UK. She is the author of *The Heroes of Attica* (1989) and *Ancient Greek Religion:*

A Sourcebook (2010), as well as articles on a wide range of subjects in Greek religion and classical literature. She is currently completing a commentary on Euripides' *Iphigenia in Tauris*, and her next project is a monograph on blood in the classical world.

Peter Kelly is Lecturer in the Classics Department in the National University of Ireland Galway, where he is currently teaching a course on the reception of Greek Tragedy. He has recently published a number of articles on the influence of Plato on Ovid in *Classical Quarterly* and *Philologus*, and he is completing the final edits of his book, titled *The Cosmic Text from Ovid to Plato*, on this theme. He is also developing a new project on the literary impact of Lucretius in Early Modern science, and his most recent article is focused on Galileo's reimagining of Lucretian metaphors.

Helen King is a historian of medicine and the body and retired from The Open University, UK, in 2017. Since then she has held a one-year post at Gustavus Adolphus College, USA, to promote interdisciplinary approaches to history. Her earlier career included visiting roles at the Peninsula Medical School, UK, and the universities of Vienna (Austria), Texas (USA), Notre Dame (USA), and British Columbia (Canada). She has published on aspects of medicine from Classical Greece to the nineteenth century, and her most recent book, *Hippocrates Now*, was published in 2019. She is currently working on a history of the female body.

Assaf Krebs is Senior Lecturer in the unit for transdisciplinary studies at Shenkar College of Engineering, Design and Art, Israel. His research focuses on the human skin in modern and ancient thought, particularly in Roman thought, from the early Republic to the high Empire. He is also engaged with modern intellectual thought and the classics and has published on classical literature and Merleau-Ponty, Deleuze, and Guattari. His current project is a monograph on skin in ancient Rome. His next projects will focus on classics and the Anthropocene, and classics in the light of Global South epistemologies.

Julie Laskaris received her doctorate in classics from the University of California, Los Angeles, USA, 1999 (BA *cum laude*, Classics, New York University, USA, 1982). Before that, she was a modern dancer. She is an associate professor in the Department of Classical Studies at the University of Richmond, Virginia, USA. Her research interests centre on ancient Greek and Roman medicine. She is most recently author of a chapter on 'Diseases', in Totelin, L. (ed.), *Antiquity* (Volume I), *Cultural History of Medicine* (Bloomsbury, 2021), and a co-editor (with R. Rosen and P. Singer) of the forthcoming *Oxford Handbook of Galen*.

Thea Lawrence is an early career researcher and received her PhD in ancient history from the University of Nottingham, UK, in 2019. Her thesis, *Odour, Perfume, and the Female Body in Ancient Rome* explores the many ways in which sources as diverse as myth, medicine, poetry, drama, and natural history interacted to construct the odour of the female body in the literary culture of Late Republican and Early Imperial Rome. She has taught on a range of subjects related to ancient history, literature, and classical reception at the University of Nottingham, UK, and the University of Lincoln, UK.

Victoria Leonard is a Research Fellow at the Centre for Arts, Memory and Communities at Coventry University, and at the Institute of Classical Studies, University of London. She is a Fellow of the Royal Historical Society. Her research focuses on the late antique and early medieval western Mediterranean. Victoria has published articles with *Vigiliae Christianae, Studies in Late Antiquity*, and *Gender & History*. Her monograph, *In Defiance of History: Orosius and the Unimproved Past*, is under contract with Routledge.

Tara Mulder is Assistant Professor in the Department of Classical, Near Eastern, and Religious Studies at the University of British Columbia in Vancouver, Canada. She has published articles on gender, sex, and medicine in ancient Greece and Rome and is writing a monograph about human reproduction in the ancient world and today. She is also working on an English translation of Metrodora's *Gynecology*.

Adam Parker is a part-time PhD student with the Open University, UK, researching the archaeology of magic in Roman Britain. He is also Associate Collections Curator at the Yorkshire Museum, UK, and Finds Liaison Officer for the Portable Antiquities Scheme. His research focuses on the materiality of ancient magic as well as sensory and phenomenological approaches.

Catalina Popescu has a PhD in classics from the University of Texas at Austin, USA, with a dissertation on memory in the tragedies about Orestes' matricide. She is interested in the function of heroic bodies as receptacles of memory in Ovid's *Metamorphoses* ('*Ta Klea Andron*, (Dis)Embodiment and Heroic Peregrinations at Ovid'). She has published papers on the literary symbolism of bodies and fluids with the University of Bucharest and the Museum of Archaeology and National History in Constanța. For the summer of 2021 she obtained an 'Eugen Lozovan' fellowship to research the transformations of heroic cultural memory in Ovid's *Metamorphoses* at the University of Copenhagen.

Irene Salvo is presently A. G. Leventis Research Associate in Hellenic Studies at the University of Exeter, UK. Her research focuses on Greek social and cultural history and material culture. She has published on the embodied and gendered dimensions of ritual practices, their emotional dynamics, pollution, and purification, as well as knowledge and education. Her latest publications include the volume co-edited with Tanja S. Scheer, *Religion and Education in the Ancient Greek World* (Mohr Siebeck, 2021).

Caroline Spearing has taught classics in a number of schools, including Winchester College, UK, and St. Paul's School, UK, where she was Head of Department. In 2017 she was awarded her PhD for a study of Abraham Cowley's *Plantarum Libri Sex* (1662 and 1668). In 2019 she was awarded a British Academy Postdoctoral Research Fellowship, based at the University of Exeter, UK, where she is working on seventeenth-century neo-Latin university verse anthologies.

Anastasia Stylianou is a Leverhulme Early Career Fellow in the History Department at the University of East Anglia, UK. Her doctoral research examined constructions of martyrs' blood in the English Reformations. She was subsequently the principal investigator for Victoria County History's project on the parish of Cradley (Herefordshire). Her current research analyses transnational Anglo-Hellenic networks, c. 1520–c. 1650, exploring how the Reformations and globalisation shaped English contacts with and constructions of Greeks and Greek Christianity. Her publications include 'Martyrs' blood in the English Reformations', *British Catholic History* (2017), Volume 33 (4), 534–60.

Laurence Totelin is Reader in Ancient History at Cardiff University, UK. Her research focuses on Greek and Roman botany, pharmacology, and gynaecology. Her key publications include *Hippocratic Recipes: Oral and Written Transmission of Knowledge in Fifth- and Fourth-Century Greece* (2009); with Gavin Hardy, *Ancient Botany* (Routledge, 2016); and edited with Rebecca Flemming, *Medicine and Markets in the Graeco-Roman World and Beyond: Essays on Ancient Medicine in Honour of Vivian Nutton* (2020).

Goran Vidović (PhD Cornell 2016) is Assistant Professor of Latin in the Department of Classics, University of Belgrade, Serbia, where he teaches various advanced Latin courses (comedy, satire, prose composition). One of his main interests is ancient drama, especially comedy, in all its variations and throughout history. His publications include works on Greek tragedy, Aristophanes, Livy, Plautus, late antique comedy, early Christian apocrypha, and the reception of Roman comedy in the twentieth-century musical. He is currently working on Plautus' comedy the *Braggart Soldier*.

John Wilkins is Emeritus Professor of Greek Culture at the University of Exeter, UK. Publications on the body include *The Boastful Chef* (2000), *Galen and the World of Knowledge* (edited with C. Gill and T. Whitmarsh, 2009) and *Galien: Sur les facultés des aliments* (2013), along with numerous chapters on food and nutrition in the ancient world. He is currently working on a translation of Galen's *Simple Medicines* and on Galen's teaching for well-being and preventive medicine.

INTRODUCTION

Mark Bradley, Victoria Leonard, and Laurence Totelin

The human body is full of fluids. Up to 60 percent of it is made up of water, nearly three-quarters in the brain and heart, more in the lungs, muscles and kidneys; even our bones are 30 percent water. As well as a container for around 5 litres of blood, our bodies are houses for mucus, milk, semen, bile, and gastric fluids and expel up to 2 litres of urine per day and half a kilogram of excrement, as well as sweat, phlegm, ear-wax, saliva, and tears. In order to produce these fluids, the human body needs to ingest between 2 and 3 litres of water per day, and cannot normally survive more than a few days without it. Bodily fluids, then, are a fundamental and vital part of our biology: they need to be carefully regulated to keep us healthy, and we need to keep a close eye (and sometimes nose) on the fluids that enter our body, as well as the ones that leave it. The body's orifices—mouth, nostrils, ear canals, tear ducts, nipple ducts, and the pores of our skin, as well as the vagina, urinary tracts, and anus—regulate those fluids and function as barriers, both potent and vulnerable, that control the physical and social integrity of the body. As well as governing our basic survival, it is little surprise that these fluids and their functions lie at the very heart of socialised behaviour, intersecting closely with approaches to gender, sexuality, emotions, and morals.

Although so often the subject of taboo, abjection, and disgust, the potency and significance of bodily fluids was not lost on the ancients. This volume will explore a wide range of ancient approaches to fluidity across a three thousand-year period of antiquity, drawing upon material from Egypt, Greece, Rome, and Byzantium, and ranging from the second millennium BCE through to Late Antiquity and beyond. It also examines the transmission and reception of ancient ideas about fluids in medical thought during the Renaissance, which in turn shaped a wide range of popular and specialist medical ideas in the modern world. Like those Renaissance physicians, ancient medical authors were acutely aware that the fluids of the body were fundamental to its health and survival and developed some sophisticated and creative theories to understand them, but—often in association with or response to this medical discourse—many other writers and thinkers took up the mantle to scrutinise fluidity across a range of different genres. A brief survey of the mid-first-century BCE Roman Neoteric poet Catullus, to take just one example, demonstrates a fascination with the proper and improper use and display of bodily fluids: across his surviving short poems—numbering just over one hundred—and eight longer poems, many of which were erotic or epigrammatic in style, nearly every bodily fluid is spilled, expelled, injected, or consumed in a way that provided to the reader an immediate and compelling assessment of the individual or individuals concerned. This was a corpus dripping

DOI: 10.4324/9780429438974-1

with tears, blood, spit, snot, semen, urine, and faeces. In a well-known poem about Furius' abstemious lifestyle (23), for example, Catullus scoffs at the dryness of the man's body: not a drop of sweat or saliva, catarrh or snot, and his monthly stools are as hard as beans or pebbles; several times—such as in poems 16 and 37—the poet promises to irrumate his opponents' mouths (a stock threat of invective, it seems), including Egnatius who, we find out in 39, *drinks* his own urine to whiten his teeth, as Catullus claims all Spaniards do; the opposite of Furius, a stereotypically unattractive girl in poem 43 has a mouth that is far from dry, inviting lewd speculation about the fluids that might have wettened it, a more subtle version of 59, where Lesbia spends her time down back alleys 'milking' the descendants of Remus; at 44, Catullus jests that Sestius' terrible books make him ill, and bring only coughs and catarrh. One of the long poems (63) tells the myth of Attis' self-castration in Phrygia, staining the surface of the land with blood, bloodstained imagery that is then repeated several times in poem 64.

Back into the short poems, Catullus complains that somebody's nasty spittle (*spurca saliva*) has befouled the lips of a pure girl (78), and then (80) reveals that Gellius' 'white lips' bear witness to the seed that he has smeared on them from his slave boy (*emulso labra notata sero*)—again, like Lesbia's milking habits, suggestive fluids are packaged up inside a metaphor. Elsewhere (97) the poet conjures up a foul, oozing sensorium in describing Aemilius' mouth, which smells as bad as his arse—inviting his audience to imagine where that mouth has been—and a smile like a summer mule's urinating vagina. And at 99, Catullus turns a poetic cliché on its head by describing a kiss he stole from the young Juventius, intended to be sweeter than ambrosia but rejected as if it were bitter poison and washed clean with water as if it were the foul spit (again, *spurca saliva*) of a foul prostitute. In all of these poems, Catullus was of course building on a long-established Greek and Hellenistic literary tradition that ascribed well-known meanings and associations to those fluids and how they were deployed, and Catullus himself was part of a rich poetic, rhetorical, and satirical culture that would influence how later Romans, Byzantines, and Europeans thought about the body and the fluids it contained.

New approaches to ancient fluidity and the body

The genesis of this volume began with a simple observation—that the fluids of the body were a pervasive presence in the ancient world, and yet much like today they were largely, and often deliberately, avoided or unobserved. An exception is the field of ancient medicine, where fluids, humours, and the balances of the body are central to ancient physiology, a connection that has been explored widely by modern scholarship. In her *Hippocrates' Woman*, Helen King argued that 'In the Hippocratic texts, organs are often of less importance than fluids' (1998: 34). This principle extends well beyond texts written (for the most part) in the Classical period; it can be applied to medical literature composed throughout antiquity, the Middle Ages, and the Early Modern period. Historians of pre-modern medicine conceive of the ancient body as an economy of fluids (see King 2012). Yet their focus of study is often on the humours, the nature, function, and number of which was a matter of debate in antiquity, until Galen's theory of the four humours—phlegm, yellow bile, black bile, and blood—became canonical (see Nutton 2005).[1] The health of the humoral body was thought to rely on a delicate balance of fluids managed through excretion and ingestion of matter. All fluids could be transformed into one another, and unbalanced humours in the body caused illness and pain. Historians of medicine have, however, paid far less attention to non-humoral liquids (such as tears, sweat, and to an extent urine) or transformed humoral fluids (such as milk and semen, which were thought to be concocted out of blood).[2]

While there remains much to be studied in ancient medical texts, this volume and the conference that it originated from were conceived of as a response to chronic inattention to fluids in wider social, political, religious, economic, and cultural contexts throughout antiquity.[3] Representations of the body and bodily fluidity are diffuse across the ancient evidence and can be masked in shadowy tangibility. This obscurity is compounded by the impulse towards euphemism and elision of bodily fluids in modern criticism. J. N. Adams' *Latin Sexual Vocabulary* (1982), which one might have expected to have been brimming with fluids, has just an appendix focusing on the vocabulary relating to bodily functions, with some focus on fluidity. Not all liquids, of course, have suffered neglect in quite the same way. The last few years, for example, have witnessed the publication of several volumes on milk (e.g. Pedrucci 2013; Penniman 2017), although even there the focus has been on the action that leads to the emission of the fluid, breastfeeding, rather than on the substance itself and its mutability—for milk, a liquid, can potentially solidify into cheese. In addition, there have been a number of individual studies on blood (especially menstrual blood) in antiquity and its significance in the realms of religion, warfare, and gender.[4] A volume from 2009 edited by Thorsten Fögen examined the role of tears in Greco-Roman literature, drama, historiography, philosophy, and religion.[5] In certain, often very specific contexts, scholars have directed their attention to ancient approaches to urine, semen, and spit.[6] Recent decades have also witnessed a flurry of interest, principally within Roman archaeology, in urban waste disposal and the evacuation of human waste products, albeit conceptually detached from the human body. Certain bodily excretions that have received limited attention in antiquity have been more fully treated in post-classical contexts.[7]

The most important precursor to *Bodily Fluids*, and one which informs and complements the current study, is the volume edited by Manfred Horstmanshoff, Helen King, and Claus Zittel, *Blood, Sweat and Tears: the Changing Concepts of Physiology from Antiquity into Early Modern Europe* (2012). This volume was concerned principally with notions of change and continuity between antiquity and the Early Modern period and made an important contribution to the understanding of bodies and its fluids within the development of the history of medicine. As well as exploring ancient and post-classical approaches to anatomy, this book contains some important chapters on blood and its relationship to sperm and milk, as well as sweat and its permeation of the barriers of the skin and tears and their relationship to bodily disorders. *Bodily Fluids* builds on the ideas and arguments of that volume, both through its individual chapters and its collective approaches, and extends them into new areas of study by considering the relationships between a wider range of fluid types and by concentrating on the permeation of ideas about fluidity across diverse ancient genres and contexts.

Another volume devoted to bodily fluids in Medieval and Early Modern culture is the recent volume edited by Anne M. Scott and Michael David Barbezat, *Fluid Bodies and Bodily Fluids in Premodern Europe: Bodies, Blood, and Tears in Literature, Theology, and Art* (2019). That volume explores the role of the body in Medieval and Early Modern culture, investigating how fluids, particularly tears and blood, both signify and explain change. Within Medieval and Early Modern bodily encounters, tears and weeping are performative modes as well as divine gifts, while blood is the mode through which the self is formed and reformed and the nature of humanity is reconfigured.

The present volume examines fluids that are produced by the human body, a body that has been the subject of a great deal of classical scholarship across the last few decades, following hard on the heels of somatic studies in the fields of anthropology and sociology. Increasingly the plurality of the ancient body is critically emphasised, and scholarship has broadened to encompass sensory and emotional approaches to the body.[8] A critical example that changed the field of fluids, flux, and the body is Dominic Montserrat's edited collection *Changing Bodies, Changing Meanings: Studies on the Human Body in Antiquity* (Routledge, 1998). Beginning from an Early Modern viewpoint, this volume focused on the ancient body through change, modification, and transition by problematising the monolithic and unvarying historiographical category of 'the body' and pushing forward to something altogether more complex and unstable. Montserrat's volume foregrounded the plurality of the body, seeking to reinstate the emotional and experiential reality of inhabiting the ancient body. As this volume does, *Changing Bodies* engaged in divergent historical contexts and periods, extending beyond the traditional confines of the classical, from Bronze Age Egypt to early Christian martyrdom and on to the reception of Classical Greek nudity in Victorian culture. At this early stage, *Changing Bodies* was not able to push back to the sticky intimacy of the queered, non-binary, and highly theorised body that is at the forefront of thought on corporeality today, but it provided a crucial stepping stone. For example, Nicholas Vlahogiannis's chapter discussed the disabled ancient body, preceding the subsequent explosion in disability studies, which are yet to explore in detail the fluids exuded by disabled and chronically-ill bodies.[9]

The field of ancient bodily studies is steadily turning away from the notion of the body as an unproblematic, discrete, and monolithic category (noted by Porter 1999: 6; Fögen and Lee 2009: 4). In her book *The Symptom and the Subject: The Emergence of the Physical Body in Ancient Greece* (2010), Brooke Holmes has convincingly argued that the concept of the physical body is not a cultural given but that it emerged over a period of time in the Greek world. The Hippocratic medical authors played a key role in this development through their focus on the notion of the symptom, 'a disruption—without obvious cause and often, though not always, painful—either to the experience of self or to the outward presentation of self' (Holmes 2010: 2).

Holmes's questioning of the very notion of the body prompted our reflection on the relationship between the fluids of the body and bodily mutability, bodily *fluidity*. Fluids exuded by the body, seeping through various orifices and beyond the boundary of the skin, usually signal change within the body and point to its fundamental instability and permeability. Does the body end with the skin? Or is it a more fluid entity that can leak, transpire, and trickle? Holmes' important study of the concept of the body in antiquity adds a classical perspective to the thriving interdisciplinary interest in bodily fluidity in recent decades in sociology, anthropology, and human geography.[10] One of these is Robyn Longhurst's *Bodies: Exploring Fluid Boundaries* (2001), which examined the concept of the 'body' and its various metaphorical associations and how ideas about the fluidity of corporeal boundaries and 'abjection' are embedded in modern approaches to politics, gender, and sexuality.[11] Traditional anthropological approaches to system, order, dirt, and disorder, epitomised by Mary Douglas's canonical study of pollution and taboo, have more recently been aligned to the central role of the body and corporeality in formulating the rules and boundaries that exist within society (Douglas 1966).[12]

One of the volume's hypotheses is that it is through fluids that the relationship between the inner body and the outside world is mediated. The collapse of corporeal boundaries signified by bodily fluids alters the perception of the body *as* fluid: it is not a solid phenomenon of flesh and bone but a container of blood, milk, urine, phlegm, sperm, and tears, which exist just below the skin and are constantly in danger of leaking out. Moreover, the orifices through which fluids pass, both into and out of the body, are frequently sites for risk, (dis)empowerment, and debate. Our objective is to explore the relationship between the fluids of the body and bodily *fluidity*—the ways in which dry, static, unchanging bodies and wet, leaking, transitory bodies were connected to ideas about social identity, morality, sexuality, and gender. Fluid bodies are changeable and uncontrollable in ways that are frequently gendered: women's bodies are overwhelmingly likely to be represented as inconsistent and unstable, even volatile and dangerous; and male bodies whose boundaries are compromised endanger their very masculinity.

This volume seeks to scrutinise the connection between fluids and gender and sex, exploring how ideas about moderation and restraint were related to the ingestion or expulsion of liquids. Why were female bodies so often connected to leakiness, and how fluid were notions of gender and sex?[13] Did the transmission of fluids in and out of the body always compromise social, political, or cultural identity, or could identity be strengthened by the generation of fluids? One overarching research question for this volume is: can we identify a broad cultural avoidance of, even aversion to, fluids in antiquity? Does a relative lack of evidence across the broad corpus of literary and visual material in antiquity allude to a situation in which bodily fluids belonged to the realm of the unspeakable, phenomena that could only in certain contexts be put into words? Or if we find that there is substantial evidence that has gone largely unnoticed and undiscussed, does that point to a set of modern sensitivities that has caused scholars to eschew the topic and brush what amounted to a much more open approach to bodies in antiquity under the carpet?

Why, then, have the fluids which those bodies contained, ingested, and expelled received so little attention? One of the answers to this is that some of the areas and discourses in which those fluids are most commonly deployed, such as the study of emotions and ideas about disgust and eroticism, are relatively new fields, and so our volume will examine how bodily fluids contribute to the expression, representation, and evaluation of emotions and related human behaviours, in those areas of study where fluids and fluidity have remained at best tangential.[14] One of the most salient areas for exploration is the relationship between different types of bodily fluid and ideas about dirt, pollution, and disgust—whether directed towards the fluids of one's own body or the bodies of others. Attitudes towards fluids are characteristically ambivalent; but why were fluids at times taboo, at times cherished? Was there a desire/revulsion hierarchy of different bodily fluids, and to what extent did that hierarchy depend on the body in question, especially pertaining to sex and gender, and the context in which bodies were employed and experienced? How were fluids coursing through and trapped inside the body perceived, and how did this perception alter when those fluids were externalised? And what role did these fluids play in the sphere of religion and ritual, where the control of the body, its functions, and its component parts was strictly regulated and policed?

Since those fluids are detected and evaluated through sight, smell, touch, taste, and even (in the case of noisy ingestion and expulsion) sound, this volume will explore—in the context of the recent flurry of scholarly interest in sensory studies—the roles the senses occupy in terms of identifying and interpreting fluids, both those within and on the perceiver's own body and those on the bodies of others. How were fluids classified and differentiated? How did physicians employ their senses to diagnose and understand the substances of the body and their relationship to its state of health or disease, and how did that medical sensory discourse become embedded in later philosophical, physiognomic, biographical, oratorical, and satirical literature?

Approaches, methods, and evidence

The present volume is part of this turn to the senses and emotions and functions as a further expansion of the study of the ancient body to include fluids and fluidity, with a methodological objective to avoid the abstract and to engage directly with physical fluids. The chapters collected here are designed to provoke further fruitful discussions; the volume is very much a point from which to begin rather than to end. It is noteworthy, however, that even in studies of eroticism and the 'lower senses' (smell, taste, and touch), where we might particularly expect to find engagement with the body's fluids, they are treated tangentially or completely overlooked. This said, our intention was not to produce a synoptic account of ancient bodily fluids or a history of scholarship on the body.[15] Instead, fluids and fluidity provide a distinct focus to specialists from a range of disciplines to illuminate how different perspectives might converge or diverge in understandings of fluids and fluidity in antiquity. To foreground a most fundamental question, then, this volume examines what ancients considered bodily fluids to *be* and what categories were used to describe them. Where did they come from, and how did they relate to the physiology and humours, and consequently to the character and temperament, of the individual who embodied them? What was it about bodily fluids that made them so often the subject of shock, disgust, or amusement? How prevalent are metaphors of fluidity in descriptions of the ancient body?

It would be naïve to assume that approaches to bodily fluids remained static across the *longue durée* of the periods represented within this volume and across the extensive geographical territories of the Mediterranean basin. Another key question that this volume will explore is how far we can detect continuities and variations in interactions with and evaluations of fluids across time and space. Contributors have been encouraged to consider the relationship between their own findings and those of other chapters in adjacent periods, territories, and fields of thought. For this reason, our volume will also explore how ancient ideas about bodily fluids were appropriated and adapted in Renaissance and Early Modern Europe, and how they have informed the way we discuss them today. Do ancient approaches to bodily fluids, we ask, complement or challenge our modern sensibilities about bodily fluids?

The evidence for ancient bodily fluids, then, is diffuse and dispersed, and fluidity often simmers (sometimes elusively) beneath the surface of ancient accounts of bodies. There are few sustained discussions within antiquity, and the majority of those are products of ancient medical discourse. These specialist treatises were written (mostly) by experts for readers training to be professional physicians—not, as we shall see, that those discussions are not revealing about ancient conceptions (and misconceptions) about the human body and how it was thought to work. Occasionally, ancient philosophers and theologians made sustained forays into bodily fluids, such as Theophrastus' treatise *On Sweat*, which adapted a range of medical theories about the body to produce an accessible outline of a single bodily fluid and its role in natural philosophy.[16] As we have seen, writers in other genres such as satirical verse and epigram, along with biography and comedy, did not miss the opportunity to explore the value of fluids as sources of humour, disgust, and salacious sexuality.

On the whole, however, the contributions within this volume—both individually and collectively—have taken the important and original step of synthesising a wide range of evidence and material that have never before been considered in a coherent way as part of the history of the human body in antiquity. This evidence comprises— among other things—ancient linguistic terminology (Coker), early Egyptian law codes, papyri and iconography (Janssen, Dobbin-Bennett), medical and gynaecological commentaries (Salvo, Fallas, Mulder, Flemming, Wilkins), philosophical treatises and natural history (Laskaris, Lawrence, Goyette, Krebs), Greek verse and drama (Kearns, Vidović), Roman elegy and epic (Burkowski, Popescu, Kelly), satire (Cent-livres Challet, Gavrielatos), sculpture and art (Parker), early Christian cultural conceptions (Totelin, Stylianou), neo-Latin poetry (Spearing), and Early Modern medicine (King). It goes without saying that these diverse forms of evidence were never intended to be considered alongside each other, and we make no claims that they help us to form a coherent and holistic perspective on the role and significance of bodily fluids in antiquity. But all ancient genres of expression were interactive and interdependent, and the volume will explore the cross-fertilisation of ideas from one context to another, and from one period or territory to the next.

The structure of the volume derives from seven central themes that offer discrete perspectives for the study of bodily fluidity while resisting the notion of a hierarchy of fluids. A thematic rather than chronological approach illuminates parallels and differences across contexts and evidence types. The themes are designed to show how bodily fluids are categorised and constructed, how fluids become imbued with didactic meaning, what the function of fluids is within the body, and how fluids are received when externalised and detached from the body. Some chapters engage in broad survey, familiarising readers with specific subjects (Janssen, Krebs) while others are more narrowly focused, opening up new fields of enquiry (Vidović, Popescu, and Burkowski).

Because the representation of fluids is often anomalous or scattered throughout the ancient evidence, and because the themes of fluids and fluidity are relatively unexplored, the temporal and spatial range covered by the chapters presented here is necessarily sweeping. This volume takes both the wide and the long view, with twenty-four chapters on fluidity or fluids, offering significant coverage of the ancient world, broadly conceived. Beginning with the ancient Egyptians (Janssen, Dobbin-Bennett) and ending with classical reception in the sixteenth and seventeenth centuries (King, Spearing), this multidisciplinary approach to past bodies and fluids is collaborative and inclusive, exploring evidence from archaic, ancient, Medieval, and Early Modern time periods. While some chapters focus on particular chronological periods (Lawrence on early Imperial Rome), others deliberately extend their analysis to transgress the barriers of periodisation (King). The volume encompasses the key geographic centres of the ancient Mediterranean basin, including Greece, Rome, Byzantium, and Egypt.

The opening theme ('The Language of Fluidity') comprises a single chapter that orientates the volume, examining how bodily emissions are classified and encoded in language. In her chapter, Amy Coker finds an inherent fluidity in the semantic boundaries between words indicating bodily substances, such as sweat, mucus, and semen, demonstrating how and why such language is characterised by an instability of meaning that can be productive, playful, and provocative in its ambiguity.

Perceptions of bodies as fluid are frequently gendered, and women's bodies are overwhelmingly likely to be represented as inconsistent and unstable, even volatile and dangerous. The second theme, 'A Woman in Flux', comprises four chapters that examine female fluidity in menstruation and sweat across a range of evidence types and contexts. Rosalind Janssen's chapter explores traces of *hsmn* or menstruation in the textual and material evidence from Deir-el-Medina. Ostraca, sherds of limestone that have been used as a writing surface, record the practice of women leaving the community to menstruate in isolation, and laundry lists detail the washing of soiled sanitary towels. Janssen synchronises disparate evidence types to show that menstruation was a public event that provided an opportunity for ritual celebration, a bodily function to be experienced in isolation, and a physical event that demanded practical intervention. Irene Salvo takes the focus on menstrual blood to the Greek Eastern Mediterranean, exploring how the transmission of knowledge about menstruation intersected both popular religious and specialist medical contexts from the Classical to the Roman Imperial period. Jane Burkowski's chapter analyses culturally constructed representations of the body in Latin love elegy; she explores the sensory body, sometimes smelly and effusive, sometimes divinely perfumed and sweet-smelling, conjured by Ovid in his didactic works. Finally, Catalina Popescu emphasises the aridity and fluid stasis of Ovid's Galatea, a flawless female creation.

The third theme, 'Erotic and Generative Fluids', includes six chapters that explore the function of internalised fluids within reproduction and the reception of sexualised fluids externalised from the body. Julie Laskaris and Rebecca Fallas examine the efficacy of fluids in Aristotelian and Hippocratic thought, and how they were understood to move through the body. Claude-Emmanuelle Centlivres Challet argues that the Roman satirist Juvenal signalled proper or improper behaviour through fluids. Issues of moisture and fluidity are then further developed by Tara Mulder, who argues that ancient medical writers such as Soranus understood foetal sex to be determined by the relative wetness of female bodies. Rebecca Flemming turns to the fluid contribution of male and female in her reassessment of ancient debates about reproduction. Finally, Adam Parker considers another way in which fluid imagery is used didactically, through the apotropaic function of Roman carved phalli ejaculating into the Evil Eye.

The fourth theme, 'Nutritive and Healthy Fluids', features four chapters that investigate corporeal fluids in their nourishment of the body or as indicators of bodily health. Emily Kearns begins by analysing blood as a signifier of death and suffering in Greek epic and tragedy. John Wilkins' chapter examines the Galenic connection between food and the four humours and how fluids emitted by the body are indicators of physiological well-being. Thea Lawrence explores a range of Roman attitudes and anxieties about breast milk. Finally, Laurence Totelin explores maternal breastfeeding in early Christian narratives.

The next theme, 'Dissolving and Liquefying Bodies', comprises three chapters that situates the permeability of the body directly in the foreground. Peter Kelly analyses weeping and bodily dissolution in Ovid's *Metamorphoses*. Michael Goyette examines how Seneca links bodily fluids and emotional fluctuations with the natural environment. Andreas Gavrielatos illustrates how Persius exploits fluid imagery for satirical purposes. Together, these chapters demonstrate how the permeable body, informed by a long tradition of Greek and Hellenistic thought, functioned as a key theme across several distinct literary genres produced within a century of each other in the city of Rome.

'Wounded and Putrefying Bodies' is a theme in which three chapters engage with damage to the body as a vessel of skin, analysing how death engenders further fluid transgression of the bounded body, where the inside flows out as a result of death and the outside is internalised, which causes death. Tasha Dobbin-Bennett's chapter concentrates on bodily decomposition and putrefaction in Egypt, arguing that putrefactive fluids exuded from the corpse were not necessarily negative but were an important part of the spiritual reconstitution of the deceased. Goran Vidović examines how the imagery of bodily fluidity was used in Aeschylus' *Oresteia* to indicate the dynamics of crime and retribution, and finally Assaf Krebs's chapter explores the nature of wounds and their symbolic function in Latin literature.

The final section, 'Ancient Fluids: Afterlife and Reception', features three chapters that explore how many of the themes discussed earlier in the volume carried an enduring influence in post-classical Europe, concentrating on the literature and medicine of the Medieval and Early Modern periods. First, Anastasia Stylianou's chapter examines the classical reception of blood in Medieval and Early Modern martyrologies. Caroline Spearing then examines menstruation in seventeenth-century England and Abraham Cowley's appropriation of the female voice in his poetic attempt to rehabilitate the toxic female body. Spearing demonstrates how Cowley's agenda is political rather than proto-feminist, but he provides valuable insight into depictions of bodily function and gender that have their roots firmly embedded in ancient ideas. Finally, Helen King's analysis of Renaissance reimaginings of notable classical women extends the scope of the volume at its close. King returns to ancient authors discussed by other contributors (such as the Hippocratic writers and Soranus) but considers fluid representation in new genres, specifically Roman mythical history (including Dionysius of Halicarnassus, Livy, and Valerius Maximus). King argues that fluids are central in the construction of gender and that anxieties about the balance and control of fluids within the body were often connected with the female sex, in antiquity and beyond. King explores fluid retention and transmission in relation to Tuccia, the Vestal Virgin who proved her chastity by carrying water from the Tiber in a sieve, and how the semantic sieve, imbued with associations of virginity and chastity, recurred in the sixteenth century in imagery of the Virgin Queen, Elizabeth I. In this motif, the themes of bodily integrity, fluids, gender, and power merged, informed and underpinned by the authority of classical antiquity.

The volume closes with a short 'Envoi' by Mark Bradley and Victoria Leonard, which summarises some of the overarching themes surrounding bodily fluids in antiquity, underscoring the use of fluids as diagnostic signs of medical, moral, and political values, and the role of fluids in exploring cultural anxieties about the human body, in all its porosity and fluidity.

Notes

1 Galen had adopted and adapted his theory of the four humours from that found in the Hippocratic text *On the Nature of Man*.
2 On tears, see Fögen 2009; Horstmanshoff 2014. The art of diagnosing illnesses through uroscopy developed into a specialised field, with treatises devoted solely to the topic, in Byzantium; see introduction in Stolberg 2016. See more later on sweat. On milk and menstrual blood, see e.g. Dean-Jones 1989, 1994; Laskaris 2008; Pedrucci 2013; on seed, see e.g. Flemming 2018.
3 Some indicative examples include Dupré i Raventós et al. 2000; Hobson 2009; Jansen et al. 2011. Most recently, see Koloski-Ostrow 2015.
4 On Greek blood(shed), see Parker 1983: 104–43; on menstrual blood, see King 1983: 109–27; King 1987: 117–26; Lennon 2010: 71–87; Cohen 1991: 273–99; Leonard 2020. On blood in general, see Lennon 2014: 90–135; cf. Clark 1998: 99–115; Gradel 2002: 235–53.
5 More recently, see Vekselius 2018 for tears in Roman politics and Alexiou and Cairns 2017 for Greek laughter and tears.
6 Bradley 2002: 21–44. On 'le lait du père', see Dupont 2002: 115–38. On spit, see McCartney 1934: 99–100; Eve 2008: 1–17. On semen, see Brakke 1995: 419–60; Petrey 2014: 343–72.
7 See, for example, Lachmann et al. 1988; Jenner 2002; Morrison 2008; Leyerle 2009: 337–56. For a broad-brush study of the symbolic and religious functions of blood, see Meyer 2005.
8 On sensory approaches to antiquity, see Ashbrook Harvey 2006; Porter 2010; Butler and Purves 2013; Toner 2014; Bradley 2015; Squire 2016; Purves 2017; Rudolph 2018. On the growing field of ancient emotions, see Kaster 2005; Konstan 2006; LaCourse Muneanu 2011; Cairns and Fulkerson 2015; Caston and Kaster 2016; Sanders and Johncock 2016; Lateiner and Spatharas 2016. For the plurality of the ancient body, see Montserrat 1998; Glancy 2010.
9 The pioneering work of Véronique Dasen (1988) is an important precedent within the field of ancient disability studies. See also Vlahogiannis 2005; Moss and Schipper 2011; Breitwieser 2012; Laes et al. 2013; Krötzl et al. 2015; Trentin 2015; Laes 2016; Draycott 2019.
10 In this context, one could note recent studies on ancient animal–human/god hybrids, as well as metamorphosis and monstrosity, which also aim at questioning bodily fluidity. See for instance King 1995; Atherton 1998; Braund and Gold 1998; Hughes 2010; Garland 2010; Aston 2011, 2014; Thumiger 2014; Lowe 2015.
11 On the idea of 'leaky bodies', see also Shildrick 1997. On bodies and boundaries in antiquity, see Fögen and Lee 2009.
12 Two significant critical reappraisals of Douglas' work which steer the focus much more onto corporeality are Meigs 1984; Valeri 1999. For discussion, see Bradley 2012: 11–40, especially 13–14.
13 Carson 1999: 87: the governing conception of woman is in every respect 'penetrable, porous, mutable, and subject to defilement all the time'.
14 Some representative scholarship on emotions: Chaniotis and Ducrey 2012; Chaniotis et al. 2017; on pollution and disgust: Parker 1983; Bradley 2012; Lennon 2014; Lateiner and Spartharas 2016; on eroticism: Rousselle 1988; Zeitlin 1999: 50–76; Vout 2013.
15 For a historical account of scholarship on the body, see Richlin 1997.
16 The second book of the pseudo-Aristotelian *Problems* also deals with perspiration.

Bibliography

Adams, J. (1982) *The Latin Sexual Vocabulary*. Baltimore, MD: Johns Hopkins University Press.

Alexiou, M. and D. Cairns (eds) (2017) *Greek Laughter and Tears: Antiquity and After*. Edinburgh: Edinburgh University Press.

Ashbrook Harvey, S. (2006) *Scenting Salvation: Ancient Christianity and the Olfactory Imagination*. Berkeley: University of California Press.

Aston, E. (2011) *Mixanthrôpoi: Animal-Human Hybrid Deities in Greek Religion*. Liège: Centre International d'Étude de la Religion Grecque Antique.

———. (2014) 'Part-animal gods', in G. L. Campbell (ed.) *The Oxford Handbook of Animals in Classical Thought and Life*. Oxford: Oxford University Press, 366–83.

Atherton, C. (ed.) (1998) *Monsters and Monstrosity in Greek and Roman Culture*. Bari: Levante.

Bradley, M. (2002) 'It all comes out in the wash: Looking harder at the Roman *fullonica*', *Journal of Roman Archaeology* 15, 21–44.

———. (2012) 'Approaches to pollution and propriety', in M. Bradley (ed.) *Rome, Pollution and Propriety: Dirt, Disease and Hygiene in the Eternal City from Antiquity to Modernity*. Cambridge: Cambridge University Press, 11–40.

——— (ed.) (2015) *Smell and the Ancient Senses*. London: Routledge.

Brakke, D. (1995) 'The problematization of nocturnal emissions in early Christian Syria, Egypt, and Gaul', *Journal of Early Christian Studies* 3(4), 419–60.

Braund, S. and B. Gold (eds) (1998) *Vile Bodies: Roman Satire and Corporeal Discourse* (special issue of *Arethusa* 31.3). Baltimore, MD: Johns Hopkins University Press.

Breitwieser, R. (ed.) (2012) *Behinderungen und Beeinträchtigungen/ Disability and Impairment in Antiquity*. Oxford: Archaeopress.

Butler, S. and A. Purves (eds) (2013) *Synaesthesia and the Ancient Senses*. Durham: Acumen Publishing.

Cairns, D. L. and L. Fulkerson (eds) (2015) *Emotions Between Greece and Rome*. London: Institute of Classical Studies, University of London.

Carson, A. (1999) 'Dirt and desire: The phenomenology of female pollution in antiquity', in J. I. Porter (ed.) *Constructions of the Classical Body*. Ann Arbor: University of Michigan Press, 77–100.

Caston, R. R. and R. A. Kaster (eds) (2016) *Hope, Joy, and Affection in the Classical World*. New York: Oxford University Press.

Chaniotis, A. and P. Ducrey (eds) (2012) *Unveiling Emotions*. Stuttgart: Franz Steiner.

Chaniotis, A., N. E. Kaltsas, and J. Mylonopoulos (eds) (2017) *A World of Emotions: Ancient Greece, 700 BC-200 AD*. New York: Onassis Foundation.

Clark, G. (1998) 'Bodies and blood: Late Antique debates on martyrdom, virginity and resurrection', in D. Montserrat (ed.) *Changing Bodies, Changing Meanings: Studies on the Human Body in Antiquity*. London: Routledge, 99–115.

Cohen, S. J. D. (1991) 'Menstruants and the sacred in Judaism and Christianity', in S. B. Pomeroy (ed.) *Women's History and Ancient History*. Chapel Hill: University of North Carolina, 273–99.

Dasen, V. (1988) *Dwarfs in Ancient Egypt and Greece*. Oxford: Oxford University Press.

Dean-Jones, L. A. (1989) 'Menstrual bleeding according to the Hippocratics and Aristotle', *Transactions of the American Philological Association* 119, 177–91.

———. (1994) *Women's Bodies in Classical Greek Science*. Oxford: Clarendon Press.

Douglas, M. (1966) *Purity and Danger: An Analysis of Concepts of Pollution and Taboo*. London: Routledge.

Draycott, J. (ed.) (2019) *Prostheses in Antiquity*. London: Routledge.

Dupont, F. (2002) 'Le lait du père romain', in P. Moreau (ed.) *Corps romains*. Grenoble: Collection Horos Jérôme Millon, 115–38.

Dupré i Raventós, X., R. Vallverdú, and J. Anton (eds) (2000) *Sordes Urbis: La eliminación de residuos en la ciudad romana*. Rome: L'Erma di Bretschneider.

Eve, E. (2008) 'Spit in your eye: The blind man of Bethsaida and the blind man of Alexandria', *New Testament Studies* 54(1), 1–17.

Flemming, R. (2018) 'Galen's generations of seed', in N. Hopwood, R. Flemming, and L. Kassell (eds) *Reproduction: Antiquity to the Present Day*. Cambridge: Cambridge University Press, 95–108.

Fögen, T. (ed.) (2009) *Tears in the Graeco-Roman World*. New York: Walter de Gruyter.

Fögen, T. and M. M. Lee (eds) (2009) *Bodies and Boundaries in Graeco-Roman Antiquity*. Berlin: Walter de Gruyter.

Garland, R. (2010) *The Eye of the Beholder: Deformity and Disability in the Graeco-Roman World*. 2nd edn. London: Bristol Classical Press.

Glancy, J. (2010) *Corporal Knowledge: Early Christian Bodies*. Oxford: Oxford University Press.

Gradel, I. (2002) 'Jupiter Latiaris and human blood: Fact or fiction?', *Classica et Mediaevalia* 53, 235–53.

Hobson, B. (2009) *Latrinae et foricae: Toilets in the Roman World*. London: Duckworth.

Holmes, B. (2010) *The Symptom and the Subject: The Emergence of the Physical Body in Ancient Greece*. Princeton: Princeton University Press.

Horstmanshoff, M. (2014) 'Tears in ancient and early modern physiology: Petrus Petitus and Niels Stensen', in D. Kambaskovic (ed.) *Conjunctions of Mind, Soul and Body from Plato to the Enlightenment*. Dordrecht: Springer, 305–23.

Horstmanshoff, M., H. King, and C. Zittel (eds) (2012) *Blood, Sweat and Tears: The Changing Concepts of Physiology from Antiquity into Early Modern Europe*. Leiden: Brill.

Hughes, J. (2010) 'Dissecting the classical hybrid', in K. Rebay-Salisbury, M. L. S. Sorensen, and J. Hughes (eds) *Body Parts and Body Whole: Changing Relations and Meanings*. Oxford: Oxbow, 101–10.

Jansen, G., A. Koloski-Ostrow, and E. Moormann (eds) (2011) *Roman Toilets: Their Archaeology and Cultural History*. Leuven: Peeters.

Jenner, M. S. R. (2002) 'The roasting of the rump: Scatology and the body politic in restoration England', *Past and Present* 177(1), 84–120.

Kaster, R. A. (2005) *Emotion, Restraint, and Community in Ancient Rome*. Oxford: Oxford University Press.

King, H. (1983) 'Bound to bleed: Artemis and Greek women', in A. Cameron and A. Kuhrt (eds) *Images of Women in Antiquity*. London: Croom Helm, 109–27.

———. (1995) 'Half-human creatures', in J. Cherry (ed.) *Mythical Beasts*. London: British Museum Press, 138–66.

———. (1987) 'Sacrificial blood: The role of the *amnion* in ancient gynecology', in M. Skinner (ed.) *Rescuing Creusa: New Methodological Approaches to Women in Antiquity*. Lubbock: Texas Tech University Press, 117–26.

———. (1998) *Hippocrates' Woman: Reading the Female Body in Ancient Greece*. London: Routledge.

———. (2012) 'Introduction', in M. Horstmanshoff, H. King, and C. Zittel (eds) *Blood, Sweat and Tears: The Changing Concepts of Physiology from Antiquity into Early Modern Europe*. Leiden: Brill, 1–24.

Koloski-Ostrow, A. (2015) *The Archaeology of Sanitation in Roman Italy: Toilets, Sewers and Water Systems*. Chapel Hill: The University of North California Press.

Konstan, D. (2006) *The Emotions of the Ancient Greeks: Studies in Aristotle and Classical Literature*. Toronto: University of Toronto Press.

Krötzl, C., K. Mustakallio, and J. Kuuliala (eds) (2015) *Infirmity in Antiquity and the Middle Ages: Social and Cultural Approaches to Health, Weakness and Care*. London: Routledge.

Lachmann, R., R. Eshelman, and M. Davis (1988) 'Bakhtin and carnival: Culture as counter-culture', *Cultural Critique* 11, 115–52.

LaCourse Muneanu, D. (2011) *Emotion, Genre and Gender in Classical Antiquity*. London: Bristol Classical Press.

Laes, C. (ed.) (2016) *Disability in Antiquity*. London: Routledge.

Laes, C., C. Goodey, and M. L. Rose (eds) (2013) *Disabilities in Roman Antiquity: Disparate Bodies*. Leiden: Brill.

Laskaris, J. (2008) 'Nursing mothers in Greek and Roman medicine', *American Journal of Archaeology* 112(3), 459–64.

Lateiner, D. (1992) 'Affect displays in the epic poetry of Homer, Vergil, and Ovid', in F. Poyatos (ed.) *Advances in Nonverbal Communication: Sociocultural, Clinical, Esthetic, and Literary Perspectives*. Amsterdam: John Benjamin, 255–69.

Lateiner, D. and D. Spartharas (eds) (2016) *The Ancient Emotion of Disgust*. Oxford: Oxford University Press.

Lennon, J. (2010) 'Menstrual blood in ancient Rome: An unspeakable impurity', *Classica et Mediaevalia* 61, 71–87.

———. (2014) *Pollution and Religion in Ancient Rome*. Cambridge: Cambridge University Press.

Leonard, V. (2020) 'The ideal (bleeding?) female: Hypatia of Alexandria and distorting patriarchal narratives', in D. LaValle Norman and A. Petkas (eds) *Hypatia of Alexandria: Her Context and Legacy*. Tübingen: Mohr Siebeck, 171–92.

Leyerle, B. (2009) 'Refuse, filth, and excrement in the *Homilies* of John Chrysostom', *Journal of Late Antiquity* 2, 337–56.

Longhurst, R. (2001) *Bodies: Exploring Fluid Boundaries*. London: Routledge.

Lowe, D. (2015) *Monsters and Monstrosity in Augustan Poetry*. Ann Arbor: University of Michigan Press.

McCartney, E. S. (1934) 'On spitting into the hand as a superstitious act', *Classical World* 27.13, 99–100.

Meigs, A. (1984) *Food, Sex and Pollution: A New Guinea Religion*. New Brunswick, NJ: Rutgers University Press.

Meyer, M. L. (2005) *Thicker than Water: The Origins of Blood as Ritual and Symbol*. London: Routledge.

Montserrat, D. (ed.) (1998) *Changing Bodies, Changing Meanings: Studies on the Human Body in Antiquity*. London: Routledge.

Morrison, S. (2008) *Excrement in the Late Middle Ages: Sacred Filth and Chaucer's Fecopoetics*. New York: Palgrave Macmillan.

Moss, C. R. and J. Schipper (eds) (2011) *Disability Studies and Biblical Literature*. New York: Palgrave Macmillan.

Nutton, V. (2005) 'The fatal embrace: Galen and the history of ancient medicine', *Science in Context* 18(1), 111–21.

Parker, R. (1983) *Miasma: Pollution and Purification in Early Greek Religion*. Oxford: Oxford University Press.

Pedrucci, G. (2013) 'Sangue mestruale e latte materno: Riflessioni e nuove proposte. Intorno all'allattamento nella Grecia antica', *Gesnerus* 70(2), 260–91.

Penniman, J. D. (2017) *Raised on Christian Milk: Food and the Formation of the Soul in Early Christianity*. New Haven, CT: Yale University Press.

Petrey, T. G. (2014) 'Semen stains: Seminal procreation and the patrilineal genealogy of salvation in Tertullian', *Journal of Early Christian Studies* 22(3), 343–72.

Porter, J. I. (ed.) (1999) *Constructions of the Classical Body*. Ann Arbor: University of Michigan Press.

———. (2010) *The Origins of Aesthetic Thought in Ancient Greece: Matter, Sensation and Experience*. Cambridge: Cambridge University Press.

Purves, A. (ed.) (2017) *Touch and the Ancient Senses*. London: Routledge.

Richlin, A. (1997) 'Towards a history of body history', in M. Golden and P. Toohey (eds) *Inventing Ancient Culture*. London: Routledge, 16–35.

Rousselle, A. (1988) *Porneia: On Desire and the Body in Antiquity*. Oxford: Basil Blackwell.

Rudolph, K. (ed.) (2018) *Taste and the Ancient Senses*. London: Routledge.

Sanders, E. and M. Johncock (ed.) (2016) *Emotion and Persuasion in Classical Antiquity*. Stuttgart: Franz Steiner.

Scott, A. M. and M. D. Barbezat (eds) (2019) *Fluids Bodies and Bodily Fluids in Premodern Europe: Bodies, Blood, and Tears in Literature, Theology, and Art*. Leeds: ARC Humanities Press.

Shildrick, M. (1997) *Leaky Bodies and Boundaries: Feminism, Postmodernism and (Bio)Ethics*. London: Routledge.

Squire, M. (ed.) (2016) *Sight and the Ancient Senses*. London: Routledge.

Stolberg, M. (2016) *Uroscopy in Early Modern Europe*. London: Routledge.

Thumiger, C. (2014) 'Metamorphosis: Human into animals', in G. L. Campbell (ed.) *The Oxford Handbook of Animals in Classical Thought and Life*. Oxford: Oxford University Press, 384–413.

Toner, J. (2014) (ed.) *A Cultural History of the Senses in Antiquity, 500 bce—500 ce*. London: Bloomsbury.

Trentin, L. (2015) *The Hunchback in Hellenistic and Roman Art*. London: Bloomsbury.

Valeri, V. (1999) *The Forest of Taboos: Morality, Hunting, and Identity among the Huaulu of the Moluccas*. Madison: University of Wisconsin Press.

Vekselius, J. (2018) 'Weeping for the *res publica*: Tears in Roman political culture', PhD dissertation, Lund University, Lund.

Vlahogiannis, N. (2005) ' "Curing" disability', in H. King (ed.) *Health in Antiquity*. London: Routledge, 180–91.

Vout, C. (2013) *Sex on Show: Seeing the Erotic in Greece and Rome*. California: University of California Press.

Zeitlin, F. I. (1999) 'Reflections on erotic desire in Archaic and Classical Greece', in J. Porter (ed.) *Constructions of the Classical Body*. Ann Arbor: University of Michigan Press, 50–76.

Part I

THE LANGUAGE OF FLUIDITY

1

FLUID VOCABULARY

Flux in the lexicon of bodily emissions*

Amy Coker

The messy edges of the body and its messy lexicon

All bodies, female and male, past and present, leak. At one level John Chrysostom, Church Father and saint, was correct when he reminded his flock that women who are beautiful on the outside are nevertheless full of blood, digestive juices, and diverse liquids which usually reside out of sight inside the body (*On Women and Beauty CPG* 4684.14 [= *PG* 63, 657–66]).[1] It is part of what Oliver Harris and John Robb (2013a: 1) speak of as the 'everyday invisible strangeness' of the body that the passing of various substances in and out of it is fundamental to the body's proper functioning but that this is a fact we nevertheless attempt to forget. There is a constant threat that the corporeal envelope may break open, either by real or imagined rupture, and remind us of our animal nature; humans are plagued by the knowledge of their own animalistic nature (see for example Critchley 2002). But when such effluvia are seen to escape, how are they conceptualised, divided up, and labelled? How were Greek and Latin used to talk about the liquids classical bodies produced, and what significance did these words have given the destabilising challenge they represent to the sacred integrity of the ordered body? Mary Beard, in a piece entitled 'Did the Romans have Elbows?' (2002) tackles an allied problem of ancient corporeal classification by looking at what we call the *arm* and talks of 'how the body's naturally unbroken surfaces were given cultural boundaries, definitions, and names' (2002: 48). This opening chapter is concerned with what happens not at the joint of upper and lower arm but when that apparently impermeable surface of the body becomes porous, and the liquids held safely within spill out into the wider Greek and Roman worlds and acquire names. For, as we will see, the inherent boundlessness and non-fixedness of liquids leads to a fluidity in the ways they are labelled and conceptualised and a lexical flux which can itself be manipulated and played with by authors and speakers.

This chapter first discusses the general non-fixedness of the vocabulary of bodily fluids before moving on to illustrate the particular lexical mobility in words for sweat, mucus, urine, and semen. In the second half, this instability of meaning is shown to be ripe for exploitation by authors and speakers who can manipulate the lexis not just to indicate an entity in the world but also to communicate social or pragmatic meaning. In this light, the evidence for inherently negatively connoted words for some fluids, and in particular for semen, is considered: if Greek had no word equivalent to English 'piss' (as we shall see), it likely did have at least one for 'spunk'. It is suggested in particular that an expression at Aristophanes' *Knights* 910 indicates more than just innocent

DOI: 10.4324/9780429438974-3

nose-blowing and that in a mock epitaph from Egypt there is concrete evidence for another crude metaphor for the same involving the brain. Overall, this chapter seeks to highlight semantic connections and porosity which from the Anglophone perspective seem bizarre, but also to show the Greek and Roman lexica of fluids show a malleability and a shiftiness which *is* familiar from English and the everyday habits of all speakers.

* * *

In working on the vocabulary of fluids there is an obvious lexicographical challenge, but also one which is cultural-historical (see Beard 2002: 51).[2] Indeed, very many of the chapters which follow in this volume are concerned at one level with how fluids are designated in language (Latin, Greek, Egyptian), and even if the focus is not specifically linguistic, language is the medium of communication and fundamentally underpins the whole enterprise. Beard talks of her piece on Roman elbows as a 'first step in a bigger project of *defamiliarization*' (2002: 48), highlighting the need to avoid both assuming that other cultures 'do things in the same way' and forcing one's own tacit preconceptions and linguistic prejudices onto foreign material.[3] Recent work on linguistic universals indeed stresses the *lack* of general principles when talking about the body (e.g. Majid 2015: 373–8). Naming patterns in languages do not always follow salient sight characteristics or 'visual parsing principles' (2015: 375) of the overall shape of the unclothed body, and indeed different senses other than sight may be involved. There are different ways of 'viewing' or 'understanding' the body and its fluids and various methods of division. For the sake of discussion these differences can be said to fall into two intersecting and mutually interdependent areas, which we might term *cultural* (the way in which the body and its effluvia are socially coded and interpreted) and *linguistic* (the partitions and descriptive units used to label the body and its effluvia in communication).

Despite the fact that the body is universally biologically shared among all members of our species, the experience of the body and its *symbolic, social, and cultural coding* (so Beard 2002: 50), perhaps especially its hidden internal mysteries, are variable (see Introduction, pp. 4–5; especially approachable are also Harris and Robb 2013a, 2013b). This difference in the reality of the body is reflected, naturally, in the distinctions and levels of complexity identified and the ways in which people talk about the body. Regardless of the labels applied to them, corporeal sheddings also represent a particularly acute existential challenge, and fluids especially, since by their very nature they are inherently mobile, transitive, and shifting. It is for this reason that the margins of the body are places to be guarded and policed carefully and fluids managed according to complex social codes and purity rules because of their potential to pollute (see Introduction p. 5, and Douglas 1966; Meigs 1978 stresses the centrality of the body and of sheddings as *decay* in particular; Rozin et al. 2008: 765 note that of bodily fluids faeces, and perhaps vomit, seem to be universal disgust-producing substances). There is therefore a basic question which needs to be asked about the extent to which the margins of the body, 'potent and vulnerable' (Introduction, p. 1), and substances associated with it are producers of vocabulary which, because of social anxieties about those aspects of the body, is inherently offensive. Talking about such topics may, according to social norms, require linguistic delicacy—which additionally has the opposite effect of making them sites of linguistic thrill and provocation; this is a topic which will be returned to in the second half of this chapter.

The local physical and natural environment, which can affect dress conventions and the degree to which the body is exposed or shaped (reported at Majid 2015: 378), or indeed the experience of the body in general, may also condition the way the body and its fluids are encoded in language. The Greek word κίκκασος (*kikkasos*) for example, recorded by the fifth/sixth-century CE lexicographer Hesychius (κ 2650 Latte) and attested only in the lexicographical tradition, may have been used to indicate (among other things[4]) the sweat which trickles down the inside of the thighs in hot climates: '*kikkasos*: the sweat [*hidrōs*] which flows from inside of the thighs [κίκκασος· ὁ ἐκ τῶν παραμηρίων ἱδρὼς ῥέων].'[5] A word for this kind of fluid is, however, not often required in northern Europe, given the rarity of the physical phenomenon in those climes. Likewise, some types of fluids are socially not salient enough to require a specific label but may do so under special circumstances. The 'frothy mixture of lube and fecal matter that is sometimes the byproduct of anal sex' does not usually have a name in English but acquired the name *santorum* in the early 2000s through linguistic activism by campaigners in the USA against American politician Rick Santorum and his views on homosexuality.[6] Indeed, *kikkasos* itself may not mean sweat but something closer to *santorum*.

Quite apart from the well-known differences between ancient and modern physiology (e.g. approaches to the circulation of the blood), important for this discussion of classical fluids is the widespread understanding of a connection between what we might call the upper and lower body and between head or neck and genitals (see e.g. Laskaris, this volume, on sex and the eyes (Chapter 6); Llewellyn-Jones 2003: 264–7 discusses the implications for female veiling). The Hippocratic text *Epidemics*, for example, identifies a sympathy between nasal mucus and semen (6.6.8): 'Those whose noses are moist by nature and whose semen—*gonē* (γονή)—is moister and more copious: they are healthy. But those with the opposite condition tend to illness' (trans. Smith 1994). Although such connections between head and penis may seem somewhat counterintuitive, this is, for example, how in much more recent times the sperm whale got its name. *Physeter macrocephalus* as it is known to specialists has a waxy substance—*spermaceti*—in its head, used in the recent past in the production of ointments, candles, and other industrial processes, but originally thought to be the animal's semen (*OED s.v. spermaceti*). Something highlighted by many contributions in this volume is the identity of some fluids which we would keep separate as the same: *we* think semen is not the same as spinal fluid, but *they* did.

Moving to consider such principles of *linguistic partitioning*, some fluids have characteristics which—at least to us—appear to make them relatively straightforwardly identifiable: faeces exit the body through the anus or rectum and are (normatively) firm and brown; urine is a thin liquid passing through a fine passage farther towards the front of the body in both sexes; freshly shed blood is red.[7] These identifications rest upon classificatory principles based around texture or colour or the points at which substances are shed from the body, although speakers rarely contemplate such substances in these terms, and some other English speakers would perhaps define them differently. As already alluded to, it is well known that languages divide up the world according to principles which may differ between languages. A series of drinking utensils in English may be labelled as either *cups* or *glasses* depending on the material from which they are made, whereas these same utensils are labelled in Russian according to their shape (e.g. Malt 2015: 331). There is no tidy correspondence between English and Russian words because the conceptual categories in each language operate with different boundaries, and languages also have different degrees of specificity, e.g. *drinking vessels* versus *cups* versus *tea cups* (see L'Homme 2015). To come back to bodily effluvia, some types of urine may look (or smell?) more like urine than others. Likewise, is an English *sneeze* noise or the expulsion of fluid? The answer likely depends on which aspects of the corporeal phenomenon are stressed each time it is mentioned and which senses are invoked (see Introduction). Compare two Latin words for crying, *lacrimo* and *fleo*. The first is specifically in reference to the shedding of liquid tears (*lacrimae*), the second can include the noise; in Greek, πταρμός (*ptarmos*) 'a sneeze' is likely onomatopoeic in origin of the noise, compared with other ways of talking about the exodus of nasal fluids, on which more shortly.

How then do we see fluids divided up, conceptualised, and labelled in Greek and Latin? We might begin with an ancient list of such cardinal points of reference, also with apparently obvious essential properties. Here is a list of bodily fluids from the famous doctor and self-professed polymath Galen; the passage comes from his enormous pharmacological work, *On Simple Medicines*, from the beginning of Book 10, introducing discussion of the use of animal products in medicine (= 12.247 Kühn; see Petit 2017 for Galen's broader introduction to this part of his treatise and its significance):

ὥσπερ δ' ἐν τῇ τῶν φυτῶν ὕλῃ καὶ περὶ τῶν ἐξ αὐτῶν γινομένων χυμῶν τὸν λόγον ἐποιησάμην, οὕτως καὶ νῦν οὐ μόνον τῶν στερεῶν μορίων ἐν τοῖς ζῴοις ἡ διδασκαλία τῆς δυνάμεως, ἀλλὰ καὶ τῶν ἐν αὐτοῖς περιεχομένων ἔσται, **φλέγματος, χολῆς, αἵματος, οὔρου, κόπρου** καὶ **τῶν ὁμοίων**.

Just as in my *Material from Plants* I also wrote about the juices which come from them, so now too here follows instruction on the properties not only of the solid parts of animals, but also on that which is contained within them: **phlegm** (*phlegmatos*), **bile** (*cholēs*), **blood** (*haimatos*), **urine** (*ourou*), **faeces** (*koprou*), and **similar substances** (*tōn homoiōn*).

Galen, *On Simple Medicines* 10

Even if we accept that faeces, urine, and blood at least exist as relatively stable conceptual categories cross-culturally (whatever their linguistic manifestations, i.e. whatever the word is to refer to them), the body produces and sheds more substances than these alone; it also produces what Galen calls *ta homoia* (τὰ ὅμοια) 'similar substances'. We might identify these in English as a range of to us disparate and often-sticky productions: *sweat, tears, spit, mucus* (from diverse orifices), *serum, pus,* and *discharges, semen, vaginal secretions* (and *menstrual blood*, and the *mess of childbirth*), *milk, ear-wax, vomit,*[8] and perhaps adding for good measure the more-solid cut or lost *hair, dandruff, nail clippings,* and *scabs*. Note that English needs two words—*vaginal secretion*—to label that substance, and at least in this author's vernacular from central England *sleepy dust* is the only phrase available to talk about that sometimes-gooey stuff you get at the corners of your eyes, known in other dialects simply as *sleep*. (I'm also sure I don't use the word *serum* 'correctly'.)[9]

This broad and unstable lexis of English is mirrored in that which we find, on the face of it, for Greek and Latin, as witnessed in the handlists on the following pages. Each word is given with a truncated version of the gloss it receives in either Liddell,

Scott, and Jones's *Greek-English Lexicon* (*LSJ*) or the *Oxford Latin Dictionary* (*OLD*, second edition). These lists make no claims to comprehensivity, nor indeed do they aim to capture corrections made by individual studies to the observations made in the main lexica:[10] they are intended only to pull together some of the diverse ways fluids were labelled in all three languages (Greek, Latin, English) and demonstrate some of the shifts and connections already observed by generations of lexicographers. These lists reflect to some extent the twin differences cultural and linguistic—but also appear to give these words a false equality, glossing over variety of use, both spatial and temporal, and obscuring distributional patterns: some of these words are very common, some exceptionally rare.

Words indicating bodily fluids and their passing in Greek

αἷμα, -ατος, τό (LSJ: '*blood* . . . II. *bloodshed* . . . III. *kin*')

ἄφεδρος, ἡ (LSJ: '*menses muliebres*')

βλέννα, ἡ (LSJ: '= μύξα, *mucous discharge*'); also πλέννα, ἡ*
   [* LSJ's πλένναι is deleted by the *Revised Supplement* in preference for πλέννα]

βρότος, ὁ (LSJ: '*blood that has run from a wound, gore*')

γόνος, ὁ (LSJ: '*offspring*; . . . IV *seed*'); also γονή, ἡ ('*offspring*; . . . II *seed*')

δακρύω (LSJ: '*weep, shed tears* . . . 2. of the eyes, *run* . . . 3. of trees, *exude gum*');
   also δάκρυ, τό, also δάκρυον, τό (LSJ: '*tear*')

ἐμέω (LSJ: '*vomit, throw up* . . . abs., *vomit, be sick*')

θολός, ὁ (LSJ [A]: '*mud, dirt*, especially in water . . . 2 *menses*. II. *ink of the cuttlefish*')

θορός, ὁ (LSJ: '*semen genitale*' ['generative seed']); also = θορή, ἡ

ἱδρώς, ῶτος, ὁ (Aeol. ἡ) (LSJ: '*sweat* . . . 2. *gum, resin*')

ἷδος, εος, τό (LSJ: '*sweat*: pl., *sweats*')

ἰδίω (LSJ: '*sweat*'); also ἀνιδίω (LSJ: '*perspire* so that the sweat stands on the surface')

ἱδρόω (LSJ: '*sweat, perspire*'); also ἀνιδρόω '*get into a sweat*'; ἀφιδρόω '*sweat off*';
   ἐνιδρόω '*sweat in*'; περιιδρόω '*sweat all over*'; συνεξιδρόω '*sweat out together*';
   προϊδρόω '*sweat beforehand*'

ἰκμάς, άδος, ἡ (LSJ: '*moisture* . . . *animal juices* or *moist secretions*'); also ἴκμαρ

ἰχώρ, ῶρος, ὁ (LSJ: '*ichor, the juice*, not blood, that flows in the veins of gods . . .
   later simply *blood* . . . II. *the watery part* of animal juices, *serum* . . . *whey* . . . 2.
   serous or *sero-purulent discharge*')

καταμήνια, τά (LSJ: *s.v.* καταμήνιος -ον, II 2, '*menses of women*')

κατάρρος (LSJ: *s.v.* κατάρρος -ον, II. subst. '*running from the head, catarrh*')

κίκκασος (LSJ: 'ὁ ἐκ τῶν παραμηρίων ἱδρὼς ῥέων, καὶ βόλου ὄνομα, Hsch.')

κορυζάω (LSJ: '*have a catarrh, run at the nose*', II. '*drivel*'); also κορύζα, ἡ (LSJ:
   '*mucous discharge from the nostrils, rheum* . . . II. metaph. *drivelling, stupidity*')

κυψελίς, ίδος, ἡ (LSJ: '= foreg. [= κυψέλιον *bee*-hive], or swallows' or sand-martins'
   nests . . . II. *wax in the ears* . . . also κυψελίτης ῥύπος, ὁ')

λέμφος, ὁ (LSJ: '= κόρυζα, μύξα')

λαγνεία, ἡ (LSJ: '*the act of coition* . . . *semen*')

λήμη, ἡ (LSJ: '*a humour that gathers in the corner of the eye, rheum*')

μύξα, ἡ (LSJ (A): '*discharge from the nose* . . . *mucus, mucous discharge*, . . . *synovial
   fluid* . . . *slime*'); also μυξάριον, τό (LSJ: 'Dim. of μύξα')

μύσσομαι (LSJ: '*blow the nose*'); also ἀπομύσσω (LSJ: '*wipe* the nose, Med. *blow one's nose*')

ὀμείχω (LSJ: '*make water*')

οὐρέω (LSJ [A]: '*make water*'); also προσουρέω '*make water upon*'; also οὖρον, τό (LSJ [A]: '*urine*')

πτυαλίζω (LSJ: '*salivate*'), also πτυελίζω; also πτυαλισμός, ὁ, also πτυελισμός, ὁ (LSJ: '*salivation*'); πτύαλον, τό (LSJ: '*sputum, saliva* . . . also πτύελος, ὁ'); also πτύελον, τό

πτύω (LSJ: '*spit out or up*'); also ἀναπτύω '*spit up or out*'

πῦον, τό (LSJ: '*discharge from a sore, matter*'); also πύος, τό; διαπύησις, εως, ἡ (LSJ: '*suppuration*'); διαπύημα, ατος, τό (LSJ: '*collection of pus*')

ῥανίς, ἡ (LSJ: '*drop* . . . 2. *semen virile*')

ῥεῦμα, τό (LSJ: III '*humour* or *discharge* from the body, *flux, rheum*')

σίαλον, τό, also σίελον, τό (LSJ: '*spittle, saliva* . . . II. *synovial fluid*')

σπέρμα, τό (LSJ: II. '*seed, semen*')

φλέγμα, τό (LSJ: II. 2 '*phlegm,* one of the four *humours* in the body')

χολή, ἡ (LSJ: '*gall, bile* . . . II. *ink of the cuttle-fish*')

χρέμπτομαι (LSJ: '*clear one's throat, hawk and spit, cough*')

χυλός, ὁ (LSJ: '*juice* in general . . . 2. of *animal juices*, 3. *juice produced by the digestion of food*')

χυμός, ὁ (LSJ: '2. animal juices, "humours"')

Words indicating bodily fluids and their passing in Latin

bīlis, is f. (*OLD²*: 1. 'the fluid secreted by the liver, bile')

cruor, ōris m. (*OLD²*: 1. 'blood [fresh or clotted] from a wound . . . 2. The shedding of blood or an instance of it . . . 3. Blood in general')

effundō, -ere (*OLD²*: '3a. to shed [tears . . . blood . . .] . . . 3b. to discharge from the body')

flētus, -ūs m. (*OLD²* 'weeping, lamentation . . . tears')

lacrima, -ae f. (*OLD²*: 'tear [usu. in pl.]')

lōtium, -ī n. (*OLD²*: 'urine')

meiō, -ere, also *mingō* (*OLD²*: 'to urinate')

mensis, -is m. (*OLD²*: 4. '[usu. pl] The menstrual discharge in a woman')

mūcor, -ōris m. (*OLD²*: 'a mucous substance')

mūcus, -ī m., also *muccus* (*OLD² s.v. mūcus* 'mucus, snot')

pītuīta, -ae f. (*OLD²*: 1. 'mucus, catarrh, phlegm' . . . 2. 'a purulent or morbid discharge' . . . 3. 'a viscous discharge (from trees)'

pūs, puris n. (*OLD² s.v. pūs¹*: 'foul matter from a sore, pus')

rōs, -ris m. (*OLD²*: 'dew [. . .], NB not always clearly distinguished from rainwater . . . 2b. (applied to tears) . . . 2c. (applied to blood)')

sanguis, -inis n. (*OLD²*: 'blood . . . blood shed in violence')

saniēs (~ēī) f. (*OLD²*: '1. watery matter discharged from a wound, ulcer, etc. . . . 2 [transf., of various other fluids]')

semēn, -inis n. (*OLD²*: '4. semen, sperm')

serum, -ī n. (*OLD²*: 'whey . . . b. . . . (transf.) . . . any similar fluid, . . . i.e. semen')

spuō, -uere (*OLD²*: 'to eject saliva, etc. from the mouth, spit') *spūtum, -ī* **n.** (*OLD²*: 'spittle')

sūdō, -āre (*OLD²*: 'to sweat, perspire')

sūdor, -ōris **m.** (*OLD²*: 'sweat (arising from heat, exertion, or sim.) . . . sweat (produced by fear, anxiety, or sim.)'

uīrus, -ī **n.** (*OLD²*: '3. a secretion (in plants or produced in the body) having medicinal, magical, etc. potency')

ūmor, hūm-, ~ōris **m.** (*OLD²*: 'moisture . . . 2. a bodily fluid or discharge')

uomitus, -ūs **m.** (*OLD²*: '1. the act of vomiting. . . . 2. vomited matter, vomit')

uomō, -ere (*OLD²*: *OLD²* '1. To be sick, vomit . . . 2. To discharge [food, etc.] through the mouth, vomit or spew out')

ūrīna, -ae **f.** (*OLD²*: 'urine')

Producing such a catalogue of terms in English, Greek, and Latin raises a number of questions regarding the homogeneity of a 'vocabulary of fluids', and the edges of these handlists themselves are messy. For a start, a large number of general and more-vague terms are excluded from these lists, e.g. Latin *excrementum* 'that which is excreted' (from *excerno* 'excrete'), be it of *oris* 'mouth' or *naris* 'nostrils', and the Hippocratic τὰ ἀπίοντα (*ta apionta*) 'the things leaving', 'excretions' (*Epidemics* 6.8.8). More fundamentally, what exactly *is* a fluid? Ear-wax can be firm or soft, sputum sticky or runny, again assuming (with an Anglophone perspective) that a fluid is defined by the presence or absence of viscosity. Greek has the word οἰσύπη (*oisupē*; also οἴσυπος, *oisupos*), which indicates the 'greasy sweat and dirt of unwashed wool' (LSJ *s.v.*), additionally borrowed into Latin (*oesypum*, also *oesopum*): while not a word likely applicable to humans but animals, such a word presents a classificatory conundrum—solid or liquid? And where does the body finish? Hair is part of the body when attached to the head, but when it drops out, presumably it is not; tears and mucus are produced by the body but easily detach from it. For the editors of Lewis and Short, English 'saliva' is a general term, but 'sputum' 'is that already spit out' (*s.v. saliva*).

Metaphors present a number of additional challenges. Some of the inclusions in the list, e.g. Latin *ros* 'dew', are only rarely used in relation to material shed from the body, whereas others, e.g. example καταμήνια (*katamēnia*) 'monthlies', for women's 'periods' (note the same metaphor in English) are very common, and so common that their original metaphorical sense is virtually lost. Conversely, there is also the risk of the latent attachment of alternate meaning where none was intended, since in the case of metaphors the mapping of linguistic expression on the object indicated is always only partial. This is particularly pressing for sexualising metaphors—and not just for fluids but in general. Henderson's *Maculate Muse*, for example, lists among what he terms the *secreta muliebria* (in the sense 'female secretions'), ἔτνος (*etnos*) and δρόσος (*drosos*) (1991: 145), two words for otherwise domestic and meteorological words which usually mean 'soup' and 'dew' respectively. The vocabulary of habitualisation ('usually') in the previous sentence is important: Henderson reads these two sexualised, corporealised meanings for these words in Aristophanes, but the words are not regularly used, so far as we can tell, with this corporealised referential content (and indeed, not everyone would agree that sexual meanings are invoked everywhere in Aristophanes where these words are used).[11] These two words are not included in the previous lists, whereas Catullus' sexualised use of Latin *serum* 'whey', on the contrary, *is* included in the list (= 'any similar fluid, . . . i.e. semen') (see also Introduction). The margins of

23

these lists must not be interpreted as concrete, despite the impression their presentation gives. More on this problem of metaphors and their interpretation in the second half.

This slipperiness of labels for fluids is less surprising when we observe the evidence that labels for even solid parts of the body are also liable to shift. It is well documented in many languages that contiguous regions of the body often share or swap names. English *hip*, for example, is used to refer to two very different features of the body, indicating both the external outward bulge of the pelvis below the waist (perhaps more usually applied to women, whose anatomy often accentuates this feature), as well as to the internal joint of the head of the femur and the socket in the innominate bone of the pelvis. The phrase '*How's your bottom?*' in English can also have a very wide range of meanings, with *bottom* used to indicate the buttocks and surrounding area and local internal workings:[12] the Latin words *culus* and *clunes* show a similar tendency to be used in this indeterminate fashion for this area of the (Roman) bottom (Adams 1981: 233, 239), as does the rare Greek word φίκις (*phikis*) (see Katz 2004, 2006). This shifting of the conceptual boundaries is altogether more pronounced where the subject of linguistic division is not the 'unbroken surface of the body' but rather those boundless liquids oozing from within it.

The boundaries of the semantic range of words within the lexicon of fluids are as porous as the body itself, despite the apparent integrity of both. The inherent visual and tactile similarity between many of these sticky, clear substances makes semantic shift in the words used to identify them likely. Put differently, these words have an unstable relationship with that which they indicate, substances which are themselves infinitely mobile. We should then perhaps not expect words for at least some bodily fluids to have as straightforward a relationship with their *denotatum* as, for example, *book* does with a set of pages with writing on them, bound into a single unit. We may speak of partitions or partonomies the body, but bodily fluids are inherently indivisible and constantly in flux. You can't nail diarrhoea to a wall.

Sweat, urine, mucus, semen

The first half of this chapter has highlighted the inherently mobile nature of words for fluids, especially those for clear fluids of the body (Galen's *ta homoia*). Discussion now turns to specific examples of shifts within and between words for sweat, semen, and urine in Latin and Greek. Any attempt to pin down meanings within this set of vocabulary must, however, take into account what might be seen as 'playful' moments where the use of words is deliberately manipulated and the author assumes that the reader will recognise that the usual range of a word is being stretched for stylistic effect. In a fragment of Hipponax (*fr.* 73 West), someone 'pissed blood and shat bile' (ὤμειξε δ'αἷμα καὶ χολὴν ἐτίλησεν). The two verbs used here—ὀμείχω (*omeichō*) and τιλάω (*tilaō*)—are used elsewhere of urination (Hesiod, *Works and Days* 727) and of excretion of a thin bowel movement, respectively (e.g. commonly in the compound κατατιλάω (*katatilaō*) 'I dribble poo on' in Aristophanes, *Birds* 1054, 1117, *Frogs* 366, etc.). For Hipponax to use them with a pair of objects of the descriptively 'wrong' bodily fluid suggests something is seriously wrong with the body in question. However, we should be careful of labelling all such incidences of apparent semantic disjunction as 'playful' when it is clear that there is such a slippage between the sphere of reference of many words for bodily effluvia. (cf. King, this volume, Chapter 24, p. 383, on 'urine is sweated out, like milk from the breast'). Original playfulness or innovation can lead to a shift in meaning, if conventionalised through use.

The difficulty comes in identifying exactly what the central, core meaning of a word is for a speaker/writer and her or his assumed audience at any one time. Without this assumption, *ipso facto* the use of a word cannot be identified as metaphorical. It is likewise very difficult to tease out whether a word is used commonly in a metaphorical sense, or whether that same word has shifted in meaning, given that it is a regular pathway for semantic change that metaphorical expressions come to replace that which they originally only indicated figuratively. For example, English *to blow the nose* is a metaphorical term, albeit one which is in such common use that it has virtually become the only way to express the action of removing mucus from the nose (*OED s.v.* blow, *v.* 1, sense 21 records its use from the middle of the sixteenth century), a nose which might at other times be *running*. To give a comparative example, French *tête* 'head' is in origin a metaphor from Latin *testa* 'pot'—but only when viewed with the benefit of historical hindsight. For speakers of French, *tête* just means 'head'. Metaphors have different degrees of lexicalisation (e.g. Crespo-Fernández 2015: especially 59ff.), and this degree is hard to quantify.

A more complex example comes from Aristophanes' Chorus in *Birds* (790–2), who describe someone undergoing an urgent need to use the toilet as 'oozing out' onto his cloak (text Wilson OCT):

εἴ τε Πατροκλείδης τις ὑμῶν τυγχάνει χεζητιῶν,
οὐκ ἂν **ἐξίδισεν** εἰς θοἰμάτιον, ἀλλ' ἀνέπτατο,
κἀποπαρδὼν κἀναπνεύσας αὖθις αὖ καθέζετο.

And if some Patrocleides among you were taken by a need to shit, he wouldn't have **oozed out** (**exidisen**) onto his cloak, but have flown away, farted, got his breath back, and sat down again.

Aristophanes, *Birds* 790–2

The compound verb here (*exidiō*) is highly unusual, found only once elsewhere much later (at Cassius Dio 44.8.3 in reference to Julius Caesar allegedly suffering an attack of diarrhoea, perhaps in imitation of the Aristophanic usage), and the simplex form ἰδίω (*idiō*) refers to sweating, not the passing of solid(-ish) matter from the anus (or gas, so Sommerstein 1999: 201). The Greek verb *idiō* 'I sweat' and family (including the nouns ἶδος [*idos*] and ἱδρώς [*hidrōs*] both 'sweat', and the later denominative verb ἱδρόω [*hidroō*] 'I sweat') are related to Latin *sudo* 'I sweat' and *sudor* 'sweat', which is strongly indicative of an original meaning of 'sweat' for this verb (reconstructable as an Indo-European root for 'sweat', *LIV* * su̯ei̯d, cf. Beekes, *IEED* 10 *s.v.* ἰδίω; Pokorny *s.v.* 2. *sueid-*). The question as to how common this compound *exidiō* was in this innovative meaning in Greek is unanswerable, given the paucity of preservation of colloquial registers, but nonetheless central to the problem: both sweating and defecation are here conceptualised as types of shedding fluid-like material from the body but through disparate routes. Determining whether Aristophanes' use here would have been seen as metaphorical depends upon whether *idiō* is viewed as having a general sense of *oozing* or a specific one of *sweating* for his audience.

Shifting from sweat back to urine, in Catullus' tabloid-esque poem 67 (for Richardson 1967 a 'Fescinnine playlet' appropriate for a wedding procession), in which a door tells of the scandalous goings on within the house he guards, we hear of a father who *minxerit* in the *gremium* 'lap' of his own son (67.29–30; trans. Lee 1990):

> egregrium narras mira pietate parentem,
>   qui ipse sui gnati minxerit in gremium.

> You're talking of astonishing devotion and the finest
>   Of fathers! **Peeing** in his own son's lap!

Catullus, 67.29–30

Latin *meio* or *mingo*[13] (perfect *mixi/minxi*) is usually used of the act of urination, but in Latin 'verbs meaning "urinate" are often used of ejaculation' (Adams 1982: 142; and cf. Catullus 97 in the Introduction, p. 2). Likewise, *urina* in Latin can refer to (English) 'semen' (Adams 1982: 92, citing Juvenal 11.170[14]), i.e. the sexual fluid, rather than the excess liquid (as we know it) filtered by the kidneys and disposed of from the body via the bladder. The use of the vocabulary of urination for an alternate kind of 'male leaking', i.e. ejaculation, is also widespread in Greek in the Classical period: Henderson (1991: 176) notes Aristophanes' *Frogs* 95 where the compound verb προσουρέω (*prosoureō*), 'urinate on', indicates sex associated with 'impotence and sterility'. Aristotle uses another compound of the same verb, ἐξουρέω (*exoureō*), of female asses who 'urinate out' or 'shed' semen (*History of Animals* 577a23 with τὴν γονήν [*tēn gonēn*], *Generation of Animals* 748a23 with τὸν γόνον [*ton gonon*], cf. Catullus 97 *meientis mulae*). The same conceptual connection was also in operation earlier, as witnessed by the development of μοιχός (*moichos*) 'adulterer', almost certainly in origin an agent noun from *omeichō* 'urinate' (Wackernagel 1916: 225 n. 1; *DELG s.v.* μοιχός; indeed, Latin *meio* and Greek *omeichō* are likely related, *OLD*[2] *s.v. meio*, although this may not go back to Indo-European). An adulterer is thus conceptualised as one who has 'pissed in' or 'watered' another man's wife (*minxerit in gremium*), problematic because of the challenge to secure paternity presented by unfettered liquid access to the female body by another man. In this connection we might also compare Catullus' *spurca saliva* of a *commictae lupae*—'the dirty saliva of a foul—or "pissed-in"—whore' (99.10), whereby 'foul' or 'polluted', as the verb is often translated, e.g. *OLD*[2] *s.v.* (*commeio*), surely arises from a sense of 'polluted with (male) fluids'.

Closer to home, a very similar pattern of shift between these various leakages can be observed in English of the late sixteenth century: Partridge has this to say about the Shakespearean phrase '*piss one's tallow*':

> Literally, to lose weight by freely sweating; hence, in *Merry Wives*, V 15–16[15] [. . .], to be so sexually excited as to experience a seminal emission. Originally, hunting slang; its appositeness to Falstaff is clear—did he not 'lard the lean earth' with his sweat?

Partridge 2001: s.v.

In Shakespeare's language we again see a transference of the vocabulary usually associated with urination ('piss') to another kind of corporeal leakage, sweating ('piss tallow'); this phrase is then additionally used to indicate ejaculation, another way in which the male body can lose fluid. Sweating as a way of losing fluid in general already has a parallel in the anal 'sweating' of Aristophanes' Patrocleides (*exidiō*); to this we might add the identity of urine and sweat according to Theophrastus (see the Envoi, this volume, pp. 399–401). While we can trace all these patterns in retrospect, the intentions of speakers and authors in deploying them are of great importance. Catullus chose to express the father's illicit sexual activity through *minxerit*, and Aristophanes makes his audience member *ooze* out, rather than expressing these ideas with an alternate selection of words. The next section sinks further into examining such negative expressions involving a particular bodily fluid, semen.

Did the Greeks have a word for 'spunk'?

Even if we were to have a perfect understanding of Greek and Roman bodies and the conceptual categories and degrees of partitioning which were at work, as already hinted at, the very idea of what a word 'means' is complex and encompasses more than just the indication of a substance or entity. Despite the centrality in traditional Western linguistics of this 'referential' or 'denotational' meaning which connects a linguistic unit and the world, words can convey socially embedded meaning of various kinds and do more than only indicate the item to which a speaker refers. Riemer (2015: 311) characterises this latter component of a word's meaning, variously described as 'expressive' (Riemer 2015: 311) or 'connotational' (Allan and Burridge 1991: *passim*) or conveying 'emotional overtones' (Geeraerts 2015: 422), as its 'penumbra'. By way of example, compare the English words *cat* and *pussy*. Both indicate felines, but the latter additionally expresses affection for the animal—and in contemporary UK English has the additional meaning of female genitalia. Each word exists for a speaker as a member of the entire lexical stock of a language (or languages) and has some kind of conceptual mental representation—but it also exists as a token used in communication, bundled up with other information about, for example, usage patterns and frequency (see Fellbaum 2015 for a sketch). As Geeraerts puts it (2015: 417), words have 'meaning potential' in that they only function as communicative tokens when understood by their audience and that understanding draws upon a rich array of knowledge about the use of any word. The shadowy penumbra of socialised meaning is usually seen as secondary to the fundamental function of a word as indicating an entity in the world, but it nevertheless exercises a strong hold over word choice ('Have you seen my cat/pussy?'). Some words for example are used only as technical terms within the parlance of the (medical) expert, others only appropriate to colloquial registers: compare the distributions of *urine* and *piss* in English. It's not just *what* you say but how you say it. And for some uses of inherently offensive words and expletives, it is the *only* principle which determines word choice, e.g. English *fuck off*, in which the semantic, referential meaning has been bleached, and it is pragmatic function alone which dominates. The phrases *fuck off* or *fuck me, that's brilliant* do not have anything to do denotationally with *making love* or *having sex*. Likewise, at Aristophanes' *Frogs* 753 we hear a slave say he, to use a current English vernacular paraphrase, 'jizzes in his pants' at something he really enjoys doing: μὰ Δί᾽ ἀλλ᾽ ὅταν δρῶ τοῦτο, κἀκμιαίνομαι (lit. 'By Zeus whenever I do this, I even *defile myself* [*ekmiainomai*]). This does not mean that extreme gossiping (in this case) actually causes the slave to ejaculate on

27

himself but is a way of expressing a pleasurable reaction to something.[16] Some words are socially coded in ways which alienate them from their referential meaning.

English has a number of highly tabooed words, in the sense of words which are usually restricted from use in everyday 'politic' conversation, which are associated with bodily functions, including some fluids, for example *fuck, shit, piss, bollocks,* and *cunt.*[17] The catalogue of offensive terms will differ between languages since it is predicated on cultural norms and cultural concerns. When working with historical languages, those anxieties must be reconstructed if a correct catalogue of offensive words is to be assembled; hence the title of this section: did the Greek language have a word which indicates semen but which most speakers would censure as part of their usual discourse? Jeffrey Henderson (1991: 35) notes that Classical Greek appears not to have had a word connotatively equivalent to English *piss.* This already highlights that we need not assume there is a universal, English-like, catalogue of *realia* which is always represented in the lexicon by terms which speakers treat as tabooed in their speech patterns (cf. Coker 2019: 191–2) and that not all words or linguistic tokens for fluids may have the same social significances as in English. A fundamental additional problem with all historical evidence of course is that some topics may be so subject to social sanction that they are not talked about at all or are restricted to groups whose language is largely unrecorded, e.g. women (see Salvo, this volume (Chapter 3) on menstruation). Ideas of dirt and pollution and disgust may thus play a role at some level in the selection of topics which are negatively marked, but any rationalisation of offensive language appealing to disgust cannot be straightforward.[18] To give one modern example, a woman not disgusted by her own genitalia or those of other members of the same sex can still call her next-door neighbour (male or female) a *cunt* (and—I think—remain a feminist). However, the same woman may be anxious about discussing genitalia openly in all contexts and use a range of different words (e.g. *bits, vagina*—see Braun and Kitzinger 2001 for a recent selection), despite her own comfort with her body. This is a reminder that offensiveness can only exist as a concept when understood within a community, since it is predicated upon the reactions of a speaker's audience and the speaker's attempts to manage their own status with the communities within which they want to be respected.[19] Words are always social signs, as well as tools which facilitate the communication of concrete facts.

To stick with semen, there are various ways of talking about this substance or *denotatum* in English, for example *semen, seed, jizz, spunk.* Each of these words has its own penumbra of features of which fluent speakers are fully aware, and which assists—or restricts—its selection and deployment. Put another way, this knowledge assists speakers or writers in choosing which of a range of denotationally similar words is appropriate for their desired aims in a particular moment in discourse. The semantic and pragmatic meanings of words are thus not so distinct as they are often described to be by traditional linguistic approaches (Allan 2007), and the boundary between descriptive and expressive meaning is indistinct. The challenge for Greek and Latin is in reconstructing the penumbra or connotations of each word as they existed for speakers (see Coker 2019 on the use of text-type as an indicator of inherently offensive connotative value). 'Semen' in Greek can be indicated by the words γόνος (*gonos*), θορός (*thoros*), θορή (*thorē*), λαγνεία (*lagneia*), ῥανίς (*rhanis*), σπέρμα (*sperma*), and σπόρος (*sporos*), among others (see also the list pp. 21–2), each of which for ancient speakers at different times and places would have been bundled up with a set of information which dictated or restricted its use. In addition, there exists a wide array of metaphorical phrases, including Archilochus' λευκὸν μένος (*leukon menos*) 'white strength' (*fr.* 196a West) and perhaps the στιλπναὶ ἔερσαι (*stilpnai eersai*) 'glistening drops' in the *Iliad* (14.351).[20] Oblique references or metaphors for male ejaculate suggest that this is something which at times required self-censorship, unlike, e.g. urine. However, metaphors do not have to be ameliorating (so Sommerstein 1999: 183–4) but can stress less-pleasant aspects of a *denotatum* for various effects, as already seen in Catullus' *minxerit in gremium* of the previous section. Kenneth Quinn's commentary on Catullus dubs the phrase 'no doubt a colloquial euphemism' (1970, *ad l. 30* = [371]) and rephrases coyly as '= "A fine father he was!"' (and for the famously prudish Fordyce [1961], poems 67, 97, and many others fall within the group of those 'which do not lend themselves to comment in English' and are thus not printed). What is questionable is, however, whether this is a euphemistic expression or something rather more negative or *dysphemistic*.[21] Catullus does not here use the kinds of Latin words which are described as having offensive content (e.g. *mentula*, *futuo*, on which see Adams 1982: 9–12, 118–22, respectively), but this does not mean the phrase he does use is ameliorating and designed to shield his readers from the act he describes through vague, obfuscating circumlocution—as it seems Tacitus (*Annals* 6.28) does in his somewhat speculative account of the Phoenix pouring out its *vis genitalis* for the purposes of procreation. Indeed, Catullus was likely intending to do precisely the opposite, stressing the physical, fluid-focused reality of illicit sex and the (damaging? amusing?) social anxieties it creates. Through the words chosen to carry a specific referential value (i.e. indicate a specific act), Catullus manipulates that referent in the eyes (or ears) of the audience, additionally conveying a negativising evaluative value: he attempts to evoke disgust or repulsion at the act and the participants in it by labelling the act the way he does. Metaphors offer an alternative way of comprehending reality, but a reality which is manipulated for the audience by the one employing that metaphor (Crespo-Fernández 2015: 61–8). The remainder of this chapter offers two such negativising examples of semen from Greek, one from the well-known Aristophanes, one from an anonymous hand from Hellenistic Egypt. These come as close as anything as being words for 'spunk' in Greek.

1. Aristophanes' *Knights* 910: 'expel your mucus'

In the middle of what is often seen as Aristophanes' most political play, *The Knights*, The Paphlagonian (a thinly veiled Cleon) and The Sausage-seller compete for the favour of the old man who is their master, Demos, or 'The People'. At the peak of a sequence of increasingly personal services both offer to Demos (871–910), running from the bottom of the body to the top, The Paphlagonian utters the following memorable line (*Knights*. 910, text Wilson 2007; trans. Henderson 1998):

ἀπομυξάμενος, ὦ Δῆμέ, μου πρὸς τὴν κεφαλὴν ἀποψῶ.

Blow your nose, Demos, and wipe your hand on my head.

Aristophanes, *Knights* 910

The emboldened word, a compound of μύσσομαι/μύττομαι (*mussomai/muttomai*), is given by LSJ (*s.v.* ἀπομύσσω, *apomussō*) as meaning in the middle voice to 'blow one's nose' (active transitive 'wipe').[22] The closely connected noun μύξα (*muxa*) and its

29

family (e.g. μυξώδης, LSJ *s.v.*, '*like mucus, abounding in it*') indicates mucus or slime of various kinds. This phrase in English 'to blow the nose' is, however, essentially a metaphorical and euphemistic term for the clearing of mucus from the nasal cavities. The way English talks about the curation of bodily fluids in this instance has obscured the full meaning of the verb as it is commonly understood by Anglophones: while 'blow the nose' conveys the correct denotational sense (the action could be re-enacted correctly, for example, using this as a translation), this phrase essentially bowdlerises the meaning by using a bland metaphor, albeit one which is in regular use. 'Blow your nose' adds a layer of vagueness which is not present in the Greek and is akin to translating Latin or Greek sex-scenes with coy metaphors, an example of the modern 'impulse towards euphemism' (Introduction p. 3), albeit one which is not in this case deliberate: the difference is that because English speakers are aware that they have an inherent anxiety about discussing sex directly, it may feel more natural to do so. A more literal translation in English of the line would be:

Expel your mucus/Snot off, Demos, and wipe yourself on my head.

Note that in the Greek text there is also no mention of the use of the hands to blow, wipe, or clear the nose—nor indeed is there mention of the nose. It is so embedded in modern Western practice to use the hands, usually in conjunction with a piece of paper or cloth, when clearing the nose, that a furtive 'hand' has crept into the first translation.

All commentaries agree that this is extreme 'sycophantic self-abasement' (Sommerstein 1981: ad loc.) because the head, the purest part of the body, is being offered up as a place to dispose of someone else's bodily emission. This is bad enough, but given what we have seen above concerning the transfers between words for fluids—and their sexualisation—this invitation for Demos to produce mucus and then clean himself up may be read as referring not to that coming from the nose but from another producer of 'mucus', the penis. Note that the verb *apopsō* (ἀπουψῶ) 'wipe' is also middle, and it is left ambiguous as to what part of the subject's own body is being cleaned: it is the same verb used for example by Carion in Aristophanes' *Wealth* to exemplify his change in fortune by now being able to use garlic leaves rather than pebbles to wipe his *bottom* (*Wealth* 817–18).[23] While this might be seen as an(other?) example of the reading of sexualised metaphors where none were intended (*viz.* the Classicist's dirty mind at work), there are a couple of compelling reasons for connecting mucus with semen in the case of *mussomai/muxa* in particular.

Firstly, there is a striking example of just this conceptual connection in a Latin graffito from Pompeii, *CIL* IV 1391. The detail of the second half of the text is the subject of continuing debate, particularly vigorous in discussions concerned with the presence of venereal syphilis in Roman Europe, but it is the first half which is of interest here (text and trans. Solin 2008: 64–5):[24]

Veneria	1
Maximo	
ment(u)la(m)	
exmuccaut	
per vindemiam	5
tota(m)	

et relinque(t) (*or relinquit*)
putr(em) ventre(m)
mucei
os plenu(m)   10

Veneria has **sucked** the cock of Maximus throughout the vintage, leaving her rotten womb empty, but her mouth full of snot.

CIL IV 1391

Even if we are not to read a form of *mucus* 'snot' in line 9, there is agreement that *exmuccaut* in line 4 refers to a sex act involving a sticky fluid: Ferri (2008) translates as ' "(she) dried (me) clean of mucus/snot" (i.e. semen).'[25] The *OLD*[2] (*s.v. exmucco, -are*) translates the verb as 'cause to discharge mucus', citing only this text, and the verb with which this form is linked, *emungo*, is given by the *OLD*[2] (*s.v.*) 'wipe mucus from [the nose, etc.]' (there is no entry for **mungo* or **mucco*). The presence of the object *mentulam* in line 3 (here translated as '*cock*') makes this sexualised meaning abundantly clear, and note that it is a tabooed Latin word for 'penis' which is used here. The same Indo-European root seems to lie behind Latin *emungo*, and Greek μύξα (*muxa*) and μύσσομαι (*mussomai*) (*LIV* *(s)meu̯k, 'slip, slide'), but it is not certain that *emungo* and *mucus* belong to this same root; *mucus* might rather belong historically with words for 'moist' (de Vaan *IEED* 7 *s.v.* mūcus), but between these groups the semantic jumps (or slips) are only small.

Secondly, whatever the direct etymological connections between these mucus words, there are in addition a few strictly etymologically unconnected—yet on the face of it similar sounding—words beginning with μυσσ-/μύζ- (*muss-/muz-*) in Greek with sexualised senses. It is suggested here that these may have reinforced a possible conceptual association between *mussomai/muxa* and sex in Greek. Of interest firstly is ἀπομύζουρις (ἡ) (*apomuzouris*), a prostitute name or term recorded in Suetonius' *On Abusive Words* 2 (ed. Taillardat 1967; also see *PCG* *192 K-A), and elsewhere found only in the scholarly tradition. Suetonius etymologises the word as a compound of οὐρά (*oura*) 'tail' and a verb for 'suck', μύζω (*muzō*) (or μυζάω [*muzaō*]); *muzō* is more regularly attested as meaning 'moan' or 'mutter', and both this sense and 'suck' are onomatopoeic in origin, meaning 'to say or do *mu*'.[26] Similarly, another word for a prostitute, μυσάχνη (*musachnē*) (also recorded in Suetonius *On Abusive Words* 2, but also rarely elsewhere)[27] is perhaps indicative of a connection between sex and pollution. *Musachnē* is likely connected with μύσος (τό) (*musos*) 'uncleanliness', which in turn may perhaps be related to the verb μυδάω (*mudaō*) 'be damp, clammy, decay' (*DELG s.v.* μύσος). Gerber (1999: 223 [= *Archil. fr.* 209]) thus translates *musachnē*, a term for a sex-worker, as 'froth of defilement'. Here might also be adduced the impurities indicated by the μυσάγματα/*musagmata* of *P. Ryl.* 3.469 (col. ii, lines 33–5), a late third-century CE papyrus letter to a Christian community in Egypt warning of the dangers of Manicheanism, be they actually physically defiling substances or abstract moral 'abominations'.[28] Finally, and very fragmentary, there are two perhaps-related words attested only in Hesychius' *Lexicon* as meaning the genitals, μύσχον (*muschon*) (μ 1987 Latte) and μυττός (*muttos*) (μ 1999 Latte), and Hesychius also records the verb μυσιᾶν (*musian*) (μ 1944 Latte) for breathing or panting during sex (ἀναπνεῖν, ἢ συνουσιάζοντα πνευστιᾶν), likely related to *muzō* in the sense 'say μυ'.[29]

What we have perhaps here are a large number of transfers and confusions: not many of these connections may be strictly etymological *per se*, but they nonetheless tell us something about the way denotata are conceptualised and connected in the minds of speakers, especially the connection between negativised sex, dampness, and decay. As Katz (2006: 182–4, especially n. 70) notes on the Greek words 'Sphinx and the anal sphincter' (Σφίγξ [*sphingx*] and σφιγκτήρ [*sphingktēr*]), any connections between them '[f]rom a strictly linguistic point of view . . . are not easily equatable', but yet 'this does not mean that there cannot be a link at the level of folk linguistics'. And it is at the level of folk linguistics which most speakers operate. Viscous fluid can be reimagined as something else, and this group of words in Greek starting with *mux-/muss-/muz-*, and Latin *muc-*, may be stained with mucus and semen, even though they have no formal etymological connection with *muxa* itself.

This discussion argues that Aristophanes' metaphor is deliberately unpleasant, conjuring the visceral and sticky aspects of sex and ejaculation. However, the same verb *apomussomai* is used by Xenophon relatively closely in time in his *Education of Cyrus* (1.2.16, c. 370 BCE) in a delicately described sketch of what we might term the 'manners' of the Persians concerned with bodily continence: here the expulsion of mucus (τὸ ἀπομύττεσθαι, *to apomuttesthai*) is presented in parallel with spitting (τὸ πτύειν, *to ptuein*) and the appearance of being full of wind (τὸ φύσης μεστοὺς φαίνεσθαι, *to phusēs mestous phainesthai*) (later we hear of the shame of being seen withdrawing 'for the sake of urinating, or anything else of that sort'). This suggests that a conventionally offensive or dysphemistic meaning did not attach to the verb *per se*—or at least not to the extent that it was unusable for Xenophon in writing an account of a foreign royal power a generation or so later. This does not mean that Aristophanes' *use* of the same was not sexualised, nor highly offensive. As highlighted in the discussion of Catullus' *minxerit in gremium*, metaphorical uses of words can be very offensive, if a speaker chooses through the metaphorical remapping of an item or action to redefine it through different and less pleasant aspects.[30] Aristophanes' trick here—assuming this verb was not associated regularly with sex—is to make his audience remap ejaculation as the removal of mucus rather than anything else. It may well be that *apomussomai* in everyday parlance in fifth-century BCE Athens did just mean the bland English equivalent of 'blow the nose', but (a) given the conceptual connections between semen and mucus, the verb was ripe for comic exploitation by a talented, crude playwright and (b) the lack of a euphemistic term in Greek for nose-blowing tells us about a comparative lack of cultural anxiety about nasal mucus when compared with the culture of the twenty-first-century western Mediterranean.

2. Clitorius' Epitaph (*SH* 975): 'ἐνκέφαλον [enkephalon]: id est semen?' (Lloyd-Jones and Parsons 1983: 488)

It might reasonably be suggested then that some ways of indicating semen were considered to be offensive by their original audiences, even those which are metaphorical. This chapter finishes by discussing one such likely example, 'he sprinkled his brain', found in a mock epitaph for a certain Clitorius, inscribed on an ostracon, in third-century BCE Egypt (*SH* 975 [= Lloyd-Jones and Parsons 1983]; *FGE* #147, lines 1686–9 [= Page 1982, 458–60]). The normalised Greek text here is the *recto* side as printed by Livrea 1987 (who restores four elegiac couplets, albeit with loss of the final pentameter). The English translation (author's own) broadly follows Livrea's under-standing of the syntax (1987: 25); the scene described seems to have taken place dur-ing a bout of wrestling, hence the reference to the 'ladder hold' in line 3:[31]

ἐνθάδε Κλειτόριος κεῖται, δρῖλον καλυκῶδες·   1
   ὡς ἔλαβ᾽ εἰς κλόπιον τὸν χαλαρὸν τὸ χρέος,
(ἦν δ᾽ ἐπὶ κλιμακίοισιν) ἰδὼν ψωλὴν ἐπέτεινε
   καὶ **ῥανίσας** εἴσω Κρούριος **ἐγκέφαλον**
ὡς ἐβόασ᾽· 'αὔεις;' < ˇ ˇ ˇ ˇ ˇ ˉ ˉ ˉ   5
   ˉ ˇ ˇ ˇ ˇ ˇ > 'μᾶτερ, ἐμοὶ βοήθει.'
'< ὦ > τέκνον, τίς ταῦτα θεῶν [ˇ ˉ ˇ ˇ ˇ ˇ
   . . . .]

Here lies Clitorius, with his bud-like prick (*drilon*).
When Crourius took hold of the soft one (*i.e. Clitorius*) for his furtive purpose
(he was in a ladder hold), on sight of his hard-on (*psōlēn*) he stiffened
and **having sprinkled his brain** (*enkephalon*) inside,
thus he shouted: 'Are you aflame? . . .
   . . .   'Mother, help me.'
'Child, who of the gods . . . these things . . . ?'
   . . .

SH 975

This short and highly sexual text has many memorable features. One is several Greek words which, by their patterns of distribution, can be fairly safely identified as sitting at the more offensive end of the spectrum: this is certainly the case for δρῖλον (line 1) (*drilon*) and ψωλὴν (line 3) (*psōlēn*), both words for the penis.[32] A second is the striking phrase in line 4, 'having sprinkled his brain (inside)'. Taillardat (1989: 207–8) interprets this as 'comic hyperbole', but Livrea (1987: 26) as ejaculation (answering Lloyd-Jones and Parsons' question reprinted as the title for this section—'ἐγκέφαλον: *id est semen?*' *enkephalon*: is it semen?' in the affirmative). Given the close connection between head and semen, the transfer of 'brain' or the contents of the head as liquid ejaculate should now be conceptually unsurprising—and indeed in English the phrase 'to screw someone's brains out' (*vel sim.*) perhaps approaches this idea of an empty-ing of the body through sexual exertion.[33] The connections between 'generative' and 'mental' strength is also found in Greek μήδεα (*mēdea*, 'thoughts' or 'genitals'), and in other languages (see Nagy 1974: 265–6).

It is difficult or impossible to make statements about the currency of such a metaphor given the scant evidence which survives from antiquity. Again, though, this metaphorical phrase 'sprinkle the brain' places a stress on the physicality of the fluids involved, with no attempt (to borrow a phrase from Allan and Burridge 1991) to shield the reader from what is being described. Indeed, on the contrary, the author forces this image upon the audience, making them complicit in the creation or understanding of the metaphor by the act of opening their eyes to what is on the page, through their attempt to map the words onto some kind of hypothetical reality. Readers and listeners always try to make meaning—nasty or nice—from the linguistic information they are presented with, filling any metaphorical gaps as necessary.

Concluding remarks

Lloyd-Jones and Parsons in their commentary on the Clitorius text give the following short assessment of it (1983: ad loc.):

epitaphium ludicrum. mortuus est Clitorius: **cetera tam obscura quam obscena.**

A playful epitaph. Clitorius is dead: **the rest is so obscure as to be obscene.**
Lloyd-Jones and Parsons 1983: ad loc.

We run into the *argumentum ex silentio* here: many expressions for unsavoury aspects of the body, and the body construed as unsavoury, were likely just not recorded or were one-off, *ad hoc* coinings. The brain-sprinkling we witness in this epigram may not have had wide currency, in the same way as the side-effects of a chesty cough are not usually described in English as '*a blob of lung butter heaved from the chest*', as they were by a caller to a national UK radio station in February 2016.[34] Both these metaphorical phrases are, however, comprehensible when interpreted through the lens of the cultural conditions under which both were made: the first makes sense since Greek physiology knew that internal conduits ran from brain to penis, and that the brain was not a solid; the second works because of a textural and chromatic similarity between a commonly consumed dairy product and phlegm—and the nasty inversion of ingesting what the body is trying to get rid of and the confusion of food and bodily waste.[35] *Enkephalon* and *muxa* might not quite have the commonplace equivalence of the English word *spunk*, but on the strength of the ways in which semen is labelled in Greek, such a similar word must have existed.

Lloyd-Jones and Parsons' comment also acts as a warning, since it is important not to assume everything which is at first look incomprehensible is a crude sexual or scatological metaphor, tempting as that may be. However, this process of defamiliarisation (so described by Beard 2002) can in turn be reconstrued as a careful *refamiliarisation*, but refamiliarisation anew with a set of principles about naming and sociolinguistically encoding aspects of the experience of the body—and anxieties which revolve around it—in ways which from our own cultural perspective we would not always expect. For Stoic Emperor Marcus Aurelius, sex can be pared back to a 'secretion of a bit of mucus' (*muxariou ekkrisis*, μυξάριου ἔκκρισις, 6.13);[36] for Philip Larkin in 1950 it could be repulsive, and 'like asking someone else to blow your own nose for you' (quoted in Motion 1993: 119). *Urban Dictionary's* 'sex wee' underlines the enduring semantic connection between clear fluids of various kinds,[37] and the boundless possibilities for indicating in words the stuff which oozes out of us, words which are chosen to manipulate liquid reality. The body is messy, but so is the language we use to talk about it.

Notes

* My thanks in particular to audiences in both Cardiff and Exeter for the many useful—and often wonderfully unpleasant—comments on this chapter, as well as to all those who have supplied intellectual, emotional, and physical support. The initial work was undertaken during the final year of a Leverhulme Early Career Fellowship in the Department of Classics and Ancient History at the University of Manchester (2013–16), during which time Professor David Langslow was a constant voice of kindness and wisdom. The production

of a final version was hampered by the unexpected: its aftermath has echoes on many of the subsequent pages.

1 *PG 63, 659* (ed. Migne): ἡ γὰρ τοῦ βλεπομένου κάλλους ὑπόστασις οὐδὲν ἕτερόν ἐστιν, ἢ φλέγμα καὶ ῥεῦμα καὶ αἷμα καὶ τροφῆς διαμασηθείσης χυλός.

2 As Beard puts it (2002: 51) '[i]t is a lexicographer's nightmare (or, maybe, delight)'—this author has experienced both.

3 Majid (2015: 378) notes attempts to force the anatomical domain of languages into the tidy order of one's own. Work in Natural Semantic Metalanguage most associated with Anna Wierzbicka (e.g. 2014) attempts to move away from these problems of Anglocentrism through the establishment of 'semantic primes', but it is not uncontroversial.

4 The same lemma states it also means a dice-throw (βόλου ὄνομα, *bolou onoma*); differently spelled κίγκασος (*kinkasos*) is also glossed as meaning a dice-throw (κ 2605 Latte). The connection between the two meanings may be negative value (compare English 'a shit/bollocks hand of cards'); other perhaps related lemmata in Hesychius which have such similar bundles of negative value are κ 2651 κίκκη (*kikkē*) 'the unpleasant smell which comes from the genitals', κ 2652 κικκίδαι (*kikkidai*) '?excrement', κ 2653 κοκκιλόνδις (*kokkilondis*) 'child excrement', and κ 2654 κικκός (*kikkos*) 'rooster; thief; excrement'.

5 In Aristophanes' *Lysistrata*, foaming sweat (*aphros*, ἀφρός, l. 1257) running down legs of the Spartans is a positive indication of their physical prowess.

6 My thanks to one of the participants from the USA at the Cardiff conference for bringing this to my attention.

7 Whichever theory of meaning is subscribed to (there is more than one), i.e. how a word maps onto an entity in the world, meaning is graded: either some entities fulfil more or fewer conditions for class membership (the 'Classical' theory) or are better or worse examples of a category (prototype theory). See Riemer (2015: 313–17) for an overview; also Geeraerts (2010), Elbourne (2011).

8 Vomit should perhaps be treated separately, as the contents of the stomach leaving the body regularly look broadly similar to what recently entered the body as comestibles. Vomit certainly featured very little, if at all, at the Cardiff conference in which this volume has its origins, despite the mouth as a prominent opening of the body (and in some respects opposite to the anus—vomit can indeed be referred to as 'mouth poo'). For new work on vomit in Latin, see Goh (2018).

9 To paraphrase a TV advert for an over-the-counter cold remedy in the UK in the winter of 2017, there is apparently a difference between a *mucus cough* and a *catarrh cough*, but at least to this author this seems to be a marketing ploy.

10 There are many fine studies on individual words: to name but a few, Jouanna and Demont (1981) on ἰχώρ (*ichōr*) and the notes in Craik (2019) on σίαλον, μύξα, μυελός, ἰκμάς, κατάρροος, δρόσος, and γονή (*sialon, muxa, muelos, ikmas, katarroos, drosos, gonē*).

11 A sexualised use of *drosos* in reference to Ariphrades' preference for performing oral sex on women at *Knights* 1285 seems certain; Henderson (1991: 145 n. 194) challenges Dover's interpretation of *drosos* as a male sexual fluid at *Clouds* 978.

12 This question was posed to the author during a spell in hospital in February/March 2018: clarity was required as to whether this was an inquiry about musculo-skeletal pain, bowel movements, or bedsores, any of which were possible in the context.

13 OLD² *s.v. meiō* states that present forms of *mingo* are doubtfully attested before the fourth century CE; see too Adams (1982: 245–6).

14 11.169–70, text and trans. Braund 2004 Loeb: *magis ille extenditur, et mox | auribus atque oculis concepta urina movetur* 'Its tension rises more and more and the next thing is that the sights and sounds make the pent-up liquid flow.'

15 *Merry Wives,* V v 13–16, FALSTAFF: 'I am here a Windsor stag, and the fattest, I think, i' th' forest.—Send me a cool rut-time, Jove, or who can blame me to piss my tallow?'

16 Sir Kenneth Dover's note on how to translate ἐκμιαίνομαι (*ekmiainomai*) (2002: 92) is worth quoting at length: '("I shoot my wad" would be too slangy a translation, since ἐκμιαίνεσθαι [*ekmiainesthai*] is a Hippocratic word). In current English "It's orgasmic!" is not uncommon as a description of (e.g.) galloping on horseback or unearthing a legible inscription.' There is in fact only a single example of *ekmiainomai* in the Hippocratic Corpus (*On Superfetation*

31), in a passage with advice as to how to conceive a male or female child; it is also found in the *Septuagint*, in Leviticus (purity rules).

17 See in general on this topic Allan and Burridge (1991, 2006).

18 Work on disgust and dirt has not yet fully engaged with how it overlaps with what some languages label 'dirty words'. While it is acknowledged that social anxiety is at the heart of the tabooing of words (Allan and Burridge 2006: 1 and *passim*), how far there are linguistic universals surrounding the production of offensive words and the body is still a desideratum.

19 The concept of 'face' is thus central to so-called 'first wave' studies of politeness; more recent approaches to (im)politeness argue for situated, socio-cognitive models, which stress the centrality of the relationship between speaker and social practice and on evaluative beliefs and expectations, e.g. Culpeper (2011), Kádár and Haugh (2013).

20 Swift (2015: 21–2) summarises interpretations of the overall scene in Archilochus *fr.* 196a—all agree this is ejaculation. On *menos* in Homer and as a part of an older IE formula indicating sex, see Van Sickle (1975: 129, 148–50); Boedeker (1984: 10–30) collects and discusses additional examples of dew and rain as impregnating and sexualised. The English idiom 'piss down with rain' also speaks to the equation of liquid from the natural environment and the body, and compare Nagy (1974: 231–7) on connections between rain and urine.

21 This term is used here *pace* R. M. Adams (1985: 44) who terms the word dysphemism 'a coinage almost as ugly as what it describes'. The phrase *dysphemistic* has the advantage of precision among a variety of earlier terms, a variety which speaks to a problem of definition.

22 Metaphorical uses of ἀπομύσσω (*apomussō*) (*s.v.*) are also given by LSJ as II '*stop his drivel*' (from Plato), and III '*snuff*' a wick. English 'snot' as a verb is also recorded (*OED s.v.* snot, *v.*) in the senses 'to snuff (a candle)', as well as to 'to sniff' or 'snivel'. Another word for snot, κόρυζα (*koruza*) can also be used of idiocy (*s.v.* II), and again echoes the English connection between a lack of control of facial mucus and a lack of brain power ('snivelling idiot'). A similarly wider metaphorical connection is found in the use of μυκτήρ (*muktēr*) 'nostril' in the pejorative sense of 'sneerer', e.g. of Socrates (by Timon of Phlius, *fr.* 25 Diels [= Lloyd-Jones and Parsons *SH* 799]).

23 *Pl.* 817–18: [..] ἀποψώμεσθα δ᾽ οὐ λίθοις ἔτι, | ἀλλὰ σκοροδίοις ὑπὸ τρυφῆς ἑκάστοτε. Note LSJ's Latin rendering of this verb (*s.v.* ἀποψάω II) '*podicem detergere* ['to wipe the anus']'. The same verb is used at *Frogs* 490 of the self-cleansing of faeces.

24 The text is discussed by Solin (2008: 64–5), and Varone (1994: 77–8), the latter reading instead for lines 7–10 *et relinque|t utr(umque) ventre | inane e(t) | os plenu*, 'leaving both their cavities empty and her mouth full'.

25 Ferri's translation of the verb is incidental to his point, which is rather on the syncopated perfect ending -*aut*, i.e. for -*avit*.

26 It is perhaps exploited punningly at *Thesmophoriazusae* 231, where the Kinsman is asked by Euripides 'τί μύζεις;' (*ti muzeis?*) 'Why are you mu-ing?' The Kinsman has just made the noise μῦ (*mu mu*) (while being shaved, as Austin and Olson 2004 *ad loc.*), but there is maybe a play on the other sense of the verb in Euripides' question. Archilochus *fr.* 42 West uses ἔμυζε (*emuze*) of the action of sucking beer through a pipe (αὐλῷ, *aulōi*), with clear allusion to the act of fellatio.

27 Pausanias (the Atticist lexicographer) also includes the word (μ 27 ed. Erbse), as does the *Suda* (μ 1470), both of whom attribute it to Archilochus (= *fr.* 209 West).

28 I have Roberta Mazza to thank for talking to me about *musagma*. The *Revised Supplement* to LSJ (*s.v.* μύσαγμα, *musagma*) gives additional examples of the word from magical papyri; at *PGM* IV.2641 it is found in conjunction with other bodily fluids including αἷμα (*haima*) (blood) and ἰχώρ (*ichōr*) (?discharge), and the embryo of a dog (κύνειον ἔμβρυον, *kuneion embruon*), which might perhaps also be considered to be mostly slippery or fluid-like.

29 Apart from here, this verb is only found once, in a passage of Cornutus' *On the Nature of the Gods* 57, which attempts to etymologise the word μυστήρια (*mustēria*) 'mysteries'.

30 Compounds are very common in the sexual and scatological vocabulary of Greek; my sense in that this may reflect a concentration on the physical details of many of the actions expressed for comical and/or offensive effect. Did this result in a specialisation into offensive

compounds vs. inoffensive simplex forms—compare English 'lick out' (often offensive, avoided by speakers) vs. 'lick' (usually inoffensive), and perhaps even Greek *miainō* 'pollute' vs. *ekmiainō* 'pollute out' > 'ejaculate'?

31 Taillardat (1989) also provides detailed discussion, based upon the line numbers of *SH*/Lloyd-Jones and Parsons. Page *FGE* is unreasonably scathing, e.g. (p. 459): 'It is impossible to explain why this gibberish was written; no effect, humorous or satirical, is discernible.' The c. 10 words on the *verso* may or may not be related to the Clitorius epigram; see again Livrea (1987) and Taillardat (1989). My thanks to Simon Pitt for discussing the niceties of this epigram with me.

32 *dril(l)os/drilon* is attested so far only elsewhere from a mosaic depicting a battles between pygmies and cranes (*SEG* 2.353) and is alluded to in a distich epigraph of Lucillius (*AP* 11.197); it likely has its origin in the word for 'worm'. *psōlē* is found in Aristophanes a handful of times (see Henderson 1991: 110) and in the *Carmina Priapea* 68; it is reasonably common in mural graffiti from across the Greek-speaking world (often in the formula 'X is a *psōlē*', e.g. *CIL* IV 1363) and is found on curse tablets, sometimes in conjunction with κύσθος (*kusthos*) 'cunt', e.g. *Def. Tab. Audollent* 77b. *psōlē* also labels the drawing in the so-called 'Obscene Proposal' letter *P.Oxy.* 3070, where it is paired with the rare φίκις (*phikis*), meaning buttocks or anus.

33 *Green's Dictionary of Slang* (*s.v.* brain, *In phrases* [= Volume I: 684–5]) gives the earliest example as 1954, glossing as 'to copulate very strenuously and poss. sadistically, usu. of a man to a woman'. Perhaps here we should also think about the warning that (male) masturbation makes you go blind.

34 Radio 6 Music, c. 0920 GMT, 22 February 2016.

35 We might compare here Lucian's anatomically bizarre Moon Men (*VH* 1.24), who also 'snot out (ἀπομύττονται, *apomuttontai*) very pungent honey' and sweat (ἱδροῦσιν, *hidrousin*) milk, from both of which they make cheese.

36 Sex is one of the nice things which the Stoic Emperor Marcus Aurelius (died 180 CE) says should be imagined for what it 'really is', along with food, which should be viewed as dead animals, and a fancy purple robe as 'a lamb's fleece dipped in a shell-fish's blood' (6.13).

37 www.urbandictionary.com/define.php?term=Sex%20Wee (Accessed February 2021).

Bibliography

Primary sources

Austin, C. and S. D. Olson (2004) *Aristophanes. Thesmophoriazusae*. Edited with Introduction and Commentary. Oxford: Oxford University Press.

Braund, S. M. (2004) *Juvenal and Persius*. Loeb Classical Library 91. Cambridge, MA: Harvard University Press.

Fordyce, C. J. (1961) *Catullus*. A Commentary by C. J. Fordyce. Oxford: Oxford University Press.

Gerber, D. E. (1999) *Greek Iambic Poetry*. Cambridge, MA: Harvard University Press.

Henderson, J. (1998) *Aristophanes. Acharnians. Knights*. Loeb Classical Library 178. Cambridge, MA: Harvard University Press.

Lee, G. (1990) *Catullus. The Complete Poems*. Oxford World's Classics. Oxford: Oxford University Press.

Lloyd-Jones, H. and P. J. Parsons (1983) *Supplementum Hellenisticum*. Berlin: Walter de Gruyter.

Quinn, K. (1970) *Catullus: The Poems*. London: Palgrave Macmillan.

Smith, W. D. (1994) *Hippocrates, Vol. VII*. Loeb Classical Library 477. Cambridge, MA: Harvard University Press.

Sommerstein, A. H. (1981) *The Comedies of Aristophanes, Vol. II, Knights*. Warminster: Aris & Phillips.

Taillardat, J. (1967) *Peri blasphēmiōn; Peri paidiōn. Extraits byzantins*. Paris: Les Belles Lettres.

Wilson, N. G. (2007) *Aristophanis Fabulae. Tomus I*. Oxford Classical Texts. Oxford: Oxford University Press.

Secondary literature

Adams, J. N. (1981) '*Culus, clunes* and their synonyms in Latin', *Glotta* 59, 231–64.

———. (1982) *The Latin Sexual Vocabulary*. London: Duckworth.

Adams, R. M. (1985) 'Soft soap and the Nitty-Gritty', in D. J. Enright (ed.) *Fair of Speech: The Uses of Euphemism*. Oxford: Oxford University Press, 44–55.

Allan, K. (2007) 'The pragmatics of connotation', *Journal of Pragmatics* 39, 1047–57.

Allan, K. and K. Burridge (1991) *Euphemism and Dysphemism: Language Used as Shield and Weapon*. Oxford: Oxford University Press.

———. (2006) *Forbidden Words: Taboo and the Censoring of Language*. Cambridge: Cambridge University Press.

Beard, M. (2002) 'Did the Romans have elbows? Or: Arms and the Romans', in P. Moreau (ed.) *Corps Romains*. Grenoble: Jérome Millon, 47–59.

Boedeker, D. (1984) *Descent from Heaven: Images of Dew in Greek Poetry and Religion*. Chico: California.

Braun, V. and C. Kitzinger (2001) ' "Snatch," "hole," or "honey-pot"? Semantic categories and the problem of nonspecificity in female genital slang', *The Journal of Sex Research* 38(2), 146–58.

Coker, A. (2019) 'How filthy was Cleopatra? Looking for dysphemistic words in ancient Greek', *Journal of Historical Pragmatics* 20(2), 186–203.

Craik, E. (2019) 'Medical vocabulary, with especial reference to the Hippocratic Corpus', in C. Stray, M. Clarke and J. Katz (eds) *Liddell and Scott: The History, Methodology and Languages of the World's Leading Lexicon of Ancient Greek*. Oxford: Oxford University Press, 141–50.

Crespo-Fernández, E. (2015) *Sex in Language: Euphemistic and Dysphemistic Metaphors in Internet Forums*. London: Bloomsbury.

Critchley, S. (2002) *On Humour*. London: Routledge.

Culpeper, J. (2011) *Impoliteness*. Cambridge: Cambridge University Press.

Douglas, M. (1966) *Purity and Danger*. New York: Routledge.

Dover, K. (2002) 'Some evaluative terms in Aristophanes', in A. Willi (ed.) *The Language of Greek Comedy*. Oxford: Oxford University Press, 85–97.

Elbourne, P. (2011) *Meaning: A Slim Guide to Semantics*. Oxford: Oxford University Press.

Fellbaum, C. (2015) 'Lexical relations', in J. R. Taylor (ed.) *The Oxford Handbook of the Word*. Oxford: Oxford University Press, 350–63.

Ferri, R. (2008) 'Review of Wright Latin vulgaire—latin tardif VIII: actes du VIIIe Colloque international sur le latin vulgaire et tardif, Oxford, 6–9 septembre 2006', *Bryn Mawr Classical Review 2009* 2(50).

Geeraerts, D. (2010) *Theories of Lexical Semantics*. Cambridge: Cambridge University Press.

———. (2015) 'How words and vocabularies change', in J. R. Taylor (ed.) *The Oxford Handbook of the Word*. Oxford: Oxford University Press, 416–30.

Goh, I. (2018) 'It all come out: Vomit as a source of comedy in Roman moralizing texts', *Illinois Classical Studies* 43, 438–58.

Harris, O. J. T. and J. Robb (2013a) 'O brave new world, that has such people in it', in J. Robb and O. J. T. Harris (eds) *The Body in History: Europe from the Palaeolithic to the Future*. Cambridge: Cambridge University Press, 1–6.

————. (2013b) 'Body worlds and their history: Some working concepts', in J. Robb and O. J. T. Harris (eds) *The Body in History: Europe from the Palaeolithic to the Future*. Cambridge: Cambridge University Press, 7–31.

Henderson, J. (1991) *The Maculate Muse: Obscene Language in Attic Comedy*. 2nd edn. Oxford: Oxford University Press.

Jouanna, J. and D. Demont (1981) 'Le sens d'ἰχώρ chez Homère (*Iliade* V, v. 340 et 416) et Eschyle (*Agamemnon*, v. 1480) en relation avec les emplois du mot dans la collection hippocratique', *Revue des Études Anciennes* 83(3–4), 197–209.

Kádár, D. Z. and M. Haugh (2013) *Understanding Politeness*. Cambridge: Cambridge University Press.

Katz, J. T. (2004) 'Sanskrit *sphij-/sphigí-* and Greek φίκις', in A. Hyllested et al. (eds) *Per Aspera ad Astericos: Studia Indogermanica in honorem Jens Elmegård Rasmussen*. Innsbruck: Institut für Sprachen und Literaturen der Universität Innsbruck, 277–84.

————. (2006) 'The riddle of the *sp(h)ij-*: The Greek Sphinx and her Indic and Indo-European background', in G. J. Pinault and D. Petit (eds) *La langue poétique indo-européenne*. Leuven: Peeters, 157–94.

L'Homme, M.-C. (2015) 'Terminologies and taxonomies', in J. R. Taylor (ed.) *The Oxford Handbook of the Word*. Oxford: Oxford University Press, 335–49.

Livrea, E. (1987) 'La morte di Clitorio', *Zeitschrift für Papyrologie und Epigraphik* 68, 21–8.

Llewellyn-Jones, L. (2003) *Aphrodite's Tortoise: The Veiled Woman of Ancient Greece*. Swansea: The Classical Press of Wales.

Majid, A. (2015) 'Comparing lexicons cross-linguistically', in J. R. Taylor (ed.) *The Oxford Handbook of the Word*. Oxford: Oxford University Press, 364–79.

Malt, B. C. (2015) 'Words as names for objects, actions, relations, and properties', in J. R. Taylor (ed.) *The Oxford Handbook of the Word*. Oxford: Oxford University Press, 320–33.

Meigs, A. S. (1978) 'A Papuan perspective on pollution', *Man* 13(3), 304–18.

Motion, A. (1993) *Philip Larkin: A Writer's Life*. London: Faber & Faber.

Nagy, G. (1974) *A Comparative Study of Greek and Indic Meter*. Cambridge, MA: Harvard University Press.

Page, D. L. (1982) *Further Greek Epigrams*. Cambridge: Cambridge University Press.

Partridge, E. (2001) *Shakespeare's Bawdy*. 3rd edn, with foreword by Stanley Wells. London: Routledge.

Petit, C. (2017) 'Galen, pharmacology and the boundaries of medicine: A reassessment', in L. Lehmaus and M. Martelli (eds) *Collecting Recipes: Byzantine and Jewish Pharmacology in Dialogue*. Berlin: Walter de Gruyter, 51–79.

Richardson Jr, L. (1967) 'Catullus 67: Interpretation and form', *American Journal of Philology* 88(4), 423–33.

Riemer, N. (2015) 'Word meanings', in J. R. Taylor (ed.) *The Oxford Handbook of the Word*. Oxford: Oxford University Press, 305–19.

Rozin, P., J. Haidt and C. R. McCauley (2008) 'Disgust', in M. Lewis, J. M. Haviland-Jones, and L. Feldman Barrett (eds) *Handbook of Emotions*. 3rd edn. New York: The Guilford Press, 757–76.

Solin, H. (2008) 'Vulgar Latin and Pompeii', in R. Wright (ed.) *Latin vulgaire—latin tardif VIII: actes du VIIIe Colloque international sur le latin vulgaire et tardif, Oxford, 6–9 septembre 2006*. Zürich: Olms, 60–8.

Sommerstein, A. H. (1999) 'The anatomy of euphemism in Aristophanic comedy', in F. De Martino and A. H. Sommerstein (eds) *Studi sull'eufemismo*. Bari: Levante, 181–218.

Swift, L. (2015) 'Negotiating seduction: Archilochus' Cologne epode and the transformation of epic', *Philologus* 159(1), 2–28.

Taillardat, J. (1989) 'L'épitaphe burlesque de Cleitorios', in M.-T. Couilloud-Le Dinahet, R. Etienne, and M. Yon (eds) *Architecture et poésie dans le monde grec: Hommage à Georges Roux*. Paris: De Boccard, 205–9.

Van Sickle, J. (1975) 'The new erotic fragment of Archilochus', *Quaderni Urbinati di Cultura Classica* 20, 123–56.

Varone, A. (1994) *Erotica Pompeiana*. Rome: 'L'erma' di Bretschneider.

Wackernagel, J. (1916) *Sprachliche Untersuchugen zu Homer*. Göttingen: Vandenhoeck & Ruprecht.

Wierzbicka, A. (2014) *Imprisoned in English: The Hazards of English as a Default Language*. Oxford: Oxford University Press.

Part II

A WOMAN IN FLUX

2

A VALID EXCUSE FOR A DAY OFF WORK

Menstruation in an ancient Egyptian village

Rosalind Janssen

In scenes reminiscent of *Desperate Housewives* or *The Archers*, Deir el-Medina, an ancient Egyptian desert settlement on the West Bank of Luxor, was a village where much was going on behind locked doors in terms of domestic violence and threats of divorce.[1] At the same time, everyone knew each other's business in relation to matters concerning hooliganism and binge drinking; sex and adultery; theft and petty crime. At its maximum extent, this was a hierarchical community comprising approximately five hundred inhabitants: men, women, and children. However, because the men were skilled artisans who worked away in ten-day shifts, it was for much of the time a village occupied solely by their families who remained behind. The menfolk were building and decorating the tombs of the New Kingdom Pharaohs in the Valley of the Kings, as well as the burial places of royal family members in the Valley of the Queens. As such, Pharaoh's remarkable village was purpose-built midway between these two valleys in c. 1570 BCE and occupied for a period of some five hundred years until c. 1070 BCE.[2]

Engaging with gender connections is one of the themes highlighted in the Introduction to this volume (p. 5), and the title to this chapter suggests that the occurrence of menstruation at Deir el-Medina was known by the men. They sometimes used the fact that their wives or daughters were menstruating as a valid excuse to absent themselves from their work. Our aim is to explore the wealth of archaeological remains and vivid literary detail, which makes it possible to delve into the minutiae of the role played by menstruation in village life. But before delving further into Deir el-Medina, we must briefly explore what the primary evidence has to tell us about both menstruation and menopause in ancient Egypt in general. The second section presents an overview of the primary evidence from the site itself, before we home in on the evidence concerning menstruation which can be derived from textual and archaeological evidence related to its professional laundry service. A third section comprises the main absence from work discussion, drawing on both primary and secondary literature.

Menstruation and menopause in ancient Egyptian thought

From the Middle Kingdom on, that is c. 2050–1650 BCE, the Egyptian word for menstruation was *ḥsmn*; the sign group usually culminates with a hieroglyphic determinative representing liquid issuing from lips, ⌐ in this case, referring to menstrual blood leaking from the vulva.[3] This determinative is interesting evidence for what Amy

DOI: 10.4324/9780429438974-5

Coker, in this volume's opening theme on ancient linguistic terminology, refers to as the widespread 'slipperiness of the labels . . . for even solid parts of the body' (p. 24). In the case of words such as 'spit', 'vomit', and 'spew out', this Egyptian determinative might refer to non-bloody flux issuing from a totally different bodily region, that is, facial, as opposed to vaginal lips.

In the next chapter (Chapter 3), on uterine bleeding in ancient Greece, Irene Salvo alludes to a connection between the flow of menstrual blood and that of rivers and water in Mesopotamian texts (p. 64). As far as Egyptian mythology was concerned, menstruation was specifically connected with the annual Nile flood (Borghouts 1978: 24). Both were linked by their anticipated regularity. It is therefore significant that, as indicated by the New Kingdom Ebers Papyrus dated to the ninth regnal year of Amenophis I, c. 1516 BCE, the irregularity of the menstrual cycle was a feature noted and treated in the medical papyri.[4] Thus, Ebers 832 references 'a woman having pain in one side of her vulva', the diagnosis of which is that 'her *hsmn* has lost its regularity'; the remedy is to bandage her pubic region with 'smashed garlic, cider and sawdust of fir tree' (Frandsen 2007: 82–3). The ancient Egyptians were also aware of amenorrhea. Papyrus Edwin Smith, a surgical treatise dating to c. 1550 BCE, records the examination of 'a woman having pain in her stomach while *hsmn* does not come for her' as 'a case of obstruction of the blood in her uterus' (Frandsen 2007: 82).[5] Such interest reflects menstrual blood as central to ancient Egyptian physiology, a point noted by the editors of this volume in relation to bodily fluids in general (p. 2).

By contrast, the concern shown with menopause is more limited, just as Maria Gerolemou (2013: 4441) acknowledges was the case in ancient Greece. It was linked to age and associated by both men and women with the inability to bear children (Janssen 2006: 5). As such, a reference to age occurs in Papyrus Ebers: 'If you examine a woman who is many years old and her menstrual period does not come.' This leads to what John Nunn (1996: 196) describes as 'a complex but innocuous remedy [which] is then prescribed to be drunk for four days'. A vivid reference to the inability to bear children occurs in the letter KBo 28.30 from the Boğazköy archive (Edel 1976: 67–75). Written in Akkadian, it was sent by Ramesses II to the Hittite king Hattušili and probably dates to the period shortly after the two monarchs had signed their famous peace treaty in c. 1258 BCE. It represents a response to Hattušili's request that Ramesses send him a man to prepare medicine to enable his sister, known as Matanazi at the Egyptian court, to bear children. The Egyptian Pharaoh patronizingly writes of her:

> See Matanazi, the sister of my brother, the king, your brother knows her. A fifty year old!! Never! She's 60! Look, a woman of 50 is old, to say nothing of a 60 year old! One can't produce medicine to enable *her* to bear children!
>
> Kitchen 1982: 92, emphasis added

Using Hittite documents, Trevor Bryce (1998: 214) has calculated that Matanazi must have been at least 58 at the time, although 'she could very well have been several years older.' He therefore concludes that Ramesses was 'speaking no more than the truth. If anything, he was being rather generous' (Bryce 1998: 215). This is certainly correct when we consider that Greek sources, such as Aristotle, were to later argue for the menopause occurring between a woman's fortieth and fiftieth year (Amundsen and Diers 1970).

Primary evidence from at Deir el-Medina

As far as Deir el-Medina is concerned, I have argued that older, post-menopausal women, including its wise woman, enjoyed increased social freedom (Janssen 2006). Jan Bremmer (1987) had previously reached the same conclusion in relation to ancient Greece. This brings us neatly to an extended introduction to the site of Deir el-Medina, which forms our main source of knowledge concerning middle class life in ancient Egypt. This is best explained by envisaging the social structure of the New Kingdom, c. 1550–1069 BCE, and indeed of ancient Egypt in general, as resembling that of a pyramid. Some 90 percent of its population were labourers and, as such, are almost entirely lost to the archaeological record. Instead, our information is largely based on evidence from the small elite class forming its apex. What happened in the middle amongst the skilled artisans is attested by what, following Barry Kemp (1989: 137–80), we might term model communities, built for a specific, short-lived purpose.[6] Of these Deir el-Medina is by far the best known, and its resultant 'history from below' representation remains unsurpassed until that afforded by Emmanuel Le Roy Ladurie's (1978) reconstruction of Montaillou, a small Medieval village in the Pyrenees. Here too we learn from textual evidence about the most intimate aspects of the daily life of its inhabitants in the early 1300s CE.

However, as indicated earlier, it is Deir el-Medina's extraordinary combination of archaeological and literary evidence that makes the site such a rich repository. Both aspects have been respectively explored in an accessible fashion by Morris Bierbrier (1982) and Andrea McDowell (1999). Systematically excavated by the French Institute from 1922 on, the extant village, comprising houses, tombs, and chapels, is still revealing its archaeological secrets during current campaigns. A notable example is the 2014 discovery of a tattooed female mummy; she was discovered amongst human remains stored in a tomb at the site by earlier excavators (Austin and Gobeil 2016).

Complementing their archaeological record, the community has bequeathed us with several thousand ostraca on limestone flakes and potsherds. These fall into three distinct types: literary and non-literary ostraca written in cursive hieratic, together with virtuoso sketches on figured (or pictorial) ostraca. The non-literary category, described by Fredrik Hagen (2007: 38) as 'drafts of administrative documents, letters and "scrap paper" for notes and records of deliveries', allow us to build up a comprehensive picture of employment conditions. Overall, there were some extremely positive aspects to being Pharaoh's tomb-builders: paternity leave, health care, and apprenticeships (McDowell 1999: 35, 53–9, 129), and retirement and state pensions (Janssen 2006). However, there were also the first recorded strikes in antiquity when the men's grain rations fell into arrears (McDowell 1999: 235–8). In addition, we learn about barter transactions, credit, and gift giving (McDowell 1999: 74–9). We can evaluate how the villagers treated both their elderly and disabled (Janssen 2006), and even their humble donkeys (Janssen 2005).

Based on his estimate of at least 30,000 literary ostraca produced at the site, Jac. Janssen (1991: 82) has calculated that 'around forty per cent' of the workmen were literate. This is a remarkable figure considering that John Baines and Chris Eyre (1983) had previously argued for a maximum 1 percent literacy rate in ancient Egypt. Moreover, of the 470 or so letters from the site, approximately one in seven were written to or from women. This, coupled with the fact that some letters are about intimate matters or even desperate situations, led Janssen (1992: 91) to conclude that 'even some of the women in the community appear to have been at least semi-literate.'

Of direct relevance to us is the ability to evaluate the life and work of the inhabitants' professional launderers, who were exclusively men until the Ptolemaic Period (Hall 1986; Janssen and Janssen 2002). They feature in the most popular work to be copied out by the Deir el-Medina schoolboys: the Middle Egyptian *Satire on the Trades* (Gasse 1992: 30). Its aim was to promote hard study as an advancement to scribal literacy, by the device of showcasing the menial professions endured by Egypt's non-literate masses. Amongst what John Foster (1999: 123) aptly terms 'the portrait gallery of the miserable' is a passage lamenting the lot of a laundryman. In William Kelly Simpson's edited edition (1973: 333–4) this begins: 'The washerman launders at the riverbank in the vicinity of the crocodile. . . . His food is mixed with filth, and there is no part of him which is clean.' The following line is crucial for our present argument in telling us that: 'He cleans the clothes of a woman in menstruation.' Paul Frandsen (2007: 100) has interpreted this 'as an indication of the low status, even social stigma' attached to this trade, and an indication that menstruation was 'regarded as negative but not dangerous'. For our purposes, it is significant that there is absolutely no indication from this text of any social taboo attached to menstruation.

Rejecting the vague 'clothes' of the Simpson translation, Terry Wilfong (1999: 430) more accurately translated *d3iw* (the kilt worn by both men and women) as 'the skirt of a menstruating woman'. He envisages this as 'a woman's skirt stained with menstrual blood' and therefore 'less likely to be a menstrual towel itself' (Wilfong 1999: 430). By contrast, surviving laundry lists from Deir el-Medina indicate that the professional all-male launderers responsible for servicing Pharaoh's village were washing actual sanitary towels. Crucial in this regard is Ostracon Deir el Medineh 30, housed in the French Institute in Cairo, which is dated to the first regnal year of Seti I in c. 1289 BCE. It records the delivery to the riverbank of ten kilts and eight loincloths, followed by five *sdw/sdy n phwy*, literally 'loincloths for the backside'. Writing many years ago, I suggested these were linen sanitary towels (Hall 1986: 55, followed by McDowell 1999: 60). It is significant that, since they are listed together with these other garments, there were seemingly no marked notions of impurity attached to such sanitary products. They are relegated to the bottom of the list merely on account of their smaller numbers. We can therefore conclude that, in what was a 'make do and mend' society, such towels were recycled from worn-out clothing. As such, 'loincloths for the backside' are attested in fifteen other Deir el-Medina ostraca (Janssen 1975: 272–7). But far more problematic is how exactly to combine this textual record by identifying such ephemeral items in the surviving archaeological record.

Absence from work at Deir el-Medina due to menstruation

As we start to consider the controversial evidence, it is important to bear in mind two points. Firstly, Thomas Buckley and Alma Gottlieb (1988: 44–5) noted that, within populations with low fertility rates, there would have been a short period of fertility characterised by a late menarche and early menopause. Amy Harris and Virginia Vitzthum (2013: 239) point out that, due to years spent pregnant and lactating, it is still the case in much of the world today that women in natural fertility populations are 'only occasionally experiencing a few sequential ovarian cycles'. As a result of their extensive literature review, these authors conclude that this amounts to an average of 'only about 40 ovarian cycles in a lifetime' (Harris and Vitzthum 2013: 239). Secondly, in relation to the onset of menarche at the comparable Middle Kingdom village of Lahun, Kasia Szpakowska (2008: 211) convincingly argues that here 'menstruation is unlikely to have been taboo or secret, neither was it likely celebrated.' She therefore asserts that for her main middle-class character, the young girl Hedjerit, 'it may simply have been an ordinary part of life' (Szpakowska 2008: 211).

We can now turn to the primary attendance registers which document reasons for the men's absence from work. Six of these were published in a seminal work by Jac. Janssen (1980). More recently, Frandsen (2007: 90–6) has provided a useful checklist of both the registers and relevant supplementary material. This chapter concentrates on the largest and most significant of these registers: Ostracon British Museum EA 5634, currently on display in the Nebamun Gallery (Figure 2.1).

Dated to year 40 of Ramesses II, which equates to c. 1239 BCE, this impressive two-sided register is compiled in black and red ink and covers a cumulative, but

Figure 2.1 Recto (a) and verso (b) of a large limestone attendance register in New Egyptian hieratic, dating to the reign of Ramesses II. Height 38.5 cm × 33 cm wide. Now in the British Museum, Ostracon British Museum EA 5634.

Source: © Trustees of the British Museum.

Figure 2.1 (Continued)

not consecutive, period of 70 working days (Janssen 1980). It comprises a list of 40 workmen, noting ten cases where a man was absent because either 'his wife' or 'his daughter' had her *hsmn*. Two of these reference the same man, and therefore probably the same woman. In relation to these infrequent and irregular absences, Jac. Janssen (1980: 141) decided that it was 'a bit unlikely' that 'all the wives and daughters of forty workmen during about seventy days only ten times had their monthly periods.' He therefore took the verb *hsmn* in its literal sense 'to purify', interpreting it as a reference to post-childbirth purification. However, both Wilfong (1999: 422) and Frandsen (2007: 84) agree that 'purification' can more readily be seen as a euphemism for menstruation, in that it was the act of ridding herself of blood that caused the woman to be purified, or restored to her normal status. This seems acceptable when we consider that in Greek, the word *katharsis* (purification) is used to refer to menstruation, with Lesley Dean-Jones (1994: 231) confirming that the discarded blood was not necessarily regarded as impure. Indeed, Velvet Yates (1998: 35) produces a convincing argument that Aristotle considered katharsis to be 'a natural and normal biological process in women, analogous to ejaculation in men'. She thereby engages with Jonathan Lear's (1988: 298) earlier throwaway suggestion that 'the preponderant use which Aristotle makes of the word "*katharsis*" is as a term for menstrual discharge.'

At the same time, Wilfong (1999: 424) believes that 'absence for *hsmn* was only permitted under certain conditions', thereby unwittingly supporting Jac. Janssen's (1980) argument that the number of absences should be higher. Highlighting two possible occasions, he suggests that this happened when the normal functioning of the household was disrupted: either menstruating women were 'absent from the home' when they went to 'the place of women' (more on this later) or were 'unable to do the necessary work in the house' because they were suffering from the incapacitating effects of dysmenorrhea or painful periods (Wilfong 1999: 424). While admitting that no estimates are available for such cases of dysmenorrhea in ancient societies, Wilfong (1999: 424) nonetheless concludes that both these situations would then have required 'the presence of the husband and/or father at home'. Frandsen (2007: 97) cynically demolishes Wilfong's unsubstantiated arguments, 'derived as they seem to be from social conditions around the year 2000 A.D.'. He effectively concludes that the limited number of absences for *hsmn* is not in fact an issue. If the man was already absent from the village, the menstruation of his wife or daughter would not have affected his ability to work and would therefore not have been recorded in a list (Frandsen 2007: 99).

Jaana Toivari-Viitala (2001: 164) argues that the reason for the specific reference to 'his wife' or 'his daughter' in Ostracon British Museum EA 5634 is because workmen's mothers had either reached menopause or died. She suggests that their sisters are similarly unmentioned because no obligations existed between siblings in relation to menstruation. In relation to the latter, Frandsen (2007: 95) more convincingly notes that by the time a man joined the workmen's crew, his sisters were likely to have already been married to other artisans. Whatever the reason, our conclusion from these six lists is that a workman was absent if his wife or daughter had her menstrual period.

Before proceeding, it is interesting to briefly consider the attitudes of two earlier male Egyptologists when discovering that menstruation featured as an excuse for absence from work in these ostraca lists. It was back in the nineteenth century when Wilhelm Spiegelberg (1895: 5) first noted the reference. However, he was, perhaps not surprisingly, particularly oblique; he refers instead to 'ein gewisses Unwohlsein' or 'a certain malaise'. In his accompanying footnote, Speigelberg reports that he had consulted Professor Gerland on this matter—presumably the anthropologist Georg Gerland, who was also at the University of Strasbourg—but without giving any further details. Such references were subsequently noticed in 1922 by Jaroslav Černý, the future doyen of Deir el-Medina administrative texts, who was laboriously writing up by hand his doctoral thesis. He wrote:

> an interesting excuse is given in the case of several workmen [he here references the Egyptian] 'his wife' or 'his mother' is menstruating. Strict prescriptions of purity did not allow the men to go to public places for the duration of their womenfolk's impurity.[7]
>
> Černý 1922: 16

Once again this reveals the contemporary attitude of a male Egyptologist towards female menstruation.

Included among the material supplementary to the six registers is an ostracon now in Chicago, which was first published by Wilfong in 1999. Dated to year 9 of Merenptah, which is c. 1204 BCE, Ostracon Oriental Institute Museum 13512 comprises a mere fragmentary three lines. It reads:

> Year 9, fourth month of the season of Inundation, day 13: the day when these eight women came out [to/from the] place of women while they were menstruating. They got as far as the rear of the house which . . . the five walls.
> Ostracon Oriental Institute Museum 13512

This text indicates that *hsmn*'ing women left the village, since the 'five walls' signify the five guard posts, which McDowell (1994: 58) suggests were designed more for the protection than the confinement of the inhabitants. While Wilfong (1999: 428) reads the women as going *to* the place of women, Toivari-Viitala (2001: 167–8) points to the possibility that they were instead coming out *from* that location, 'although they probably were supposed to stay there (?).' The guard posts then take on a new significance as she suggests that 'this was the place where the villagers could call various matters to the attention of the administrators' (Toivari-Viitala 2001: 168). This would explain the lacuna in the text: the event that happened after the women had reached the rear of the house, and which had something to do with the five walls.

Ostracon Oriental Institute Museum 13512 contains the only mention of *s.t hm.wt* ⌸⌸⌸⌸ , or 'the place of women' in Egyptian texts. Such sites of seclusion are widely attested in traditional cultures. In their introductory chapter, which challenges the existence of a universal menstrual taboo, Buckley and Gottlieb (1988: 13) argue: 'we do find many suggestions that seclusion is not always onerous'; moreover, it could be indicative of 'enhanced rather than lowered status'. This is well expressed in Anita Diamant's (1997) best-selling novel *The Red Tent*. Based on the biblical story of the rape of Dinah, it imagines menstruating women congregating together in a sanctuary type structure which symbolises women's empowerment and social control. However, Buckley and Gottlieb (1988: 14), rightly remind us that such notions of status are always 'culturally variable and specific'.

The archaeological location of 'the place of women' is completely unknown. The fact that Raphael Ventura (1986: 120–44) suggested that the 'five walls' were to the south of the village, whereas Frandsen (1989: 121) later argued that they were to the north, certainly does not help. Moreover, any notion of what are generically called 'menstrual huts' may well be completely misleading. As we have seen earlier, Diamant's concept of a tent envisages a type of structure that would not survive in the archaeological record. Therefore, all we can say at present is that such a 'place' outside the village was clearly spacious enough to accommodate at least eight women.

Moreover, we may further need to query whether such a 'place of women' was necessarily outside the village. Lynn Meskell (1998: 235–7) has suggested that spaces at the rear of several of the larger Deir el-Medina houses could provide evidence for this practice. More tangible in terms of associated material culture is the tiny room found under the stairs of house N49.21 at New Kingdom Amarna which contained two model pottery beds, two fragmentary female figurines, and a stela depicting a woman and girl before Taweret, the hippopotamus goddess of childbirth (Kemp 1989: 305). This led Dominic Montserrat (1996: 48) to describe Pharaonic women's lives as similarly comprising 'frequent childbirth and death punctuated by squatting in the cupboard under the stairs during menstruation'. Szpakowska (2008: 210–11) is highly sceptical of what she describes as Montserrat's 'dark and dubious reconstruction' and sees this type of confinement to a specific space as 'an unlikely scenario'. Instead, considering 'these kind of areas' as 'typically inviting to children', she prefers to see the votive objects as a child's 'secret stash' in what was once a den (Szpakowska 2008: 211). On the face of it, Montserrat's reconstruction does read as the more likely, since the artefacts are ones specifically associated with women and childbirth.

Moreover, Montserrat's (1996: 48) 'cupboard under the stairs' theory is supported by an Aramaic document from Elephantine, dated to 404 BCE (Porten 1996: 237–8). Greek and Demotic house sale contracts from the Hellenistic to the Byzantine Period also detail the structure. For example, in Demotic Papyrus Louvre 2424 and 2443, mention is made of the *hrr.t* or 'women's space' (Zauzich 1968: 17–21, 21–6). According to Colin (2001), this was usually situated under the stairs of multi-story Theban houses. By comparing the written sources with the archaeological evidence, Ada Nifosi (2019) has recently been able to describe this room in detail for the first time. Concluding from the dimensions that it was not a comfortable place and therefore not intended to be used for an extended period of time, she interprets its use as a household latrine. A toilet seat and adjacent jug of water could have converted it into a temporary 'women's space' for menstrual ablutions. An exciting and plausible interpretation, this may be of relevance to Dynastic Egypt.

Since such references to the women's space are never negative, Wilfong (1999: 431) concluded that Ostracon Oriental Institute Museum 13512's record of the 'trip' by the eight women to the 'place of women' might 'well have been a positive one'. Toivari-Viitala (2001: 168) goes so far as to interpret the Deir el-Medina's 'place of women' as the site of a rite for a group of girls at first menstruation. She derives her evidence from Ostracon Oriental Institute Chicago 9, which mentions a man bringing something to a woman 'because she came *hsmn*'ing'; her suggestion is that this is indicative of a ritual involving gifts and/or food provisions (Toivari-Viitala 2001: 165). Comparable evidence for transition rites at the time of menarche in ancient Greece is rather scant (King 1998: 75–88). Although more certain from an anthropological perspective (Britton 1996: 649–50), we need to be extremely cautious about the presence of any such rite in ancient Egypt.

Returning to Ostracon Oriental Institute Museum 13512, all we can ultimately be certain about is that it details a village event important enough to have been recorded. For our purposes, its crux is that, while the men stayed away from work, the women left the home. Frandsen (2007: 99) reaches the conclusion that 'menstruating women should ideally leave the village before their men were "contaminated"', and that 'men stayed at home only if the women failed to reach the "women's place" in time.' This aligns with two of the questions highlighted in the Introduction to this volume: the possible avoidance of menstruation in the ancient world and ideas about bodily fluids and pollution (p. 5).

Spatial concerns are fundamental and lead us to a third theme foregrounded by the editors: that of fluids in the sphere of religion and ritual (p. 6). Frandsen (2007: 100) suggests that the reason why the artisans had to avoid contact with their menstruating women was because their work in the tomb was connected with death and that the women therefore needed to leave the necropolis area by passing the walls of the village. Ultimately, therefore, it was 'the fertile menstruating women who were seen as vulnerable and in need of protection' (Frandsen 2007: 100). This neatly corresponds to Deborah Winslow's (1980: 615) anthropological findings in relation to Catholics in Sri Lanka, where menstruation signalled a sign of women's *vulnerability* to threats imposed by the cosmos and society. By contrast, for Sri Lankan Buddhists, menstruation was evidence of women's *threat* to cosmic purity and therefore to society.

The editors of this volume pose a hypothesis 'that it is through fluids that the relationship between the inner body and the outside world is negotiated' (p. 5). This is neatly addressed through an additional argument which connects women and the tomb as 'instruments of regeneration' (Frandsen 2007: 100). In ancient Egypt, the womb was thought of as a container in the form of a jar, as Helen King (1998: 33–5) confirms was also the case with the use of the word *angos* in the Hippocratic texts of ancient Greece. Similarly, both the sarcophagus and the tomb were containers that functioned as a womb. Paralleling the gestation process inside the mother's womb, the process of rebirth took place in the tomb. Here, the preserved body returned to the uterus of the sky-goddess Nut, to be reborn at sunrise as a renewal of creation. In her chapter later in this volume (Chapter 19), Tasha Dobbin-Bennett stresses the need for a lifelike, fully functioning body in the afterlife, as promoted by ancient Egyptian religious texts (p. 306).

However, just as Papyrus Ebers demonstrates a 'clear opposition between menstruation and birth', so it was necessary to keep menstruation separate from such funerary regeneration (Frandsen 2007: 103). This was because it implied absence of pregnancy and a resultant lack of fertility. In this reading, it was once again the artisans who formed the intermediaries between women and the tomb, thereby corresponding to a further theme of this volume, whereby female bodies were seen in antiquity as posing a danger to men's masculinity (p. 5).

While all this may be true on a lofty religious level, there is perhaps an alternative explanation for men staying at home. If we turn to the Hebrew Bible, Genesis 31 relates the story of Rachel's unexplained theft of the teraphim or household gods. Hiding them in the saddlebag taken from her camel, she proceeds to sit on this within the confines of her tent. According to the translation of verse 35 in the New Revised Standard Version, Rachel eludes her father Laban in his search for the precious teraphim by telling him: 'Let not my lord be angry that I cannot rise before you, for the way of women is upon me.' The verse then simply continues: 'So he searched, but did not find the household gods.' Esther Fuchs (1988: 80) argues that Laban's total lack of challenge to her 'way of women' trick is due to the fact that knowledge of menstruation is solely under female control, and that, because it can be faked, is associated with sexual cheating. Beverly Straussmann (1996: 306) indeed records rare instances of Dogon women in Mali, West Africa, who were faking menstruation in cases of adultery. The deception took the form of either 'going to the [menstrual] huts when not menstruating or not going when menstruating' (Straussmann 1996: 306). However, in her critical response to Fuchs, Mieke Bal (1988: 151) perceptively argues that she has failed to take into account the fact that Rachel's appeal to her menstrual period became for Laban 'a sign of male inferiority, of male fright, a fright that blinds'. Bearing in mind the previously quoted Harris and Vitzthum (2013: 239) and their description of the 'only about 40 ovarian cycles in a lifetime' experienced by women in natural fertility populations, we can easily see how menstruation was an occasion for male fright. I suggest it could have been fright sufficient to have resulted in women leaving the village, and their men taking time off work.

Wilfong (1999: 426) controversially asserts that Ostracon Oriental Institute Museum 13512 provides a 'rare ancient account of menstrual synchrony'. Citing a series of chronological texts—most notably our Ostracon British Museum 5634—which detail the names of women menstruating at the same time over most of the year, Wilfong (1999: 426) concludes that 'it may be possible' that we can attest to such menstrual synchrony at Deir el-Medina over a forty-year time period. He draws on Martha McClintock's 1971 study, which was based on a group of 135 women living in dormitories at the prestigious all-women Wellesley College. In arguing for an increase in menstrual synchrony, McClintock asserted that their menstrual synchrony was because the women spent time together. Wilfong's argument is similarly based on the close physical proximity of Deir el-Medina and, as indicated by his Ostracon Oriental Institute Museum 13512, public awareness outside the household of women's menstrual cycles. Citing only supportive studies, he chooses to completely ignore the wealth of subsequent literature debunking McClintock's (1971) idea of menstrual synchrony. For instance, Wilfong (1999: 428) references a later study by Kathleen Stern and McClintock (1998), but this was summarily dismissed by Beverly Straussmann (1999) in terms of its statistical and methodological flaws.

Particularly valuable for our purposes are Straussmann's studies (1997, 1999) of the Dogon. Comprising the only testing of the concept of menstrual synchrony in a natural fertility population, they undermine the common belief among anthropologists that it occurred in pre-industrial societies. More recently, in their review of the literature for and against menstrual synchrony, Harris and Vitzthum (2013: 238–9) conclude that 'it seems there should be more widespread doubt than acceptance of this hypothesis.' We can therefore say that there is no evidence to support Wilfong's (1999: 432) conclusion that 'explicit and implicit records of menstrual synchrony . . . are now definitely identified among the Deir el-Medina documentation.' Rather, it is all part of a menstrual myth.

At Deir el-Medina there was neither menstrual synchrony nor a general taboo regarding contact with menstruating women. However, corresponding to the overarching question posed by this volume (p. 5), we can in this ancient Egyptian community identify an avoidance of menstrual blood. This was due to spatial religious and ritual concerns. As long as the artisans were away in the Valley of the Kings or that of the Queens, they were not polluted by contact with their menstruating women. As Frandsen (2007: 103) puts it, they posed 'no threat to the construction work nor did the Tomb, through them threaten the fertility of their absent womenfolk'. However, if the men were at home, this picture automatically changed. In what was a symbiotic relationship, the menstruating women were then 'both vulnerable to harm from vicarious contact to the Tomb and a menace to the potential cosmic fertility intended for that sacred construction' (Frandsen 2007: 103–4). The end result was that this physical event demanded a practical intervention: their menfolk could not go to work. I have additionally argued that we might simultaneously move away from this lofty religious interpretation to a simplistic explanation from daily life: that of male anxieties when faced with what was the rare occurrence of their menstruating wives and daughters. Drawing together a wide range of evidence, what is abundantly clear is that the behaviour of men and women at Deir el-Medina was intricately interconnected by means of the menstrual cycle. While menstruation as a bodily function may have been experienced by women in isolation, it was advertised in the administrative documents as a public event to be ritually celebrated.

As such the *hsmn*'ing of their wives and daughters, constituted a valid excuse for men at this ancient Egyptian village to take a day-off work.

Notes

1 Egyptologists have always referred to Deir el-Medina ('Monastery of the City') by its Arabic name, whereas the ancient inhabitants called their village Set Maat ('The Place of Truth').
2 Pharaonic dates are based on king-lists, dated inscriptions, and astronomical records. This chapter uses the chronology provided by Ian Shaw (2000: 479–83), who states that 'In the New Kingdom the margin of likely error is about a decade.'
3 It is D26 in Sir Alan Gardiner's (1957: 453) Sign List, the standard reference work.
4 Papyrus Ebers is best accessed in Paul Ghalioungui's (1987) English translation.
5 Papyrus Edwin Smith was published by James Henry Breasted (1930a, 1930b). Online versions of the two volumes are available respectively from https://oi.uchicago.edu/research/publications/oip/edwin-smith-surgical-papyrus-volume-1-hieroglyphic-transliteration (Accessed February 2021) and https://oi.uchicago.edu/research/publications/oip/edwin-smith-surgical-papyrus-volume-2-facsimile-plates-and-line-line (Accessed February 2021).
6 The other examples of such model communities are the workers' villages at Old Kingdom Giza, Middle Kingdom Kahun, and New Kingdom Amarna.
7 My thanks are due to Dr Hana Navratilova, who kindly provided me with a translation from the original Czech.

Bibliography

Primary sources

Borghouts, J. F. (1978) *Ancient Egyptian Magical Texts*. Leiden: E. J. Brill.
Breasted, J. H. (1930a) *The Edwin Smith Surgical Papyrus, Vol I: Hieroglyphic Transliteration, Translation, and Commentary*. Chicago: University of Chicago.
———. (1930b) *The Edwin Smith Surgical Papyrus, Vol II: Facsimile Plates and Line for Line Hieroglyphic Transliteration*. Chicago: University of Chicago.
Gardiner, A. H. (1957) *Egyptian Grammar: Being an Introduction to the Study of Hieroglyphs*. Oxford: Oxford University Press.
Ghalioungui, P. (1987) *The Ebers Papyrus: A New English Translation, Commentaries, and Glossaries*. Cairo: Academy of Scientific Research and Technology.
Porten, B. (1996) *The Elephantine Papyri in English: Three Millennia of Cross-Cultural Continuity and Change*. Leiden: Brill.
Simpson, W. K. (1973) *The Literature of Ancient Egypt: An Anthology of Stories, Instructions, and Poetry*. New Haven: Yale University Press.

Secondary literature

Amundsen, D. W. and C. J. Diers (1970) 'The age of menopause in classical Greece and Rome', *Human Biology* 42, 79–86.
Austin, A. and C. Gobeil (2016) 'Embodying the divine: A tattooed female mummy from Deir el-Medina', *Bulletin de l'Institut Français d'Archéologie Orientale* 116, 23–46.
Baines, J. and C. Eyre (1983) 'Four notes on literacy', *Göttinger Miszellen* 61, 65–96.
Bal, M. (1988) 'Tricky thematics', *Semeia* 42, 133–55.
Bierbrier, M. (1982) *The Tomb-Builders of the Pharaohs*. London: British Museum Publications.
Bremmer, J. (1987) 'The old women of ancient Greece', in J. Blok and P. Mason (eds) *Sexual Asymmetry: Studies in Ancient Society*. Amsterdam: J. C. Gieben, 191–215.

Britton, C. J. (1996) 'Learning about "the curse": An anthropological perspective on experiences of menstruation', *Women's Studies International Forum* 19(6), 645–53.

Bryce, T. (1998) 'How old was Matanazi?', *Journal of Egyptian Archaeology* 84, 212–15.

Buckley, T. and A. Gottlieb (1988) 'A critical appraisal of theories of menstrual symbolism', in T. Buckley and A. Gottlieb (eds) *Blood Magic: The Anthropology of Menstruation*. Berkeley: University of California, 3–54.

Černý, J. (1922) *Život dělníků thébské nekropole v nové říši (1300–1000 př. Kr.)*. Archive of the Charles University in Prague, dissertations, no. 1148: Prague, 16.

Colin, F. (2001) 'Un espace réservé aux femmes dans l'habitat de l'Égypte hellénistique d'après des papyrus grecs et démotiques', in I. Andorlini, G. Bastianini, M. Manfredi, and G. Menci (eds) *Atti del XXII congresso internazionale di papirologia: Firenze 23–29 agosto 1998*, vol. I. Florence: International Congress of Papyrology 'G. Vitelli', 259–68.

Dean-Jones, L. (1994) *Women's Bodies in Classical Greek Science*. Oxford: Clarendon Press.

Diamant, A. (1997) *The Red Tent*. New York: St. Martin's Press.

Edel, E. (1976) *Ägyptische Ärtze und ägyptische Medizin am hethitischen Königshof: Neue Funde von Keilschriftbriefen Ramses' II. aus Bogazköy*. Opladen: Westdeutscher Verlag.

Foster, J. L. (1999) 'Some comments on Khety's instruction for little Pepy on his way to school', in E. Teeter and J. A. Larson (eds) *Gold of Praise: Studies on Ancient Egypt in Honor of Edward F. Wente. Studies in Ancient Oriental Civilization 58*. Chicago: The Oriental Institute, University of Chicago, 121–9.

Frandsen, P. J. (1989) 'A word for "causeway" and the location of "the five walls"', *Journal of Egyptian Archaeology* 75, 113–23.

———. (2007) 'The menstrual "taboo" in ancient Egypt', *Journal of Near Eastern Studies* 66(2), 81–106.

Fuchs, E. (1988) ' "For I have the way of women": Deception, gender, and ideology in biblical narrative', *Semeia* 42, 68–83.

Gasse, A. (1992) 'Les ostraca hiératiques littéraires de Deir el-Medina: Nouvelles orientations de la publication', in R. J. Demarée and A. Egberts (eds) *Village Voices: Proceedings from the Symposium 'Texts from Deir el-Medîna and their Interpretation' Leiden, May 31- June 1, 1991*. Leiden: Centre of Non-Western Studies, Leiden University, 51–70.

Gerolemou, M. (2013) 'Menopause', in R. S. Bagnall, K. Broderson, C. B. Champion, A. Erskine, and S. R. Huebner (eds) *The Encyclopedia of Ancient History*. Malden: Wiley-Blackwell, 4441.

Hagen, F. (2007) 'Ostraca, literature and teaching at Deir el-Medina', in R. Mairs and A. Stevenson (eds) *Current Research in Egyptology 2005: Proceedings of the Sixth Annual Symposium*. Oxford: Oxbow Books, 38–51.

Hall, R. (1986) *Egyptian Textiles*. Princes Risborough: Shire Publications.

Harris, A. L. and V. J. Vitzthum (2013) 'Darwin's legacy: An evolutionary view of women's reproductive and sexual functioning', *The Journal of Sex Research* 50(3–4), 207–46.

Janssen, J. J. (1975) *Commodity Prices from the Ramessid Period: An Economic Study of the Village of Necropolis Workmen at Thebes*. Leiden: E.J. Brill.

———. (1980) 'Absence from work by the necropolis workmen of Thebes', *Studien zur altägyptischen Kultur* 8, 127–52.

———. (1992) 'Literacy and letters at Deir el-Medina', in R. J. Demarée and A. Egberts (eds) *Village Voices: Proceedings from the Symposium 'Texts from Deir el-Medîna and their interpretation' Leiden, May 31-June 1, 1991*. Leiden: Centre of Non-Western Studies, Leiden University, 81–94.

———. (2005) *Donkeys at Deir el-Medîna*. Leiden: Nederlands Instituut voor het Nabije Oosten.

Janssen, J. J. and R. Janssen (2002) 'The laundrymen of the Theban Necropolis', *Archiv Orientální* 70(1), 1–12.

Janssen, R. (2006) 'The old women of Deir el-Medina', *Buried History* 42, 3–10.

Kemp, B. J. (1989) *Ancient Egypt: Anatomy of a Civilization*. London and New York: Routledge.

King, H. (1998) *Hippocrates' Woman: Reading the Female Body in Ancient Greece*. London: Routledge.

Kitchen, K. A. (1982) *Pharaoh Triumphant: The Life and Times of Ramesses II*. Warminster: Aris and Phillips.

Lear, J. (1988) 'Katharsis', *Phronesis* 33(3), 297–326.

Le Roy Ladurie, E. (1978) *Montaillou: Cathars and Catholics in a French Village 1294–1324*. London: Scolar Press.

McClintock, M. K. (1971) 'Menstrual synchrony and suppression', *Nature* 229, 244–5.

McDowell, A. G. (1994) 'Contact with the outside world', in L. H. Lesko (ed.) *Pharaoh's Workers: The Villagers of Deir el-Medina*. Ithaca, NY: Cornell University Press, 41–59.

———. (1999) *Village Life in Ancient Egypt: Laundry Lists and Love Songs*. Oxford: Oxford University Press.

Meskell, L. (1998) 'An archaeology of social relations in an Egyptian village', *Journal of Archaeological Method and Theory* 5(3), 209–43.

Montserrat, D. (1996) *Sex and Society in Graeco-Roman Egypt*. London: Kegan Paul.

Nifosi, A. (2019) *Becoming a Woman and Mother in Graeco-Roman Egypt: Women's Bodies, Society and Domestic Space*. Abingdon: Routledge.

Nunn, J. F. (1996) *Ancient Egyptian Medicine*. London: British Museum Press.

Shaw, I. (ed.) (2000) *The Oxford History of Ancient Egypt*. Oxford: Oxford University Press.

Spiegelberg, W. (1895) *Arbeiter und Arbeiterbewegung im Pharaonenreich unter den Ramessiden (ca. 1400–1100 v. Chr.): eine kulturgeschichtliche Skizze*. Strasbourg: Karl J. Trübner.

Stern, K. and M. McClintock (1998) 'Regulation of ovulation by human pheromones', *Nature* 392, 177–9.

Straussmann, B. I. (1996) 'Menstrual hut visits by Dogon women: A hormonal test distinguishes deceit from honest signaling', *Behavioral Ecology* 7(3), 304–15.

———. (1997) 'The biology of menstruation in *Homo sapiens*: Total lifetime menses, fecundity and nonsynchrony in a natural-fertility population', *Current Anthropology* 38, 123–9.

———. (1999) 'Menstrual synchrony pheromones: Cause for doubt', *Human Reproduction* 14(3), 579–80.

Szpakowska, K. (2008) *Daily Life in Ancient Egypt: Recreating Lahun*. Oxford: Blackwell.

Toivari-Viitala, J. (2001) *Women at Deir el-Medina: A Study of the Status and Roles of the Female Inhabitants in the Workmen's Community during the Ramesside Period*. Leiden: Nederlands Instituut voor het Nabije Oosten.

Ventura, R. (1986) *Living in the City of the Dead: A Selection of Topographical and Administrative Terms in the Documents of the Theban Necropolis*. Freiberg: Universitätsverlag.

Wilfong, T. G. (1999) 'Menstrual synchrony and the "place of women" in ancient Egypt (OIM 13512)', in E. Teeter and J. A. Larson (eds) *Gold of Praise: Studies on Ancient Egypt in Honor of Edward F. Wente*. Chicago: The Oriental Institute, University of Chicago, 419–34.

Winslow, D. (1980) 'Rituals of first menstruation in Sri Lanka', *Man* 15(4), 603–25.

Yates, V. (1998) 'A sexual model of *catharsis*', *Apeiron: A Journal for Ancient Philosophy and Science* 31(1), 35–57.

Zauzich, K.-T. (1968) *Die ägyptische Schreibertradition in Aufbau, Sprache und Schrift der demotischen Kaufverträge aus ptolemäischer Zeit*. Wiesbaden: Harrossowitz.

3

UTERINE BLEEDING, KNOWLEDGE, AND EMOTION IN ANCIENT GREEK MEDICAL AND MAGICAL REPRESENTATIONS*

Irene Salvo

With the 2017 campaign 'Blood normal', the brand of feminine hygiene products Bodyform was the first in the UK to show in an advertisement red blood poured out from a test tube instead of an artificial blue liquid: the advertisement won the Grand Prix Glass Lions at Cannes in 2018 for its contribution to dismantling the shame that still surrounds periods.[1] In contrast, an advertisement from 1894 took a much more covert approach to the matter: 'For private parcels of Southhall's Sanitary Towels by post securely packed—with private address labels, quite free from anything to attract observation; write to—The Lady Manager, 17 Bull Street, Birmingham—this department being entirely managed by Ladies' (quoted in Andrews and Lomas 2018: 34). In the nineteenth-century advert, discretion was imperative, and gender boundaries were clearly demarcated. These advertisements reveal perceptions of bodily fluids embedded within their socio-historical contexts and demonstrate a shift over time in the assumptions that underlie the commercial activity of monetising women's bodily fluids. Southall's marketisation of sanitary towels seems to be built on shame over buying pads and the wish to conceal the purchase. By contrast, in what has been described as 'a love letter to periods', Bodyform speaks to the revision of female understandings of menstruation by subverting prudish standards of behaviour and revealing usually invisible bodily matters.[2] Women have had to manage menstruation, and any other bleeding from the womb, throughout all periods of human history. How common was the discourse on bleedings, and to what extent was it visible and emotionally neutral? Rosalind Janssen's chapter in this volume (Chapter 2) has shown that contact with menstruation in ancient Egypt was not regarded as a taboo. This chapter focuses on cultural attitudes to uterine bleeding in the Greek Eastern Mediterranean from the late Classical to the Imperial period, examining how information about how gendered female emissions could be controlled through magico-religious rituals when they cross bodily boundaries. In the first part, it will recall basic principles of ancient physiology, highlighting debates in Hellenistic Alexandria that testify to teaching medical knowledge. The second part focuses on gems and their properties in controlling bleeding from the uterus. The final section will explore the use of menstrual blood in particular as a magical ingredient. This chapter does not cover exhaustively the topic of gendered fluids in Greek ritual; rather, it emphasises the connections between specialist medical knowledge of bodily fluids, ritual knowledge, social visibility of fluids, and emotions,

DOI: 10.4324/9780429438974-6

particularly shame, of menstruation, as well as of not conforming to the model of a fertile mother with no reproductive dysfunctions.

Representations of physiology: some basic concepts

Physiological understandings of gendered bodily emissions were well developed in antiquity: according to the Hippocratic writers, regular menstrual periods were essential to women's health (see King 1998: 2013), while for Herophilus of Alexandria, menses could sometimes be harmful for certain women, and Soranus thought that they were always harmful—especially for fragile bodies (Herophilus T 204 von Staden; Soranus, *Gynaecology* 1.29.4, translation Temkin 1956; cf. menstrual blood in Pliny, on which see Lawrence, Chapter 14 in this volume). Blood was derived from digested food, and only women expelled it because their flesh was spongier and looser than that of men (Dean-Jones 1994: 55).

The Hippocratic writers understood that the production of a monthly flow was due to an alteration in temperature that let the fluids move out of the body:

> First blood is stirred up in their body of necessity each month because one month differs greatly from another month, both in its coldness and its heat, and a woman's body senses this, since it is moister than a man's; as the blood is stirred up and fills her vessels, some of it separates off, which I suppose has its origin in nature.
>
> Hippocratic Corpus, *On the Nature of the Child* 4,
> edition Potter 2012

Other bleeding from a woman's body was an indication of disease, such as preterm rupture of membranes, if pregnant, or vaginal lesions (Dean-Jones 1994: 52, 213), which could be remedied with potions: 'If a woman suffers pain in her lower belly during sexual intercourse and fresh blood appears, pound linen, green rushes and goose grease, mix these together, dissolve them in dilute white wine, and give to drink' (Hippocratic Corpus, *On Barren Women* 34, edition Potter 2012).

Ancient medical knowledge connected menses to breast milk (see Lawrence and Totelin in this volume, Chapters 14 and 15). The Hippocratics, Aristotle, and Herophilus thought that milk came from a surplus of nourishment: the menstrual blood not used by the foetus ascended into the chest, and heat turned it into milk.[3] If milk results from fluids moving upwards, the excess of blood that does not contribute to the growth of the foetus moves downwards and is expelled after childbirth in the form of lochia, at first stimulated by the movements of the baby during labour (Hippocratic Corpus, *On the Nature of the Child* 7). To preserve a woman's health, the lochia must flow after childbirth for about 25–42 days after giving birth to a girl and 20–30 days in case of a boy; lochial emissions are essential after abortions as well (Hippocratic Corpus, *On Diseases of Women* 1.72, edition Potter 2018; on lochial blood, see Dean-Jones 1994: 213–15). If lochia do not flow, serious illnesses may arise, such as happened to Phrontis, a female patient who after self-examination reported to the doctor that she had an obstruction in her vagina: a prompt treatment of the condition made her healthy and fertile again, avoiding a potentially cancerous ulcer.[4]

Bodily fluids—and remedies to treat them, such as fumigations—should not encounter obstructions, from the uterus to the mouth via the vagina (Bonnard 2013: 28; on women and fumigations, see Gourevitch 1999; Totelin 2009: 253–6). Sexual intercourse might have been necessary for the correct flowing of menstrual blood, and it was deemed essential for preserving women's health (King 2005: 157; cf. Steinert 2013 on gynaecological pathologies and obstructions to the flow of bodily fluids in Mesopotamian texts).

Alongside sexual difference, age was also particularly significant for understanding human anatomy, since it altered the level of moisture as well as the quantity of blood—younger women were moister and with more blood, and seasons and the environment had an impact too—cold weather stimulated larger flows (Hippocratic Corpus, *On Diseases of Women* 1.1). In another Hippocratic treatise, menstrual blood is understood as foam appearing in adolescent girls because the bodily humours have more space to be agitated in wider channels; the same mechanism produces the semen in adolescent boys (Hippocratic Corpus, *On Generation* 2, edition Potter 2012), and Aristotle upholds it;[5] in *Generation of Animals* (726b10–11, edition Peck 1942), he outlines the analogy between menstruation and seminal fluid. The latter is 'a residue from that nourishment which is in the form of blood'. It appears for the first time in boys at the time when menses appear in girls, and both fluids lose their generative power or cease in old age (Aristotle, *Generation of Animals* 727a3–11). Heat transforms blood into semen, while the colder nature of women does not allow a complete process (Aristotle, *Generation of Animals* 725a11–22; 726b31–727a1; Dean-Jones 1994: 60–1; cf. also Boylan 2015: 66; Flemming 2018). Herophilus followed Aristotle's theory of spermatogenesis, although his knowledge of anatomy from dissections enabled him to include the testes and the spermatic ducts in the production of seed (von Staden 1989: 288–96; see Laskaris and Flemming in this volume, Chapters 6 and 10).

Fluid discourses and medical knowledge

Regarding age, Herophilus emphasises that milk, menses, and sperm distinguish adults from children. Zeuxis, a second-century BCE Empiricist physician and commentator of the Hippocratic Corpus whose thought is transmitted to us via Galen, reports the following: 'Herophilus, too, was content to call [children] of such an age [sc. up to puberty] "infants", inasmuch as he says: "In "infants" seed, milk, menstruation, foetus, and baldness do not occur"' (T267a6–8 von Staden; trans. von Staden 1989: 436). Bodily fluids linked to the generative process, then, clearly mark out not just genders but more particularly stages of life. The assertion may sound obvious, and indeed Herophilus was ridiculed for having highlighted this, but Zeuxis clarifies the matter in his defence, as the text continues:

For, he [Herophilus] does not mean that these things do not occur in [children] until they reach the aforementioned age, i.e. right from the first moment of their birth, which is exactly the [erroneous] interpretation [of Herophilus's words] that some people—among them Callimachus, too—accept and for which they ridicule Herophilus, as though he were teaching things that are recognized by all. But [Herophilus in fact means] children up to puberty, [and he makes this point] because some people assume that these things occur in them, too.

Herophilus T267a-9–14 von Staden; trans. von Staden 1989: 436–7

This passage merits attention. Herophilus is presented as a teacher of anatomical knowledge, who is derided by his pupil Callimachus.[6] Herophilus' teaching seems to state something that is already well known, namely that children do not yet produce fluids such as menstrual blood, sperm, and milk. It is a piece of information available to everyone, it does not need to be acquired after specific educational training or through careful observation and analysis. Nonetheless, there were people who wrongly assumed the presence of reproductive fluids before puberty, and Herophilus deemed it important to clarify the correct knowledge on infant anatomy. The debate that Zeuxis presents on knowledge about fluids in children selects the Greek word *nēpios* instead of *pais* (child). The etymology of the substantivised adjective *nēpios* is unclear, and the idea that it means 'infant not able to talk' is not supported by any ancient tradition (Chantraine 1968–80: *s.v.*). But it unquestionably indicates the first stage of human life as characterised by a state of immaturity, and its semantics also comprises the meaning of 'being childish, foolish, ignorant'. From a medical point of view, it indicates children up to puberty. The aforementioned Hellenistic debate focuses on opposing the level of information held by experts with those depicted as ignorant persons, possessing silly misconceptions about the human body and its development. Moreover, the debate continues among the experts, Herophilus and Callimachus, about the suitability of this subject as a teaching topic. Is it worth dedicating time to teaching something that can be considered as triviality? According to Herophilus, the answer is affirmative, since the appearance of gendered bodily fluids at a certain age does not seem to be correctly and clearly understood by non-specialists.

To advance their knowledge further, doctors resorted to instruments, appliances, knives, and, most importantly, their hands (Hippocratic Corpus, *Decorum* 8, edition Jones 1923). Palpating and touching the patient was central to the Hippocratic diagnosis and to the doctor–patient relationship (on the doctor's body as a means of communication, see Bodiou 2016). Not just body parts, but also bodily fluids were touched and handled by the healer—probably without changing attitude when visiting a male or a female patient (Kosak 2016: 251–2). This gives an indication as to why practical experience was considered essential to the advancement of knowledge, as explicitly underlined again in the Hippocratic treatise *Decorum*: 'And if any of them do not know many things, they are brought to understanding by the facts of actual experience' (Hippocratic Corpus, *Decorum* 18). The stress on the importance of practical experience and direct examination seems, in the end, to be in accordance with Herophilus' epistemological method. The observation and analysis of human bodies made clear that gendered fluids did not appear before puberty in children (on infants' physiology, see also Dasen 2011: 293), and this information was worth spreading, especially among those who assumed otherwise.

Liquids, blood, and gems: the Tantalus-*historiola*

The circulation of opinions on fluids and puberty in Alexandria and the discussion on medical theories and practical experience demonstrate the multifaceted different levels of knowledge about the body. This variety of perspectives needs to be further examined by surveying the data beyond medical writings. What can other types of sources tell us about the diffusion of knowledge on bodily fluids, their visibility, and reactions on them? In the following two paragraphs, I will turn my attention to fluid discourses in non-literary documents, gems and papyri in particular, to highlight the manipulation of blood from the womb when magic and rituals tried to mediate the exchange between the inner and outer physical worlds (see Introduction to this volume, p. 5).

Knowledge about physiology and pathologies of bodily fluids was essential to the functioning of magical gems. By magical gems, I refer to semi-precious stones with a design and an inscription not retrograde and of a particular chosen colour (for example, red or green jasper), which were worn in close contact with the body—on rings or as necklace pendants—to avert evil or gain good fortune. They were produced primarily in Egypt and the Near East from about the second century CE until the fourth century CE, and the majority of them come from the antiquarian market, therefore without a clear archaeological context (for an introduction to gems, see Faraone 2018: 16, with previous bibliography). Alongside various purposes such as attracting a lover or prosperity, one of the most popular scopes for these gems was healing diseases. For the present discussion, I would like to focus on two aspects in particular: first, amulets for controlling bleeding from the womb, and, second, the relationship between magical gems and learned information on bodily fluids.

Gemstones as a field cross-fertilised by magical as well as medical knowledge have been the subject of numerous studies in recent years (see Bonnard et al. 2015, with previous bibliography; on religion and medicine, especially in the Classical period, see Nutton 2013: 104–5; on healing magic, most recently, Edmonds III 2019: 116–48). A specific stone seems to be linked to bodily fluids and gender-related illnesses: haematite. Haematite is relevant here for having been considered as antihaemorrhagic and efficacious in cases of uterine pathologies. Attilio Mastrocinque has stressed how the properties attributed to semi-precious stones related strictly to their colour; in the case of haematite stone, its dark black shade and globular form 'resembled coagulated blood, but when pulverized assumed the colour of living blood' (Mastrocinque 2011: 66, see also 62–3). Both, then, in its solid state as an amulet to be worn and as a liquid to be drunk, haematite resembled blood, as other stones were associated to other bodily fluids, like galactite to breast milk (Faraone 2018: 79).

An explicit reference to the fluidity of blood comes from a series of nine haematite amulets in which recur the same mythological elements charged with magical powers (*historiola*).[7] These gems are probably from Egypt and the Eastern Mediterranean and are dated to the second/third century CE. They present a wing-shaped inscription (*pterugōma*; each following line loses the first or the last letter) and underneath either an Ares figure (identified with the Syrian god Hadad) or an animal-headed demon with bound arms (see an example in Figure 3.1a). On the reverse there is an upright amphora, interpreted by modern scholars as the representation of a woman's uterus, surrounded by lines resembling snakes and magical words (*voces magicae*), such as the Greek vowel series and Jewish divine names (e.g. *Adōnai, Sabaō, Iaō*; see Figure 3.1b). (On iconographies of the womb in gems, see Dasen and Nagy 2019: 429–31, with previous bibliography.) The inscription on the obverse orders Tantalus to drink blood, and it reads (with variations): 'You are thirsty, Tantalus, drink blood.'[8] These amulets have parallels in other iatromagical incantations, whose text is written in a triangular- or heart- or grapes-shape, also on different media, such as silver lamellae or papyri (see Faraone 2012, 2018: 216–20). The progressive disappearance of the letters signifies the exorcism of a demon inflicting the disease (*deletio morbi*, elimination of a disease). On the Tantalus gems, a haemorrhage is seen as a demon threatened by the command to the mythologically always-thirsty Tantalus—if the blood-demon does not stop flowing, he will be extinguished by Tantalus (Michel 2005: 150–1), or, in other words, as soon as Tantalus tries to drink blood, the flow will dry up (Barb 1952: 272), in accordance with the mythical story best described by Homer (*Odyssey* 11.582–7, translation Wilson 2018). Alphonse A. Barb noted that the haemostatic properties of these amulets were aimed at preventing miscarriage and therefore at supporting the fecundity of pregnant women (Barb 1952: 279).

Figure 3.1 (a) Mars Ultor and Tantalus formula, obverse of Michel 2001: no. 383; (b) amphora-shaped womb with divine names and vowels, reverse of Michel 2001: no. 383, gem in black haematite, third century CE. Now in the British Museum, OA.10717.

Source: © Trustees of the British Museum.

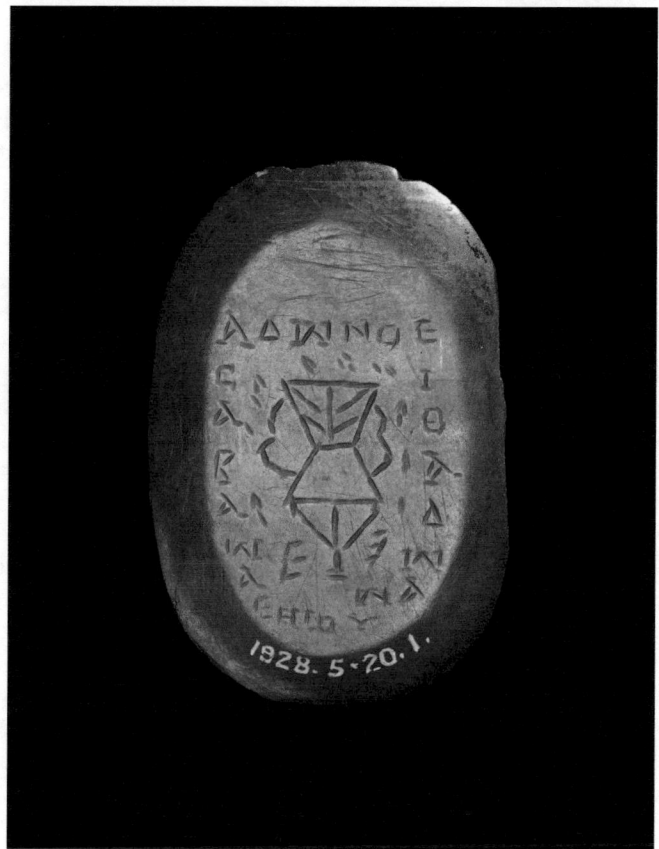

Figure 3.1 (Continued)

Christopher A. Faraone has studied these 'vanishing' incantations in great detail, and he has offered a fresh interpretation of the texts, iconographies, and functions of the nine gems with the Tantalus *historiola* (Faraone 2009, 2012: 35–49, 2018: 94–7). He thinks that the vanishing wing-layout of the text can reverse the command to Tantalus and promote bleeding as well. These amulets, therefore, could either stem or encourage uterine bleeding depending on the clinical circumstances and the desired effects; as the command vanishes, blood reappears (Faraone 2009: 265–71). Among other evidence, he takes as a parallel some haematite gems with representations of the womb as a round octopus-shaped cup over a key that could open or close the organ, either to support conception, or menstrual flow, or contraception (see on these gems Salvo 2017: 141–3, with previous bibliography, and Dasen 2018b). Attilio Mastrocinque develops this idea, adding that other iconographic elements hint at the 'magic' production of liquid: Ares could provoke bleeding wounds; the bite of a snake (the *haïmorrous*) could cause heavy haemorrhages, while another could induce insatiable thirst; in India, Tantalus was venerated as a benefactor of humanity, whose cup, as the gods' cupbearer, was instantaneously full again once emptied (Philostratus, *Life of*

Apollonius 3.25; 30; 32, edition Jones 2005); and finally, the bound demon was not able to drain the mythical water (Mastrocinque 2018: 116–18).

Although these contributions have significantly advanced our understanding of these amulets, we should further elaborate on Barb's interpretation. Several arguments can be put forward in favour of a primarily anti-abortifacient action. Firstly, one key element of the myth that must be maintained in the amuletic application is the fact that Tantalus remains thirsty, existing liquids recede in his presence, no new liquid is produced, and his thirst is not quenched (see also Mastrocinque 2000: 137; Faraone 2009: 271: 'cruel-hearted' command; and Veltri 2015: 192–3: it is an *adunaton-*formula). The bleeding could not be stopped if it is Tantalus that has to drink the liquids (see Mastrocinque 2018: 118). The invitation to drink in the formula should be interpreted *ad absurdum*: it is impossible that Tantalus will drink blood because the womb protected by the gem will not leak any fluid (cf. Steinert 2013: 4 for Mesopotamian incantations to stop haemorrhages or flowing blood during pregnancy).

Although no actual drinking is envisaged, uterine bleeding seems to be associated with water (see King, Chapter 24 in this volume, on water and virginity as well as on the importance of timing in keeping fluids in or out of the body). Tantalus would drink blood as he would drink water (on watery connotations of blood, see Boylan 2015: 33). The drawing of the womb has the shape of a drinking cup. Similarly, in Mesopotamian incantations against haemorrhages, the uterus is compared to a fermenting vessel and a leaking waterskin:

([The blood] . . . from the vagina of the young woman . . . drips and flows constantly) li[ke] [a waterlogged] meadow whose dike is not holding back (the water), [like] a fermenting vessel whose stopper does not block (the outflow), like a waterskin whose knot is not strong, whose drawstring is untrustworthy.

SpTU 4, No. 129 i 20–2; edition von Weiher 1993,
trans. from Steinert 2013: 9

Watery associations recur also in the conception of menstrual blood according to Chinese medicine. The Qing dynasty scholar Fu Qing-zhu (1607–84 CE) writes: 'menstruation is not blood but heavenly water or the *tian gui*. Originated in the kidneys, it is the essence of consummate yin (the kidneys), but possessed of the qi of consummate yang (the heart)' (Fu et al. 1995: 52). Furthermore, menstruation is controlled by the Penetrating Vessel (*Chong Mai*) and Liver-Blood, which supply blood to the womb (Maciocia 2011; on parallels between the ancient Greek and Chinese vascular theories, see Craik 2009). A connection between uterine bleeding and the liver seems to emerge from the Greek gems as well. The Ares on the Tantalus gems has been interpreted as a protective figure (Faraone 2009: 262) or as a god who could help in provoking bleeding wounds (Mastrocinque 2018: 116), but, most importantly, it recurs in another haematite gem in which Ares heals hepatic illnesses (Barb 1952: 280; Michel 2001: no. 385–6). According to Galen (*The Doctrines of Hippocrates and Plato* 6.3.7, edition De Lacy 1981; see also Nutton 2013: 239), the liver was a common area for the vegetative, nutritive, and reproductive faculties. In particular, it was the source of nutrition and growth and the seat of desires (Galen, *The Doctrines of Hippocrates and Plato* 8.1.11). The location of the appetitive soul in the liver leads us to further consider the figure of Ares in the gems, who also appears in relation to love magic in gems as well as magical papyri (Michel 2005: 152–3; *Papyri Graecae Magicae* [hereafter *PGM*] IV.296–466, XII.401–44). In the case of the Tantalus gems, we might venture to imagine that he could have aided conception and fertility as a symbol of the male semen and as the strongest figure within the love triad of Ares, Aphrodite, and Eros. Alternatively, he could have functioned as guarantor of the *historiola* as a planetary and astrological reference (see also Barb 1952: 280; Michel 2005: 153; *PGM* XIII.220, 646–734, 1031).

A further formula is worth analysing. On one of these gems, there is the Tantalus wing-inscription and Ares on its obverse, while on the reverse, above the womb and two snakes, four lines of letters read: 'The hidden lord will heal the hidden things (*ho kurios ho apokruphos iasetai ta apokrupha*)' (Festugière 1961: pl. I; trans. Faraone 2009: 259). This addition has been interpreted as a reference to the womb and menstruation (Faraone 2009: 259). In my opinion, it refers more specifically to preventing a miscarriage or a preterm delivery of the foetus in gestation. This gem carries the best representation of a womb that most resembles terracotta body parts votive offerings (on these see Dasen 2015: 120–5; Flemming 2017). Snakes are often associated with the womb, and the snake-god Chnoubis protects the correct working of the womb in the gems where deities are depicted above the octopus-shaped uterus (see Mastrocinque 2018: 213–14). A face with seven snakes protects the womb on a silver Byzantine pendant from Asia Minor, with a formula commanding the black *hustera* to eat and drink blood (Spier 1993: no. 34, pl. 3b; on Byzantine amulets protecting pregnant women, see also Congourdeau 2009: 42–3). Furthermore, the association of divine agents with 'hidden things' occurs mostly in relation to the Christian God in the Church Fathers: He has the power to reveal knowledge that is obscure, invisible, deep, and hidden (see e.g. Origen, *Philocalia* 23.5.12, edition Lewis 1911; Leontius, *In Job*, *Homilia* IV 144, edition Allen and Datem 1991).

The figure of Tantalus himself has a link with procreation and children. He notoriously killed and cooked his son, Pelops, who was then rescued by the gods. In evoking precisely this mythical story, the magical gem seems to summon divine agents to rescue an unborn child in danger. A poignant epitaph in Rome, coeval with the gems, beautifully displays the interplay between children, fathers, danger, and the myth of Tantalus. The text reads:

> (Face A) I know in myself that I have never done anything evil, but encountering evil misfortune and bitter bewitchment, I suffered such things as no one ever has (suffered). For I, who became wealthy in regard to children, was deprived of them along with their mother, my honourable and chaste wife. The children which I had, see them on the tomb: I was their father and as I grow old I grieve terribly. . . . (Face B) I, once wealthy in regard to children, having an evil warden in my misfortune, am now being punished like Tantalus.
>
> *Inscriptiones Graecae Urbis Romae [IGUR]* III 1490
> (Rome, Italy, third century CE); edition Moretti 1968–90,
> trans. Horsley and Llewelyn 1987, slightly modified

Although there is no indication of Christian or Jewish influence, the story reminds of the biblical Job grieving the loss of his children (Horsley and Llewelyn 1987: 31).

However, while Job accepts God's will as the source of his sufferings, in the Roman epitaph the cause of the destruction of the happiness bestowed to a fertile couple is identified in misfortune and bewitchment. The result is a suffering compared to that of Tantalus, who was punished for his *hubris*. Reversing the sequence of fertility, bewitchment, and Tantalic suffering, the story of Tantalus in the gems could help in averting the Evil Eye that puts the life of small children in danger (on Evil Eye and body parts, in particular 'double pupils', see Laskaris, Chapter 6, this volume).

The use of amulets to avert evil influences to the normal course of pregnancy, resulting in pregnancy loss or pre-term delivery, is ethnographically attested by contemporary medical anthropological research (see Cecil 1996). In a study conducted on Qatari Muslim women, 33 percent of respondents explained miscarriage as provoked by the Evil Eye casted by infertile and envious women (Kilshaw et al. 2017: 5). Although younger and educated women are less likely to believe in it, older family members request them not to tell anyone about the pregnancy and to conceal the bump from the eyes of others for as long as possible. For protection and healing, there is a counter-formula, *ruqyah*, that is, reciting certain verses of the Quran (Kilshaw et al. 2017: 5). Pregnancy and Evil Eye attacks between women are also intertwined in the story of Sarah and Hagar in Rabbinic commentaries (Ulmer 1994: 112). Greek cursing prayers from the Imperial period similarly attest to magical assaults and rivalry around fertility (see Salvo 2017: 132–7; on illness, gems, and the Evil Eye, see also Giannobile 2006). Low expectations of a successful outcome reinforce beliefs in magical danger during pregnancy.

Considering these connections, the Tantalus gems might have protected the wearer from haemorrhages caused by someone else. Their use as a control measure for menstrual bleeding does not seem entirely convincing. This last idea may be supported by two Latin recipes against flowing blood, especially one in which Tantalus is invited to drink blood, as the amulet does not touch the earth (see Barb 1952: 271). Faraone argues that this points to the ambiguity of menstruation, and if the patient does not wear the amulet, healthy bleeding returns (Faraone 2009: 263, 2012: 42). However, the specification that the amulet must not touch the earth seems to be more an indication for how to wear it rather than a way for stimulating bleeding at wish. Furthermore, a corrupted line in a gem of this series seems to hint at gastric haemorrhages, a kind of bleeding that certainly was not intentionally induced (Sternberg Auction 19–20 Nov. Zürich 1990, no. 459, l. 2: ΔΝΣΤωΜΑω (ἐν στομάχῳ), Mastrocinque 2000: 137). Faraone also argues that the incantation could be used to induce the flow of blood given its disappearing layout (Faraone 2009: 269, 2012: 49). However, thinking of the oral dimension of amulets, emphasised by Faraone, one might reach right the opposite conclusion: like an echo, as long as at least one letter of the formula is heard, it continues to be efficacious, to exercise its power, and it even reaches places where it could not have been heard in full.

The geometrical layout, in which each letter of the text is almost always in the same column on the gem surface, adds further efficacy (see e.g. Mastrocinque 2014: no. 353; on the *pterugma* scheme see also Giannobile 2006; Martín Hernández 2012; Mastrocinque 2018: 115). These complex compositions were inspired by Greek and non-Greek philosophical schools and religious doctrines and absorbed various realms of knowledge—from medicine to theurgy (Mastrocinque 2000: 138; Michel 2005: 142). The portability and potential invisibility of the amulet guaranteed protection from the Evil Eye. At the same time, the visualisation of internal organs in images on gems contributed to the transmission of oral knowledge and memory about the body (see also Dasen 2018a: 132 on 'the transmission and transformation of magical knowledge as a lived religious experience'). Learned information about pathologies and cures as represented on portable objects like the magical gems could have opened the way to talking about female bodies and the dangers around the womb, reducing fears and any potential sense of shame arising from a failed fertility.

Bodily fluids as magical ingredients

The body of knowledge behind the magical gems is shared by the *Papyri Graecae Magicae* (hereafter *PGM*), a collection of Greek magical papyri dated from the second century BCE to the fifth century CE (Michel 2005: 144: 'common intellectual background' between gems and papyri; see also Vitellozzi 2018, with previous bibliography). Gems might have been engraved on the basis of a papyrus (Vitellozzi 2018: 183). However, gendered fluids are treated in the magical handbooks under an additional perspective. While in the gems uterine bleeding emerges mostly as a pathological symptom to be cured, in the papyri it may be used as an ingredient for magical recipes. The number of attestations is limited, but those extant reveal an unashamed acquaintance with the manipulation of female and male emissions.

Menstrual blood could have been used as ink to write a spell, as attested by this fragmentary text:

> Let the genitals and the womb of her, Ms-so-and-so, be open, and let her become bloody by night and day. And [these things must be written] with menstrual blood, and recite before nightfall, the offerings (?) . . . she wronged first . . . and bury it near a sumac, or near . . . on a tablet.
> *PGM* LXII.76–106; trans. from Aubert 1989: 430

These instructions could be aimed at provoking menstrual bleeding as a healing therapy for amenorrhea. However, they have also been interpreted as an abortifacient spell in redress for a suffered injustice (Aubert 1989: 434–5). We can speculate that the user and the target of this aggressive magical ritual both could have been women. A third possibility might have been to employ the charm as a contraceptive (see Salvo 2017: 139).

Menstrual blood or the vaginal discharge (*ichōr*) of a virgin who died an untimely death is listed twice in the Greek Magical Papyri as part of a spell. The first occurrence is in an attraction spell:

> She NN, is burning for you, Goddess, some dreadful incense and dappled goat's fat, blood and filth, the menstrual flow of virgin dead, heart of one untimely dead, the magical material of dead dog, woman's embryo, fine-ground wheat husks. . . . And this is sacrilege! She placed them on your altar. . . . A dog-faced baboon now is born whene'er there's menstrual cleansing. . . . Attract her NN to me very quickly, I myself will clearly convict her of everything, goddess, which she had done while sacrificing to you.
> *PGM* IV.2574–621; trans. Betz 1986

Slander is a common technique to persuade the divine agent to bring the beloved target to the agent of the curse (on persuasion strategies, Salvo 2016). In this spell, the transgression committed by the victim consisted of defiling the altar of the goddess Selene placing on it prohibited substances, among which were vaginal fluids of a dead virgin together with the heart of a girl who died an untimely death and an embryo. The birth of the animal sacred to the Egyptian god Thoth coincides with menstrual periods (see Daniel and Maltomini 1990–92: 252 for the dog-faced baboon). Interestingly, the vocabulary for menstrual period recalls the monthly purge of the Hippocratic writers, testifying that the expression was widespread and long-lasting. These references to female fluids further underline the scope of the spell, that is, to control the sexuality of the victim. Moreover, it is not menstrual blood *per se* that is presented as a powerful magical agent, but virginity and untimely death amplify its properties, as in the second instance: 'For you the woman NN burns some hostile incense, goddess; the fat of dappled goat, and blood, defilement, embryo of a dog, the bloody discharge of a virgin dead untimely' (*PGM* IV.2646; trans. Betz 1986). These two analogous spells emphasise the inappropriate displacement of bodily fluids over sacred altars. The expected result is fulfilment of desires and attraction of the target. In modern Greek folklore, menarcheal blood can be kept by a girl's mother to use in love potions that guarantee the fidelity of her future husband—sometimes adding pulverised hair from the girl's vulva, while men give to their favourite lover sun-dried sperm as a love potion (Paradellis 2008).

As there is continuity between ancient and modern Greek magic in using menstrual blood in love magic, the same can be noted in highlighting the abortifacient or contraceptive properties of menstruation. In modern Greek folklore, menstruation is the antithesis of conception; it must not stay in contact with consecrated bread and wine, and it represents sexuality as opposed to motherhood like Eve to the Virgin Mary (see Paradellis 2008). In ancient Greek magic, there is a contraceptive recipe that includes menstrual blood:

> A contraceptive, the only one in the world: Take as many bittervetch seeds as you want for the number of years you wish to remain sterile. Steep them in the menses of a menstruating woman. Let them steep in her own genitals.
>
> *PGM* XXXVI.321–4; trans. Betz 1986

A frog then swallows the seeds, and the prescription continues with the fabrication of an amulet with animal substances (on this text see Gordon 2007: 129–34; De Haro Sanchez 2015: 165–6 with reference to *P. Oslo* 1.1). The ritual procedure for manufacturing this contraceptive amulet is quite complex, especially if compared with two other contraceptive recipes in the *PGM*, which merely advise taking a bean with a hole and putting it in a piece of mule skin (*PGM* LXIII.24–5, 26–8; on the mule related to amulets and fertility, Soranus, *Gynaecology* 1.63). Similar procedures in other texts are also simpler, with the recurring seeds—barley this time—soaked in menses and placed in a container made of mule skin (Aelius Promotus 775.4–8, edition Rohde 1901, see Faraone 2018: 283). It is worth remembering that these amulets could have been ambiguously used on oneself for avoiding conception as well as on others for impeding their fertility in an action of aggressive magic (Gordon 2007: 132; Salvo 2017), although the latter intention might have involved a more forceful movement to put the amulet on the body of another person.

Uterine bleeding occurs in a prescription of the Alexandrian physician Aelius Promotus (second century CE), who suggest this procedure to accelerate labour (*ōkutokia*): 'It is profitable, if she holds a date or the testicles of a living weasel. Burn the stalk of a cabbage, mix it with her menses (*tois emmēnois autēs*) and tie it on' (Aelius Promotus in Wellman 1908: 774, ll. 24–7 and 775, ll. 1–4; trans. Faraone 2018: 283). Cabbage was considered to induce menstruation (Cyranides 1.1.119, edition Kaimakis 1976, see also Waegemann 1987; Dioscorides, *De Materia Medica* 4.183, edition Osbaldeston and Wood 2000). However, it seems unclear how a parturient woman could have had menses. Although magical recipes often include bizarre elements, they could have an internal logic. In this case, we could speculate that it may be the case of bleeding during labour rather than menstrual blood. Vaginal discharge and blood-stained mucus could signal the onset of labour. Otherwise, antepartum or intrapartum haemorrhages could be symptomatic of pathologies such as the premature separation of the placenta from the uterine wall (*abruptio placentae*) or ruptured uterus or a placenta praevia (when it is positioned over the cervix). In any case, at least in the Hippocratic treatises, the terminological and ontological difference between menses, lochia, and pathological uterine haemorrhages was not so precise (see King 1998: 90). Moreover, a speculative hypothesis that is hard to prove might contemplate the storage of menstrual blood for future use in magical rituals.

Finally, menstruating women themselves could have been considered as possessing magical powers. Explaining how dreams work, Aristotle mentions that 'if a woman chances during her menstrual period to look into a highly polished mirror, the surface of it will grow cloudy with a blood-coloured haze' (*On Dreams* 460a, edition Hett 1957). This might recall a magical belief, but Aristotle finds the cause of this phenomenon in the greater quantity of blood vessels in the eye during the menstrual period, a change that also involves the action of sight, the atmosphere through which it operates, and the surface observed (see Laskaris in this volume, Chapter 6, for more on this passage, p. 115).

Visibility, knowledge, and shame

Gendered fluids and leaks from the uterus emerge from the medical writings as well as from the amuletic gems and recipes in papyri. This presence in texts and objects can be read as facilitating the circulation of knowledge about these emissions. Informed discourses on physiological mechanisms and pathologies could have made women and men alike more familiar with menses and any other form of uterine bleeding. The question of how gender, class, and level of education changed the effects of this knowledge on the individual remains open. It is plausible to assume that shared discourses about the womb could have lessened a sense of shame around menstruation and female fertility. An example from contemporary cultures might help to frame the connections between female fluids, knowledge, and emotion. Although the practice has been illegal since 2005, menstruating girls and women in rural Nepal sleep in a hut (*Chhaupadi goth*), exposed to any danger—from poor hygiene to wild animals and rape (Standing and Parker 2018: 158). Adolescent girls perceive menstruation as disgusting and shameful, and the physiological process is not well understood: education is crucial to making menstruation a less frightening experience (Standing and Parker 2018: 159–60). The dissemination of medical knowledge and education among women, girls, midwives, and doctors is considered by the World Health Organization, UNIceF, and the United Nations Population Fund to be one of the most effective tools for improving health conditions in disadvantaged communities (Mathai and Sanghvi 2017). The link between shame and lack of information is increasingly evident in modern contexts, with awareness-raising functioning as an antidote to embarrassment. On a psychological level, a study has shown how women that self-objectify themselves as sexed objects can more frequently develop emotions like shame and disgust towards their leaking bodies and tend to conceal menstruation (Roberts 2004; see more references to feminist research in Johnston-Robledo and Chrisler 2011).

Both ethnographical and psychological data prompt us to think of the magical and medical management of fluids from the womb as a process that implicated a net of connections between knowledge, self-perception, and emotion. A way to pin down ancient personal and individual states is to look at the various representations of uterine bleedings here selected. A widespread distribution of detailed anatomical knowledge is hard to imagine, if we think of the debate on children and puberty at Alexandria mentioned earlier. However, the circulation of empirical knowledge on uterine bleeding seems to be attested by ancient Greek representations of bodily fluids, the visualisation of internal organs in gems, and their use in magical spells. These sources seem to give a different picture from modern times. Female bodies were objectified in a gem to protect them from danger. Knowledge and use of fluids depended from where and when they became visible. One of the main aims of knowledge about the body seems to have been to diagnose eventual illnesses and find a remedy. Shame emerges in magical papyri as the consequence of a desecration with menstrual blood, but in the case of other uterine bleedings, shame might have been caused by fear of a failed fertility. Healthy or pathological emissions from the womb might not have been concealed in shame, since their functionalisation either in medical or magical purposes might have been more vital.

Notes

* I would like to warmly thank Laurence Totelin, Victoria Leonard, and Mark Bradley for their valuable suggestions to improve this chapter. My research at the University of Goettingen was supported by the German Research Council (DFG), at the Collaborative Research Centre 1136 *Education and Religion*, sub-project C01 'Aufgeklärte Männer—abergläubische Frauen? Religion, Bildung und Geschlechterstereotypen im klassischen Athen'. Unless otherwise indicated, all translations are from Loeb Classical Library editions.

1 'Periods are normal, showing them should be too', www.bodyform.co.uk/our-world/blood normal/ (last accessed 11 September 2018); www.adweek.com/creativity/check-out-all-30-grand-prix-winners-from-cannes-lions-2018/; www.campaignlive.co.uk/article/blood-new-normal-why-bodyform-libresse-took-marketing-risk-campaign-periods/1447989 (Accessed February 2021). Bodyform is known in other countries by brands such as Libresse, Nana, or Nuvenia.

2 www.marketingweek.com/how-bodyform-took-the-toxic-shame-out-of-periods/. (Accessed February 2021).

3 Hippocratic Corpus, *Epidemics* 2.3.17: διὸ τὰ γάλακτα, ἀδελφὰ τῶν ἐπιμηνίων, therefore milk is sister of the menses (omitted in the Loeb edition by Smith 1994); *On the Nature of Women* 1.73; *On the Nature of the Child* 10; Aristotle, *Generation of Animals* 776b4–5, 29–32; Herophilus T191 von Staden. See von Staden 1989: 292, 297; Dean-Jones 1994: 215–23.

4 Hippocratic Corpus, *On Diseases of Women* 1.40. Phrontis is the only female patient whose name is mentioned in the Hippocratic gynaecological works, see Totelin 2009: 116; King

1998: 48–9 on the relationship between doctors and women who know their bodies and have experience about their health.

5 The equation of female menstrual blood and male semen, however, does not often recur in other Hippocratic works, since the male semen is generally juxtaposed to the female seed (Dean-Jones 1994: 49; see further Flemming in this volume, Chapter 10).

6 On testimonia about Callimachus, see von Staden 1989: 480–3: he was very critical of Herophilus' school and gained independence from his master, remaining however on the 'rationalist' side against empiricism.

7 Bonner 1950: no. 144; Delatte and Derchain 1964: no. 364; Auction Sternberg 19–20 nov. Zürich, Wolfe and Sternberg 1990: no. 459; Michel 2001: no. 382, 383, 384, 93, 2004: no. 28.12.b_1; Festugière 1961: 287 no. 1.

8 Bonner 1950: no. 144: Διψᾷς Τάνταλε, αἷμα πίε. See Faraone 2009: 258–60 for the textual variants in the nine gems (e.g. ἔμα for αἷμα). Bonner (1950: 276) reads διψάς as a substantive (LSJ: 'venomous serpent, whose bite caused intense thirst'), not as a form of the verb διψάω, and translates 'Tantalus-viper, drink blood.'

Bibliography

Primary sources

Allen, P. and C. Datem (1991) *Leontius: Presbyter of Constantinople: Fourteen Homilies*. Leiden: Brill.

Betz, H. D. (1986) *The Greek Magical Papyri in Translation*. Chicago: University of Chicago Press.

Bonner, C. (1950) *Studies in Magical Amulets, Chiefly Graeco-Egyptian*. Ann Arbor: University of Michigan Press.

Daniel, R. W. and F. Maltomini (1990–92) *Supplementum magicum*. 2 vols. Opladen: Westdeutscher Verlag.

De Lacy, P. (1981) *Galen. On the Doctrines of Hippocrates and Plato*. 2nd edn. Berlin: Akademie-Verlag.

Delatte, A. and P. Derchain (1964) *Les intailles magiques gréco-égyptiennes de la Bibliothèque Nationale*. Paris: Bibliothèque Nationale de France.

Festugière, A. J. (1961) 'Pierres magiques de la Collection Kofler (Lucerne)', *Mélanges de l'Université St. Joseph* 37, 287–93.

Hett, W. S. (1957) *Aristotle. On the Soul. Parva Naturalia. On Breath*. Cambridge, MA: Harvard University Press.

Jones, C. P. (2005) *Philostratus. Apollonius of Tyana, Vol I: Life of Apollonius of Tyana, Books 1–4*. Cambridge, MA: Harvard University Press.

Jones, W. H. S. (1923) *Hippocrates. Prognostic. Regimen in Acute Diseases. The Sacred Disease. The Art. Breaths. Law. Decorum. Physician (Ch. 1). Dentition*. Cambridge, MA: Harvard University Press.

Kaimakis, D. (1976) *Die Kyraniden*. Meisenheim am Glan: Anton Hain.

Lewis, G. (1911) *The Philocalia of Origen*. Edinburgh: T. & T. Clark.

Mastrocinque, A. (2014) *Les intailles magiques du Département des monnaies, médailles et antiques*. Paris: Bibliothèque Nationale de France.

Michel, S. (2001) *Die magische Gemmen im Britischen Museum*. London: The British Museum Press.

Moretti, L. (1968–90) *Inscriptiones Graecae Urbis Romae [IGUR]*. 4 vols. Roma: Instituto italiano per la storia antica.

Osbaldeston, T. A. and R. P. A. Wood (2000) *Dioscorides Pedanius. De materia medica*. Johannesburg: Ibidis.

Peck, A. L. (1942) *Aristotle. Generation of Animals*. Cambridge, MA: Harvard University Press.

Potter, P. (2012) *Hippocrates, Vol X. Generation. Nature of the Child. Diseases 4. Nature of Women and Barrenness.* Cambridge, MA: Harvard University Press.

———. (2018) *Hippocrates, Vol XI. Diseases of Women 1–2.* Cambridge, MA: Harvard University Press.

Rohde, E. (1901) 'Aelius Promotus', in E. Rohde (ed.) *Kleine Schriften.* Tübingen: Mohr Siebeck, 381–410.

Smith, W. D. (1994) *Hippocrates. Epidemics 2, 4–7.* Cambridge, MA: Harvard University Press.

Temkin, O. (1956) *Soranus' Gynecology.* Baltimore, MD: Johns Hopkins University Press.

von Staden, H. (1989) *Herophilus: The Art of Medicine in Early Alexandria.* Cambridge: Cambridge University Press.

von Weiher, E. (1993) *Uruk: spätbabylonische Texte aus dem Planquadrat U 18. Teil IV [SpTU].* Mainz am Rhein: von Zabern.

Waegemann, M. (1987) *Amulet and Alphabet: Magical Amulets in the First Book of Cyranides.* Amsterdam: Gieben.

Wellmann, M. (1908) 'Aelius Promotus. Ἰατρικὰ φυσικὰ καὶ ἀντιπαθητικά', *Sitzungsberichte der Königlich Preussischen Akademie der Wissenschaften* Juli-Dec. 1908, 772–7.

Wilson, E. (2018) *Homer. The Odyssey.* New York: W. W. Norton.

Secondary literature

Andrews, M. and J. Lomas (2018) *A History of Women in 100 Objects.* Stroud: The History Press.

Aubert, J.-J. (1989) 'Threatened wombs: Aspect of ancient uterine magic', *Greek, Roman, and Byzantine Studies* 30, 421–49.

Barb, A.-A. (1952) 'Bois du sang, Tantale', *Syria* 29, 271–84.

Bodiou, L. (2016) 'Le corps du médecin hippocratique: Média, instrument, vecteur sensoriel', *Histoire, Médecine et Santé* 8, 31–46.

Bonnard, J.-B. (2013) 'Male and female bodies according to ancient Greek physicians', *Clio* 37, 21–39.

Bonnard, J.-B., V. Dasen, and J. Wilgaux (2015) 'Les technai du corps: La médecine, la physiognomonie et la magie', in F. Gherchanoc (ed.) *L'histoire du corps dans l'Antiquité: Bilan historiographie.* Besançon: Presses Universitaires de Franche-Comté, 169–90.

Boylan, M. (2015) *The Origins of Ancient Greek Science: Blood, a Philosophical Study.* London: Routledge.

Cecil, R. (ed.) (1996) *The Anthropology of Pregnancy Loss: Comparative Studies in Miscarriage, Stillbirth, and Neonatal Death.* Oxford: Berg Publishers.

Chantraine, P. (1968–80) *Dictionnaire étymologique de la langue grecque: Histoire des mots.* Paris: Klincksieck.

Congourdeau, M.-H. (2009) 'Les variations du désir d'enfant à Byzance', in A. Papaconstantinou and A.-M. Talbot (eds) *Becoming Byzantine: Children and Childhood in Byzantium.* Washington, DC: Dumbarton Oaks Research Library and Collection (HUP), 35–63.

Craik, E. M. (2009) 'Hippocratic bodily "channels" and oriental parallels', *Medical History* 53, 105–16.

Dasen, V. (2011) 'Childbirth and infancy in Greek and Roman antiquity', in B. Rawson (ed.) *A Companion to Families in the Greek and Roman Worlds.* Oxford: Wiley-Blackwell, 291–314.

———. (2015) *Le sourire d'Omphale: Maternité et petite enfance dans l'Antiquité.* Rennes: Presses Universitaires de Rennes.

———. (2018a) 'Amulets, the body and personal agency', in S. McKie and A. Parker (eds) *Material Approaches to Roman Magic: Occult Objects and Supernatural Substances.* Oxford: Oxbow, 127–35.

———. (2018b) 'Uterine amulets from Graeco-Roman Egypt', in R. Flemming, N. Hopwood, and L. Kassell (eds) *Reproduction: From Antiquity to the Present Day*. Cambridge: Cambridge University Press.

Dasen, V. and Á. M. Nagy (2019) 'Gems', in D. Frankfurter (ed.) *Guide to the Study of Ancient Magic*. Leiden: Brill, 406–45.

Dean-Jones, L. (1994) *Women's Bodies in Classical Greek Science*. Oxford: Clarendon Press.

De Haro Sanchez, M. (2015) 'Magie et pharmacopée: L'utilisation des végétaux dans les papyrus iatromagiques grecs', *Mythos* 9, 149–72.

Edmonds III, R. G. (2019) *Drawing Down the Moon: Magic in the Ancient Greco-Roman World*. Princeton, NJ: Princeton University Press.

Faraone, C. (2009) 'Does Tantalus drink the blood, or not? An enigmatic series of inscribed haematite gemstones', in U. Deli and C. Walde (eds) *Antike Mythen: Medien, Transformationen und Konstruktionen*. Berlin: Walter de Gruyter, 248–73.

———. (2012) *Vanishing Acts on Ancient Greek Amulets: From Oral Performance to Visual Design*. London: University of London, Institute of Classical Studies.

———. (2018) *The Transformation of Greek Amulets in Roman Imperial Times*. Philadelphia, PA: University of Pennsylvania Press.

Flemming, R. (2017) 'Wombs for the gods', in J. Draycott and E.-J. Graham (eds) *Bodies of Evidence: Ancient Anatomical Votives Past, Present and Future*. London: Routledge, 112–30.

———. (2018) 'Galen's generations of seeds', in N. Hopwood, R. Flemming, and L. Kassell (eds) *Reproduction: Antiquity to the Present Day*. Cambridge: Cambridge University Press, 95–108.

Fu, S., S.-Z. Yang and D. Liu (eds) (1995) *Fu Qing-zhu's Gynecology*. Boulder: Blue Poppy Press.

Giannobile, S. (2006) 'Malanni fisici e malanni spirituali nelle iscrizioni magiche tardoantiche', in R. Marino, M. Cassia, C. Molè, and A. Pinzone (eds) *Poveri ammalati e ammalati poveri: Dinamiche socio-economiche, trasformazioni culturali e misure assistenziali nell'occidente romano in età tardoantica*. Catania: Edizioni del prisma, 335–63.

Gordon, R. (2007) 'The coherence of magical-herbal and analogous recipes', *MHNH Revista Internacional de Investigación sobre Magia y Astrología* 7, 129–34.

Gourevitch, D. (1999) 'Fumigation et fomentation gynécologiques', in I. Garofalo, A. Lami, D. Manetti, and A. Roselli (eds) *Aspetti della terapia nel Corpus Hippocraticum*. Florence: Olschki, 203–17.

Horsley, G. H. R. and S. Llewelyn (1987) *New Documents Illustrating Early Christianity*, 4. North Ryde: Ancient History Documentary Research Centre, Macquarie University.

Johnston-Robledo, I. and J. Chrisler (2011) 'The menstrual mark: Menstruation as social stigma', *Sex Roles* 68, 1–10.

Kilshaw, S. et al. (2017) 'Causal explanations of miscarriage amongst Qataris', *BMC Pregnancy and Childbirth* 17, 250.

King, H. (1998) *Hippocrates' Woman: Reading the Female Body in Ancient Greece*. London: Routledge.

———. (2005) 'Women's health and recovery in the Hippocratic Corpus', in H. King (ed.) *Health in Antiquity*. London: Routledge, 150–61.

Kosak, J. (2016) 'Interpretations of the healer's touch in the Hippocratic Corpus', in G. Petridou and C. Thumiger (eds) *Homo Patiens: Approaches to the Patient in the Ancient World*. Leiden: Brill, 247–64.

Maciocia, G. (2011) *Obstetrics and Gynecology in Chinese Medicine*. 2nd edn. London: Elsevier.

Martín Hernández, R. (2012) 'Reading magical drawings in the Greek magical papyri', in P. Schubert (ed.) *Actes du 26e Congrès international de papyrologie Genève, 16–21 août 2010*. Geneva: Droz, 491–8.

Mastrocinque, A. (2000) 'Studi sulle gemme gnostiche', *Zeitschrift für Papyrologie und Epigraphik* 130, 131–8.

———. (2011) 'The colours of magical gems', in C. Entwistle and N. Adams (eds) *'Gems of Heaven': Recent Research on Engraved Gemstones in Late Antiquity c. AD 200–600*. London: British Museum Research Publications, 62–8.

———. (2018) 'Amuleti e medicina: Rimedi per problemi mestruali', in L. Verderame and M. E. Couto-Ferreira (eds) *Cultural Constructions of the Uterus in Pre-Modern Societies, Past and Present*. Cambridge: Cambridge Scholars Publishing, 109–20.

Mathai, M. and H. Sanghvi (2017) *Managing Complications in Pregnancy and Childbirth: A Guide for Midwives and Doctors*. 2nd edn. Geneva: World Health Organization, Department of Reproductive Health.

Michel, S. (2005) '(Re)interpreting magical gems, ancient and modern', in S. Shaked (ed.) *Officina Magica: Essays on the Practice of Magic in Antiquity*. Leiden: Brill, 141–70.

Nutton, V. (2013) *Ancient Medicine*. 2nd edn. London: Routledge.

Paradellis, T. (2008) 'Erotic and fertility magic in the folk culture of modern Greece', in J. C. B. Petropoulos (ed.) *Greek Magic: Ancient, Medieval, and Modern*. London: Routledge, 125–34.

Roberts, T.-A. (2004) 'Female trouble: The menstrual self-evaluation scale and women's self-objectification', *Psychology of Women Quarterly* 28, 22–6.

Salvo, I. (2016) 'Emotions, persuasion, and gender in Greek erotic curses', in E. Sanders and M. Johncock (eds) *Emotion and Persuasion in Classical Antiquity*. Stuttgart: Franz Steiner Verlag, 263–80.

———. (2017) 'Owners of their own bodies: Women's magic and reproduction in Greek inscriptions', in M. Dillon, E. Eidinow, and L. Maurizio (eds) *Women's Ritual Competence in the Ancient Mediterranean*. Farnham: Ashgate, 131–48.

Spier, J. (1993) 'Medieval Byzantine magical amulets and their tradition', *Journal of the Warburg and Courtland Institutes* 56, 25–62.

Standing, K. and S. L. Parker (2018) 'Girls' and women's rights to menstrual health in Nepal', in N. Mahtab (ed.) *Handbook of Research on Women's Issues and Rights in the Developing World*. Hershey, PA: IGI Global, 156–68.

Steinert, U. (2013) 'Fluids, rivers, and vessels: Metaphors and body concepts in Mesopotamian gynaecological texts', *Le Journal des Médecines Cunéiformes* 22, 1–23.

Totelin, L. M. V. (2009) *Hippocratic Recipes: Oral and Written Transmission of Pharmacological Knowledge in Fifth- and Fourth-Century Greece*. Leiden: Brill.

Ulmer, R. (1994) *The Evil Eye in the Bible and in Rabbinic Literature*. Hoboken, NJ: Ktav Publishing House.

Veltri, G. (2015) *A Mirror of Rabbinic Hermeneutics. Studies in Religion, Magic, and Language Theory in Ancient Judaism*. Berlin: Walter de Gruyter.

Vitellozzi, P. (2018) 'Relations between magical texts and magical gems: Recent perspectives', in S. Kiyanrad, C. Theis and L. Willer (eds) *Bild und Schrift auf 'magischen' Artefakten*. Berlin: Walter de Gruyter, 181–253.

Wolfe, L. A. and F. Sternberg (1990) *Antike Münzen, Griechen – Römer – Byzantiner. Phönizische Kleinkunst – Objekte mit antiken Inschriften. Geschnittene Steine und Schmuck der Antike. Auktion XXIV, Zürich, 19–20 November 1990*. Zürich: Sternberg Frank AG.

4

PUELLAE GENTLY GLOW

Scent, sweat, and the real in Latin love elegy and Ovid's didactic works

Jane Burkowski

Latin love elegy is a genre rich in sense imagery: densely elaborated suggestions of the sights, sounds, and feel of the elegists' contemporary surroundings, as well as of imagined landscapes, give many of their poems evocative, quasi-realistic settings and characters. This richness in sense imagery often includes scent—a topic with which that of bodily fluids is inextricably related. But the 'smellscape' of elegy is, for the most part, redolent of scents of a different, external kind: exotic perfumes, spices, and incense.[1] References to these fragrant substances help to build atmosphere, but they also bring with them a variety of connotations, suggesting sanctity, or foreign luxury, exoticism, and sensuality, depending on the context. The latter set of connotations are, in elegy, not typically spun in a negative way (see Griffin 1985: 10–11), but in at least one prominent case they do carry the more negative implications with which they are often charged in other genres, of deceit, luxury (in a negative sense), and moral laxity, sexual and otherwise.[2] This passage comes at the opening of an elegy by Propertius (c. 55–c. 15 BCE), where expensive perfume is listed among the luxuries that act as an unnecessary mask for the natural beauty of the speaker's mistress:

> What good is there, darling, in going out with your hair styled, and gliding about in the delicate folds of a Coan [i.e. silk] dress? What good is there in drenching your locks in Orontean [i.e. Syrian] myrrh (*Orontea crines perfundere murra*), and 'selling yourself' in(/for) foreign gifts, losing your natural glory in purchased adornments, and not allowing your limbs' natural advantages to shine forth? Believe me, nothing can improve your appearance; Love, naked himself, has no love for those who tamper with beauty.
>
> Propertius 1.2.1–8[3]

Though the speaker's moral standpoint grows more ambiguous as the poem develops, this opening passage, at least, is clear enough in presenting the simple and natural as not only aesthetically but also morally superior to the complex and artificial.[4] Where the visual and tactile appeal of adornments such as elaborate hairstyles, rich jewels, and silk clothing are concerned, it is easy enough for the reader to imagine the simple or natural alternative that Propertius' speaker is endorsing. In the case of perfume, though (*Orontea . . . murra*, line 3), especially in an era before the widespread use of soap or the invention of effective anti-perspirants, and in which perfumed oils and

DOI: 10.4324/9780429438974-7

powders played an important role in hygiene and in masking the smell of sweat (see Stewart 2007: 53–5), this is rather more difficult. It is, then, perhaps unsurprising that the speaker passes over the subject in silence, and the reader is left to infer that elegiac *puellae* (like Victorian ladies) do not sweat at all, but 'gently glow'.

This suggestion is not entirely facetious, because the idealising solution to the problem of what the unperfumed *puella* might smell like does get some explicit confirmation in another Propertian passage; in 2.29a, the speaker, wandering drunk at night, is waylaid by a gang of Cupids, who encourage him to head to Cynthia's house, where she is getting ready for bed. When she removes her headdress, they tell him, 'scent will drift over [him]: not that of the plant of Arabia [i.e. myrrh again, most likely], but that which Love has made with his own hands' (*afflabunt tibi non Arabum de gramine odores,/sed quos ipse suis fecit Amor manibus*, 2.29.17–18).[5] What, then, is this 'perfume made by the hands of Love'? It is tempting to turn to a partly parallel passage in Catullus (c. 84–c. 54 BCE) for help in answering this question:

> For I shall bring you the perfume (*unguentum*) that Venus and the Cupids gave to my girl, and as soon as you catch a whiff of it, you will ask the gods to turn you, Fabullus, into one great nose.
>
> Catullus 13.11–14

In the Catullan passage, though, the nature of this unguent—that is, whether it is to be imagined as an actual perfumed oil or as a scent given off by the girl herself—is left equally unclear, prompting scholars to put forward a great variety of suggested interpretations.[6] As Shane Butler explains, though, the simplest and perhaps most likely solution is that the ambiguity itself is the desired effect: that both of the ideas implied, of a perfumed oil and of the natural scent of the woman's body, freely interact and augment one another. Is the same thing happening in Propertius 2.29a? To a certain extent, perhaps—and it is altogether likely that Propertius is consciously alluding to the Catullan passage, encouraging us to 'read it in' to his poem. However, the two passages are not perfectly parallel. In the case of Propertius 2.29a, unlike in Catullus 13, a direct and explicit contrast is set up between the fragrance wafting off Cynthia's hair and that produced by expensive imported perfumes. We are thus encouraged to imagine them in tandem and compare them, allowing the connotations of each to interact one with the other, but by no means to conflate them fully, in the way that the Catullan passage would have us do. It therefore seems to me hard to resist the idea that in the case of Propertius 2.29a the Cupids are much more straightforwardly hinting at Cynthia's own natural scent.[7] In the fantasy scenario described, of course, this is a fragrance so divinely appealing that it equals or surpasses artificial perfumes in its attractive effect. This is not by any means a concept that originates with Propertius, or with Catullus; the idea that women, and especially beautiful and/or virtuous women, are either scentless or pleasant-smelling by nature crops up in Greek and Roman literature of various genres, while the divinely irresistible scent of one's beloved in particular (whether male or female)—of their breath, or their hair, or their sweat—is a common trope in ancient love poetry.[8] If anything, it is comparatively rare in extant Latin elegy, appearing in this passage from Propertius 2.29a alone.

Given the relative rarity of the trope in this genre, it is telling that, in the one passage in which it does appear, it is Cynthia's loosened hair in particular that gives off this attractive scent. As compared with the idea of honey-sweet breath, or floral-scented sweat, this has some plausibility to it. Most of us can probably relate to the idea that the smell of a loved one's hair has a special appeal, and the particular substance to which it is here compared (a hair-oil scented with a desert plant, probably myrrh) is oil-based and musky rather than light and sweet, and thus makes for a closer, more 'realistic' sensory parallel than other common comparanda in such passages, like apples, or honey, or fresh flowers might. Of course, the same might have applied to a scent emanating from Cynthia's body more generally; the decision to cite Cynthia's hair as the source of her magically attractive scent, though, has the added effect of preserving a certain level of decorum. Compared with any of the other areas from which Cynthia might have wafted her musk at Propertius, the hair is relatively free of lingering associations with less appealing odours—or, for that matter, with the perhaps more earthy idea of the attractive power of the smell of a lover's body—and so is less prone to evoke them in the reader's mind. Consider, by contrast, some of the more explicitly sexual interpretations that have been suggested for the less specifically localised *unguentum* of Catullus 13.11–14, including the suggestion that it refers to Catullus' girlfriend's vaginal secretions or to an artificial lubricant intended for use in anal sex (see Littman 1977; Hallett 1978, respectively). The Propertian passage, by clearly siting Cynthia's 'perfume made by the hands of Love' up around her head, ensures that it invites such speculation much less readily. Certainly, hair (and especially women's hair) was an important locus of messages about women's sexuality and held a lot of erotic cultural 'baggage' for the Romans, but it did so less by direct association than by a process of metonymy (see Bartman 2001: 1–4; Myerowitz Levine 1995: 85–90 and Eilberg-Schwartz's introduction to the same volume). Thus, letting the smell of Cynthia's hair stand in for the attractive pull of the rest of her body does not entirely sanitise the image, but it does help it to occupy a liminal zone (as hair does itself) between the bodily and the symbolic. It also taps into some even more specifically Roman cultural associations: taken together with the Cupids that provide both the 'perfume' and the description of it, the image calls up both the general association between the gods and pleasant scents (see Clements 2015: especially 56–9) and the more specific connections between Venus, her hair, pleasant scents, and attractive power.[9] Several effects are thus carefully balanced with each other in a highly concise way: the mythicised perfection of the speaker's fantasy Cynthia is beautifully conveyed, but a foothold in reality is maintained, and the passage as a whole avoids straying either into implausible hyperbole or into vulgarity.

As this passage from Propertius 2.29 is the only explicit mention that we get of the smell of the unperfumed elegiac mistress (for good or ill), it would be wrong to describe it as representative. But it is in keeping with the broader features of the genre, as a highly evocative, quasi-realistic, but essentially idealised image, intimate enough to be sensual but not so frank and bodily as to come across as crude. It makes sense that this should be the sort of limitations that we see put on references to the body in elegy; the idealising nature of the genre, together with the 'softness' that Propertius in particular continually identifies as its defining feature, means that a certain level of decorum must be maintained, and that some aspects of reality, including such 'up close and personal' aspects of the body as its smell, must therefore be either partly or completely elided, or cloaked in elegant flattery.[10] There is no blanket ban on bodily fluids in elegy; blood makes frequent appearances, and tears are everywhere: they and their connotations are generically appropriate.[11] But sweat and the smell of sweat have a very different set of connotations, which are bound up in ideas of gender, class, social status, and literary tone (see Bradley 2015b and in the Envoi to this volume). Sweat, just like spittle, semen, urine, and any other bodily fluids that are considered inconsistent with elegy's particular poetic tone, is thus largely absent—which does not go without saying, as there are plenty of poetic genres to which such subject matter is suited; see for example Centlivres Challet's and Gavrielatos's chapters (Chapters 8 and 18) in this volume, on satire. On those rare occasions when any of these substances do appear, they are relegated to particular contexts. That is, those in which their more unpleasant connotations are irrelevant—for example, when the brow of a fearful lover breaks out in a cold sweat—or in the context of magical rituals, curses, or invective, where their ability to produce a visceral reaction in the reader is being put to a generically appropriate purpose.[12] Such, then, is the limited part that the 'baser', more taboo bodily fluids play in love elegy: if they contribute in any way towards reinforcing the characteristics of the genre, they do so mostly by their conspicuous absence.

Having laid down this background principle, we can proceed to the departure from the pattern represented by Ovid (43 BCE–c. 17 CE), especially in his didactic works, the *Art of Love* (*Ars amatoria*) and *Cures for Love* (*Remedia amoris*), and the ways in which his handling of the subject helps to complicate these works' generic status. Even in Ovid's *Amores* (*Loves*), which nominally belong to the same genre as Propertius' and Tibullus' elegies, the realistic and even sordid details of life and love that had generally been elided or avoided in their works had begun to intrude into Ovid's descriptions of his speaker's relationships. A level of decorum sufficient to let the *Amores* qualify as love elegy remains, but even this is sometimes suspended to a degree that makes Ovid's elegy come close to parodying that of his predecessors, or at least to stepping far enough beyond its world to comment on it from the outside (on the relationship between the *Amores* and earlier elegy, see especially Conte 1994: 44–8, 46, who notes that 'Ovid's poetry tries to look at elegy instead of looking with the eyes of elegy'). One illustrative example that is particularly relevant in the context of this volume is *Amores* 3.7, in which the speaker has experienced an episode of impotence and spends the eighty-four-line poem dwelling on his feelings of frustration and humiliation. Its special relevance here lies in the fact that the final couplet, though admittedly obscure, seems to contain a reference to bodily fluids that are usually off limits: the speaker's partner, he tells us, felt compelled to hide the 'shameful' non-consummation of their evening from her slaves, 'having fetched a little water (*sumpta . . . aqua*)' (*Amores* 3.7.83–4). No details about what she does with it are provided, but A. A. R. Henderson (I think, rightly) suggests that she must be imagined as sprinkling it on her bed to fabricate a 'damp patch', suggestive of messy lovemaking (Henderson 1979 *ad* Ovid, *Remedia* 431–2). Others, based on linguistic parallels for *sumpta aqua*, suggest that the phrase alludes to sham post-coital douching (Brandt 1963: ad loc.; Butrica 1999: 137). In either case, Ovid would seem to be sneaking subject matter more characteristic of epigram into a love elegy almost by stealth. Protected by his own vagueness and the degree to which he relies on the reader to fill in the blanks—to the point where, at our cultural remove, we cannot even be sure of what he means—Ovid rides as close as possible to the boundaries of the genre without quite crossing them, bringing the poem to an end with what is, on the one hand, the most shockingly intimate bodily detail to appear in a poem that turns on such details but is on the other hand even more coy in tone and expression than what had come before. The passage thus makes an interesting foreshadowing of Ovid's treatment of bodily fluids in his didactic works, in which he exploits the humorous and poetic possibilities of this kind of generic mixing more often and more daringly.

In the *Ars amatoria*, as in its companion the *Remedia amoris*, Ovid has much freer play than he does in the *Amores*, as the work's generic status is ambiguous to start with. The three-part work is (on the face of it) a didactic poem, as its overall structure, illustrative digressions, allusions to earlier didactic works, and addresses to an audience of pupils all suggest.[13] Its elegiac metre, though, is a surprising feature for a didactic poem, where we would expect to find hexameters alone, and we should not underestimate the impact that this choice of metre will have had on the ancient reader's experience of the poem.[14] Indeed, in the quasi-apologia for his didactic works that Ovid inserts into the *Remedia amoris*, he explicitly tells us to adjust our expectations and to judge his intentions based on his poems' metre ('but you, whoever you are, whom my licence offends,/if you are wise/have taste (*si sapis*), measure each [work] by its own numbers', *Remedia* 371–2; this is surely a disingenuous excuse, if it implies that the didactic works' elegiac metre excuses their relative 'licence': see Holzberg 2002: especially 50–3). The speaker, too, the so-called *praeceptor amoris* ('Instructor of Love'), is half jaded elegiac lover and half parody-didactic faux expert. As Volk notes (2002: 163–6), the speaker's dual status as both lover and poet, though typical of elegy, is unusual for didactic, in which the speaker is not typically characterised as a practitioner of the art that he is teaching. This multifaceted combination of didactic tone and approach with elegiac attitudes and content is the basis for much of the humour in the *Ars Amatoria*, and the poem ends up having as much in common with satire or epigram as it does with either of the genres to which it ostensibly belongs. It does not, however, stray fully into lampooning elegy, or even into straightforwardly parodying didactic. Instead, it carves out for itself an entirely unique generic niche. Ovid's treatment of the realities of life and love where the body is concerned plays a subtly important role in that process.

To return to sweat, then: it is not a subject to which the *praeceptor amoris* devotes much time, but neither does he shy away from it or hide it with flattery in the way that the 'straight' love-elegists do. In advising his male pupils, he straightforwardly warns them that, if they want to attract women, they must remember not to neglect their personal grooming: 'let not the breath of your ill-smelling mouth be dreadful,/ and let not the husband and father of the herd (i.e. the billy goat) offend the nose' (*Ars* 1.521–2). This couplet encapsulates several features of Ovid's more general approach in his didactic works. The actual practical advice offered ought to be self-evident: the poem's real-life readership will not, one assumes, have needed the *praeceptor* to tell them to bathe, or to explain why they need to do so. But this obviousness is interesting in itself; it reminds us that Ovid, as the poet behind his *praeceptor* persona, has chosen to throw in these realistic but prosaic details, familiar from everyday life but foreign to the love elegy in which the work is heavily rooted, even though the context does not strongly demand them. The result is somewhat jarring, but this is surely the aim: to amuse the reader who expects others to be surprised or put off by such mixing of conventions and disregard of generic decorum. This effect is bolstered by the comparison of the smell of the unwashed lover to that of the billy goat: a classic symbol of crude sexuality as well as repellent smells, evocative of satyrs and of all that is traditionally set in opposition to urbanity and refinement (see Lilja 1972: 133–7; Nicholson 1997: 253–4, especially n. 8). This feeds into one of the dominant themes of the *Ars*—the importance of *ars* and *cultus* ('art/artificiality/technique' and 'cultivation', in all its senses) in all aspects of life and love—but it also underlines the strong sensory imagery of the couplet, describing one pungent and unpleasant smell by comparison with an even stronger one. The phrasing, meanwhile, is mock-serious and exaggeratedly poetic. By referring to the goat in high-flown and allusive terms ('the husband and father of the herd'), Ovid maintains at least a surface level of decorum; in essence, of course, though this coyness may masquerade as a concession to propriety, its true purpose is to magnify the humour of having brought up the subject of body odour at all, by drawing attention to its taboo nature. In short, the apparent goals and attitude of the speaker and of the poet are very much at odds, and it is the tension between the two, working together with the tension created by the defeat of the reader's expectations where the generic conventions of elegy are concerned, that creates much of the couplet's humour.

To give this kind of advice on basic personal hygiene to male lovers may be surprising in general (and generic) terms, but it is perhaps not all that incongruous when viewed in the context of the physical nature of the elegiac lover. The male speakers of love elegy do tend to claim that their identity as lovers and poets grants them a special status over other mere mortals, but this idealisation is largely spiritual; in bodily terms, though they understandably do not go so far as to describe themselves as unwashed or smelly, they are more likely to depict themselves as pale and sickly than as idealised physical specimens.[15] The elegiac speakers' beloveds, however, as we have seen, are depicted as divinely perfect where their physical selves are concerned. It is thus significant that Ovid's advice to the women to whom book three of the *Ars* is addressed runs along much the same lines as his advice to their prospective lovers:

> I was about to advise you not to let the savage billy-goat (*trux caper*) appear in your armpits, and not to let your rough legs bristle with coarse hairs! But I'm not advising girls brought up in the Caucasus.
>
> Ovid, Ars 3.193–5

Viewed in a certain light, there is a concession to convention and politeness inherent in the *praeceptor*'s assumption that his female addressees already know not to let their body odour offend others, especially as he made no such gallant assumption in the case of his male addressees. This concession is, however, very slight, and is offset both by the speaker's patronising tone and by his phrasing. Whereas, for the men, the billy goat had been alluded to in coyly allusive and even rather grand terms, as a sort of dignified patriarch, here he is simply and straightforwardly *caper*—and not just a billy goat, but a fierce, wild one (*trux*). The effect is much harsher, even in terms of how these strongly consonantal words strike the ear; it is a way of describing body odour (and perhaps armpit hair, too, given the reference to depilation that follows it) that we might sooner expect to encounter in invective or satire: compare Catullus 69.6 (perhaps the inspiration for Ovid's *trux caper* lurking in armpits, as the same image and wording are used), 71.1; Horace, *Epodes* 12.7–9, *Satires* 1.2.30, *Epistles* 1.5.29; Juvenal 6.132, 11.172ff.; Palatine Anthology 11.239, 11.240 (Lucillius); Martial 4.4,

6.93, 3.93.11, 11.22.7, many of which feature comparisons with goats.[16] Of course, Ovid immediately explains that he is referring not to the armpits of sophisticated urban courtesans but to those of rustic, mountain-dwelling tribes. But the speaker is, again, having it both ways: by asserting that his addressees already know that they must keep themselves smelling sweet, he is at the same time reminding his readers that this is something that women do need to think about. In doing so, he is in a sense informing the reader that the *puellae* so carefully idealised by the elegiac speaker are, in the invisible world behind the scenes of elegy, only ordinary women after all and liable to smell just as goaty as unkempt barbarian ones if they neglect their beauty routines. In other words, both the advice and the way in which it is couched imply that the two groups differ only in cultural terms, and that, in natural terms, they are essentially similar.[17] As often, the *praeceptor* is found to be hinting at his own 'inside knowledge' of the world of women by revealing to his other, implied audience of men that the perfection that *puellae* show to their lovers only masks completely normal human women, who by no means exude only 'perfume made by the hands of Love'. In the context of passages like that from Propertius 2.29a examined earlier, all of this also has wider poetic implications: by reminding us of the reality that lurks behind the idealised picture presented by elegy, Ovid reveals the artificiality of that genre's conventions.

It is important, however, to stress the fact that Ovid is not taking shots at elegy from the safety of the clearly defined borders of any other genre. The image of his female addressees that he here presents is not that of the caricatured prostitutes and 'loose women' that populate satire and epigram, who are often characterised as smelly and off-putting (as in some of the passages from these genres cited previously), but only of the average that lies somewhere between this extreme and that of the divine beauties of elegy—in other words, its potential for humour lies not in exaggeration but in its recognisable reality. Ovid's version is not even quite in the same vein as true didactic, as far as we can determine. We know that there were other love manuals extant at the time (see Volk 2002: 158–9), but as they are now largely lost to us, it is impossible to say how Ovid's treatment of the more basic, bodily, or down-to-earth aspects of his advice differed from theirs. We can, however, compare Ovid's approach to that of at least one extant didactic poet who did cover similar material: namely, Lucretius (c. 94–c. 50 BCE). Certainly enough borrowings and references to his work *On the Nature of Things* (*De rerum natura*, henceforth *DRN*) have been identified in Ovid's works that he must have had Lucretius in mind in composing his own didactic, especially those passages in which their subject matter overlaps.[18] It is therefore telling that Lucretius' references to sweaty and smelly *puellae* are much closer in tone to satiric ones than to Ovid's. In explaining to his pupils that buying women expensive clothes is a waste of money, for example, Lucretius describes how an expensive 'sea-blue dress' will simply wear out and end up 'drinking the sweat of lovemaking [lit., of Venus]' (*DRN* 4.1127–8); elsewhere, he describes how the lover, blinded by his affection and apt to interpret faults as endearing quirks, sees his beloved as simply 'unkempt' (*acosmos*) when others would call her 'filthy and stinking' (*immunda et foetida*, *DRN* 4.1160).[19] Ovid's playful comparisons to goats and the barbarian women of the Caucasus, much as they may push the generic boundaries of elegy, seem mild by contrast with Lucretius' studied bluntness. Both poets are lampooning the idealisation of the beloved that is characteristic of love poetry by reminding their readers that bodies are bodies, and no one's is appealing at all times and in all its aspects, and both quite naturally use sweat and body odour as examples.[20] Lucretius, though, takes aim at love poetry and its conventions from the outside, maximising the physically off-putting force of his descriptions.[21] Ovid instead does so essentially from the inside, with affectionate parody of generic and social conventions, rather than pointed satire (see Conte 1994: 54–6; Dalzell 1996: 146–51).

Though I have so far used sweat as the main pivot of this analysis, it is by no means the only generically unexpected subject to be alluded to in Ovid's didactic works. The practical and realistic approach that he shows in the case of sweat and body odour applies also to yet more taboo bodily fluids. Like his impotence poem in the *Amores*, Ovid's didactic works acknowledge the nitty-gritty of sex, not stopping at those aspects which make their way into elegy proper—breasts, thighs, kisses, and love-bites—but getting right down to bodily secretions. Lucretius' discussion of sex in book four of the *DRN* focuses heavily on the role of both male and female 'seed' (see especially *DRN* 4.1035–60, 1210–77), and so it would evidently not be outside the generic boundaries of didactic poetry to talk frankly about semen or about vaginal secretions. As noted earlier, however, Ovid's didactic is no ordinary didactic; the *praeceptor* is less a Lucretius figure, with a focused, scientific approach and a polemically philosophical outlook, than he is an alternative version of the speaker of the *Amores*, who claims still to feel bound by the demands of propriety: conventions which he implies are societal ones but which are clearly also generic (elegiac) ones. As in the final couplet of the impotence poem, then, his references to taboo sexual subjects are coy and allusive: they are clouded by double meanings or hidden behind claims of modesty and embarrassment. But, as with sweat and body odour, the fact that he mentions them at all is striking enough. In one of the books of the *Ars* addressed to men, for example, he describes the woman who agrees to sex because her partner wants it, but without much enthusiasm of her own, as 'dry' ('I hate a woman who puts out because it is necessary to do so,/and is dry (*sicca*), thinking about her wool-working', *Ars* 2.685–6). The word can be taken in metaphorical terms, to mean restrained and sober in habits, but in the context of a discussion of women's sexual pleasure, the literal meaning of 'not wet' must surely come to the fore, with the metaphorical sense at best providing the poet with plausible deniability of crudeness.[22] In his advice to women on the same topic, he warns that faking pleasure and orgasm can be more difficult than it sounds, because 'ah! embarrassing as it is, that part has secret signs of its own' (*Ars* 3.804). He here characteristically signposts the risqué nature of his material, not only by the modesty of his wording ('that part', *pars . . . ista*) and through calculated vagueness (he does not, after all, tell us what its 'secret signs' are), but by exclaiming at how embarrassing it is to have to think of such things. He thus amusingly creates the fiction of a narrator who feels forced by his duty to his pupils to enter into subjects that test the boundaries of decorum, as a mask for the poet who rejoices in playing with those very restrictions.

The same trick is exploited in the *Remedia amoris*, where it is pushed even further. This is typical of the *Remedia*, the overall theme of which—instruction in how to cure oneself of love—lends itself even more readily than the *Ars amatoria* to a Lucretius-style demystification of love and lust. To this end, the *praeceptor* explains how even sexual encounters themselves can furnish opportunities to cure yourself, proving his assertion through examples (including a clearer reference to the dreaded damp patch in the bed that may be alluded to in *Amores* 3.7):

> One man, because he had seen the private parts on an uncovered body while he was in the act, fell out of love at once: another because, as his girl rose from the act of love, he saw shameful traces (*signa pudenda*) in their soiled bed. Ah, but you're not serious if that kind of thing can dissuade you: wan fires have enflamed your heart! When Cupid has drawn his strained bow with more force, those whom he has wounded will need a greater cure.
>
> Ovid, *Remedia* 429–36

The *praeceptor*'s assertion that the strategy of disgusting yourself with the physical realities of sex will work only if you are not very deeply in love is particularly reminiscent of Lucretius, who stresses lovers' irrational compliance with the dictates of their bodily fluids, and the blindness that pleasure causes both to the ordinariness of one's beloved and to the grossly physical and animalistic nature of sexual behaviour (*DRN* 4.1037–207). Ovid's speaker, though, is not sufficiently far removed from the underlying principles of elegy to fully embrace Lucretius' suggestion that simply examining the realities of love and lust will be sufficient to dispel both. Ovid's proposed solution to this problem—the strategy that he suggests for the lover who is too deeply in love to be dissuaded by the physical details of sex—is typically irreverent: why not watch your beloved go to the toilet?

> What of him who hid as his girl did her nasty business (*reddente obscena puella*), and secretly saw what custom forbids us to see? Gods forbid I should advise anyone to do such a thing! Granted, it works, but it's just *not done*.
>
> Ovid, *Remedia* 437–40

The speaker is once again self-conscious about his own boundary-pushing, claiming that he would never advise such practices, though he has essentially just advised them. The passage as a whole is negotiated without the use of a single dirty word and yet manages to incorporate references to women's genitalia, the sexual secretions of both partners, and a man spying on a woman at her chamber pot, all hedged with faux embarrassment and claims to be bound by generic and social convention—which, strictly speaking, he has been, though he has consistently twisted these same conventions to make them act entirely contrary to their usual purpose, using them not genuinely to curb himself but to add extra humour and surprise value to his treatment of them.[23]

Thus, Ovid's handling of the subject of bodily fluids, though it certainly distinguishes his didactic works from the elegy in which they are based, does not place them within the conventions of any other clearly defined genre either. He departs sharply from the idealisation and decorum of elegy but steers clear of the exaggeration, obscene vocabulary, and targeted cruelty of satire or the philosophical justification, genuine didacticism, and relatively serious tone of Lucretius. The final question is, then, where does this leave the didactic works on the spectrum of reality, from idealisation on one side to satiric hyperbole on the other? Ovid, in his exilic works, is at pains to emphasise the fictionality of his *praeceptor* role and suggests that the didactic works' potential for being taken seriously contributed to the ruin of his life and career (see *Tristia* 1.1.111–16; *Tristia* 2, especially 349–60). It should also be noted that the apparent frankness and plausibility which the speaker's approach to bodily matters helps to create has been a factor in the tendency of scholars to quote extensively from the *Ars* and the *Remedia* in sourcebooks and social-historical accounts of Roman life of the period—often in contexts in which it is impossible fully to take into account their shifting tones, the complex dynamic at play between the speaker's attitudes and those of the poet, and the wider literary goals by which these have been shaped. The question is thus neither a trivial one in terms of the poems' own historical context nor without significance for our own broader understanding of Roman life and attitudes. As we have seen, many of the central effects of Ovid's didactic works rely on reintroducing a dose of the real into elegy—that is, into a poetic world whose rich outer texture ordinarily serves to distract from its idealisation and elision of the prosaic or unsavoury aspects of reality. Mentioning bodily fluids is an economical and, in the way that Ovid does it, an amusingly shocking way of doing so. In the end, though, the pose of a speaker who is telling things as they are, which is only made more plausible by his feigned embarrassment and his nominal concessions to decorum, acts together with the poems' defiance of clear generic classification to confuse more than to clarify, creating a layered message in which it is often difficult to tell where the speaker's motives end and where the poet's begin. The didactic works thus play not only with the conventions of elegy but with their readers, leaving them unsure quite how to respond. In reading elegy, we can either join the speaker in appreciating the beauty of his idealising fantasies, or we can see the pathetic side of their hopelessness (which he often explicitly stresses), but we must see how they differ from reality. In satire, we can see that we are being presented with a hyper-real world filled with hyper-real characters and can respond with amusement, disgust, or both, as required. In Lucretius' didactic, too, it is all too clear how the speaker wants us to interpret the world he sets before us, for he tells us. In Ovid's didactic, though, this is much less well defined; faced with a work whose position on the reality spectrum is left just as playfully ambiguous as its categorisation in the familiar list of genres, the reader is cut adrift—or, perhaps, is liberated, and left to make of it what they will.

Notes

1 See Tibullus 1.3.61–2; 1.5.11, 36; 1.7.51, 53; 1.8.70; 2.2.3–6; [Tibullus] 3.4.28; 3.6.63–4; 3.11.9; Propertius 1.2.3; 2.4.5; 2.13.23, 30; 2.19.13; 2.29.17–18; 3.10.19–22; 3.13.8; 3.14.28; 3.16.23–4; 3.17.27, 31; 4.7.32; 4.8.86; Ovid, *Amores* 2.13.23; 3.3.33. Flowers, especially bound into garlands, also make frequent appearances. On the sense of smell and its greater prominence in the culture and literature of pre-modern societies than in contemporary Western society, see Lilja 1972; Classen et al. 1994, especially 13–50; Stevens 2002; Mark 2011; Bradley 2015a.

2 On moralising and satirical attitudes to perfumes in Rome of the late Republic and early Empire, see Lilja 1972: 71–3, 79–81; Potter 1999: 175–80 ('nature was the standard of propriety'); Stevens 2002: 59–61; Stewart 2007: 95–7, 101–41. Propertius cites perfumes among the expensive luxuries that have corrupted modern morals also at 3.13.8.

3 Translations are my own. Where the Latin is quoted, the texts referred to are the *OCT* editions of Heyworth (2007a) for Propertius and of Kenney (1961, reprinted with corrections 1995) for Ovid.

4 On the moral and economic complexities of Propertius 1.2, see Watson 1982: 238–9; James 2001: 245–6; Gibson 2009: 287; Burkowski 2012: 17–22.

5 I follow most editors in dividing '2.29' into two poems after line 22. See Butler and Barber 1933: 242; Enk 1962: 368–9; Heyworth 2007b: 238 for the primary reasons for doing so.

6 For some suggested interpretations of the Catullan passage, see Quinn 1973: ad loc.; Littman 1977; Hallett 1978; Witke 1980; Kilpatrick 1998; Nappa 1998; Karakasis 2005: 105–13; Butler 2010: 98–9, 106–7, 2015: 81–2, 87.

7 This is the position taken (e.g.) by Paley 1872: ad loc.; Shackleton Bailey 1956: ad loc.; Lilja 1972: 77. Enk (1962: ad loc.) and Camps (1967: ad loc.) both suggest that it refers to a literal perfumed oil, so lovely that it seems 'divine', and Butler (2010: 106, n. 50; 2015: 88, n. 72) implies the same but allows for the possibility that it is the combination of this scent with Cynthia's own that creates its attractive effect.

8 On the natural smell of women as more pleasant or neutral than that of men, see for example Xenophon, *Symposium* 2.3; Plautus, *Mostellaria* 273 ('a woman smells right when she smells of nothing', *mulier recte olet ubi nihil olet*); Cicero, *Letters to Atticus* 2.1.1. On the natural fragrancy of women as compared to men in Greek thought, see Totelin 2014. For examples of sweet-smelling beloveds, see Horace, *Odes* 2.8.21–4; Martial 3.65; 5.37; 11.8; Achilles Tatius 2.38.3; Palatine Anthology 5.13 (Philodemus); 5.15.3, 5.18.5 (Rufinus); 5.144.6, 5.197.2 (Meleager); 12.7 (Strato).

9 Although the relative dating makes a direct allusion to the divine scent that wafts off Venus' hair at Virgil, *Aeneid* 1.402–4, doubtful (the latter half of 'book two' of Propertius seems to have been composed contemporaneously with the *Aeneid*, judging from Propertius 2.34.60–6), the two passages at least rely on a shared set of cultural associations; see Butler 2010: 102–3. On Venus/Aphrodite's hair as a symbol used in love magic, see Ficheux 2006.

10 For elegy as 'soft' (*mollis*), see Propertius 1.7.19; 2.1.2; 2.34.41–2; 3.1.19–20; 3.3.17–18. On the 'ideologies' underlying elegy more generally, see Conte 1994: 37–43.

11 For blood, see Tibullus 1.2.39; 1.6.47–54; 2.3.38; 2.6.40; Propertius 2.4.22; 2.7.14; 2.8.26, 34; 2.9.11, 40; 2.12.16; 2.17.2; 2.24.37; 2.30.21; 3.9.1; 3.11.34, 40; 3.15.41; 3.16.19; 4.8.65; Ovid, *Amores* 1.7.51, 60; 1.8.11–12; 1.12.11–12; 2.1.23; 2.5.12; 2.10.32; 2.12.6; 2.14.29, 32; 2.16.40; 3.5.35; 3.8.10, 16, 54; 3.9.64; 3.14.38. For tears and weeping, see Tibullus 1.1.52, 61–6; 1.2.76; 1.3.8, 14; 1.4.71–2; 1.5.38; 1.6.83–4; 1.8.53–4, 67–8; 1.9.29, 37–8, 79; 1.10.55–6, 63; 2.4.37; 2.5.77; 2.6.32, 41–3; [Tibullus] 3.2.25; 3.8.17–18; 3.10.21–2; Propertius 1.3.46; 1.4.23; 1.5.14–15, 30; 1.6.24; 1.7.17–18; 1.9.7; 1.10.2; 1.12.15–16; 1.13.16; 1.15.40; 1.16.4, 13, 31–2, 47–8; 1.18.6, 15–16; 1.19.18, 23; 1.21.6; 2.5.8; 2.7.2; 2.8.2; 2.13.51; 2.14.6, 14; 2.16.31, 54; 2.17.18; 2.20.1–8; 2.26.8; 2.27.7; 3.6.9–10, 17, 37; 3.7.55; 3.8.24; 3.20.4, 29; 3.25.5–8; 4.7.28, 69; Ovid, *Amores* 1.6.18; 1.7.4, 16, 22, 57–8, 60; 1.8.83–4, 110–11; 1.12.1; 1.14.51; 2.2.36, 59; 2.14.30; 2.18.7; 3.6.57, 60, 67–8; 3.9.1–3, 11–12, 46, 49; 3.10.15. For one take on weeping in elegy, see James 2003. On tears in Ovid's works more generally, see Kelly, Chapter 16, this volume.

12 See Propertius 2.22.12; 2.24.3 (sweat on the fearful lover's brow); Tibullus 1.2.54, 96 (spittle, used in magic); Tibullus 1.8.37–8 (wet kisses); Tibullus 1.5.49–50 (blood-soaked food and gall, in a curse); Tibullus 2.4.57–8 (mare's mucus, used in magic; cf. Ovid, *Amores* 1.8.8); Propertius 3.6.27 (toad's pus, in a magical context); Propertius 4.5.68 (bloody sputum, in invective directed at a witch-like bawd). The milk of the she-wolf that nursed Romulus and Remus is also referred to at Propertius 4.1.55–6 in an epic vein, and properly belongs in the same category as blood and tears (see previous note).

13 See Dalzell 1996: 21–4; Volk 2002: 36–41 on these as the major defining features of didactic poetry and Conte 1994: 50–2 and Dalzell 1996: 138–9 on their role in the *Ars*. It is not clear whether didactic was traditionally considered a distinct genre, clearly separate from other types of epic, but Volk (2002: 60–8) cautiously concludes that it would have been considered so by the period in question.

14 See Conte 1994: 52–3; Dalzell 1996: 137–8; Morgan 2010: 17–24; and more generally Morgan 2001.

15 Most often, the elegiac speaker's self-proclaimed special status manifests itself either as a unique insight or as divine protection from harm. See for example Tibullus 1.2.15–30; 1.3.57–66; 1.4.21–6, 77–80; 1.5.57–8; 1.8.1–6; 2.5.113; [Tibullus] 3.10.15; Propertius

1.7; 1.9.5–8; 2.17.9; 2.27.11–12; 2.34.54–8; 3.16.11–20; Ovid, *Amores* 1.6.9–14. For the elegiac speaker as pale and sickly, see for example Tibullus 1.3; Propertius 1.1.22; 1.5.21–2; 1.9.17; 3.16.19; Ovid *Amores* 1.6.5–6; 2.9.13–14; 2.10.23–4. Incidentally, it would not have been entirely beyond the pale for the elegists to depict their lover-poet personas as unkempt or even unwashed, judging from Horace's comic description of 'mad' poets in *The Art of Poetry* (*Ars poetica*): 'the better part of them take no care of their nails,/or their beard; they seek secluded places, and avoid the baths' (297–8); this is, then, perhaps another omission informed by generic convention.

16 On body odour as appropriate subject matter for comedy and satire, see Lilja 1972: 133–7 (cf. also 224–5, where she notes the general prevalence of scent imagery in satire and epigram as compared to other genres); Gibson 2003: *ad* 3.193–4 (who describes how tone, style, and content work together in the passage to stress how *in*appropriate it is to the quasi-elegiac genre of the *Ars*).

17 The power and significance of *cultus*, in the twinned senses of culture and of adornment, form a central theme of the *Ars*, and especially of its third book. See especially Watson 1982.

18 On Lucretius' influence on Ovid, see for example Sommariva 1980; Wheeler 1995; Miller 1997; Schiesaro 2014; and, most relevant to the topic at hand, Shulman 1981; Conte 1994: 64–5.

19 As Brown points out (1987: ad loc.), Lucretius is in this line playing with the same tradition in love poetry of idealising and championing natural beauty to which Propertius contributes in 1.2. The broader passage to which the line belongs (*DRN* 4.1149–70) has much in common with Ovid, *Ars* 2.657–62 and *Remedia* 323–30, both of which clearly rely on the reader's familiarity with Lucretius for part of their impact.

20 Lucretius makes this point explicitly, and at length; see *DRN* 4.1171–91, a passage which (as many have noted) heavily influences Ovid's discussions of the unpleasantness of the process of applying makeup (*Ars* 3.209–16; *Remedia* 351–6).

21 On Lucretius' approach here as similar to that of satire in terms of style, content, and purpose, see Brown 1987: 137–9 (cf. also his note on Lucretius' use of coarse vocabulary and tone, *ad* 4.1108).

22 cf. *OLD s.u. siccus* 7b, 3a. For parallel uses of *sicca*, see Ovid's own *Heroides*, in which Sappho thus describes the effect that sexual fantasies about the absent Phaon have on her: 'all is accomplished, and I take pleasure in it, and cannot remain dry (*siccae non licet esse mihi*)', *Heroides* 15.133–4, on which see Thorsen 2014: 14–16, and Martial, 11.81.2, describing a woman left unsatisfied by either of her two lovers: 'and the girl lies there, dry (*sicca*), in the middle of the bed.'

23 For a fuller discussion of this passage and its relation to the apologia that precedes it, see Holzberg 2002: especially 52.

Bibliography

Primary sources

Brandt, P. (1963) *P. Ovidi Nasonis Amorum libri tres*. Hildesheim: Georg Olms.

Butler, H. E. and E. A. Barber (1933) *The Elegies of Propertius*. Oxford: Clarendon Press.

Camps, W. A. (1967) *Propertius Elegies Book 2*. Cambridge: Cambridge University Press.

Enk, P. J. (1962) *Sex. Propertii elegiarum liber secundus*. Leiden: Sijthoff.

Gibson, R. K. (2003) *Ovid: Ars amatoria Book 3*. Cambridge: Cambridge University Press.

Henderson, A. A. R. (1979) *Remedia amoris*. Edinburgh: Scottish Academic Press.

Heyworth, S. J. (2007a) *Sexti Properti Elegos critico apparatu instructos edidit S.J. Heyworth*. Oxford: Oxford University Press.

Kenney, E. J. ([1961] 1995) *P. Ovidi Nasonis Amores, Medicamina faciei femineae, Ars amatoria, Remedia amoris*. Oxford: Oxford University Press.

Paley, F. A. (1872) *The Elegies of Propertius, with English Notes*. London: Bell and Daldy.

Quinn, K. (1973) *Catullus: The Poems*. London: Palgrave Macmillan.

Shackleton Bailey, D. R. (1956) *Propertiana*. Cambridge: Cambridge University Press.

Secondary literature

Bartman, E. (2001) 'Hair and the artifice of Roman female adornment', *American Journal of Archaeology* 105, 1–25.

Bradley, M. (2015a) *Smell and the Ancient Senses*. London: Routledge.

———. (2015b) 'Foul bodies in ancient Rome', in M. Bradley (ed.) *Smell and the Ancient Senses*. London: Routledge, 133–45.

Brown, R. D. (1987) *Lucretius on Love and Sex*. Leiden: Brill.

Burkowski, J. M. C. (2012) 'The symbolism and rhetoric of hair in Latin elegy', DPhil thesis, University of Oxford, Oxford.

Butler, S. (2010) 'The scent of a woman', *Arethusa* 43(1), 87–112.

———. (2015) 'Making scents of poetry', in M. Bradley (ed.) *Smell and the Ancient Senses*. London: Routledge, 74–89.

Butrica, J. L. (1999) 'Using water "unchastely": Cicero "Pro Caelio" 34 again', *Phoenix* 53(1), 136–9.

Classen, C., D. Howes and A. Synott (1994) *Aroma: The Cultural History of Smell*. London: Routledge.

Clements, A. (2015) 'Divine scents and presence', in M. Bradley (ed.) *Smell and the Ancient Senses*. London: Routledge, 46–59.

Conte, G. B. (1994) *Genres and Readers: Lucretius, Love Elegy, Pliny's Encyclopedia*. Baltimore, MD: Johns Hopkins University Press.

Dalzell, A. (1996) *The Criticism of Didactic Poetry*. Toronto: University of Toronto Press.

Ficheux, G. (2006) 'La chevelure d'Aphrodite et la magie amoureuse', in L. Bodiou, D. Frère, and V. Mehl (eds) *L'expression des corps: Gestes, attitudes, regards dans l'iconographie antique*. Rennes: Presses Universitaires de Rennes, 181–94.

Gibson, R. K. (2009) 'The success and failure of Roman love elegy as an instrument of subversion: The case of Propertius', in G. Urso (ed.) *Ordine e sovversione nel mondo greco e romano*. Pisa: Edizioni ETS, 267–87.

Griffin, J. (1985) *Latin Poets and Roman Life*. Chapel Hill, NC: Duckworth.

Hallett, J. P. (1978) 'Divine unction: Some further thoughts on Catullus 13', *Latomus* 37, 747–8.

Heyworth, S. J. (2007b) *Cynthia: A Companion to the Text of Propertius*. Oxford: Oxford University Press.

Holzberg, N. (2002) 'Staging the reader response: Ovid and his "contemporary audience" in the *Ars* and *Remedia*', in R. Gibson, S. Green, and A. Sharrock (eds) *The Art of Love: Bimillennial Essays on Ovid's Ars Amatoria and Remedia Amoris*. Oxford: Oxford University Press, 40–53.

James, S. L. (2001) 'The economics of Roman elegy: Voluntary poverty, the *recusatio*, and the greedy girl', *American Journal of Philology* 122(2), 223–53.

———. (2003) 'Her turn to cry: The politics of weeping in Roman love elegy', *Transactions of the American Philological Association* 133(1), 99–122.

Karakasis, E. (2005) '*Totum ut te faciant, Fabulle, nasum*: Catullus' XIII reconsidered', *Studies in Latin Literature and Roman History* 12, 97–114.

Kilpatrick, R. S. (1998) '*Nam unguentum dabo*: Catullus 13 and Servius' note on Phaon (*Aeneid* 3.279)', *Classical Quarterly* 48(1), 303–5.

Lilja, S. (1972) *The Treatment of Odours in Antiquity*. Helsinki: Societas Scientiarum Fennica.

Littman, R. J. (1977) 'The unguent of Venus: Catullus 13', *Latomus* 36, 123–8.

Mark, S. R. (2011) 'Follow your nose? Smell, smelling, and their histories', *The American Historical Review* 116(2), 335–51.

Miller, J. (1997) 'Lucretian moments in Ovidian elegy', *Classical Journal* 92(4), 384–98.

Morgan, L. (2001) 'Metre matters: Some higher-level metrical play in Latin poetry', *Cambridge Classical Journal* 46, 99–120.

———. (2010) *Musa Pedestris: Metre and Meaning in Roman Verse*. Oxford: Oxford University Press.

Myerowitz Levine, M. (1995) 'The gendered grammar of ancient Mediterranean hair', in H. Eilberg-Schwartz and W. Doniger (eds) *Off with Her Head! The Denial of Women's Identity in Myth, Religion, and Culture*. Berkeley: University of California Press, 76–130.

Nappa, C. (1998) 'Place settings: *Convivium*, contrast, and persona in Catullus 12 and 13', *American Journal of Philology* 119(3), 385–97.

Nicholson, J. (1997) 'Goats and gout in Catullus 71', *Classical Weekly* 90(4), 251–61.

Potter, D. S. (1999) 'Odor and power in the Roman empire', in J. I. Porter (ed.) *Constructions of the Classical Body*. Ann Arbor: University of Michigan Press, 169–89.

Schiesaro, A. (2014) '*Materiam superabat opus*: Lucretius metamorphosed', *Journal of Roman Studies* 104, 73–104.

Shulman, J. (1981) '*Tu quoque falle tamen*: Ovid's anti-Lucretian didactics', *Classical Journal* 76(3), 242–53.

Sommariva, G. (1980) 'La parodia di Lucrezio nell'*Ars* e nei *Remedia* ovidiani', *Atene e Roma* 25, 123–48.

Stevens, B. (2002) 'The scent of language and social synaesthesia at Rome', *Classical Weekly* 101(2), 159–71.

Stewart, S. (2007) *Cosmetics and Perfumes in the Roman World*. Stroud: Tempus.

Thorsen, T. S. (2014) *Ovid's Early Poetry: From His Single Heroides to His Remedia Amoris*. Cambridge: Cambridge University Press.

Totelin, L. M. V. (2014) 'L'odeur des autres: Femme et odeurs à l'intersection de la pratique hippocratique et de la pratique religieuse', in J. Jouanna and M. Zink (eds) *Hippocrate et les hippocratismes: Médecine, religion, société*. Paris: Académie des Inscriptions et Belles-Lettres, 83–99.

Volk, K. (2002) *The Poetics of Latin Didactic: Lucretius, Vergil, Ovid, Manilius*. Oxford: Oxford University Press.

Watson, P. A. (1982) 'Ovid and cultus: Ars amatoria 3.113–28', *Transactions of the American Philological Association* 112, 237–44.

Wheeler, S. M. (1995) 'Ovid's use of Lucretius in *Metamorphoses* 1.67–8', *Classical Quarterly* 45(1), 200–3.

Witke, C. (1980) 'Catullus 13: A reexamination', *Classical Philology* 75(4), 325–31.

5

OVERFLOWING BODIES AND A PANDORA OF IVORY

The pure humours of an erotic surrogate

Catalina Popescu

My chapter discusses the presence of fluid bodies and the embodiment of Galatea (the conventional name for Pygmalion's statue) in Ovid's *Metamorphoses* (10.243–97, editions Magnus 1892; Simmons 1929) and uses an ancient medical perspective to explain the myth of Pygmalion. Using the theories of Aristotle (*Generation of Animals* and *On the Soul*) and the Hippocratic Corpus (*On Diseases of Women*), I will argue that Pygmalion's Galatea represents a new type of erotic partner that opposes through her dynamic of humours the previous feminine models described in Ovid's *Metamorphoses* and Hesiod's *Works and Days* and *Theogony*.

In my view, there are more Pygmalion episodes in the *Metamorphoses*. The first book abounds in cosmic sculptural acts, where water, clay, and stone morph into bodies through the will of a conscious artist. Ovid's *Metamorphoses* describes the genesis of early humans first as the work of an experimental demiurge (1.76–86), then as spontaneous generation involving Earth's parthenogenetic fertility and sexual multiplication (1.381–415; see Popescu 2016: 148). The first model of genesis implies earth and water and divine moulding performed by a skilled *artifex* (Feldherr 2010: 123, 128). After the Flood, the act of shaping people is left to humans. While Andrew Feldherr (2010: 112) believes that the first gods are in a certain sense Daedalian artisans, 'in the forging of the crafted cosmos' and 'their very capacity for duplicitous imitation', Douglas Bauer (1962: 2) asserts that the story of Deucalion and Pyrrha anticipates the central episode of Pygmalion through 'a complete catalogue of the vocabulary to be used throughout and a forecast of the countless permutations the image will undergo'. Bauer (1962: 2, 13) actually assimilates Deucalion's efforts to those of Pygmalion because the former also mollifies stone in order to fashion new living bodies. Thus, while Deucalion and Pyrrha are not skilled in any form of carving, their work bears sculptural undertones, turning them into unwitting demiurges, as much as their gesture of tossing stones turns them into the first sowers of earth. Agriculture, craftsmanship, art, and sexuality are thus connected, being diverse forms of the same process: insemination of life (Popescu 2016: 148).

As Patricia Salzmann-Mitchell (2008: 308–10) pointed out, the entire poem teems with restless bodies constantly articulating a single giant *corpus*. Thus, there is a natural connection between earth, stone (a sterile version of soil), and human bodies, up

DOI: 10.4324/9780429438974-8

to the point of fusion: 1.427–9, 'some bodies [were] unfinished . . . and often in the same body (*eodem in corpore*) one part is alive (*vivit*) and another part is still rough earth (*rudis tellus*).'[1] Life and its imitation constantly exchange places: stone, water, and earthly 'veins' (1.409–10) stand for flesh, blood, and its vessels. If body and earth/stone are similar in nature, then what is body can be treated as stone and what is stone can also be used as body.

In my view, the metamorphoses are possible because the bodies are fluid, or imbued with restless fluids whose dynamic is similar to that of human humours (see Introduction, p. 8 for the fluidity and mutability of bodies). Throughout the entire poem, there is a subtle thread between human physiology with its humoral flow and the cosmic cycle with its transmutations. Liba Taub (2011: 43–8) shows that such ideas of analogy between human physiology and natural phenomena were permeating the Greek and Roman world (Aristotle, *History of Animals* 486b19–22). In *Meteorology* (366b14–19), Aristotle compares the storm in nature to the wind trapped inside the human body; in his view, earthquakes correspond to the tremors and the shivers of the human body. In Ovid too, the Cosmos, like a real organism, secretes, exudes, overflows, and drains. In the first act of creation in the *Metamorphoses*, nature rapidly arrange itself according to Hippocratic (and Aristotelian) principles of cold and hot, humid and dry, light and heavy. The amorphous mass of the beginning suffers stratification and layers, and the balance between these elements dictates its future health and viability. Ovid names the original Chaos 'a body':

> quia **corpore in uno**
> frigida pugnabant calidis, umentia siccis,
> mollia cum duris, sine pondere, habentia pondus.

> Because **in one body** the cold parts were fighting with the warm, the humid with the dry, the soft with the tough, the weightless with the heavy.
>
> Ovid, *Metamorphoses* 1.18–20

The first god rearranges this Chaos and allows for healthy *humoral* circulation: 'At last, a liquid flowing around (*circumfluus umor*) holds and surrounds the solid earth' (*Metamorphoses* 1.30–1).

If the first creation refers to physiology of fluids in nature, Ovid strengthens the idea that life itself is an exchange of fluids when the stone thrown by Pyrrha and Deucalion become first humid and then alive. In the newly created humans, the water principle stirs up the possible divine seeds concealed in the earth. Through observation of the correspondence between the earth's mass and the human body, Deucalion is able to repopulate the planet with more bodies after the Flood: 'I believe that the stones (*lapides*) are called "bones" (*ossa*) inside the earth's body (*in corpore terrae*)' (*Metamorphoses* 1.393–4). The separation of fluids from solids creates order and life inside these new human bodies, just like the original mass turned into a viable universe once it collected its humours.

> quae tamen ex illis aliquo pars **umida suco**
> et terrena fuit, versa est in **corporis** usum;

quod **solidum** est flectique nequit, mutatur in **ossa**,
quae modo **vena** fuit, sub eodem nomine mansit.

From these, what part was **wet with any moist** and earthy, turned into **body**,
what was **solid** and unbendable, turned into bones, what was only **vein**,
stayed under the same name.

Ovid, Metamorphoses 1.407–10

Nevertheless, humans are often sickened by humoral misbalance, and a similar condition causes earth to become deadly ill. The problematic humour is blood. Ovid uses sanguine metaphors to link physical illness with moral decay. Immoral behaviour pairs with disruption of sanguine flows inside of a sickened body:

protinus inrupit **venae** peioris in aevum
omne nefas: **fugere pudor** verumque fidesque;

At last, all evil bursts in the age of baser **vein: shame**, truth, and loyalty **ran away.**

Ovid, Metamorphoses 1.128–9

The versatility of the symbol of blood makes it an excellent signifier for various key themes: decency is the ability to blush (*pudor*, 1.129; 10.241), blood is nurturing life force when inside the body ('redness, blush' *rubor*, 2.450; see also 10.241), discarded blood is a sign of death and turpitude, and lack of shame is bruising and lividity. When sickened by crimes, the pressure builds up until the vessels of the world pulsate (*pulsant*, 1.42), begging for relief. From high above, the father of gods examines the body of the world throughout the times of Giants and the age of human decay: 'that war was threatening to start from the same body (*ab uno corpore*) and the same source' (*Metamorphoses* 1.185–6). Like a doctor, Jupiter diagnoses crimes as humoral disturbances and concludes that there is no other cure than surgery. As a result, he decides to cut open the faulty body to excise the evil and purge the sickness 'through flows', *per flumina*.[2] The earth is drenched in blood, its veins bursting under the pressure of depravity. Outside its body, healthy 'blood', *sanguis*, degenerates in monstrous 'gore', *cruor*: 'they say that the Earth, splashed with much blood (*sanguinem*) from her offspring, was fully drenched and drew new life from the warm gore (*cruorem*)' (*Metamorphoses* 1.157–8). The gods withdrew in the ethereal, dried and hotter areas, while the world is left to the mercy of humoral purges which degenerate into flood and lifeless dryness.

The initial separation of fluids is followed by instability, purge, and fossilisation. Unfortunately, the Cosmos is not the only victim of overflow and death by desiccation. A similar phenomenon happens to humans. A particular case is the story of the Propoetides, who first abhor Aphrodite and deny her divinity and later receive their punishment by being the world's first prostitutes. Because they made their bodies 'public', the Propoetides turn the natural openness of the female body into a life-threatening vice when their wasteful body turns dry, stiff, and discoloured (see Introduction, p. 9 for the socio-medical and ethical implications of fluids). Similar to the vices in the Iron Age society, promiscuity translates as sanguine loss and physical dysfunction. Their body follows their mind: unable to feel shame (*pudor*), they lose their *rubor* as well. As a result, their blood dries out or coagulates in their cheeks, turning them into stones.

> Sunt tamen obscenae Venerem Propoetides ausae
> esse negare deam; pro quo sua numinis ira
> corpora cum fama primae **vulgasse** feruntur,
> utque **pudor** cessit, **sanguisque induruit oris**,
> in rigidum parvo silicem discrimine versae.

> However, the shameless Propoetides dared to deny that Venus was a goddess; because of this, by her divine anger, they are said to have been the first to **disgrace [make public/prostitute]** their bodies together with their reputation, and since they ceased from feeling **shame, their blood dried in their cheeks** and they turned into stiff flint, since they were no different.
>
> Ovid, *Metamorphoses* 10.238–40

The stone metamorphosis is thus a manifestation of a sexually overactive female. Women in general were suspected of being imbalanced, cold and leaky, wasteful of their humours, and prone to promiscuity (van't Land 2011: 368–71, quoted in Popescu 2016: 157, note 21). Being 'bound to bleed' (King 1993: 113–16, on Hippocrates *On Diseases of Women* 2.133, 3.213; *Epidemics* 6.8.32), they needed replenishment, which led to further waste/corruption of precious fluids through their insatiable sexuality. Nevertheless, taking into consideration that the Propoetides' original sin was detesting Aphrodite, one might imagine that their problem was initially excessive abstinence (see Bauer 1962: 2 for their 'insensitivity to love' translated as petrifaction). The Hippocratic author saw the danger of strangulation and dryness in the females who refuse to allow their body a normal sexuality and a healthy fluid discharge (see also King 1993: 116; Bonnard 2013: 30–3). In her research, Helen King also associates this stiffening and lividity with the ancient perception regarding the supposed pathology of abstinence and the cravings of an otiose womb (1993: 116–17). Page Du Bois (1988: 107) also specifies: 'This [stone] is another aspect of the female body, the asexuality of presexual and postsexual female.'

Both conditions (frigidity and promiscuity) were regarded as morbid. A leaky or sexually voracious female was in ancient view a danger not only to social norms but also to her own somatic well-being (see Introduction, p. 8).[3] Rogue nature was supposed to be deadly. This impulsive, sexual excess translates in humoral misbalance, and the Propoetides turn from warm and moist bodies back into the stones originally cast by Pyrrha.[4]

This case would seem unusual unless one finds out that the medical text of Hippocrates and Aristotle portrayed women as anomalous by nature, in particular when it came to humours and sexuality (see also Introduction, p. 8 for the gendered biases against bodily fluids and their economy):

> The paradox of the feminine consists precisely in its being an unstable object, always in need of being brought into equilibrium and stability through the

reestablishment and then sudden loss of that biological medial state which the man possesses naturally.

Manuli 1980: 402 (see also Du Bois 1988: 184;
Arthur-Katz 1989: 169–70)

Pygmalion, too, indirectly criticises the faulty female nature when he despises womankind and takes refuge in his sculpture of a pure maiden: 'disgusted by the vices that nature (*natura*) gave in abundance to the female mind' (*Metamorphoses* 10.244–5). Although the mythical sculptor constructs Galatea in order to replace the faulty females of earth, he creates a statue which strangely imitates the birth of the first female of Greek myth, the arch-villainess whom poets love to hate: Pandora. Ovid's Galatea bears subtle ties to Hesiod's 'all-gifted' curse, feminine prototype, and erotic companion. Both creations are virginal and endowed with enchanting beauty and both have the touch of Aphrodite. In Hesiod's *Works and Days* (65–6, edition Evelyn-White 1914), the goddess of love gives Pandora her charisma ('and golden Aphrodite spread charm (χάριν) around her head/and painful desire (πόθον) and limb-loosening worries', while in Ovid's poem, the same goddess contributes to Galatea's animation (10.270–9). Both dolls/statues are made to be adorned. For Pandora, all the deities offer her presents: she has skills and precious ornaments (*Works and Days* 81–2). Depending on the poem, she even has a jar of 'gifts' (*Works and Days* 94–5). Ovid's Galatea is no different. She, too, is adorned by her creator with amazing clothing, coloured jewellery, and gifts pleasing to the girls (10.260–5). In addition, while Pygmalion desires her to be pure, his amorous gestures prove that, much like Pandora, Galatea is intended to satisfy a frustrated erotic need.

However, Pandora and Galatea have many differences. One is the prototype, the other is the enhanced model. One is worked out to be a tool of punishment for humankind, the other to redeem the faulty nature of the prototypical female (see also McCurdy 2011: 37). One represents the earth with all its misfortunes and 'gifts', the other is the opposite of whatever earthly nature can produce (Ovid, *Metamorphoses* 10.243–6). There is another dissimilarity between Galatea and Pandora. While Galatea's harmless limbs are mollified by love, Pandora softens men's bodies with 'painful desire and sorrows that **gnaw limbs**' (*Works and Days* 66: καὶ πόθον ἀργαλέον καὶ γυιο**βόρους** μελεδώνας). Pandora conceals her cheap inside, or her 'clay core' (Sebesta 1995: 136) under beguiling layers of cosmetics and clothes. She is herself a layered product of craftsmanship, never meant to be stripped or opened without impunity. In opposition, Galatea comes nude into existence, and while later dressed, she is nonetheless meant to represent the perfect simplicity of a truly precious essence ('neither did she appear less beautiful (*formosa*) naked (*nuda*)' (*Metamorphoses* 10.266; see Sharrock 1991: 45 for *nuda* as both 'erotically naked and simple').

Beside the divergent intentions of their builders, Galatea and Pandora imply different media and have a different 'hydrodynamic equilibrium'. Pandora is moulded of earth humidified by water (both in *Works and Days* and *Theogony*), the same clay that corresponds to the faulty original creation in *Metamorphoses*.[5] Thus, Pandora stands for the humid and corruptible beginnings of humankind. Additionally, her creation recapitulates that original cosmogony and geogony: she stands as a living embodiment of the very earth she is made of, with creatures of all sorts crowning her forehead: 'He [Hephaestus] created on it [her crown] many intricate works, wonders to behold,/creatures as many as the land and the sea nurture' (κνώδαλ᾽, ὅσ᾽ ἤπειρος πολλὰ τρέφει ἠδὲ θάλασσα) (*Theogony* 581–2).

Pandora has a jar, and in some other interpretations of the story she represents a jar herself. Du Bois believes that Pandora is a representation of the earthen vase, a metaphor of opening up the earth, opening wine jars, or performing sexual acts: all are acts which warrant caution lest the liquid contents of such vessels might overspill.[6] According to Jeffrey M. Hurwit (1995: 185), who compares this maiden statue to the goddess Athena Parthenos and even emphasises the mythical presence of a divine, Earth-like Pandora (1995: 173), her existence raises fear of the revelation of feminine sexual mysteries:

> If Pandora is *parthenos*, it is in the sense that she is a maiden ready for marriage—'sexually available', one who has not yet lost her virginity—which, of course, Athena Parthenos is not. As the representative Woman, she is (potentially) promiscuous whereas Athena is virtually asexual.
>
> Hurwit 1995: 185

Indeed, the revelation of her femininity, in this case the physiology of her fluids and bleeding events (defloration or even periods) can be quite troublesome. In the context we discuss, Pandora's body is problematic because she can be associated with excessive 'openness' or *vulgare* (like the Propoetides). Judith Lynn Sebesta (1995: 136) emphasises that there are similarities between Pandora, wives in general, vases, and *hetairai*, as they are all commodities:

> Each exhibits craft/tiness: the *hetaira*, wife and Pandora as craftswomen, the vase as product of craftsman's skill. . . . Each is superficially charming: the *hetaira*, wife, and Pandora whose liveliness conceals their dangerous sexuality, and the vase whose gleaming lovely exterior hides its clay core.
>
> Sebesta 1995: 136

Even when she is not presented as an opener of a jar, Pandora is associated with a gaping jar/belly/womb.[7] The prototypical woman is meant to be 'filled' up but might also 'spill' out her goods.[8] She does not produce but eats up the substance provided to her belly by her hard-toiling husband. Whether uterus or stomach, a belly is meant for consumption of fluids, in both physiological acts of either copulating or eating/drinking (see Vidović in this volume, p. 329, for similarity between *nedus* and bladder). The note is even more prominent when we realise that Pandora is associated with drones sipping up the honey of the working bees. In Du Bois' vision (1988: 59), honey also stands for semen, further proving that eating and making love are both forms of liquid replenishment/waste. Her nature is thus not far away from that of a wasteful vessel. She is bound to be filled up and she is bound to leak out, oozing her goods and losing them all over the earth.

By comparison, Galatea's flawlessness and incorruptibility is due to the absence of the damp and humid principles characterising the other females. Her reaction to eroticism is different precisely because of her initial lack of humours. In my view, the success of her (later) monogamous sexuality is the result of her pre-existent dry composition. She is the very contradiction of the pre-existent models, born leaky and cold, promiscuous and insatiable (van't Land 2011: 370, Aristotle, *Generation of Animals* 416a). What makes her special is her inability to bleed, her lack of life, lack of colour, warmth, and lack of orifices. In creating her, Pygmalion imitates the original demiurgic acts of the gods but avoids any of the defects. She is not born out of clay mollified with water, which later dries out and cracks, but out of dry and stable ivory (Popescu 2016). For some scholars, ivory appears warmer than regular stone.[9] It is, no doubt, nobler and rarer, and it requires a lot of hard work to put together. Salzmann-Mitchell (2008: 297–310) sees the conception of the ideal woman as a painstaking patchwork of tiny ivory sheets in a material with natural warmth. While the scholar appreciates that this organic material is actually closer to flesh (being essentially a giant tooth), she believes Pygmalion sacrificed a tremendous amount of time articulating these small and frustrating ivory strips in order to obtain something incorruptible.[10] In my view, the outcome of Pygmalion's labour is an impermeable doll which cannot be penetrated from inside or outside. While the prototype in Greek myth comes hollowed and vase-like, Galatea was not shaped like a vessel: one should not forget that the ivory was not hollowed, but underneath its smooth surface there was always a skeleton of rods or a filling of wood (Salzmann-Mitchell 2008: 297–8). While Pygmalion's touch carries undertones of penetration (*insidere*), it only probes her flesh as deep as smooth ivory would yield, and for the most part even such insertion remains imaginary: 'and he **believes** that his fingers **dip into** the touched limbs' (*et credit tactis digitos insidere membris*) (*Metamorphoses* 10.257). It never seemed to dig deeper than the surface. Her virgin body is void of any entrance.

This is the most striking difference between Pandora and Galatea: the former was created to be a real sexual partner; the latter's purpose is to offer Pygmalion a shelter from the overt sexuality of 'natural' women.[11] Simply put, Pygmalion does not want her to fulfil the role of usual sexual surrogates. Alison Sharrock (1991: 45) even believes that the chromatic fluctuation between white and red in Galatea's description betrays anxiety 'or the uneasiness between sex and purity, life and lifelessness'. Disgusted with Propoetides and their petrified blood, which caused them not to blush (10.240–1), he wants his statue to remain pure. The reader might notice that when he handles his snowy, unblemished statue, once more Pygmalion fears his statue internally bleeding or bruising: 'and he believes that his fingers dip into (*insidere*) the touched limbs and he fears lest bruising (*livor*) may come into the pressed joints' (*Metamorphoses* 10.257–8). It is the same lividity (*induruit sanguis*) that petrified the Propoetides.[12] Such a bleeding event could announce the arrival of unwanted feminine physiology, whether it may be menstruation, bleeding during defloration, or birth (King 1993: 116–17). The concerns that his 'probing fingers' might bruise the gentles members of Galatea show that he fears actually awakening her dormant sexuality under his ardent touch (or 'probing').[13] In the story of Pandora, Du Bois (1988: 59) associated the unopened jar with a virgin body and its opening or breakage with intercourse and its inherent dangers: 'Opening it up may unleash dangerous substances, and it may anger the one opened, the female body of the earth who is thus split.'[14] It looks as if the enamoured artist postpones opening such a jar, fearing her penetrability and haemorrhage—both morbid events to his mind.

What makes Galatea so special and pure is also what makes her useless as a feminine bed companion. What represents her perfection (impenetrability and purity) is also what defeminises her: she is only a simulacrum of a woman. In this new light, *livor* feared by Pygmalion stands once more for a virginal disease, where blood solidifies due to the *pnix* of abstinence, the hysteric choking due to a blockage of menses caused by the absence of sexuality, according to the Hippocratic school (King 1993: 116–18, on Plato, *Timaeus* 91c and Hippocratic passages such as *On Diseases of Women* 1.7). In Du Bois's view (1988: 107) as well, stone and petrifaction are a metaphor for stubborn virginity or post-menopausal sterility. Either of these presupposes dryness or bareness, 'the recesses of the woman's body', and 'emphasizes untouchability, unavailability'. The idea stands well with that expressed by Michael Paraskos (2008: 32, 34), who believes that rejection of Aphrodite is the cause of petrifaction, death, and sterility. Bauer (1962: 2–7) also argues that mollification or petrifaction in Ovid are strongly related to the fluctuations of feminine feelings of love (Echo, Aglauros, Narcissus, Cyparissus, Galatea): erotic insensitivity is stony death.[15] As a result, brought to the verge of erotic exasperation, Pygmalion begs Aphrodite to give Galatea life.

Ovid is indeed delicate with the sensuous aspect of Pygmalion's touch, but ascetic quest and sexual fulfilment meet under his caresses. With scholars divided between seeing the episode of Galatea's animation as gaze and artistic exercise (Freedberg 2008: 326) or gentle groping (Paraskos 2008: 32–4; Stoichita 2008: 14) and sensuous touch (Scobie and Taylor 1975: 49–54; Elsner 2007: 116–17; Feldherr 2010: 263; Sharrock 1991: 46–9), my opinion stays with the latter.[16] While Aphrodite is performing the actual miracle, a reader has no problem understanding that Pygmalion's ardent passion is actually the agent of her animation. Victor Stoichita (2008: 14) and Salzmann-Mitchell (2008: 309) even emphasise the importance of sensuality in the revitalisation and humanisation of Galatea. In this case, the entire act of sculpting was erotic in nature, a metaphor of sexual fulfilment, even when born out of a paradoxical need for abstinence (Sharrock 1991: 47). Paraskos (2008: 31, 34), too, understands the animating power of touch:

> By connecting this story specifically with that of 'Echo and Narcissus', therefore, we see that the means of escaping from one's stony origin is not by human sight, which failed to save Echo and Narcissus, but through licit human touch—through tactility.
>
> Paraskos 2008: 34

The scholar even accentuates that tactility is what creates realities; sensuality and art reunite in the gesture of touch: 'this tactility is of the outmost importance, as it is the touch of the objects that gives them reality, not simply their sight.'[17]

If sculpture shares traits with sensuality, the reciprocal is also true. Sexuality and siring are acts of shaping, moulding, or sculpting. Jean-Baptiste Bonnard (2013: 26–33) and Karine van't Land (2011: 371) insist: 'When creating the embryo, the male semen forced form on the female matter' (see Aristotle, *Generation of Animals* 740b-41a, also van't Land 2011: 364; Leitao 2012: 35, on Aristotle's view on the sensitive soul existent only in male semen). In an Aristotelian sense, there is no difference between creation as art and creation as insemination of the female by the male. Any act of siring is a 'building' act:

> Aristotle, in the *Generation of Animals*, in the context of his larger argument against *pangenesis*, would claim that male seed ought to come not from all parts of the body but only from the one part that functions as 'craftsman' (δημιουργοῦντος) and 'builder' (τέκτονος, 723b29–30).
>
> Leitao 2012: 51

The revitalisation of Galatea through art and loving caress places her animation under the double sign of sexual insemination, because love making is just as much caressing as it is shaping into existence. Galatea is both a sculpture and an erotic surrogate, both an embryo being shaped and the womb receiving the seed of life.[18]

In the act of touching and shaping, a physiological transformation takes place within Pygmalion. The poet describes the sculptor's amorous thrill both in emotional and somatic terms. According to Kathryn McKinley (2001: 99), the epigones of Ovid even saw Galatea as a sex doll 'inseminated' by her creator with both sperm and a principle of soul, in Aristotelian fashion (*spermatizabat*). Ovid's Pygmalion has no visible seed, but his chest burns with love (Sharrock 1991: 47 for *ignes* as sexual 'heat'): 'Pygmalion is amazed and hides in his chest (*pectore*) a fire (*ignes*) for the imitated body' (*Metamorphoses* 10.252–3). In my view, this 'fire' of Pygmalion also has Aristotelian echoes and brings once more into discussion the dynamics of blood. Men were believed to be hotter and able to aerate their blood into the most nourishing humour (sperm). For Aristotle (*Generation of Animals* 726a26–7), semen results 'from the blood—or rather, the hot part of the blood, aerated like foam' (Bonnard 2013: 28–9; see also van't Land 2011: 369–71; Leitao 2012: 35).[19] Additionally, seed was both warm and airy, capable of offering *pneuma* to a foetus, which stands both for 'breath' and 'soul' (Leitao 2012: 35). David Leitao (2012: 34) argues that in such tradition, 'male semen is the most logical vehicle for the transmission of the soul because its fiery heat, *pneuma*, is similar to the *aithēr* of which the soul itself is composed' (*Generation of Animals* 736b27–737a1).[20] Thus, women are receptacles (Reeder 1995: 248; Zeitlin 1995: 49–57), and men fill them with fluid, aerated life.[21] It is such nurturing fire that consumes Pygmalion's heart and travels through his kisses to the object of his desire: Galatea. Under his ardent touch, Pygmalion's erotic surge brings warmth and softness to the cold ivory. At verse 10.281, the poet specifies that 'she seemed warm (*tepere*).' Further sexual undertones are present here, according to Sharrock (1991: 47–8), since her skin finally 'yields' to his caress, and his fingers eventually dip into her permissive flesh:

> temptatum mollescit ebur **positoque rigore**
> **subsidit** digitis **ceditque**.

> The ivory softens when touched, **gives way**, and **yields** under his fingers, **losing its rigidity**.
>
> Ovid, *Metamorphoses* 10.283–4

97

Life implies a soft and pliable skin, but the essential metamorphosis takes place deeper than the surface: the hollowed insides are finally teaming with fluid life. In other words, her creator does not only 'impress' her superficially but also deeply fluidifies her, working on the circulation inside her body: her throbbing pulse announces promptly the arrival of blood in her ivory cheeks (10.280–93).

Galatea and Pygmalion finally consummate a union which remains mostly elusive. It is a perfect embrace that unifies chastity and love, sexuality and life, in opposition to the spasmodic copulation of older creatures. His *ignes* (sexual heat) surreptitiously turn into the force of life, and the response to his caress is a heartbeat. Instead of soul as *pneuma* (Aristotle, *Generation of Animals* 726a26–7; *On the Soul* 405a21–5), Galatea gains blood, which both sexualises and embryonises her by replenishing and fostering her, as if she is both an erotic partner and a *foetus* in an Aristotelian sense (*Generation of Animals* 740b–741a). Announced by warmth, pulse, and redness, the sanguine flow regains its lost status as a noble humour of life. Blood is once more *sanguis* and not *cruor*. As mentioned in a previous paper, for Galatea, Ovid chooses the infusion of real blood over water, the inferior serum used for the creation of Pandora (Popescu 2016: 154).

Unlike the other women, her body does not corrupt or exhaust any of the humours of her lover, simply because her ivory nature does not allow for fluid exchange but permanently absorbs his 'transfusion'.[22] Moreover, Pygmalion seals her up in the very act of love and animation. Her skin is soft like Hymettan wax and he moulds that wax all over her surface: 'as the Hymettan **wax** softens under the sun' (*ut Hymettia sole/cera remollescit*) (*Metamorphoses* 10.284–5). On one hand, the metaphor of warm wax bears undertones of closure of orifices, as if Pygmalion moulds heraldic seals all over her body.

On the other hand, the imagery of wax stands for the final touch of the bee filling the comb cells with nurturing fluids. The image of wax brings us back to Hesiod, to the metaphor of the bee in the story of Pandora (see *Theogony* 590–612, edition Evelyn-White 1914, and *Works and Days* 85–95). Like a drone in a hive, Pandora wastes the fluids of her husband, while he works like an industrious bee to fill her with nourishing foods.[23] These vital fluids of Pygmalion are more ambiguous. In Bronwen Wickkiser's view (2010 : 570) and Du Bois' (1988 : 59), honey is even close to the idea of semen, 'since honey may perhaps be seen as a bodily product, analogously connected to seminal fluid.' Food and livelihood are the link between honey and semen: 'Just like milk, the menses, or fat, *the seed*, passing through the intermediate stage of blood, is a "useful residue" (*perittôma*) of food' (Bonnard 2013: 29). Here, Pygmalion seals up his most precious fluid in the body of Galatea, like a bee seals it in a honeycomb filled to the brim.

While earlier fearful of mysterious bleeding events (menstruum or bruises), the lover finally rejoices to see Galatea's veins pulsating with fresh blood: 'the veins jump (*saliunt*) when touched by the thumb (*temptatae pollice*)' (*Metamorphoses* 10.289)'; 'she felt it and blushed (*sensit et erubuit*)' (*Metamorphoses* 10.293). It is significant that Galatea's life source is just as invisible and reclusive as the unseen fluid 'fires' in Pygmalion's heart. Her life remains an inner mystery: she is warm, she throbs with heart beat deep inside, and ultimately she feels the touch and blushes. Galatea's blood rests concealed in the tiny vessels of her body, as a source of feminine fertility which is best left hidden from the male eye (regarding the fear of menstruation, see King 1993:

114–17; Laskaris 2008: 460–1 and note 30; on breastfeeding, see Pedrucci 2015: 45). Thus, her blood no longer signifies the frightening aspect of death or the pollution of menstruation but the reassurance of the hidden life force. Blood represents redness of youthful, maidenly cheeks; blood is once more health and modesty. As a virgin and then a monogamous wife, Galatea restores through her blushing the healthy mystery of the female life, in opposition to the sickening *livor* of the Popoetides. *Rubor* is *pudor* once more.

There is one element that betrays Galatea's openness to the world: her eyes raised to the skies. Unlike the Propoetides, who open up their bodies to many men, Galatea simply raises her eyelids and only to contemplate her lover–creator (10.293–4). While silent and almost action-less, 'Pandora exerts her irresistible power simply by being seen' (Francis 2009: 16). As explained in a previous essay, Galatea too preserves the instinct to watch the sky, an instinct also present in the early clay humans created by the original demiurge (Popescu 2016: 155).[24] Nevertheless, her gaze is harmless and does not wander past her lover, because Pygmalion's image is finally fused with the divine sky of pure beginnings.[25] Unlike the leaky and unsatisfied creatures born from moistened soil, this maiden needs to search no more, because her lover reunites both the Father–Sky (10.294) and the Hymettan Sun of her creation (10.283–4). He is the only source she needs for warmth. Sharrock (1991: 49) specifies that 'Pygmalion has created the perfect woman whose whole world is her lover.'

Thus, the imperfections of the wet, leaky, and unstable life form in permanent need of copulation are answered by the creation of a dry creature that is once and for all filled with warm fluids through definitive intercourse with her animator. The fluid cycle is thus completed, with a new mixture of elements. The result is a hierogamy (Law 1932: 337; Frazer 1961: 173; Reinhold 1971: 316).[26] Eros (*amans*) and Cosmos (*caelum*), reunited in a ring composition, go back to the original creation. The unstable world is pacified with this micro-cosmic union followed by legitimate generation (the birth of Paphos, 10.296–7).

Notes

1 The translations from Greek and Latin are my own unless otherwise shown.
2 Ovid portrays Jupiter as a surgeon about to amputate a malignant section (1.191–2): 'but the incurable (*immedicabile*) body must be cut off (*recidendum*) by the blade for healing (*curae ense*), lest the healthy part be taken.'
3 Douglas L. Cairns (2002: 81) explains that women's dress code was a symbol of their containment, segregation, and permanent liminality: it was deemed to be for their own good. Pandora herself is presented as veiled, in the sexually contained position of a virgin bride.
4 Paraskos (2008: 34) believes that Deucalion and Pyrrha's 'story makes one immediately aware that all human beings begin as dead (that is, as stone)'. In his view, 'the process is incomplete, or at the very least that the humans created by Deucalion and Pyrrha are forever on the verge of reverting to their origins to a non-human death state, like stony statues.'
5 Jane E. Harrison (1900: 108) draws a parallel between Pandora as a Gaia figure and the festivities of *Pithoigia* and *Choes*. 'The Pithoigia, the opening of graves, existed no doubt before the earth became anthropomorphized into a goddess.'
6 François Lissarrague (1995: 98) argues that a certain similarity between women and boxes, as confined household areas, exists through metaphor. There is also a connection between 'the word for basket, *kistē*, and the word *kustos*, a vulgar expression for the female sexual organs.' Also, Ellen Reeder (1995: 195–298) associates Pandora with the container she supposedly misuses. Judith Lynn Sebesta (1995: 136) emphasises the necessity to veil Pandora

as if she is a boxed commodity, a 'clay core' wrapped in feminine fabric. Du Bois (1988: 59), believes that the femininity itself starts as earth, originally free and parthenogenetic, and ends as earthen empty container, under the control of men.

7 As Du Bois (1988: 59) emphasises the connection with Anthesteria, 'the analogy made among vase and body and earth, and the Pandora's opening of the *pithos* may be associated with her connection to Epimetheus, who foolishly receives her as a gift, presumably opens her body and reaps the consequences.'

8 As we see in Hesiod (*Theogony* 598–9), Pandora was the initial model for women 'who staying indoors, like the drones down into the sheltered hive, harvest the labour of others (ἀλλότριον κάματον) into their own bellies (σφετέρην ἐς γαστέρ')'.

9 Salzmann-Mitchell (2008: 308–10) sees here 'a fetishistic "dismemberment" of the female body for male sexual pleasure': 'While the visual artist Pygmalion makes his perfect woman out of pieces of ivory turned into body parts, the elegiac poet/lover creates his mistress with words as a collection of body parts.'

10 According to Feldherr (2010: 261), Pygmalion criticises also the artistic flaws behind the creation of the first 'natural' women. Barolski and D'Ambra (2009: 20), however, believe that it is a matter of rejecting humankind and the previous generations.

11 Justin Glenn (1977: 179–85) discusses at length the fact that the sexuality of Pandora seems to be her problem. Hurwit (1995: 185) sees this sex-toy in stark opposition with the virginity of Athena Parthenos. In addition, Pandora 'is unknowing whereas Athena is wise. She is artifice whereas Athena is artificer. She is passive whereas Athena is active. Pandora is, in effect, the Anti-Athena.'

12 Sharrock (1991: 43) reads Livor as malicious gossip or envy feared in elegies.

13 Caroline Spearing suggested to me this really useful term ('probing'): Pygmalion's caress is actually a frustrated sexual act. Sharrock (1991: 46–9) reads all the verbs of touching, feeling, attempting, and warming as sexual gestures leading to arousal and intercourse. Sharrock even connects *temptare* to the imagery of sexual touch in elegiac poems.

14 See Giula Sissa (1990) for the Greek idea of virginity without the hymen and sexuality without the modern concept of defloration.

15 Bauer (1962: 2) believes that the driving themes of the poem are (1) the stone metamorphosis and (2) the prevalence of love as a driving cosmic force of life and animation.

16 Michael Goyette suggested to me the works of Scobie and Taylor and Salzmann-Mitchell, and I would like to thank him for his kindness.

17 Paraskos (2008: 32) discusses the relationship between tactile sense and reality, that 'without an embodied and tactile sense of self, and equally an embodied and tactile sense of nonself, we risk ceasing to be conscious, as though we can revert back to the unconscious stones that Deucalion and Pyrrha cast over their shoulders.'

18 Paraskos (2008: 29) believes that the association between tactile sense and sexuality might be responsible for the discomfort modern viewers have when asked to interact with art by touching.

19 While Aristotle believed that women convert menstrual blood into food for the embryo, while uncooked 'food' is eliminated as after-birth (*Generation of Animals* 726a26–7; Bonnard 2013: 22). As Haak (2012: 295–304) explains, semen was indeed an excellent derivative of blood (better than maternal milk) and capable of nourishment.

20 Further Leitao (2012: 29) explains that the students of Aristotle supported this idea 'that warmth is the primary constituent of the human body because "seed (σπέρμα) is warm, and this is what creates the living being (κατασκευαστικὸν τοῦ ζῴου), and the place (τόπος) into which seed is deposited—this is the womb—is rather warm and similar to it [seed]".'

21 McKinley even discusses Aristotle's theory where he associates semen with the source of life and soul (Aristotle, *Generation of Animals* 729a11-b23, 737a8–34). While the scholar (2001: 99) does not envisions Galatea and Pygmalion engaged in sexuality, she explains that the epigones of Ovid might have even seen the process of animation as physiological insemination, 'an explicit sexual gloss', and quotes Giovanni's use of the verb *spermatizabat*.

22 According to Ovid, people are more or less statues or descendants of animated simulacra. In Hesiod, women are all imitation of life. See Francis (2009: 16): 'As Hephaestus' metallic maids demonstrate, there is no clear line between an image of life and life itself.'

23 Wickkiser (2010: 570) points out the paradox of female as bee and female as drone. Also, while Pandora was a delightful creature, 'the race of women descended from her are like lazy, gluttonous drones who exploit the labor of their worker bees (*Op.* 590–612).' Sharrock (1991: 47–8) sees in the malleable wax a further form of sexual pliability in this one-of-a-kind doll (Galatea) whose 'life and sexuality are a gift from her lover-creator'.

24 In 1.85–6 Ovid specifies that 'he ordered them to look at the sky (*caelum videre*) and lift their raised gaze to the stars.'

25 These beginning are visible in the first creation. See 10.294: 'she saw her lover together with the sky' (*pariter cum caelo*).

26 Frazer (1961: 173) asserts: 'The story of Pygmalion points to a ceremony of a sacred marriage in which the king wedded the image of Aphrodite, or rather of Astarte.' Bauer (1962: 16, quoted in Popescu 2016: 154) and Sharrock (1994: 171) believe that the worship is more of a romantic/sensuous kind.

Bibliography

Primary sources

Evelyn-White, H. G. (1914) *Hesiod. Works and Days. Theogony. The Homeric Hymns and Homerica.* Translated by H. G. Evelyn-White. Cambridge, MA: Harvard University Press.

Magnus, H. (1892) *Ovid. Metamorphoses.* Edited by H. Magnus. Leipzig: Teubner.

Simmons, C. (1929) *Ovid. Metamorphoses.* Edited by C. Simmons. London: Palgrave Macmillan.

Secondary literature

Arthur-Katz, M. (1989) 'Sexuality and the body in ancient Greece', *Mètis: Anthropologie des mondes grecs anciens* 4(1), 155–79.

Barolski, P. and E. d'Ambra (2009) 'Pygmalion's doll', *Arion: A Journal of Humanities and the Classics* 17(1), 19–24.

Bauer, F. D. (1962) 'The function of Pygmalion in the *Metamorphoses* of Ovid', *Transactions and Proceedings of the American Philological Association* 93, 1–21.

Bonnard, J.-B. (2013) 'Male and female bodies according to ancient Greek physicians', *Clio* 37(1), 19–37.

Cairns, D. (2002) 'The meaning of the veil in ancient Greek culture', in L. Llewellyn-Jones (ed.) *Women's Dress in the Ancient Greek World.* Swansea: The Classical Press of Wales, 73–95.

Du Bois, P. (1988) *Sowing the Body: Psychoanalysis and Ancient Representations of Women.* Chicago: The University of Chicago Press.

Elsner, J. (2007) *Roman Eyes: Visuality and Subjectivity in Art and Text.* Princeton, NJ: Princeton University Press.

Feldherr, A. (2010) *Playing Gods: Ovid's Metamorphoses and the Politics of Fiction.* Princeton, NJ: Princeton University Press.

Francis, J. A. (2009) 'Metal maidens, Achilles' shield, and Pandora: The beginnings of "Ekphrasis"', *The American Journal of Philology* 130(1), 1–23.

Frazer, J. G. (1961) *The Golden Bough.* Garden City: Doubleday.

Freedberg, D. (2008) *The Power of Images.* 2nd edn. Chicago: University of Chicago Press.

Glenn, J. (1977) 'Pandora and Eve: Sex as the root of all evil', *The Classical World* 71(3), 179–85.

Haak, H. L. (2012) 'Blood, clotting and the four humours', in M. Horstmanshoff, H. King, and C. Zittel (eds) *Blood, Sweat and Tears: The Changing Concepts of Physiology from Antiquity into Early Modern Europe.* Leiden: Brill, 295–304.

Harrison, J. E. (1900) 'Pandora's box', *The Journal of Hellenic Studies* 20, 99–114.

Hurwit, J. M. (1995) 'Beautiful evil: Pandora and the Athena Parthenos', *American Journal of Archaeology* 99(2), 171–86.

King, H. (1993) 'Bound to bleed: Artemis and Greek women', in A. Cameron and A. Kuhrt (eds.) *Images of Women in Antiquity*. 2nd edn. London: Routledge, 109–27.

Laskaris, J. (2008) 'Nursing mothers in Greek and Roman medicine', *American Journal of Archaeology* 112(3), 459–64.

Law, H. H. (1932) 'The name Galatea in the Pygmalion myth', *The Classical Journal* 27(5), 337–42.

Leitao, D. D. (2012) *The Pregnant Male as Myth and Metaphor in Classical Greek Literature*. Cambridge: Cambridge University Press.

Lissarrague, F. (1995) 'Women, boxes, containers: Some signs and metaphors' in E. D. Reeder (ed.) *Pandora: Women in Classical Greece*. 2nd edn. Princeton, NJ: Princeton University Press, 91–101.

Manuli, P. (1980) 'Fisiologia e patologia del femminile negli scritti ippocratici dell'antica ginecologia greca', in M. D. Grmek (ed.) *Hippocratica: Actes du colloque hippocratique de Paris*. Paris: Éditions du CNRS, 393–408.

McCurdy, C. G. (2011) 'Ovid's Pygmalion myth: Conceptions of the image in Greek myth and philosophy', CLAS Honors thesis, Washington and Lee University, Lexington.

McKinley, K. L. (2001) *Reading the Ovidian Heroine: 'Metamorphoses' Commentaries 1100–1618, Mnemosyne Supplements*. Leiden: Brill.

Paraskos, M. (2008) 'Bringing into being: Vivifying sculpture through touch', in P. Dent (ed.) *Sculpture and Touch*. 2nd edn. New York: Routledge, 17–34.

Pedrucci, G. (2015) 'Baliatico, αἰδώς e malocchio: capire l'allattamento nella Grecia di epoca arcaica e classica anche con l'aiuto delle fonti romane', *Eugesta* 5, 27–53.

Popescu, C. (2016) 'Privind cerul: jocul şi truda în destiul omenesc în *Metamorfozele* lui Ovidiu' in A.-C. Halichias and M.-L. D. Oancea (eds) *Joc. Joacă. Jucării*. Bucureşti: Editura Universității din Bucureşti, 147–59.

Reeder, E. D. (1995) 'Containers and textiles as metaphors for women', in E. D. Reeder (ed.) *Pandora: Women in Classical Greece*. 2nd edn. Princeton, NJ: Princeton University Press, 195–298.

Reinhold, M. (1971) 'The naming of Pygmalion's animated statue', *The Classical Journal* 66(4), 316–19.

Salzmann-Mitchell, P. (2008) 'A whole out of pieces: Pygmalion's ivory statue in Ovid's *Metamorphoses*', *Arethusa* 41(2), 291–311.

Scobie, A. and J. Taylor (1975) 'Agalmatophilia, the statue syndrome', *The Journal of the History of the Behavioral Sciences* 11(1), 49–54.

Sebesta, J. L. (1995) 'Visions of gleaming textiles and a clay core', in E. D. Reeder (ed.) *Pandora: Women in Classical Greece*. Princeton, NJ: Princeton University Press, 125–42.

Sharrock, A. (1991) 'Womanufacture', *The Journal of Roman Studies* 81, 36–49.

———. (1994) *Seduction and Repetition in Ovid's Ars Amatoria 2*. Oxford: Clarendon Press.

Sissa, G. (1990) *Greek Virginity*. Cambridge, MA: Harvard University Press.

Stoichita, V. (2008) *The Pygmalion Effect: From Ovid to Hitchcock*. Chicago: University of Chicago Press.

Taub, L. (2011) 'Physiological analogies and metaphors in explanations of the earth and the cosmos', in M. Horstmanshoff, H. King, and C. Zittel (eds) *Blood, Sweat and Tears: The Changing Concepts of Physiology from Antiquity into Early Modern Europe*. Leiden: Brill, 41–60.

van't Land, K. (2011) 'Sperm and blood, form and food: Late Medieval medical notions of male and female in the embryology of *membra*', in M. Horstmanshoff, H. King and C. Zittel (eds)

Blood, Sweat and Tears: The Changing Concepts of Physiology from Antiquity into Early Modern Europe. Leiden: Brill, 363–92.

Wickkiser, B. L. (2010) 'Hesiod and the fabricated woman: Poetry and visual art in the "Theogony"', *Mnemosyne* 63(4), 557–76.

Zeitlin, F. (1995) 'The economics of Hesiod's Pandora', in E. D. Reeder (ed.) *Pandora: Women in Classical Greece*. Princeton, NJ: Princeton University Press, 49–56.

Part III

EROTIC AND GENERATIVE FLUIDS

6

THE EYES HAVE IT

From generative fluids to vision rays

Julie Laskaris

The centrality to sexual reproduction of the head, the brain, and the vessels connecting them with the rest of the body is evident in numerous Greek texts. Not sufficiently explored, however, is the special emphasis given in some texts to the role of the eyes in the formation of generative seed. Following is a summary of relevant anatomical and physiological concepts and a discussion of related theories and pathological conditions, with a particular focus on the eyes. Among the texts to which I shall be referring are the treatises of the Hippocratic Corpus, a collection of approximately sixty medical texts written or compiled in the late fifth and early fourth centuries BCE by many authors, all of them unknown. Beginning in antiquity, they came to be attributed to the famous physician, Hippocrates.

Seed formation

Reproductive seed and cerebral spinal fluid, called *myelos* in Greek,[1] are closely aligned in early Greek thought, as noted by Richard Onians (1951: 108–15) and Elizabeth Craik (2009: 111), who confine their discussions to male semen. Onians argues that though the Greeks, with their practice of animal castration, clearly understood the testes' centrality to male sexual potency and fertility, they believed that the testes were merely a delivery system: the seed itself was produced in the head, they thought, and travelled to the testes and then to the penis (1951: 109–10 with n. 4). I will argue in this paper that reproductive seed in women was believed to have its origin in *myelos*, too, and that because of the presence of *myelos* in them, the eyes—and consequently vision itself—were related to reproductive matters in Greek thought.

Myelos is frequently translated as 'marrow', generally inappropriately, since *myelos* in Greek medical texts most often refers not to bone marrow, the producer of blood cells found in the spongy part of bones, but to cerebral spinal (or cerebrospinal) fluid (Craik 2009: 108). Cerebral spinal fluid is the clear colourless fluid found within and surrounding the brain and spinal cord; it functions as a shock absorber for the central nervous system, providing buoyancy for the brain, which floats in it, and circulating nutrients to the central nervous system (the brain and spinal cord) and providing it with immunological protection. *Myelos* was thought by numerous Greek and Roman medical writers (e.g., the authors of the Hippocratic treatises *On Glands, On Places in Man, Coan Prognoses,* and *On Internal Affections* and, hundreds of years later, Celsus and Galen) to be carried in vessels running from the brain to the rest of the body, with some vessels going from the brain to the eyes (Craik 2009: 111) and a crucial one

DOI: 10.4324/9780429438974-10

running from behind the ear to the groin. Cerebral spinal fluid does in fact bathe the optic nerve, whose connection to the brain was very likely first noticed by Alcmaeon, who wrote of passages connecting the brain with the eyes (Diels and Kranz 1996; 24 A 5 = Theophrastus, *On the Senses* 25).

Alcmaeon was followed in this observation by several Hippocratic authors and Aristotle (Craik 2006: 20). These texts also assert that the health of the eye—and, indeed, of the body as a whole—could be affected by a fluid present in the head and the area of the eye: if the fluid was in a pure state, all would be well; but if there was too much of it, or if it became corrupted, it could cause problems for the eyes and head and even travel to the lower regions of the body, causing blockages (Craik 2006: 20). Illnesses involving *myelos* were generally considered by Greek medical writers to be intractable and were often linked to overindulgence in sex (Craik 2009: 111).

With regard to the association of *myelos* and male semen, Elizabeth Craik has called particular attention to *On the Nature of Bones* 11–18,[2] a Hippocratic treatise dated most probably to the late fifth century (Craik 2015: 229), that offers detailed insight into the understanding of at least some Greek doctors regarding the physical structures underpinning sexual reproduction. This text is closely aligned with several others, a fact that, while it provides difficulties for textual critics, perhaps attests to the acceptability of its views.[3] Its complicated reconstruction of the vascular system—the most complete of extant Classical Greek texts—describes a single original, fundamental, or primary vessel.[4] The primary vessel makes a complex circuit throughout the body, in effect connecting every part with every other, and serving as the 'trunk' from which numerous large and small vessels branch out. Among these are fine vessels that take root in the brain and cover the entire head (12). The primary vessel itself then travels towards the back of the head and along the spine, sending out more fine vessels to form a network within the 'spinal *myelos*', where 'the thinnest and purest' fluid (presumably, cerebral spinal fluid), collected from all parts of the body by other vessels, is released (14). From the spine, the primary vessel runs down the body, rooting itself in various organs along the way and often sending out fine vessels that enmesh the organs. Among these organs are the kidneys. The fine vessels enmeshing the kidneys, however, are themselves compressed into a sinewy vessel that travels to the groin. In men, this vessel implants the sphincter, bladder, testicles, and epididymides with fine vessels; its widest and straightest part then turns back from the groin and sends up a stalk that is the penis (14–15).

Before returning to male anatomy and physiology, the text briefly describes how the primary vessel runs a course in women parallel to the one in men:

> In females, it [sc. the primary vessel] extends to the uterus, bladder, and urethra. From there it goes straight on: in women it is bound around the uterus; in males, it is coiled around the testicles. Because of its nature, this vessel also collects the most seed: for drawing nourishment from the most numerous and purest parts—and being of scant blood, hollow, thick and sinewy, full of *pneuma*, and tautened by the penis, it violently compresses the smaller vessels located in the spine. The vessels compressed, as if by a cupping instrument, empty into the upstream vessel. An influx into the vessel does occur from the rest of the body's limbs also, but the greatest amount, as already stated, is collected from the *myelos*.[5]
>
> *On the Nature of Bones* 15

Thus, while seed comes from all parts of the body—including presumably the brain—it is collected mainly from the 'spinal *myelos*' and travels via the primary vessel along the spine to the penis or uterus. We shall see that this basic scheme persists in the three main theories concerning reproductive seed, regardless of where the given theory locates the origin of the seed.

Myelos appears in the three theories most prevalent in Greek thought concerning where reproductive seed is formed. The oldest of the three theories holds that the brain and the *myelos* are the site of the generation of reproductive seed; it is called the encephalogenetic theory. *Myelos*, however, plays a part in the other two theories, as well, even though the brain has no special function in them. These are, chronologically, the pangenetic theory, in which seed is thought to be produced by every part of the body of both males and females; and the haematogenetic theory, in which seed is said to result from the concoction of the blood.[6]

The Hippocratic treatise *On Places in Man*, probably written in the first half of the fifth century BCE (Craik 1998: 29), describes the anatomical structures as conceived by the advocates of encephalogenesis and mentions the vessel that runs behind the ear (3.5). According to this text, seed-bearing vessels run behind the ears to the groin and inner thighs, ending at the big toes; if these vessels are cut, the person (*anthropos*, which can signify 'man' or 'woman') becomes infertile (*akarpos*; literally, 'without fruit').[7] Concern for the crucial vessel behind the ear is reinforced when the author advises cupping for it and not phlebotomy, with a warning not to break the skin (12.1).[8] That sterility and not impotence is the negative outcome mentioned and that the gender-neutral 'person' (*anthropos*) is used and not 'man' (*anēr*), is worth noting, as the condition can thus apply to both sexes, just as the primary vessel of *On the Nature of Bones* runs from the brain to either the penis or the uterus. Furthermore, in the case of any men suffering damage to this vessel, impotence may have been assumed, since sterility and impotence were not often sharply differentiated because the same fluids thought to carry or produce the seed were also thought to swell the penis.[9] We see this very clearly at *Epidemics* 7.105 (edition Smith 1994), which mentions the more or less opposite problem of priapism, which causes men to have persistent erections and swellings behind the ear. Presumably the excess of fluid that causes the excessive number of erections also clogs the crucial vessel behind the ear; or perhaps it is the other way around and the constant erections do not result from an excess of fluid but rather cause it. The following discussion, at any rate, will suggest that bodily fluids related to reproduction apparently functioned in a 'demand-based' economy.

The best-known passage warning against phlebotomising the vessel behind the ear comes, contrary to expectation, not from a text advocating for encephalogenesis, as one might expect, but pangenesis: *On Airs, Waters, and Places* (edition Jones 1923a). This rather early Hippocratic text is concerned with the impact of climate and geography on human health. One portion, often referred to as the 'ethnographic' section, describes the physiological traits imagined of peoples who live in extreme climates. Among these are the Scythians, nomads who inhabited the central Eurasian steppes. According to *On Airs, Waters, and Places*, one group of Scythians was the Anarieis, a name that approximates to the Greek word for 'unmanned'. The unfortunate Anarieis are appropriately named since, according to the Hippocratic text, they suffer from impotence resulting from their practice of cutting the vessel behind the ear to treat the joint problems that they themselves attribute to long hours spent on horseback (22).

The author, perhaps aware that attributing impotence to cutting only this one vessel puts him at odds with the pangenetic theory he otherwise espouses, offers an additional cause for the impotence: the wearing of trousers (22). This strange custom, which the author alleges did not permit Scythian men sufficient opportunity to handle their genitals, pairs with the priapism passage discussed earlier to show that the production of male seed was considered a demand-based operation.[10] As we shall see, the production of breast milk was also believed to function in the same way (as indeed it does).

Another pangenetic text, *On Generation* 2, also raises the alarm about phlebotomising behind the ear but maintains that the damage is less drastic:

> Those who've been phlebotomized beside the ears have sex and ejaculate, but it [their seed] is scanty, weak, and sterile; this is the case because most seed flows from the head, alongside the ears, and into the spinal *myelos*. This path through gets blocked owing to the scarring that arises from the cut.
>
> *On Generation* 2, edition Potter 2012

The claim that intercourse and ejaculation can still occur has a contentious ring to it, as if a tacit argument is being made against the strict encephalogenetic view, in which impotence and failure to ejaculate would be a necessary result of phlebotomising behind the ear. Our author cannot support that view and be consistent with the theory of pangenesis, and yet for him the head, ear, and *myelos* still hold great importance for seed production and the flow of seminal fluid. Like the author of *On Airs, Waters, and Places*, he finds a way to adjust his explanation of the outcome of cutting the vessel behind the ear, so that he both incorporates elements of the older theory and retains the possibility of performing phlebotomy on that particular spot. Heinrich von Staden has called attention to the lack of consistency in these texts with regard to these two theories of spermatogenesis and notes that at least *On Generation* attempts to reconcile the views (1989: 290). Perhaps by the time of these texts the belief was widespread that the vessel behind the ear was related to fertility and potency and that getting bled there could have dire consequences for them both, so to pacify their patients, doctors steered clear of the practice regardless of their own theoretical allegiances. On the other hand, perhaps we are witnessing one of the many examples of 'the agon between old and new' (von Staden 1989: 289–90) in the essentially conservative tradition of ancient medicine.

On the Sacred Disease (edition Jones 1923b) is a text rich for the present discussion in that, like *On Generation* and *On Airs, Waters, and Places* (with which latter it famously shares some similarity of wording, especially concerning seed production), it advocates for pangenesis while retaining relics of encephalogenetic theory. In an eloquent passage sometimes referred to as the 'Elegy of the Brain', *On the Sacred Disease* celebrates the brain as the dominant organ of the body and as the locus of the emotions, perception, and intelligence and argues strongly against those who believe that the heart is the central organ (17–20; for discussion, see Laskaris 2002: 127–31). The author, despite holding this view, states that reproductive seed comes not from the brain, but from all parts of the body: healthy seed from healthy parts and diseased from diseased. But since the text attributes so many vital functions to the brain and considers it to be the source of all the 'most serious illnesses' (6), the brain retains a special connection to spermatogenesis, being productive of the seed that will result either in a healthy brain that, as such, will not be the source of the gravest diseases, or in an unhealthy brain that will lead to vulnerability to such serious illnesses as the sacred disease of the title. The text evokes the descending phlegm theory and asserts that the sacred disease is brought about when phlegm descends from the brain, blocking off crucial passages and so leading to seizures and other problems (8–14). The prominence of the brain in the encephalogenetic tradition very likely contributed to the development of the descending phlegm theory and to the dominant role attributed to the brain in this treatise, even as it subscribes to the newer pangenetic theory of seed formation.

Aristotle is the most influential proponent of the third of the main theories of seed production—haematogenesis. He proposed that the heart is the central organ of the body and that both male and female seed arise from the blood (*Generation of Animals* 721a26–727b30). In males, the seed is said to be fully concocted blood and to be generative. In females, the seed is menstrual fluid; it is not generative but rather nutritive, providing nourishment and substance to the foetus. While the head and the eyes and the moisture in them should not, with these premises, be related to sexual reproduction in Aristotle, they in fact are. For example, in *History of Animals* we see that moisture in the head is correlated with sexual maturity and causes luxuriant hair growth, and that sexual activity can cause hair loss, presumably by draining moisture from the head:

> Hair grows naturally—in greater or lesser amounts—in other places, too, but especially on the head and chin—and especially if it is fine. The eyebrows get shaggy in some older people to the point of needing to be cut, because they are placed where bones grow together—bones that separate and give off more moisture as people age. But when people start to have sex, the hair of the eyelashes does not grow, but rather falls off, and more so for those who have more sex. These hairs turn grey very slowly.
>
> Aristotle, *History of Animals* 518b5–12, edition Louis 1964–9

The eyes and fertility tests

In a passage from *Generation of Animals* concerned with fertility testing for men and women, Aristotle makes explicit the connection between the head, brain, eyes, and seed. Men's seed, he says, is tested by placing it in water (fertile seed sinks to the bottom) (747a4–10). Testing women is more complicated:

> They test women both with vaginal suppositories, to see whether their odours make their way upwards from below to the breath in the mouth, and with coloured ointments in the eyes, to see if the ointments colour the saliva in the mouth. If these things do not happen, obviously the passages in the body through which residue is excreted have been filled up and have grown shut. For of all the parts of the head, the area around the eyes is the most productive of seed. This is obvious from the fact that in sexual intercourse it alone manifestly changes form, and because the eyes of people who take part in sex too frequently are obviously sunken.
>
> Aristotle, *Generation of Animals* 747a14–20

111

The use of suppositories of wool or linen, infused with medicinal substances and inserted into the vagina, was a fairly common drug delivery system in Greek antiquity and later. In this case, the suppositories were not intended as medicinal but, being imbued with some heavily scented substance, were used to test whether the passages in the woman's body were open and therefore available for receiving the man's seed. If one could smell the odour on the woman's breath, then the passages must be clear, and she will be able to produce a child. A similar story occurs with the eyes: if when coloured ointments are smeared on the eyes, the colour shows up in the saliva, that is proof that the crucial seed-bearing passages in the head—and especially those around the eyes—are open.

Versions of this fertility test are found in an earlier and in a later work. According to the Hippocratic text, *On the Nature of Women* 99 (edition Potter 2012), which probably took its final form in the mid-fourth century BCE (Craik 2015: 217), a woman's fertility can be assessed by seeing whether a drug called 'the red stone' 'enters' or 'passes through' after it has been rubbed into the eye. There is no mention of saliva in this text, but the concern for clear passages in the head and particularly around the eyes is apparent. The later reference comes from the Pliny the Elder in the first century CE; as in Aristotle, saliva is involved: 'We accept as a prognostic of fecundity in women that when the eyes have been anointed with a medicament, the saliva is tinted with it' (*Natural History* 7.67, edition Rackham 1942).

In the view of all three sources, the passages in the head, particularly in the area of the eyes, must be kept open for conception to take place. Thus, in all three theories of seed formation, the head, the brain, and the areas around the ears and eyes remain of central importance.

The eyes and pregnancy

Pregnancy tests may also reflect the connection between the eyes and the uterus, as in this case: one of a short list from the Hippocratic text, *On Barren Women*:

> Another test if you do not know by some other means if a woman is pregnant: The eyes are sunken and very hollow, and the whites of the eyes do not have their natural whiteness, but have become rather dark red, if she is pregnant.
> *On Barren Women* 3, edition Potter 2012[11]

The passage seems intended to aid in determining pregnancy in its earlier stages, since the assessment is offered as useful 'if you do not know by some other means' whether or not the woman is pregnant—an uncertainty that for obvious reasons does not exist in the later stages of pregnancy. The rationale is not spelled out, but it is likely that the embryo, in drawing nourishment from the menstrual fluid, was thought to deplete the supply of it produced in the area of the eye, creating a concavity there. The greater redness perhaps reflects the greater volume of menstrual fluid in circulation as it is demanded by the growing embryo—another example of the 'demand-based economy' of reproductive fluids. Another possibility is that the embryo's drawing of nourishment may be thought to create suction, resulting in a concavity and the constriction of the fine blood vessels of the eye, which leads in turn to a build-up of blood and the reddening of the whites of the eye. Recall that suction is the mechanism in

On the Nature of Bones for male ejaculation, when smaller vessels become forcibly compressed as the vessel in the penis draws heavily on the *myelos* collected in the spine, and that in Aristotle the eyes of those who are overly enthusiastic about sex are sunken.

On Superfetation, a Hippocratic text closely related to *On Barren Women*, brings the breast into the eye–uterus pathway, stating that on whichever side of the uterus the embryo is lying, that breast will be larger and that eye larger and brighter:

> One should find out which of a woman's breasts is larger, for on that side is the embryo. And the same with the eyes: for that eye will be larger and brighter on the whole within the eyelid, on whichever side is also the larger breast.
>
> *On Superfetation* 19, edition Potter 2010

This passage, where the fact of pregnancy is already established, may be aimed at determining the sex of the child, since girls were generally thought to develop on the left side of the uterus and boys on the right.[12] That the eye here becomes larger and brighter, and not red and sunken as in *On Barren Women*, may be accounted for by the different context: here nourishment from fully concocted menses (i.e., milk) is implied, whereas in *On Barren Women*, concerned as it is with the earlier stages of pregnancy, nourishment for embryos was thought to come from unconcocted menses.[13] Milk production was generally believed to begin during the course of the pregnancy (colostrum, in fact, is produced during pregnancy), with opinions varying as to precisely when and how (see Chapters 14 and 15 by Lawrence and Totelin, respectively, in this volume). Aristotle believed in a very early start to milk production and here correlates the cessation of menstruation with it:

> Menses occur for some time in most women after conceiving: in the case of female embryos, usually for thirty days, and around forty for males.[14] After birth, as well, the menses generally return in the same number of days, though not with equal precision for all women. After conception, however, and the [numbers of] days just mentioned, they [occur] no longer naturally, but turn to the breasts and become milk.
>
> Aristotle, *History of Animals* 583a26–34

In *On Superfetation*, the underlying thought in connecting the breast with the eye and the uterus appears to be that the foetus is being nourished on menstrual fluid produced in the area of the eye—fluid that is now being concocted into milk and stored in the breast.[15] The greater size and brightness of the eye is from the greater production of the fully concocted—i.e., white—menstrual fluid on the given side of the body. This interpretation aligns with the 'demand-based economy' we have seen above with regard to sexual function in men and the production of generative seed.

The eyes and pathology

In the Hippocratic text *On Diseases of Girls*, the build-up of menstrual fluid and its putrefaction—a risk for unmarried girls and women—could cause many serious problems, including derangement brought about by visual hallucinations (Flemming and Hanson 1998, lines 4–16, 29–37; Potter 2010: 358–62). The latter is consistent with what we have seen concerning the role of the eyes in the generation of reproductive seed and with the passages connecting the eyes with the uterus: if the woman is bearing babies, the menses nourish the growing foetus and are eventually transformed into milk, and the woman's eyes will reflect these processes. In *On Diseases of Girls*, we have the converse: the woman remains unmarried or childless, and the built-up blood in her head causes terrible murderous visions—literal bloody-mindedness. This is not to say that absorption of excess blood and moisture by the rest of the body's tissues does not also happen and to a problematic degree; that it does is indeed clearly stated and repeatedly (King 1998: 28–9). The references to deranged visions, however, rely on the eye–uterus connections already mentioned and to some of the concepts related to optical theory, discussed in the next section.

Less drastic problems with sight can also be related to menstrual fluid. The Hippocratic text *Coan Prognoses* states that 'Occluded vision is cured by the appearance of plentiful menses', (chapter 541 in Potter 2010: 240). Celsus, a first-century CE follower of Hippocratic ideas, reports a condition in which people can see well during the day but not at all at night. This condition, he maintains, does not befall women with regular periods (6.38, edition Spencer 1938). In these cases, built-up menses apparently cloud the vision and can even obscure it entirely at night.

Optical theory

The beliefs expressed in our texts concerning some, at least, of the pathological conditions discussed may be influenced by contemporary ideas concerning the nature of vision. The belief that vision resulted from emanations from the eyes can be found in a variety of Greek and Roman authors. Plato posited that the eyes were 'light-bearing', with streams of fire flowing through and emanating from them as the source of vision (*Timaeus* 45b2–46a2). Alcmaeon may have prefigured Plato: he is reported to have stated that there is a 'gleaming element' in the eye that contains fire (DK 24 A 5 = Theophrastus, *On the Senses* 26; see von Staden 1989: 200–55, 570–4; Craik 2006: 83; Nutton 2013: 48–9). More fully elaborated theories concerning vision emerged from these earlier ideas, and an extramission theory, which held that vision was produced from the emanation of vision rays from the eyes, was current in intellectual circles from Euclid through to the Early Modern period (Keyser 1993; Hub 2016).

The pathological conditions described in the preceding section, while they admittedly do not refer explicitly to emanations, depend implicitly on the notion that the blood that is so prevalent in and around the eyes can affect the eyes' emanations and/or that the emanations can have a material (and bloody) component; indeed, there is no reason to think that Plato's streams of fire or Euclid's vision rays were other than material. Some of the beliefs about menstruation that tend to be characterised as superstitious have similar implications. To take the lustre off ivory and the edge off a sword-blade with her mere gaze were among the more than a dozen powers attributed to the menstruating woman by Pliny (*Natural History* 7.15.64–6). Included in the list was the woman's ability to dim the bright surface of a mirror by looking into it. Pliny was neither the first nor last to make use of the 'bloody mirror motif': it had appeared already in Aristotle's *On Dreams* (following), and it had staying power through to the Early Modern period:

> In the case of highly polished mirrors—whenever women who are having their menstrual periods look into a mirror, the surface of the mirror becomes like a bloody cloud. If the mirror is new, it is not easy to wipe away such a stain, but it is easy if it is older. The reason, as we said, is that the eye not only is affected by the air, but also does something to it and stirs it, just as bright objects do. . . . For logically the eyes are in the same state as any other part [of the body] during menstruation; then, too, they are highly vascular by nature. Therefore, when menstruation comes about because of the upheaval and swelling of the blood, change in the eyes occurs, though unseen by us, (for the nature of the seed and of the menses is the same), and the air is stirred by the eyes and it has the same effect on the air continuous with the mirrors as it itself underwent. And that air affects the surface of the mirror.
>
> Aristotle, *On Dreams* 459b29–460a13, edition Ross 1955

Of central concern for us in this passage is the emphasis on the eyes with mention of their many blood-bearing vessels being swollen by menses, with the obvious implication that the gaze itself contains a material element, namely blood.

The attribution of this text to Aristotle is doubted by some on the grounds that it contradicts the optical theory of *On the Soul* and *On Sense and Sensibilia*, where Aristotle 'maintains that the process that triggers perception consists solely of a mediation of the colours of an object *to the eye* through the intervening medium' and overtly rejects the extramission view (Hub 2016: 3–4). On the other hand, Aristotle does adopt the extramission view in other works (Hub 2016: 1 with n. 1, 4–5). More relevant for present purposes than questions of authorship, however, is the fact that the extramission theory had the wider currency by far in antiquity, arguably because it aligned reasonably well with the older concepts and established beliefs discussed earlier concerning the powers of menstruating women and the visual hallucinations to which childless women and girls were prone and to the power of the gaze for expressing love and desire, as depicted in countless works of art and literature (see, for example, Toscano 2013).

An interesting related belief was that those persons, and particularly women, with a 'double pupil' were especially able to cast the 'Evil Eye' (on the 'performative gaze' and optical theory, see Hub 2016). At least in some cases, the 'double pupil' was probably coloboma of the iris (McDaniel 1918: 345–6), a congenital defect or hole in the iris. The coloboma is dark like the pupil, and depending on how it is placed in relation to the pupil, it may give the impression that the pupil is of a greater than normal size and is shaped irregularly or that there is a second pupil entirely. In the latter case, where the coloboma is distinct from the pupil, double vision may arise which, given the extramission theory, may have contributed to the notion that persons with this affliction had the power of the Evil Eye, as they were able to send out double rays, or perhaps that the extra 'pupil' was for transmitting harmful rays. Pliny tells us that Cicero considered the glance of all women with 'double pupils' to be injurious (*Natural History* 7.2.18). Pliny's remark comes during his discussion of the Evil Eye and of those ethnic groups who were thought to be prone to it, including those with double pupils; but Cicero speaks categorically of women with double pupils, even Roman women—and his allegation is against women only. Since the concept of the Evil Eye is tacitly predicated on the notion that emanations can pass through the pupils of the eye to their target, those who appeared to have double pupils would have been deemed doubly dangerous.

Conclusion

The role of *myelos*, with its origin in the head and eyes, in Greek ideas concerning spermatogenesis in women was significant and is reflected in medical texts. Understanding the eyes as spermatogenetic in Greek thought elucidates some of the obscure passages about them in the Hippocratic Corpus and elsewhere. That these ideas cohered with ubiquitous notions concerning the powers of menstruating women and of the Evil Eye no doubt contributed to their longevity, even as they were folded into the theories of pangenesis and haematogenesis, alternate views of seed formation with which they should have been at odds, and possibly informing—or sharing a conceptual ancestor with—optical theory.

That the head and the eyes in particular should have gained such prominence in beliefs and theories concerning spermatogenesis may seem counter-intuitive to present-day readers of these texts. But when we consider the immediate arousal of sexual desire that often occurs upon catching sight of an attractive person—alluded to throughout Greek literature and present in many works of Greek art (see Toscano 2013)—we should not be surprised that a connection is made. Then again, we may wish to consider the research being conducted on the evolution of the human eye, with its unusually elongated shape and broad expanse of white sclera (the 'whites of the eyes')—both unique in the animal kingdom.[16] Current theories for these phenomena centre on the greater expressiveness for communicating intention and emotion that these attributes confer and for their aiding in mate selection, as white sclera may reveal the health status and so the desirability of a potential mate (see e.g., Mayhew and Gómez 2015). Our ancient sources were of course not aware of evolutionary theory. But if the eyes have had this function in human life, then that fact could have been absorbed into the fundamental conceptions of Greek society and rationalised by philosophical and medical authors. In addition, the custom of veiling may have made the eyes of women seem even more expressive, or perhaps made the women more adept at expressing themselves with their eyes when wearing their veils in public. In a conceptual system that assumed that reproductive fluids were produced based on demand, as I have argued, the desire expressed by the eyes, or felt by the body when the eyes caught sight of someone desirable, could easily have been related to the production of that purest of fluids, *myelos*, which was so prevalent in the head and the area of the eyes.

Notes

1 I have opted for the transliteration *myelos* (rather than *muelos*) as readers may be familiar with the *myel-* stem from medical and scientific writings.
2 I will generally cite the Loeb editions of the Hippocratic texts, which have convenient facing-page English translations. Readers of Greek may wish to consult the editions published by Littré or Budé, which may number the texts differently.

3 See Duminil (1998: 75–115); Potter (2010: 11–12); Craik (2015: 224–30) for discussion of the relationship of *On the Nature of Bones* to other works. As noted by them, *On the Nature of Bones* 11–19 is very likely the work Galen refers to as appended to *Instruments of Reduction* and entitled *On Vessels* (19.114, 128 Kühn).

4 I am accepting 'ἀρχαίη', the reading of M, with Duminil (1998) and Craik (2009: 109) over Harris's conjecture of 'παχείη', which is accepted by Potter; Berrey also reads 'ἀρχαίη', translating it as 'primary' (2014: 288).

5 The last few lines of this passage contain vocabulary particularly apt for discussing the movement of fluids through the vessels: τὰ δὲ βιαζόμενα | ὥσπερ σικύη ἐς ἑωυτὰ πάντα ἐκδιδοῖ ἐς τὴν ἄνω φλέβα· συλλείβεται δὲ καὶ ἐκ τῶν ἄλλων μελῶν τοῦ σώματος ἐς ταύτην· τὸ δὲ πλεῖστον, ὥσπερ εἴρηται, ἀπὸ τοῦ μυελοῦ τοῦτο συναλίζεται. I am translating ἄνω as 'upstream' (*s.v. LSJ* I for this meaning of ἄνω in the context of rivers); it is here qualifying the primary vessel which, being closer to the source, is the 'upstream vessel'. Such a meaning is in keeping with other words in the vicinity that can refer to the movement of fluids: ἐκδίδωμι can be used of rivers emptying into the sea (*s.v. LSJ* II), and συλλείβω's primary meaning refers to the movement of fluids. These are apt words for the context: the movement of *myelos* and other fluids through the vessels.

6 See von Staden (1989: 288–96) for discussion of spermatogenesis in men.

7 As Totelin notes, plants could be considered 'male' or 'female', and though usually fruit-bearing was associated with 'female' plants, there were some fruit-bearing plants considered 'male' (2018: 59). See, too, Flemming's discussion of words for infertility that were restricted to women (*atokos, aphoros*) (2013: 577). That a different word is chosen here supports my argument that both sexes are under discussion.

8 See Craik on cautery of the *myelos*-bearing vessels of the head in other texts (2009: 114).

9 See Berrey 2014 for thorough discussion, and cf. *On the Nature of the Child* (9), where in men fluids agitated during sex descend from the head.

10 See Berrey (2014) for the role of friction in medical theories of potency.

11 οἱ ὀφθαλμοὶ εἰλκυσμένοι καὶ κοιλότεροι, καὶ τὰ λευκὰ τῶν ὀφθαλμῶν οὐκ ἔχει τὴν φύσιν τῆς λευκότητος, ἀλλὰ πελιώτερα, ἢν κύη. The translation of this passage brings us to the perennial problem of understanding Greek words for colours: *pelios* generally refers to extravasated blood and can indicate either the dark red of venous or dried blood or the 'black-and-blue' of bruises (*LSJ* ad loc., where it is translated as 'livid', a word that bears the same ambiguity in English as *pelios* does in Greek). 'Dark red' is my choice here, as it seems the more likely colour for the whites of the eyes to turn.

12 The idea that the right testicle produced males and the left females is attributed by Aristotle to Anaxagoras and other unnamed natural philosophers (*Generation of Animals* 763b30 = Diels and Kranz 59 A 107). See, too, *On Superfetation* (31), where men are advised to bind the right testicle before sex as tightly as they can tolerate it if they want to produce a girl (presumably so that the left testicle will be the productive one) and the left one if they want to produce a boy. That the right testicle is usually larger than the left may be relevant (McManus 2004), though see Lloyd for a thorough discussion of the powerful symbolism of right and left in Greek thought broadly (1991: 27–48; originally published 1962, reprinted 1973).

13 See Craik (2006: 83) on adjectives denoting brightness as indicators of the healthiness of the eye. The use of milk as a topical medicine may contribute to the positive significance milk had—quite apart from its obvious benefits. That milk, and human milk especially, has powerful antimicrobial properties and would have been effective for some of the conditions to which it was applied would have raised its value further; that it was frequently used in treating eye infections in babies is especially appropriate given the present discussion (see Laskaris 2008).

14 The greater number of days for male embryos can perhaps be understood as contributing to their larger and more robust bodies, since Aristotle considered menstrual fluid to supply the embryo with substance.

15 This is not to say that the area of the eye is the only one in which menstrual fluid is being produced, concocted into milk, and diverted to the breasts, but the passage does seem to

reflect the view that this part of the body is especially connected with spermatogenesis. For a discussion of the diversion of menstrual blood into milk, see King (1998: 34–5, 143–4).

16 Some other animals have white sclera, but no others reveal such a broad expanse of it.

Bibliography

Primary sources

Craik, E. M. (1998) *Hippocrates. Places in Man*. Edited with a translation and commentary by E. M. Craik. Oxford: Oxford University Press.

———. (2006) *Two Hippocratic Treatises: On Sight and On Anatomy*. Edited with a Translation and Commentary by E. M. Craik. Leiden: Brill.

Diels, H. and W. Kranz (1996) *Die Fragmente der Vorsokratiker*. 6th edn. Zürich: Weidmann.

Duminil, M.-P. (1998) *Hippocrate, Vol VIII: Plaies, nature des os, coeur, anatomie*. Paris: Belles Lettres.

Flemming, R. and A. E. Hanson (1998) 'Hippocrates' *Peri Partheniōn* ("Diseases of Young Girls"): Text and translation', *Early Science and Medicine* 3(3), 241–52.

Jones, W. H. S. (1923a) *Hippocrates*, vol. I. Cambridge, MA: Harvard University Press. (This volume includes *Airs Waters Places*).

———. (1923b) *Hippocrates*, vol. II. Cambridge, MA: Harvard University Press. (This volume includes *The Sacred Disease*).

Louis, P. (1964–9) *Aristote. Histoire des animaux*, vol. I–III. Paris: Les Belles Lettres.

Potter, P. (2010) *Hippocrates*, vol. IX. Edited and translated by P. Potter. Cambridge, MA: Harvard University Press. (This volume includes *Nature of Bones*; *Superfetation*; and *Girls*).

———. (2012) *Hippocrates*, vol. X. Edited and translated by P. Potter. Cambridge, MA: Harvard University Press. (This volume includes *Generation*; *Nature of Women*; *Barrenness*).

Rackham, H. (1942) *Pliny. Natural History*, vol. II. Translated by H. Rackham. Cambridge, MA: Harvard University Press.

Ross, W. D. (1955) *Aristotle. Parva naturalia*. Edited with a commentary by W. D. Ross. Oxford: Oxford University Press.

Smith, W. D. (1994) *Hippocrates*, vol. VII. Edited and translated by W. D. Smith. Cambridge, MA: Harvard University Press. (This volume includes *Epidemics 7*).

Spencer, W. G. (1938) *Celsus. De Medicina*, vol. II. Translated by W. G. Spencer. Cambridge, MA: Harvard University Press.

Secondary literature

Berrey, M. (2014) 'The Hippocratics on male erotic desire', *Arethusa* 47(3), 287–301.

Craik, E. M. (2009) 'Hippocratic bodily "channels" and oriental parallels', *Medical History* 53(1), 105–16.

———. (2015) *The 'Hippocratic' Corpus: Content and Context*. London: Routledge.

Flemming, R. (2013) 'The invention of infertility in the Classical Greek world: Medicine, divinity, and gender', *Bulletin of the History of Medicine* 87(4), 565–90.

Hub, B. (2016) 'Aristotle's bloody mirror and natural science in Medieval and Early Modern Europe', in N. Frelick (ed.) *The Mirror in Medieval and Early Modern Culture: Specular Reflections*. Turnhout: Brepols, 31–71.

Keyser, P. T. (1993) 'Cicero on optics ("Att." 2.3.2)', *Phoenix* 47(1), 67–9.

King, H. (1998) *Hippocrates' Woman: Reading the Female Body in Ancient Greece*. London: Routledge.

Laskaris, J. (2002) *The Art is Long: On the Sacred Disease and the Scientific Tradition*. Leiden: Brill.

———. (2008) 'Nursing mothers in Greek and Roman medicine', *American Journal of Archaeology* 112(3), 459–64.

Lloyd, G. E. R. (1991) *Methods and Problems in Greek Science*. Cambridge: Cambridge University Press.

Mayhew, J. A. and J.-C. Gómez (2015) 'Gorillas with white sclera: A naturally occurring morphological trait linked to social cognitive functions', *American Journal of Primatology* 77, 869–77.

McDaniel, W. B. (1918) 'The *pupula duplex* and other tokens of an "evil eye" in the light of ophthalmology', *Classical Philology* 13(4), 335–46.

McManus, J. C. (2004) 'Right–left and the scrotum in Greek sculpture', *Laterality: Asymmetries of Body, Brain and Cognition'* 9(2), 189–99.

Nutton, V. (2013) *Ancient Medicine*. 2nd edn. London: Routledge.

Onians, R. B. (1951) *The Origins of European Thought about the Body, the Mind, the Soul, the World, Time, and Fate*. Cambridge: Cambridge University Press.

Toscano, M. M. (2013) 'The eyes have it: Female desire on Attic Greek vases', *Arethusa* 46(1), 1–40.

Totelin, L. M. V. (2018) 'Animal and plant generation in classical antiquity', in N. Hopwood, R. Flemming and L. Kassell (eds) *Reproduction: Antiquity to the Present Day*. Cambridge: Cambridge University Press, 53–66.

von Staden, H. (1989) *Herophilus: The Art of Medicine in Early Alexandria*. Cambridge: Cambridge University Press.

7

'INFERTILE' AND 'SUB-FERTILE' SEMEN IN THE HIPPOCRATIC CORPUS AND THE BIOLOGICAL WORKS OF ARISTOTLE

Rebecca Fallas

Some individuals have much semen (*polusperma*), some little (*oligosperma*), some none at all (*asperma*).

Aristotle, *Generation of Animals* 725b29–31;
trans. Peck 1942: 85

As Julie Laskaris demonstrated in the previous chapter, bodily fluids are fundamental to all the theories of conception put forward by the ancient medical writers, and as Rebecca Flemming notes in Chapter 10, semen was viewed as a crucial bodily fluid. In order to generate a child a woman needed to produce menstrual fluid and a man (and depending on the theory, a woman) semen.[1] If these fluids were not produced in the correct amounts at the correct time then this could result in a person being considered infertile. The ancient medical writers put forward two main factors for male reproductive failure: impotence and the inability to produce fertile semen.

This chapter focuses on the descriptions of male infertility due to a lack of fertile semen, as discussed by Aristotle in his biological works[2] and the authors of the Hippocratic Corpus (a collection of around 60 medical treatises largely dating from the fifth and fourth centuries BCE written by multiple unknown authors; see Craik (2015) for details on the authorship and content of individual treatises). As the quotation from Aristotle that opens the chapter suggests, there was an understanding in the ancient world that not only did some men produce more semen than others but also that there were some who produced no semen at all. There are many ways in which semen is described by both Aristotle and the Hippocratic writers as non-generative (*mē gonimon* e.g. Aristotle, *Generation of Animals* 718a24), less fertile (*agonōteroi* e.g. 726a3), and, in the Hippocratic Corpus, small in amount, weak, and infertile (*oligon de kai asthenes kai agonon*, *On Generation* 2.8–11, edition Potter 2012). In modern medicine, semen analysis takes place on both a macroscopic and microscopic level. The former can include measuring the volume, appearance, colour, coagulation, and viscosity of the semen and on a microscopic level accessing the sperm concentration, motility, and viability (Agarwal and Said 2010: 16–17). The concentration of sperm is one of the key values, and terms such as oligozoospermia (low sperm count) or azoospermia (zero sperm count) are often used to describe the fertility of semen.

DOI: 10.4324/9780429438974-11

However, individual sperm cells were only identified under a microscope in 1677 by the Dutch biologist Antonie van Leeuwenhoek.[3] Therefore, the question arises: what did the ancient medical writers believe caused semen to be 'less fertile', 'infertile', or 'weak'?

Although fertility in the ancient world has been widely discussed over the last thirty years, infertility has only recently become a focus of scholarship (see for example Flemming 2013; Berrey 2014; Fallas 2015; Senkova 2015; Totelin 2017).[4] It is alongside this attention to infertility that interest in male fertility has developed. However, what has not received any detailed attention is the nature of infertile semen.

This chapter seeks to address this first through an exploration of how the differences in the theories of conception and semen production by the Hippocratic writers and Aristotle have an effect on their understanding of what makes semen 'infertile', 'less fertile', or 'weak'. This will be followed by discussions on the key factors described by these authors as having an effect on the fertility of semen, including age, the amount and quality of semen produced, physical traits, and damage to the body. Finally, I consider how ancient authors thought that if a man with lower quality semen fathered a child, the effect of this could be seen in the child produced.

Producing semen

There are two main words used for semen in ancient Greek: *to sperma* and *hē gonē*.[5] The most used Greek-English dictionary, Liddell and Scott's *A Greek-English Lexicon* (LSJ, 1996), defines *sperma* as 'seed' either in plants or animals but when referring to animals it is also translated as 'semen'.[6] *Gonē* is usually translated as 'seed' but is sometimes also translated as 'semen' and is also used to refer to the offspring of an animal. Their use by the ancient authors seems to be mostly interchangeable.[7]

As is often the case in the Hippocratic Corpus, we do not find one complete theory which is the same across all the treatises and which we could therefore classify as the definitive Hippocratic theory. However, the explanation of semen production in the text *On Generation* provides us with the most detailed account. This theory includes four components, the male semen, the female semen, breath (*pneuma*), and menstrual blood. The author explains that 'a man's seed comes from all the moisture in his body' (*On Generation* 1.1–2; trans. Potter 2012: 7). In this theory the seed is drawn from the whole body, and the author states that it comes from the 'solid parts and the soft parts and from moisture' in the body, with the four types of moisture here being blood, bile, water, and phlegm (*On Generation* 3.1–5; trans. Potter 2012: 11).

The movement of the seed through the body was theorised to occur during intercourse itself, as opposed to semen being stored in the testicles prior to intercourse. The Hippocratic author describes this movement in the following way: as the penis is rubbed during intercourse and as the man moves, the vessels and cords throughout the body, all of which ultimately lead to the penis, are warmed, and the moisture in the body is warmed. This turns the moisture into a liquid which, agitated by the movement, changes to foam. The most powerful and fattest part of this foam is secreted and goes into the spinal marrow, and from here, it passes to the kidneys and through the middle of the testicles to the penis, where it is ejected through a different tube from the urine and moves into the uterus (*On Generation* 1.1–26). Therefore, in this Hippocratic theory, the semen moves through the body and is ejaculated due to the combination of heat and movement during intercourse. On entering the uterus, the male seed would mix with the female seed and conception would take place.

The theory put forward by Aristotle differs substantially from the viewpoint put forward by the Hippocratics. Aristotle thought that semen was a residue from nourishment (*Generation of Animals* 727a31–2). To explain this process from the beginning, Aristotle's theory suggests that, when nourishment, essentially in the form of food, enters the body, it undergoes several processes of concoction; the concoction changes the nature of the food by a process which can be likened to the ripening of food or to baking, processes completed by heat (*Parts of Animals* 650a3–8, edition Peck 1937). Having entered the body, the food travels to the stomach, where it undergoes the first stage of concoction. The nourishment then passes to the heart, where it is concocted again into blood, which then provides nourishment to the rest of the body; this nourishment is used for growth and maintenance of the body (*Parts of Animals* 647b5; cf. 666a8). The leftover nourishment can undergo further concoction and produce a variety of parts, such as marrow (*Parts of Animals* 652a5), fat (if there is a substantial amount left over) or nails, and hair (*Parts of Animals* 651.20–5). Some of the leftover nutriment is further concocted into semen, milk, and menstrual fluid (*Generation of Animals* 726a26; 728a36). Semen is described by Aristotle as having undergone the ultimate concoction, that is to say, it is produced at the last stage of concoction. This, according to Aristotle makes it the most potent residue, whereas menstrual fluid is described as un-concocted semen.

Aristotle, like the Hippocratics, supposed that the semen moved through the body during intercourse.[8] In his theory, the penis was moved during intercourse, the semen collected, and it then moved forward (*Generation of Animals* 717b23–6) and that to emit the semen the man needed to hold his breath (718a2–4). Once emitted, the semen would make its way into the uterus, where it would mix with the menstrual blood already in the uterus and conception would take place.

The 'infertility' at the beginning and end of life

Having explained how both Aristotle and the Hippocratic writers understood semen to be produced, we can now examine their theories about fertility problems related to semen. Until puberty occurs, a person is effectively infertile, and the accounts of the ancient authors as to why this is the case help us begin to understand how they thought about the fertility of semen. For the Hippocratics, the focus was on the passageways through which the semen moved around the body during intercourse, which first needed to be wide enough for the fluid to travel through. The author of the text *On Generation* (2) suggests that narrow and solid passageways are why children do not emit semen, and it is only as the child grows and the passages widen that semen can be emitted.[9]

Conversely, Aristotle was more concerned with the ability of the body to produce semen. He describes how children do not produce semen because their bodies are growing, and therefore all the nutriment is used up in this process, leaving no surplus from which semen can be produced (*Generation of Animals* 725b20–6). In *History of Animals*, Aristotle describes how the beginning of procreation in males is marked by their emission of seed but notes that 'they are not fertile immediately (*out' archomenōn gonima*) when the emissions begin, nor any longer when they become few (*oligōn*) and weak (*asthenōn*)' (*History of Animals* 585a35-b5; trans. Balme 1991: 451; cf. *History of Animals* 544b15, edition Peck 1970). Although the emission of seed was thought to be a sign of impending fertility, a person was not considered fully fertile by Aristotle until they reached the age of 21. Aristotle (*History of Animals* 582a17–29) says that, although males begin to produce semen at 14, it is largely infertile until they reach the age of 21. Furthermore, he notes that, if a child is conceived with either parent being under 21, the child will be small and imperfect, with the birth also presenting difficulties for the mother. Aristotle describes women as reaching their peak at 21, whereas men continue to improve. Of course, this improvement would only occur for so long, and Aristotle noted that fertility declined after the age of 60, although he recognised that some men were still able to procreate into their seventies (*History of Animals* 585b6–8; cf. 545b27–31). The reason Aristotle gave for fertility declining with age was based on his understanding that people were born with a certain amount of internal heat, left over from conception, and this heat constantly diminished throughout a person's life, so that by the time they reached old age it had almost been extinguished (*Generation of Animals* 748a33–5, 766b30–2).

The idea of fertility changing over a person's life is important, as it demonstrates that a person's fertility status was not thought to be static but could change over time. In this case, it is age that is the underlying issue, but as the rest of this chapter will show, it was not the only factor that influenced a person's fertility.

'Copious and weak' or 'plentiful and strong'?

The two key factors in assessing the fertility of semen in the ancient world were 'quantity' and 'quality'. As noted earlier, Aristotle describes the emission of semen in ageing males as 'few and weak', and the idea of weak semen is found repeatedly in both Aristotle and the Hippocratic texts. However, something which was open to debate was whether a large emission was a sign of a high level of fertility or a weakened one. There are several mentions of men experiencing seminal incontinence (e.g. *Epidemics* 6.3.9, edition Smith 1994; *On Internal Affections* 43, edition Potter 1988b). In one case it is noted that when a man passes urine or stools and brings forth a copious amount of moist semen (*proerchetai hoi thoros polus kai hugros*), he will not be able to produce children (*On Diseases* 2.51, edition Potter 1988a). Here, 'copious moist semen' is linked to reduced fertility. However, in other texts there appears to be some conflicting opinions on whether 'moist and copious semen' is a positive or negative trait. In the texts *Epidemics* and *Aphorisms*, we find two almost identical sentences which appear to come to different conclusions:

> Those whose noses are moist by nature and whose [seed] (*gonē*) is moister (*hugrotere*) and more copious (*pleiōn*): they are healthy. But those with the opposite condition tend to illness.
>
> *Epidemics* 6.6.8; trans. Smith 1994: 251

> Those whose nostrils are naturally watery, and whose seed (*gonē*) is watery (*hugrē*), are below the average when in health; those of an opposite character are above the average when in health.
>
> *Aphorisms* 6.2; trans. Jones 1931: 181[10]

123

While the second passage does not mention fertility directly, it is noted that seed being watery is linked to below-average health so it would seem this is considered a negative. The common consensus does seem to be that 'watery' semen was a bad sign for fertility.

Aristotle also links thin seed to infertility (e.g. *Generation of Animals* 747a4; *History of Animals* 581a30–3 and 582a30), leading to one of the only tests for male infertility from ancient Greece, stating:

> The water-test is quite a fair one for infertility in the male semen (*agonon*), because the thin, cold semen (*en tō hudati*) quickly diffuses (*diacheitai*) itself on the surface, whereas the fertile semen (*gonimon*) sinks to the bottom (*buthon*).
>
> Aristotle, *Generation of Animals* 747a3–7; trans.
> Peck 1942: 247

The denseness and heat in the fertile semen allows it to sink to the bottom, whereas infertile semen, being cold and thin, diffuses on the surface. It was not just the relative thinness or thickness of the seed which could be used to determine fertility but also its texture, with Aristotle noting that fertile semen was 'granular' (*chalazōdē*, *History of Animals* 581a30). Although 'watery' semen might be a problem, semen that was too concentrated could also be an issue. In the Hippocratic *Aphorisms* (5.64), the author states that a man's fertility would be affected if he did not have enough 'breath' (*pneuma*) in his body to force the seed through and out of the body, which can happen if the body is too dense. Alternatively, the seed itself may be too dense, due to cold-ness, to move through to the penis.

As noted in the previous examples from *Epidemics*, a link was sometimes made between 'fertility' and 'ill health' in the ancient world. Being infertile did not always mean being unhealthy. In the case of women, Helen King (2005: 158) has demon-strated that, although conception and childbirth could be seen as signs of health, hav-ing an unhealthy womb was not always a sign of general ill health. Similarly, having infertile semen did not mean a man was viewed as unhealthy, but it was recognised that diseases in other parts of the body and general ill health could cause problems with the quantity and fertility of the semen. Aristotle (*Generation of Animals* 746b39; cf. 725a9–10) notes that diseases can occur elsewhere in the body that leave the semen emitted fluid and cold, although he does not state the exact nature of these diseases. However, in another passage he states that those in bad health may emit other residues that are morbid along with the semen. This means that the emission (*apochōroun*) is infertile because it contains so little semen (*spermatikon*, *Generation of Animals* 725b14–17). This suggests that there was a recognition that fluid ejaculated could be made up of more than just 'the seed'.

The idea of a distinction between 'semen' and 'seed' in Aristotle's semen theory has only recently being made. Marwan Rashed (2018: 114) argues that Aristotle (*Genera-tion of Animals* 736a13–21) states that the reason for the whiteness of semen (*tou spermatos*) is that the seminal fluid (*hē gonē*) is foam (*aphros*) and that his terminol-ogy is 'precise and univocal' on this matter and makes the distinction between semen and seminal fluid, with the foam being the latter. Rashed (2018: 115) supports his theory with another passage from *Generation of Animals* (736b33–7), where Aristo-tle notes that semen (*en tōi spermati*) contains an agent, heat, that makes semen (*ta spermata*) fertile and that this heat is not fire but the 'breath' which is stored up in semen (*en tōi spermati*) and in the frothy material (*kai en tōi aphrōdei*).[11] In Aristotle's theory of conception, it is 'breath' that is an essential ingredient for conception occurring. While I am not entirely convinced that Aristotle always makes this distinction between *sperma* and *gonē*, Rashed's argument is persuasive. It also fits with the ideas presented here that thick semen is more fertile than thin semen: Rashed further notes that at 735b8–13 Aristotle states that foam (*ho aphros*) becomes thicker and whiter, with the bubbles becoming smaller and more indistinguishable.

It was not just illness that could have a weakening effect on the body, which was then replicated in the semen. The Hippocratic author of *On Regimen* suggests that ingesting certain foods could also have an effect on semen, noting that 'Mint warms, passes easily by urine, and stops vomiting; if eaten often it melts the seed and makes it run (*tēn gonēn tēkei hōste rhein*) preventing erections (*enteinein kōluei*) and weakening the body' (*On Regimen* 2.54; trans. Jones 1931: 331). Here the seed is described as melting (*tēkei*), thus becoming runny, which as we have seen is linked to lowered fertility. It is noted that the seed being thin could contribute to another problem associated with fertility—that of impotence; the link between the seed and impotence will be discussed in detail later in this chapter.

Whether due to disease, age, or the ingestion of certain foodstuffs, it seems clear that for both Aristotle and the Hippocratic writers, observations of the appearance of semen, particularly the amount and consistency of what was emitted, could help assess its fertility. Aristotle even provides us with an empirical test for the potency of semen, by observing its behaviour when added to water, which for him confirms the observations being made.

Physical traits determining the fertility of semen

For Aristotle it was not just diseases which caused thin, cold semen. And he speaks with relative frequency about the physical characteristics which could lead to a lack of fertility.[12] For example, he suggests that baldness occurs in men who have plenty of semen, whereas women, children, and eunuchs are not affected due to their lack of semen (*Generation of Animals* 783b35–784a10; cf. *History of Animals* 632a5–6). The colour of the flesh was also thought to be an indication of fertility, with Aristotle stating that men who are fair produce more semen than those who are dark (*History of Animals* 583.7–10).

An important physical trait which according to Aristotle has an effect on fertility is the size of a man's penis. In *Generation of Animals*, he notes that men who have a large penis are less fertile than those who have a moderately sized one, explaining that this is because the semen cools off as it is transported over a greater distance, and cold semen is not generative (*mē gonimon*, *Generation of Animals* 718a24; cf. 725a10 and *Problemata* 871a23, edition Mayhew 2011). The idea that a larger penis results in the semen losing a greater amount of heat when transferred into the female during intercourse fits with Aristotle's theory of heat playing an important part in the production of fertile semen. Here the man has the required heat in his body to produce semen which is fertile, but the loss of heat as it moves through the penis causes it to become ungenerative.

For the most part, the physical traits affecting fertility mentioned by Aristotle are natural ones: people are born with a particular hair or skin colour. However, there is one trait which Aristotle mentions that could be potentially changed: a person's weight. Aristotle suggests that too much fat on the body could have a direct effect on semen. He describes why this is the case in the following passage:

> Some individuals have much semen (*polusperma*), some little (*oligosperma*), some none at all (*asperma pampan*); and this is not due to any bodily weakness, but in some cases, at any rate, it is due to the opposite: the available supply gets used up to benefit the body; as an example of this we have men in sound health putting on rather a lot of flesh and getting a bit fat: these emit less semen (*proientai sperma*) and have less desire for sexual intercourse than is normal (*hētton epithumousi tou aphrodisiazein*).
>
> Aristotle, *Generation of Animals* 725b29–34; trans. Peck 1942: 85

The reason why being overweight affects fertility is that the nutriment that the body receives via food, which would normally go into producing semen or menstrual blood, is instead diverted to sustain this extra fat, and this is why those who are overweight are not able to produce the essential reproductive fluids (cf. *Generation of Animals* 726a3–7; 746b25–9).[13] The effect of fat on fertility was not necessarily believed to be a permanent state but could be reversed by losing the additional weight (*Generation of Animals* 726a3–5)

Physical damage to the body

Whereas Aristotle focuses on physical characteristics, in the Hippocratic Corpus physical damage to the body represents the main causes of male infertility, and invariably this is linked to a disturbance in the passageway of the seed through the body. In *On Generation* (2) we are informed that, both in eunuchs and in those whose testicles are crushed, not only is the passageway for the seed through the testicles disrupted but also the cords by which the penis is raised and lowered are affected, so such men are therefore unable to have erections.[14]

One of the most repeated causes for male infertility in the Hippocratic Corpus is the act of making an incision beside the ear. In *On Generation* the author describes how those who have this done are able to have intercourse and to ejaculate but their seed is 'small in amount' (*oligon*), 'weak' (*asthenes*), and 'infertile' (*agonon*) (2.8–11)—the reason being because the seed has to flow past the ears into the spinal marrow, and the scarring from the cut causes the passageway to become solid. The incision beside the ear does not cause a direct problem with the reproductive system, and these men are still able to have intercourse. However, as I described at the beginning of this chapter, the Hippocratic author thought that semen was drawn from the entire body and it worked its way through the body during intercourse. The incision behind the ear interrupted this passage through the body, weakened the semen, and made it infertile.

One of the most famous examples of male infertility in the Hippocratic texts is that of the Scythians outlined in the text *On Airs, Waters, and Places*. While this also features a cut behind the ear, the explanation for their infertility is somewhat different to that given in *On Generation*.[15] In *On Airs, Waters, and Places* (22) the author states that as the Scythians ride astride their horses, they have swelling in their joints (*kedmata*)[16] and become lame and develop sores on the hips. This in itself does not cause their infertility, but the attempted cure does. The vein behind the ear is cut and bleeds until they feel faint and they sleep. On awakening, some find themselves cured. The effect on the semen is described thus:

> Now, in my opinion, by this treatment the seed is destroyed (*iēsei diaphtheiresthai ho gonos*). For by the side of the ear are veins, to cut which causes impotence [infertility] (*agonoi*), and I believe that these are the veins which they cut.
>
> *On Airs, Waters, and Places* 22.13–17; trans. Jones 1923: 127–9

In this passage, cutting behind the ear destroys the seed and leads to the man becoming infertile. W. H. S Jones chooses to translate *agonoi* as 'impotence' in this passage, although *agonos* generally means infertile. Later in the same passage, the author notes that when these men attempt intercourse with women, they find they cannot do so. It is not explicitly stated they are impotent, but this is implied by the fact they cannot complete the act of intercourse.[17] The author of *On Airs, Waters, and Places* goes on to say that these men, after two or three attempts at intercourse, assume that they have offended the gods and begin to dress like women and perform women's work.[18] Whether this type of incision caused erectile dysfunction differs depending on the Hippocratic text. While the author of *On Airs, Waters, and Places* suggests that there is a correlation, the author of *On Generation* suggests otherwise. Marquis Berrey (2014: 294) states the reason for the difference is because, while the author of *On Generation* theorised that it is the tendons connecting the penis and the testicles which creates an erection, the author of the other text sees the downward flow of the semen itself as being responsible. Therefore, without semen there cannot be an erection. This seems a plausible analysis of the situation, which, although not mentioned by Berrey, is supported by the assertion given in *On Regimen* (2.54) that mint causes the seed to melt, which means erections cannot occur. However, I still believe that the description given by the author of *On Airs, Waters, and Places* remains somewhat ambiguous because of his choice of language. There is only a limited amount of discussion by Aristotle on physical damage to the body causing problems with fertility, but in *Generation of Animals* (728a15–18) he describes men whose generative organs have been destroyed as not being able to concoct semen, which leads to them having loose bowels due to the unconcocted semen being secreted into their intestines.

The effect of 'sub-fertile' semen on the child

A problem with semen production did not always mean that a man was completely incapable of procreating but may have meant he would find it more difficult to do so, what in modern medicine is sometimes described as being 'sub-fertile'. In both Aristotle and the Hippocratic texts we find descriptions of men with weakened semen and fertility, which not only makes the man less likely to impregnate his partner but can also have an effect on any child which is conceived.

The effects described are twofold.[19] Firstly, it could affect the sex of the child conceived. The author of *On Generation* (6) explains that, if strong semen (*ischuroteron*) comes from both parents, then a male child will result, whereas, if weak (*asthenes*) semen is provided by both parents, a female will be engendered. If there is a greater amount of weaker semen than of strong semen, the stronger seed is effectively overcome, and a female child is produced. Conversely, if there is more of the stronger semen than of the weaker, then the result is a male child. The idea that a child was generated from 'weak' semen then the result would be a female makes several appearances in the Hippocratic Corpus. Advice is given in the Corpus for activities and regimens a couple should undertake should they wish to conceive a child of a particular sex. This includes the timing of intercourse in the menstrual cycle, eating or refraining from certain foods or drinks, and tying up one testicle, the right for a girl and the left for a boy, during intercourse (*On Superfetation* 31, edition Potter 2010). What is interesting in these regimens is that, while the advice for conceiving a male child, such as eating hot and dry foods and drinking dark wine (*On Regimen* 1.27), were similar to advice for improving general fertility, the regimens suggested for producing a girl—for example eating moist foods and white wine (*On Regimen* 1.27)—seem to contradict such advice.

It was not just the sex of the child conceived which could be affected but also the health of the subsequent child:

> But when some disease befalls the moisture (*hugrou*) from which the sperm (*sperma*) is formed, the four kinds of substances that are naturally present in this part do not produce a complete seed (*tēn gonēn ouch holēn parechousin*), but one weaker (*asthenesteron*) to the degree that it is maimed (*pepērōmenon*); thus it does not seem any wonder to me that this offspring is maimed (*pērōthēnai*) like its parent.
>
> *On Generation* 11; trans. Potter 2012: 23[20]

As the semen is drawn from each part of the body in the Hippocratic theory, it makes sense that a problem with one part of the body would not necessarily lead to all of the semen becoming unproductive but could still affect the particular part in the child produced in the same place as it did their parent.

Aristotle too thought that weakness in the semen could affect the child. He notes that when semen production begins at puberty, what is secreted is mostly infertile, and if a child is conceived, it is often small and weak (*History of Animals* 544b15). Aristotle also suggests that a problem at conception or in the development of the child in the womb, what we might refer to as a congenital abnormality, could affect the fertility of that child in adulthood.

> Some of these deformities are curable, some are not; those, however, who have become deformed during the original constitution of the embryo, have a special tendency to remain infertile (*agona*) throughout; thus, masculine-looking women are produced in whom the menstrual discharges do not occur, and effeminate men whose semen (*sperma*) is thin (*lepton*) and cold (*psuchron*).
>
> Aristotle, *Generation of Animals* 746b33–7a3;
> trans. Peck 1942: 247

Therefore, we find that it was not only thought that weakened semen could directly affect the health of the child produced by it, but it could also potentially have an effect on whether the child itself would be able to conceive a child of their own when the time came.

Conclusion

By bringing together this evidence from the Hippocratic Corpus and the biological works of Aristotle, we see that there were not only many ideas surrounding the causes of 'weak' or 'infertile' semen, but also that the results of such semen could differ, too. Having 'weak' or 'infertile' semen did not automatically mean that a man would not produce children, but it did make him less likely to do so; however, even if a child was produced, there was a chance that this child would have health problems of its own, including fertility problems in later life.

The way in which semen was assessed for fertility in the ancient world was not dissimilar to the macroscopic tests performed today. To be considered fertile, it needed to be of sufficient volume, its consistency was assessed to ensure it was neither too runny nor too thick or 'granular', and observations were made about its viscosity. Many things were thought to affect the fertility of semen, including disease, consumption of foodstuffs, weight, and congenital abnormalities. Certain physical features such as penis size and body size could be used as a guide as to whether a man was more likely to have weakened semen. Due to the fundamental differences between the Hippocratic writers and Aristotle in the understanding of how semen was produced, it is unsurprising that their understanding of the causes of 'weak' semen differ. Since the Hippocratic understanding was that semen was not only produced during intercourse but also made its way through the body through a series of passageways, many of the problems associated with semen are linked to a disruption of its passage through the body. On the other hand, in his theory of semen production, Aristotle focuses on problems associated with heat and nourishment.

Of all the bodily fluids, semen is perhaps the only one which is judged not only on its effect on the body it originates from but also on another body it enters. The quality of menstrual blood and female semen (where the theory suggests this exists), like male semen, can be judged on their ability to produce a child, but this takes place within the body where the fluids originate. To be judged as fertile, male semen not only has to be produced in the correct amounts, be of sufficient quality, and be unimpeded in its movement through the body, but also be transferred successfully into a new body to fulfil its potential of creating a new life.

Notes

1 In the Hippocratic text *On Generation* (3–4) the author states that both men and women produce semen. See Rebecca Flemming's chapter in this volume (Chapter 10) for the discussion surrounding the nature of female 'semen'. As there is no description of female reproductive failure due to problems in their production of semen, here I concern myself only with male semen. All citations from the Hippocratic Corpus in the chapter are from the Loeb editions; other editions, including that by Littré (1839–61), may have slightly different chapter numbering.

2 Here I concentrate on the texts *Generation of Animals* and *History of Animals* as they constitute the bulk of the evidence for Aristotle's opinion on such matters from a biological

perspective. Although book 10 of *History of Animals* does offer good descriptions of infertility, because of the disputes over authorship of this work, I have excluded it from this study. For a good overview of this problematic text see Philip van der Eijk (1999) and Lesley Dean-Jones (2012).

3 It should be noted that van Leeuwenhoek's observation was only the beginning towards an understanding of the role 'sperm' played in reproduction and many theories were put forward. For a good overview of the development of these ideas, see Vienne (2018).

4 While Flemming (2013) provides the first detailed study of infertility, her focus is not on male fertility. Senkova (2015) and Fallas (2015) discuss male infertility in more detail. Although Berrey's (2014) focus is on erotic desire rather than fertility, his argument is intrinsically linked to fertility, as will be demonstrated in this chapter. Aristotle's ideas on semen have received more attention, although much of the scholarship is focused on the ideas regarding *pneuma* and the soul, particularly from a teleological and mechanical viewpoint, neither of which are my focus here. In addition to those referenced in this chapter, see for example Peck 1953; Matthen 1989; Bos 2009; Connell 2016.

5 Other words are used in relation to semen, but these are rare and often relate to animals rather than humans. For example, *thoros* is used in the Hippocratic text *On Diseases* (2.51) to describe male semen. Aristotle uses the same word in *History of Animals* (568b12) to describe fish milt.

6 See Totelin (2018: 60–2) for the use of *sperma* in relation to animals and plants.

7 In this chapter I translate *gonē* as 'seed' and 'semen' when *sperma* appears in the ancient texts. Although the terms are generally interchangeable, often the ancient writers have a preferred term. For example, in the Hippocratic text *On Diseases of Women*, *gonē* is used more often than *sperma*. In the text *On Generation*, the author seems to favour *sperma* for semen before it enters the uterus and then *gonē* once the seed enters the uterus and conception takes place. Rashed (2018) does put forward an argument for Aristotle distinguishing between semen (*sperma*) and seminal fluid (*gonē*); this distinction will be explored later in this chapter.

8 The exact nature of the differences between where Aristotle believes semen is drawn from in the body compared to the Hippocratic view has been debated in modern scholarship. Many of the debates have surrounded the ideas of pangenesis. This term was introduced by Charles Darwin in the nineteenth century, but while the same term is used for the theory given in the ancient texts because of its similarity to Darwin's, they are not in fact the same theory. In pangenesis, the seed is drawn from individual areas of the body, contrasting with the view that it is drawn from the body as a whole. Generally, it is accepted that the Hippocratic theory is based around the idea of pangenesis, whereas Aristotle subscribed to the alternative school of thought. This has been challenged by Andrew Coles (1995) who argues that the two theories are closely related and that Aristotle can be read as subscribing to a similar view as the Hippocratic Corpus on this point. This in turn has been disputed by Daryl Mcgowen Tress (1999), who argues that Coles comes to this conclusion because he is approaching the two authors as both being physicians as opposed to a physician and a philosopher.

9 The Hippocratic authors do not give exact ages for when puberty or a decline in fertility would occur. Darren Amundsen and Carol Diers (1969: 125–6) suggest that in *Coan Prognoses* (502, edition Potter 2010), the statement that some diseases do not appear before puberty but only between 14 and 42 is an indirect statement that puberty occurred at 14 and a decline in fertility around the age of 42.

10 The Greek in both the passage from *Aphorisms* and the one from *Epidemics* differ slightly in the various manuscripts. Although the overall meaning appears to be the same, the possibility of corruption of the text as a possible reason for the different conclusions should be considered here.

11 This is only one part of Rashed's theory of the nature of semen in *Generation of Animals*, and Rashed himself points out that there are problematic parts in these passages. However, these fall outside the constraints of this chapter.

12 Physical characteristics affecting male fertility are not mentioned in the Hippocratic texts. However, the effect of all the physical characteristics Aristotle mentions are reported as factors of female infertility by the Hippocratic writers (e.g. *Prorrhetic* 2.24, edition Potter 1995; *Diseases of Women* 2.111, edition Potter 2018; and *Aphorisms* 5.46). Why the Hippocratics do not mention these with regards to male fertility is not clear.

13 Aristotle notes that in general, animals that have more fat produce less semen than lean ones (*Generation of Animals* 727a-32–5).

14 Berrey (2014: 294) suggests that in the case of the genitals being crushed, only ejaculation is prevented, not erections. However, if eunuchs were believed not to be able to achieve an erection because of the cutting of the cords that raise and lower the penis, then surely the hardening of these same cords, which results in them not being able to contract or relax, must mean that the author thought both types of men to be impotent.

15 An incision beside the ear causing infertility is also found in *Places in Man* (3, edition Potter 1995). In *Epidemics* (6.5.15), an incision behind the ear for *kedmata* is described, but there is no mention of any effect on fertility. Here I focus on the Hippocratic writer's view of the causes of impotence among the Scythians. Lieber (2003) gives a good account of the various attempts to retrospectively diagnose the impotence of the Scythians, including suggesting this may be due to the group of diseases called haemochromatosis or hereditary disorders.

16 The exact nature of the disorder *kedmata* is difficult to determine, although it seems to involve swelling of the joints. See Lieber (2003: 355–6 n.14) for a summary of the different ideas surrounding this disease.

17 As Berrey (2014: 294 n.15) has noted, in *On Airs, Waters, and Places* (22), the author describes the Scythian men as *mē hoioi t' eōsi chrēsthai sphisin* (are not able to have sex with women). In *On Generation* (2), eunuchs are said to be *ou chrēstoi* (not able to have intercourse). The similarity in language suggests that the Scythians too suffered from erectile dysfunction.

18 Herodotus too describes the impotence of the Scythians, but in his account, this condition is retribution from the gods for the looting of shrines (1.105.2–4). See Thomas (2000: especially 54–71) and McMahon (1998: 69–73) for a summary of the differences between the two accounts.

19 See also Flemming, this volume, Chapter 10, for further discussion on this matter.

20 The idea that healthy seed comes from the healthy parts or the body and diseased from the parts that are disease is also mentioned in *On the Sacred Disease* (2, edition Jones 1923).

Bibliography

Primary sources

Balme, D. M. (1991) *Aristotle. History of Animals*, vol. III: Books 7–10. Edited and translated by D. M. Balme. Cambridge, MA: Harvard University Press.

Jones, W. H. S. (1923) *Hippocrates*, vol. I. With an English translation by W. H. S. Jones. Cambridge, MA, Harvard University Press.

———. (1931) *Hippocrates*, vol. IV. With an English translation by W. H. S. Jones. Cambridge, MA: Harvard University Press.

Littré, E. (1839–61) *Œuvres complètes d'Hippocrate*. 10 vols. Paris: Baillière.

Mayhew, R. (2011) *Aristotle. Problems*, vol. I: Books 1–19. Edited and translated by R. Mayhew. Cambridge, MA: Harvard University Press.

Peck, A. L. (1937) *Aristotle. Parts of Animals*. Edited and translated by A. L. Peck. Cambridge, MA: Harvard University Press.

———. (1942) *Aristotle. Generation of Animals*. Edited and translated by A. L. Peck. Cambridge, MA: Harvard University Press.

———. (1970) *Aristotle. History of Animals*, vol. II: Books 4–6. Edited and translated by A. L. Peck. Cambridge, MA: Harvard University Press.

Potter, P. (1988a) *Hippocrates*, vol. V. Edited and translated by P. Potter. Cambridge, MA: Harvard University Press.

———. (1988b) *Hippocrates*, vol. VI. Edited and translated by P. Potter. Cambridge, MA: Harvard University Press.

———. (1995) *Hippocrates*, vol. VIII. Edited and translated by P. Potter. Cambridge, MA: Harvard University Press.

———. (2010) *Hippocrates*, vol. IX. Edited and translated by P. Potter. Cambridge, MA: Harvard University Press.

———. (2012) *Hippocrates*, vol. X. Edited and translated by P. Potter. Cambridge, MA: Harvard University Press.

———. (2018) *Hippocrates*, vol. XI. Edited and translated by P. Potter. Cambridge, MA: Harvard University Press.

Smith, W. D. (1994) *Hippocrates*, vol. VII. Edited and translated by W. D. Smith. Cambridge, MA: Harvard University Press.

Secondary literature

Agarwal, A. and T. M. Said (2010) 'Interpretation of basic semen analysis and advanced semen testing', in E. S. Sabanegh (ed.) *Male Infertility: Problems and Solutions*. New York: Springer, 15–22.

Amundsen, D. and C. Diers (1969) 'The age of menarche in classical Greece and Rome', *Human Biology* 41(1), 125–32.

Berrey, M. (2014) 'The Hippocratics on male erotic desire', *Arethusa* 47(3), 287–301.

Bos, A. P. (2009) 'Aristotle on soul and soul-"parts" in semen (GA 2.1, 735a4–22)', *Mnemosyne* 62, 378–400.

Coles, A. (1995) 'Biomedical models of reproduction in the fifth century BC and Aristotle's Generation of Animals', *Phronesis* 40(1), 48–88.

Connell, S. M. (2016) *Aristotle on Female Animals: A Study of the Generation of Animals*. Cambridge: Cambridge University Press.

Craik, E. M. (2015) *The 'Hippocratic' Corpus: Content and Context*. London: Routledge.

Dean-Jones, L. (2012) 'Clinical gynaecology and Aristotle's biology: The composition of HA X', *Apeiron* 45(2), 180–99.

Fallas, R. (2015) 'Infertility, blame and responsibility in the Hippocratic Corpus', PhD thesis, Open University, Milton Keynes.

Flemming, R. (2013) 'The invention of infertility in the classical Greek world: Medicine, divinity, and gender', *Bulletin of the History of Medicine* 87(4), 565–90.

King, H. (2005) 'Women's health and recovery in the Hippocratic Corpus', in H. King (ed.) *Health in Antiquity*. London: Routledge, 150–61.

Liddell, H. D. and R. Scott (1996) *A Greek-English Lexicon*. Oxford: Clarendon Press.

Lieber, E. (2003) 'The Hippocratic "Airs, Waters, Places" on cross-dressing eunuchs: "natural" yet also "divine"', in M. Golden and P. Toohey (eds) *Sex and Difference in Ancient Greece and Rome*. Edinburgh: Edinburgh University Press, 351–69.

Matthen, M. (1989) 'The four causes in Aristotle's embryology', *Apeiron* 22(4), 159–79.

McGowan Tress, D. (1999) 'Aristotle against the Hippocratics on sexual generation: A reply to Coles', *Phronesis* 44(3), 228–41.

McMahon, J. M. (1998) *Paralysin Cave: Impotence, Perception, and Text in the Satyrica of Petronius*. Leiden: Brill.

Peck, A. L. (1953) 'The connate pneuma: An essential factor in Aristotle's solutions to the problems of reproduction and sensation', in E. A. Underwood (ed.) *Science, Medicine, and*

History: Essays on the Evolution of Scientific Thought and Medical Practice Written in Honour of Charles Singer. Oxford: Oxford University Press, 111–21.

Rashed, M. (2018) 'A latent difficulty in Aristotle's theory of semen', in A. Falcon and D. Lefebvre (eds) *Aristotle's Generation of Animals: A Critical Guide.* Cambridge: Cambridge University Press, 108–29.

Senkova, M. (2015) 'Male infertility in classical Greece: Some observations', *Graeco-Latina Brunensia* 20(1), 121–31.

Thomas, R. (2000) *Herodotus in Context: Ethnography, Science and the Art of Persuasion.* Cambridge: Cambridge University Press.

Totelin, L. M. V. (2017) 'Whose fault is it anyway? Plant infertility in antiquity', in G. Davis and T. Loughran (eds) *The Palgrave Handbook of Infertility in History.* London: Palgrave Macmillan, 57–75.

———. (2018) 'Animal and plant generation in classical antiquity', in N. Hopwood, R. Flemming, and L. Kassell (eds) *Reproduction: Antiquity to the Present Day.* Cambridge: Cambridge University Press, 53–66.

van der Eijk, P. J. (1999) 'On sterility ("HA X"), a medical work by Aristotle?', *Classical Quarterly* 49(2), 490–502.

Vienne, F. (2018) 'Eggs and sperm as germ cells', in N. Hopwood, R. Flemming, and L. Kassell (eds) *Reproduction: Antiquity to the Present Day.* Cambridge: Cambridge University Press, 413–26.

8

SAY IT WITH FLUIDS

What the body exudes and retains when Juvenal's couple relationships go awry

Claude-Emmanuelle Centlivres Challet

Fluids in the satires

Over the past decades, Roman satire has been generally put to use by social historians to 'isolate areas of anxiety for the male elite about society and its norms' and 'about standards of male and female behaviour' (Braund and Raschke 2002: 66). While the specific intentions of Juvenal are debated, it can be argued that bodily fluids are used by the satirist as another tool to emphasise his discourse (Gavrielatos, this volume, Chapter 18).[1] Fluids are markers, extensions of the body expressing its inner workings, mirroring the moral and emotional processes of an individual. As such, fluids are used by the satirist to signpost the issues that he considers need addressing and to complement his descriptions of what he finds criticisable: bodily fluids are additional evidence, as valuable as actions or speeches, on which the satirist bases his analysis of people and their morals.[2] This chapter investigates what the exudation or retention of certain fluids tells of the way Juvenal sees male–female couple relationships and what role they play in the satirist's conveying of his message concerning men's and women's expected and distorted roles within the conjugal relationship. It will start with a general overview, then focus on *Satire 6*.

The sixteen satires of Juvenal, written between the end of the first century and the beginning of the second century, contain sixty mentions of twelve different bodily fluids. There appears to be a coherence in the vices or virtues attributed to each fluid depending on the circumstances in which it is exuded or retained. Nine of the twelve bodily fluids appearing in Juvenal's satires are not specifically associated with either sex—tears, nasal mucus, saliva, sweat, bile, stomach contents, blood, urine, and faeces—and three of them are linked to one of the two sexes—sperm, vaginal secretion, and milk.[3] Juvenal uses occurrences of bodily fluids either in a context of scorn or anger, where the presence or absence of the fluids in question is meant to denounce something, twice as many times as when he uses them in a context where he passes no judgement, or when the fluids in question are positively connoted.

Tears are physiologically produced by both males and females, and Juvenal has four occurrences of women shedding tears (6.8, 273, 276–7; 10.260–1), seven occurrences of men shedding tears (1.168; 3.101 (two different people); 5.158–60; 6.539; 10.32; 16.27), and three neutral occurrences (1.48–50; 13.133–4; 15.132–7). In all cases tears denote sadness, with various nuances ranging from compassion to guilt, and neither men nor women are ridiculed for crying; they are only mocked when they

DOI: 10.4324/9780429438974-12

cry fake tears. So Juvenal, despite his satirical streak, does not use crying as a sign of weakness. He states that nature has given men a tender heart and tears to testify to it, and that tenderness is the best of human feelings.[4] Male emotions are not denied, nor is their display through tears, a way of expressing grief used by men as well as by women: the satirist speaks of how it is natural to be moved by pity by a friend on trial or when a ward with ambiguous feminine locks and wet cheeks brings a defrauder into court (15.134–7) and tells of the natural instinct of weeping at the funeral of a virgin or a child (15.138–40).[5]

Nasal mucus dripping is ridiculous in a man (10.199–201) as well as in a woman (6.147–8). Saliva is associated with ridicule and pollution in men five times (5.127–9; 6.623; 7.111–12; 9.35–6; 13.212 [lack thereof]) and in women once (6.O14). Sweat symbolises ridicule, guilt, or fear in four men (1.28, 166–7; 10.178; 13.220) and two (6.259, 420) or three women (11.186–8); one case is not clear-cut, and we will come back to the ambiguous dampness in question.

Bile is the symbol of anger in men three times (5.158–60; 11.186–8; 13.143) and once perhaps emerges in a case of real nausea (6.432). Stomach contents are once vomited by a woman when they should not be and once when they should (6.100–1, 432): this fluid is used in a different capacity depending on the message of the satirist—we will come back to that as well.

All occurrences of blood are meant either to ridicule sterile violence or to signify good or bad health. Blood appears twice in a sex-neutral way, once as a symbol of violence (8.136), and once as a marker of health when tinting complexion (11.54). Six occurrences concern the violence of men (8.241–3; 10.120–1, 185–6; 13.178–9; 15.58, 92) or their health (6.46; 10.217–18), and three occurrences associated with women symbolise violence (6.525–6) or life (4.8–10; 8.217–19), while one is related to abortion (6.595–7).

The evacuation of urine is depicted as a natural activity for men in two cases (1.129–31; 3.107), but for women it is an activity associated negatively with lust (6.64; 11.169) and ridicule (6.264, 310–13) four times. Last, faeces are mentioned twice in relation to men to ridicule them (9.43–4; 14.199–200).

As for fluids associated with one or the other sex, vaginal secretion is criticised, as it is associated with unlawful sexuality (6.318–19; 10.321–2; perhaps 11.186–8); however, sperm (6.366–8) and milk (6.9, 398–401, 592–4) are not connoted negatively, and their absence marks out-of-bounds behaviour, as we will see.

Of the sixty occurrences of bodily fluids making their appearance in the Juvenalian corpus, a third appear in the famous *Satire 6*, in which all types of fluids are present except faeces. The topic of *Satire 6* is debated; I have argued elsewhere that *Satire 6* is neither a rant against women in general nor a pamphlet against marriage as an institution, but, to put it briefly, a reminder that it takes both partners for a marriage to function, that both male and female traditional gender roles should be heeded in public while leeway and room for manoeuvre should be given in private, and that a union without affectionate feelings cannot function properly (Centlivres Challet 2013: 113–50). Within this thematic framework, which pervades the whole corpus of the satires, Juvenal uses the presence or absence of bodily fluids to show how each gender should behave. The fluids that are used to express approval or criticism of gender roles are milk, sperm, vaginal secretion, stomach contents, urine, and tears, and will thus be explored in more detail.

Fluids associated with one of the sexes

Milk

The most traditional gender role of Roman women is as reproductive agents, and among the tasks included in this role is that of nurturing the child. Despite the fact that texts show that elite women often shunned this task and resorted to wet-nurses, there were authors that advocated maternal breastfeeding, among them Juvenal, who pits breastfeeding women against women whom he considered frivolous and self-centred.[6] He says the following of the wife of mountain people (*montana uxor*, 6.5), the rough mother of prehistoric times: 'The opposite of you, [Propertius's] Cynthia, and of you [Catullus's Lesbia], whose sparrow's death clouded the shining eyes, she presents breasts destined to feed large babies' (*Haut similis tibi, Cynthia, nec tibi, cuius / turbavit nitidos extinctus passer ocellos, / sed potanda ferens infantibus ubera magnis*, 6.7–9).[7] Primitive women had the proper, good instincts, according to Juvenal, contrary to contemporary women, whose lifestyle is perverted. The satirist then contrasts the nurturing practices of women of the lower classes with those of women of the elite. Of poor women he says: 'But these submit themselves to the hazard of bearing a child and endure all the strains of nursing as their fate demands, but on a golden bed rarely lies a mother in childbirth' (*Hae tamen et partus subeunt discrimen et omnis / nutricis tolerant fortuna urguente labores, / sed iacet aurato vix ulla puerpera lecto*, 6.592–4).

Still worse is the absence of maternal milk, if this lack is associated with a behaviour that runs counter to other expected female traditional roles: 'But let her sing rather than fly through the whole city, presumptuous and able to endure men's meetings, daring to address herself to generals in their commander's cloaks in her husband's presence, a picture of righteousness with dry breasts' (*Sed cantet potius quam totam pervolet urbem / audax et coetus possit quae ferre virorum / cumque paludatis ducibus praesente marito / ipsa loqui recta facie siccisque mamillis*, 6.398–401).

The opposition between the husband, who plays his role in his rightful place, among other men, and the wife, who transgresses the limits imposed by the traditional separation of the gendered spheres of action, is highlighted by an anatomical detail: the wife has 'dry breasts' (*siccisque mamillis*). In other words, she does not have the attribute of a woman who behaves within the scope of her gender role: the breasts with no milk symbolise her wandering away from motherhood.[8] James Duff (1970: 234) comments on 'dry breasts', saying 'like an unsexed creature', as Edward Courtney does (1980: 314); but this does not take into account the underlying issue of gender boundaries made clear by Juvenal: the know-all trespasses on male territory and is thus physiologically, and in a purposefully derogatory way, compared to a man and not to an unsexed creature, which would neither make sense nor have the intended subtextual weight in this context. George Ramsay's 'hard breast'(1940) is vague enough to cover both the concepts of motherhood and femininity in a way that his chaste times perhaps demanded. Niall Rudd (1991) has 'with no milk in her breasts', a matter-of-fact interpretation. Franco Bellandi (2003: 32) proposes to understand *siccisque mamillis* as *sine sudore pudoris*, 'without the perspiration of shame', and associates it with *recta facie*, 'with a straight face': the woman would not even show the expected signs of emotion and embarrassment that her behaviour should cause; this idea of the shameless busybody fits the context but weakens the contrast with the generals in their commander's cloaks. Not only does the dry-breasted woman stray from the maternal realm—epitomised by nurturing characteristics—and thence from her roles as mother, carer, and housewife, but she has also lost the characteristics of the wife and sexual object, since her 'dry breasts' evoke something shrivelled, suggesting not only the lack of milk of traditionally nurturing breasts but also a flatness of breast that is more male than female.

Sperm

King (2008: 154–6) highlights sperm as well as a beard as the epitomes of masculinity in antiquity. As the quintessential attribute of the Roman man, sperm can be seen as the equivalent of milk for women, and according to Juvenal, like milk it should not be retained within the context of matrimony.[9] Its mention, or rather the mention of its absence, is used to criticise women as well as men: 'Some women are delighted by innocuous eunuchs, smooth kisses, and the absence of a beard, because there is no need for an abortifacient' (*Sunt quas eunuchi inbelles ac mollia semper / oscula delectent et desperatio barbae / et quod abortivo non est opus*, 6.366–8).

In *Satire 6*, female adulterous behaviours of all kinds point to the intention of obtaining sexual satisfaction, and this is corroborated by the mention of women who like the attentions of eunuchs, with whom relations do not have a reproductive aim, and thanks to whom they do not have to worry about abortions. Eunuchs as sexual partners give women sexual freedom and control over their bodies, which husbands do not offer; contrary to the latter, the former present the advantages of offering both sexual satisfaction and contraception. The absence of sperm highlights the absence of the dominant male; the relationship is dysfunctional since there is no male conjugal authority to keep women's behaviour in check. The absent and redundant husband is here ridiculed.

Such a useless male spouse is hinted at by Juvenal when he compares a weak cuckold to a worm or to an ant: 'You believe that it is love, you are then satisfied with yourself, you worm, and you dry her tears with your kisses' (*Tu credis amorem, / tu tibi tunc, uruca, places fletumque labellis / exorbes*, 6.275–7). 'Their fearful husbands sometimes provide against cold and hunger in imitation of the ant' (*Prospiciunt aliquando viri, frigusque famemque / formica tandem quidam expavere magistra*, 6.360–1).

As it happens, *uruca* is really 'caterpillar' rather than 'worm'—but 'worm' conveys a pejorative nuance in English which is used to better effect in this context—thus a sexually immature creature. In any case, to the untrained eye, a caterpillar looks asexual. As for the ant, the queens can retain the male ants' sperm for several years (Baer et al. 2006), making the latter superfluous during that time. What is more, the ant looks asexual to those who do not know that males are winged and/or hidden in the anthill; therefore, both these insects make for interesting comparison. Admittedly, Juvenal may not have known the specifics of ants and caterpillars, but to amateurs, these insects lack the decisive male features of a visually clear sexual dimorphism. And what is more important, after emission sperm is under the control of the females, in the case of the ants, or there is none, in the case of the caterpillars, as is the case with eunuchs.

Vaginal secretion

There are two relatively clear occurrences of vaginal secretion in the *Satires*, and both are negatively connoted. The first one appears during the rites of the Bona Dea, when women get excited at the prospect of having some energetic interactions with men: 'What a great torrent of sheer lust flows down their drenched legs!' (*Quantus / ille meri Veneris per crura madentia torrens!* 6.318–19).

The second occurrence concerns a woman who would give everything to a young, handsome lover, because, as the satirist says, once vaginal secretion is produced and flows, there is nothing to stop a woman's libido: 'No woman indeed will say no to her moist pudenda' (*Quid enim ulla negaverit udis / inguinibus*, 10.321–2).

A third occurrence is open to debate, because its understanding is not so straightforward: Juvenal condemns the shameless adulteress coming back home with a damp dress, *umida . . . multicia* (11.188), from a rendezvous: one can interpret this dampness as sweat, as mentioned earlier, as vaginal secretion, or even as unlawful semen.

So while milk and sperm are associated with conjugal relationships, vaginal secretion is not. Milk and sperm are positively viewed by the satirist not only for their intrinsic qualities but because they are associated with a woman's and a man's traditional roles, that is, nurturing and engendering, respectively, while vaginal secretion is associated with inappropriate sexual appetite and as such is a fluid used by Juvenal as a marker to signpost deviance.

Fluids that are not specifically male or female

We have just seen that vaginal secretions flow when an adulteress encounters a lover, not her husband, that milk disappears when a wife misbehaves in front of her husband in a manner that belittles him, and that sperm is avoided in the female quest for sexual satisfaction. So what types of exudations occur when spouses directly and actively interact with each other? The answer is, paradoxically, fluids that are not associated with one sex only.

Stomach contents

When an adulteress boards a ship to accompany her lover on a rough sea, she is not sick; however, she whom her husband asks to come with him is sick: 'She who follows her lover has a strong stomach. That other one vomits all over her husband' (*Quae moechum sequitur, stomacho valet. Illa maritum / convomit*, 6.100–1). The dutiful wife following her husband is not able to hold in her stomach contents. An adulteress is happier when following her lover, and it gives her body the strength to function properly. The spilled stomach contents of the dutiful wife tell of the unhappy state of the conjugal relationship.

Another lady behaves in an untraditional fashion, which this time results in a gross and unfeminine display: the wife for whom famished guests have been waiting and who comes back from the baths, where she was intimately taken care of by a masseur, only to guzzle in a disgusting fashion litres of wine that she immediately regurgitates in front of the appalled company and sickened husband: 'She drinks and vomits. Thus the husband feels sick and, eyes closed, holds down his bile' (*Bibit et vomit. Ergo maritus / nauseat atque oculis bilem substringit opertis*, 6.432–3).[10] She behaves in ways that run counter to her traditional gender role as the perfect housewife, but the husband manages to keep his bile down, meaning both that he manages to stay calm and that his stomach contents stay where they belong, unseen from the rest of the company. Danielle Gourevitch shows that the Roman *stomachus*, as the seat of anger, worry, and ill-being in general, 's'oppose aux idées raisonnables, au jugement, *ratio* et *animus*' (1977: 73): the husband acts righteously; thus, his stomach retains its contents and does not show his disapproval of the situation; the wife's stomach, on the contrary, empties itself, and the fact that it does so exposes her immoral behaviour.

Maurizio Bettini argues that, since there is evidence that the god Liber presided over seminal fluid in the animal kingdom, wine might be perceived as a direct parallel to semen (1995). In the aforementioned image of the gross lady, the latter could be seen as playing with semen, appropriating it, and then discarding it, having no need of a fluid that does not add anything to her status, since she already behaves like a dominant individual. Even if we do not go that far into a metaphorical interpretation that Juvenal might not have intended, it is still worth considering the connection that was understood to exist between female wine-drinking and sexuality: drinking wine could lead to adulterous behaviour, something that society was keen on curbing (Bettini 1995; Laurence 1998). Marcel Durry explains this prohibition as springing from the early notion that wine drinking had contraceptive and abortive properties, another reason to prevent women from having access to a beverage that would imperil their essential role as child bearers (1955). Nicholas Purcell reflects on the issue of female independence at a historical time when the proper management of the household was at the centre of the elite's concerns (1994). The woman misbehaving in front of her guests commits offences that taint every aspect of the ideal life she should lead: she seeks sexual pleasure outside wedlock, neglects her role as a housewife and host, publicly behaves in an indelicate and thus unfeminine way, appropriates the male prerogatives of drinking wine and dominating the scene, and endangers her reproductive functions.

Urine

Despite the fact that urine is common to both man and woman, it takes on certain gender associations when it is evacuated through the genital area. Its evacuation by women in a public setting is used by Juvenal not only to mock them but also to point to gender role issues. Here is what he writes about the female gladiator:

> Aspice quo fremitu monstratos perferat ictus
> et quanto galeae curvetur pondere, quanta
> poplitibus sedeat quam denso fascia libro,
> et ride positis scaphium cum sumitur armis.

> See how she snorts when she performs the thrusts that she has been shown,
> and how the weight of the helmet bends her, see the size of the bands—the

density of their fibre—that sit on her knees, and laugh when she resorts to the chamber pot once she has laid down her weapons.

Juvenal, *Satires* 6.261–4

The audacity of the female gladiator's untraditional behaviour is made clear by the narrative contrast between her activity in the arena, traditionally purely male, and the way she relieves herself, in a purely female fashion. The difference between the genders is signified by the sexual characteristics of the woman, the genitals symbol-ised by the chamber pot.[11] The bodily element puts the woman back into the place that she should not have left, her female realm, where her husband is still the one in charge.

Juvenal has another wife urinating in an inappropriate setting, and her out-of-bounds behaviour puts her husband not only in a shameful but also in a redundant position:

> Noctibus hic ponunt lecticas, micturiunt hic
> effigiemque deae longis siphonibus implent
> inque vices equitant ac Luna teste moventur,
> inde domos abeunt: tu calcas luce reversa
> coniugis urinam magnos visurus amicos.

> Here at night they [women] stop their litter, here they urinate and spray with long gushes the effigy of the goddess [Pudicitia], then they ride each other in turn, moving under the eye of the Moon, and then return home: you, in the morning, on your way to see your powerful friends, you tread in your wife's urine.

Juvenal, *Satires* 6.309–13[12]

The wife's distortion of the traditional performance of sexuality is an essential denial of her gender as she behaves against traditional ideals by interacting sexually with a woman rather than a man. But she also negates men's gender role since she makes her husband sexually redundant and a passive witness of her misbehaviour. Her activities here thus run counter to both genders' ideals, and both husband and wife are implicitly criticised.

Tears

For this last example of fluid we have to go back to a passage that we have encoun-tered before:

> Tum gravis illa viro, tunc orba tigride peior,
> cum simulat gemitus occulti conscia facti,
> aut odit pueros aut ficta paelice plorat
> uberibus semper lacrimis semperque paratis
> in statione sua atque expectantibus illam,
> quo iubeat manare modo. Tu credis amorem,

tu tibi tunc, uruca, places fletumque labellis
exorbes, quae scripta et quot lecture tabellas
si tibi zelotypae retegantur scrinia moechae!

She is then a nuisance to her husband, now worse than a tigress deprived of her young, when she feigns laments, consciously hiding some misdeed, or she abominates his boys or weeps over some imaginary mistress, tears always in abundance and always ready at their post, waiting for her to order how they should flow. You believe that it is love, you are then satisfied with yourself, you worm, and you dry her tears with your kisses; what letters, how many messages would you read if you were to open the drawers of the jealous adulteress!

Juvenal, *Satires* 6.270–8

This is an example of the fake tears that Juvenal hates in woman and man alike; indeed, tears are only criticised by the satirist if they are fake outward signs of sadness or guilt.[13] In this case, they are the fluid at the centre of the conjugal interaction: since the woman is suspected of being an adulteress, she cries to ward off suspicions; her husband is thus convinced of her innocence and kisses away her tears. He is thereby guilty of weakness, of being easily deceived, and of falling prey to his wife's control. While so far he was vomited upon or treading into his wife's urine in a rather passive way of interacting with her through fluids, this time he is willingly soiled.

Conclusion

As we see throughout this volume, bodily fluids are not only symptomatic of strong emotions and reactions but also trigger them. Juvenal uses them as props to move his audience; they emphasise what he says of the men and women whom he stages: the fluids, as bodily expressions of the protagonists' morals, are meant to disgust or provoke laughter, allowing the audience to experience the satirical message on another, more sensory level: the audience can feel in their guts, as it were, what Juvenal means.

Juvenal shows throughout his corpus, and in a focused way in *Satire* 6, that men and women who do not behave in accordance with traditional gender rules as they are imposed on them by social ideology imperil couple relationships. Women who humiliate their husband in public by out-of-bounds behaviour as well as men who do not act as dominant and controlling figures are criticised by the satirist; marriages cannot function harmoniously if men and women do not play, in public, ideal roles that are strictly defined by gender-segregating rules. Bodily fluids which are exuded or retained mirror the erring ways of their owners: through these, Juvenal highlights dysfunctional relationships to denounce the lack of respect for traditional gender roles and its corollary consequences on spouses' happiness and marital harmony.

Juvenalian conjugal couples are not, paradoxically, shown in the presence of each other exchanging sexed fluids or witnessing the other exuding a positive fluid. Semen, the male fluid by definition, is absent when it should flow to show husbands' traditional reproductive masculinity and dominant position; and, conversely, vaginal secretion, associated with sexual excitement, flows to point to a behaviour that does not correspond to wives' expected traditional reproductive role. There is, to put it in a nutshell, not enough procreating sperm and too much lustful vaginal secretion for Juvenal's couples to function. Milk, as a symbol of wives' traditionally most sought-after quality, child-bearing, is also conspicuous by its absence, which Juvenal highlights and criticises.

Wives are shown exuding and interacting through fluids common to both men and women that should not flow when and where they flow. The disgusting stomach contents, the shameful urine, and the fake tears are all exuded in a context where the wife behaves contrary to the ideal rules imposed on her by tradition, and concomitantly, husbands are shown as not being able to counteract by resuming the dominant position that is expected of them: they are passive and are soiled without a word, themselves behaving contrary to male traditional gender rules. Fluids that are not sexed are used by the satirist to make his point regarding the functioning of the couple: having too many of them outside the body taints the conjugal relationship and prevents harmony. Juvenal thus mocks the dysfunctional wedlock within which one spouse humiliates the other in front of guests without consequence, or within which the only way of getting in touch with one's spouse is through a puddle of urine. According to Juvenal, conjugal relationships have gone seriously wrong because gender roles are not heeded, and what better way than to say it with fluids?

Notes

1 On satire as a genre and its debated aim(s), see Freudenburg 2001, 2005; Plaza 2006; Keane 2006; Jones 2007; Rosen 2007; Braund and Osgood 2012.
2 See the Introduction to this volume, pp. 1–2 for the 'assessment of individuals' provided by Catullus through their association with specific fluids.
3 This categorisation does not take into account the fact that lactation can happen in men following hormonal imbalance; for examples and explanation of human male lactation, see e.g. Marieskind 1973.
4 *Mollissima corda / humano generi dare se natura fatetur, / quae lacrimas dedit. Haec nostri pars optima sensus*, 15.131–3. The Latin text used for quotations from Juvenal's *Saturae* is the 1992 edition by Clausen.
5 On Roman approval of exaggerated grief, see MacMullen 1980; on male and female tears as a way of expressing virtue in Latin texts in general, see Hostein 2006; Rey 2015.
6 Soranus, *Gynaecology* 2.18; Favorinus *apud* Gellius, *Attic Nights* 12.1.4–7; Musonius Rufus 3; Tacitus, *Germania* 20; Plutarch, *Cato the Elder* 20.3. For more on the issue of breastfeeding and male views on it, see Centlivres Challet 2017. See also Lawrence, and Totelin, Chapters 14 and 15, this volume.
7 All translations are my own.
8 Milk is all the more associated with women's reproductive role, since in antiquity breast-milk was associated with menstrual blood: Pedrucci 2013.
9 On ejaculated sperm as a positive, apotropaic fluid, see Parker, Chapter 11, this volume; on sperm and fertility, see Fallas, Chapter 7, this volume.
10 On the Hippocratic therapeutic cleansing use of vomiting, see Demand 2002.
11 For references about the *scaphium* being a woman's chamber pot, in opposition to the *lasanum* used by men, see Courtney 1980: 292.
12 On this scene as an 'exemplary profanation' and on the cultural value of urine, see Lentano 1995. Urine and amorous relationships are associated in recipes of potions supposed to break love bonds, from Greco-Roman times to the modern era: Griffiths 1965.
13 For instance, a man lies shamelessly about some disappeared sum of money and sheds tears while swearing on the head of his son (13.76–85, 174–5).

Bibliography

Primary sources

Clausen, W. V. (1992) *A. Persi Flacci et D. Iuni Iuvenalis saturae.* 2nd edn. Oxford: Clarendon Press.

Secondary literature

Baer, B., S. A. O. Armitage, and J. J. Boomsma (2006) 'Sperm storage induces an immunity cost in ants', *Nature* 441, 872–5.

Bellandi, F. (2003) '*Siccis . . . mamillis* (Giovenale 6, 401)', *Paideia* 58, 32–9.

Bettini, M. (1995) 'In vino stuprum', in O. Murray and M. Tecusan (eds) *In Vino Veritas.* London: British School at Rome, American Academy in Rome, Istituto Universitario Orientale Napoli, Università di Salerno, and Svenska Institutet i Rom, 224–35.

Braund, S. M. and J. Osgood (eds) (2012) *A Companion to Persius and Juvenal.* Malden: Wiley-Blackwell.

Braund, S. M. and W. J. Raschke (2002) 'Satiric grotesques in public and private: Juvenal, Dr Frankenstein, Raymond Chandler and Absolutely Fabulous', *Greece and Rome* 49(1), 62–84.

Centlivres Challet, C.-E. (2013) *Like Man, Like Woman: Roman Women, Gender Qualities and Conjugal Relationships at the Turn of the Century.* Oxford: Peter Lang.

———. (2017) 'Roman breastfeeding: Control and affect', *Arethusa* 50(3), 369–84.

Courtney, E. (1980) *A Commentary on the Satires of Juvenal.* London: The Athlone Press.

Demand, N. H. (2002) 'What is normal? Vomiting as a health measure in Hippocratic medicine', in A. Thivel and A. Zucker (eds) *Le normal et le pathologique dans la collection hippocratique: Actes du Xème colloque international hippocratique (Nice, 6–8 octobre 1999).* Nice: Publications de la Faculté des lettres, arts et sciences humaines de Nice-Sophia Antipolis, 499–508.

Duff, J. D. (1970) *D. Iunii Iuuenalis, saturae XIV, fourteen satires of Juvenal.* 3rd edn. Cambridge: Cambridge University Press.

Durry, M. (1955) 'Les femmes et le vin', *Revue des Études Latines* 33, 108–13.

Freudenburg, K. (2001) *Satires of Rome: Threatening Poses from Lucilius to Juvenal.* Cambridge: Cambridge University Press.

———. (ed.) (2005) *The Cambridge Companion to Roman Satire.* Cambridge: Cambridge University Press.

Gourevitch, D. (1977) '*Stomachus* et l'humeur', *Revue de Philologie, Littérature et Histoire Anciennes* 51, 56–74.

Griffiths, J. G. (1965) 'A modern Welsh anti-love charm with ancient antecedents', *Anthropos* 60, 108–12.

Hostein, A. (2006) '*Lacrimae principis*: les larmes du Prince devant la cité affligée', in M.-H. Quet (ed.) *La 'crise' de l'Empire romain de Marc Aurèle à Constantin: Mutations, continuités, ruptures.* Paris: Presses de l'Université Paris-Sorbonne, 211–34.

Jones, F. (2007) *Juvenal and the Satiric Genre.* London: Duckworth.

Keane, C. C. (2006) *Figuring Genre in Roman Satire.* Oxford: Oxford University Press.

King, H. (2008) 'Barbes, sang et genre: Afficher la différence dans le monde antique', in V. Dasen and J. Wilgaux (eds) *Langages et métaphores du corps dans le monde antique.* Rennes: Presses Universitaires de Rennes, 153–68.

Laurence, P. (1998) 'Ivresse et luxure féminines: Les sources classiques de Jérôme', *Latomus* 7(4), 885–99.

Lentano, M. (1995) 'Le matrone e il simulacro: Giovenale 6.303–310', *Bollettino di Studi Latini* 25(1), 74–89.

MacMullen, R. (1980) 'Romans in tears', *Classical Philology* 75, 254–5.

Marieskind, H. (1973) 'Abnormal lactation', *Journal of Tropical Pediatrics* 19(2), 123–8.

Pedrucci, G. (2013) 'Sangue mestruale e latte materno: Riflessioni e nuove proposte. Intorno all'allattamento nella Grecia antica', *Gesnerus* 70, 260–91.

Plaza, M. (2006) *The Function of Humour in Roman Verse Satire: Laughing and Lying*. Oxford: Oxford University Press.

Purcell, N. (1994) 'Women and wine in ancient Rome', in M. McDonald (ed.) *Gender, Drink and Drugs*. Oxford: Berg Publishers, 191–208.

Ramsay, G. G. (1940) *Juvenal and Persius*. 2nd edn. Cambridge, MA: Harvard University Press.

Rey, S. (2015) 'Les larmes romaines et leur portée: Une question de genre?', *Clio* 41, 243–64.

Rosen, R. M. (2007) *Making Mockery: The Poetics of Ancient Satire*. Oxford: Oxford University Press.

Rudd, N. (1991) *Juvenal, the Satires*. Oxford: Clarendon Press.

9

FLABBY FLESH AND FOETAL FORMATION

Body fluidity and foetal sex differentiation in ancient Greek medicine

Tara Mulder

According to medical texts in the Hippocratic Corpus, both male and female parents could influence the sex of the foetus at the moment of conception or during development in the uterus based on the amount of fluidity in their bodies, the makeup of their semen and menstrual fluid, and their regimens of diet and exercise. The writer of *On Regimen* (late fifth to early fourth century BCE) even proposes a spectrum of what we would today call gender, resulting from the combination, at varying levels of potency, of male and female seed in utero.[1] Then, starting with Aristotle (384–322 BCE) and finding fuller expression in the Greek texts of the second century CE, medical writers ascribed to the female parent more responsibility for foetal sex differentiation, due largely to the paired ideas that female parents contributed more constructive matter (menstrual fluid) to the generative process and that they had more influence over the developing foetal body for a longer period of time than the male parent.

Ancient medical writers indicate that people in the ancient world had control over the fluidity of their bodies, changeable by means of diet and exercise. Since foetal sex was seen as largely determined by the temperature and fluidity of the parents' bodies, this somatic mutability went some way towards explaining culpability for the sex of the foetus. I say 'culpability' because, as Rebecca Fallas shows in Chapter 7 in this volume, there was a correlation in these ancient Greek medical texts between the sort of regimen that would lead to both increased fertility *and* to the likelihood of conceiving a male child. Thus, we see two main themes emerge from these texts: first, that to produce a female foetus was to have achieved a poor, if teleologically necessary, reproductive outcome, and second that, because of their protracted contact with the developing foetuses, female parents were inherently more culpable for this outcome.

Principles of foetal sex differentiation

There were three basic principles of foetal sex differentiation in the ancient Greek medical texts (for the theories of generation more generally, see Julie Laskaris and Rebecca Flemming, Chapters 6 and 10 in this volume, respectively, as well as Connell 2016: 107–8). One theory held that maleness was associated with the right side of the body and femaleness with the left; another that males were associated with hot, dry things and females with cold, wet things; and the third, that there was male and female

DOI: 10.4324/9780429438974-13

seed which could come from both parents and whichever sex of seed gained mastery in the process of generation would determine the foetal sex. I paint with broad strokes here, but later in this chapter I will unpack some of the nuances of each theory. These theories do not seem to have been mutually exclusive, perhaps because understanding foetal sex differentiation was almost certainly an elusive undertaking in the ancient world. Even today, we know that there are many factors that interact in complicated ways which cause foetal bodies to present as female, male, or intersex, including chromosomes, hormones, and environmental factors. But particular medical writers in the ancient world did have their preferred explanations.

Males on the right, females on the left

In *On Superfetation* (mid-fourth century BCE), the writer advises a man who wishes to engender a female child to bind up his right testicle as tightly as he can so as to prevent sperm from coming from that testicle. If he wishes to engender a male child, he should bind up the left testicle (*On Superfetation* 31). He binds in this way because male sperm was thought to come from the man's right testicle and female sperm from his left testicle. While this advice is ostensibly value neutral, the author of *Epidemics 2* (c. 400 BCE) connects the right side of the body more broadly to maleness and strength and the left side to femaleness and weakness, explaining that the right side of the body is stronger and more forceful because males are engendered on the right side of the body and vice versa.[2] Similarly, the writers of *Epidemics 6* (c. 400 BCE), *Aphorisms* (c. 400 BCE), and *Prorrhetic 2* (end of the fifth century BCE) explain that a male foetus develops on the right side, or in the right chamber, of the uterus, and a female foetus develops on the left side or chamber (*Epidemics* 6.2.15, edition Smith 1994; *Aphorisms* 5.48, edition Jones 1931; *Prorrhetic* 2.24, edition Potter 1995).

The association of the male with the left and the female with the right side of the body also appears in the fragments of Parmenides (late sixth to mid-fifth century BCE) and Anaxagoras (510–428 BCE) (Parmenides fr. B17; Anaxagoras fr. A107). Aristotle, who believed that the male represented 'perfect' or completed generation, while the female was 'imperfect,' but necessary, tells us that 'the right is naturally better than the left.'[3] And Galen (129–after 199 CE) too thought that male foetuses came from the right side of the body and female foetuses from the left.[4] In *On the Usefulness of the Parts of the Body*, he develops his theory that male foetuses form within the hotter, drier, right side of the uterus, while female foetuses take shape in the colder, wetter, left side of the uterus (*Usefulness of the Parts* 14.2; trans. May 1968).

Males are hot and dry; Females are cold and wet

We can see from Galen's theory in *On the Usefulness of the Parts of the Body* that for him the right and left sides of the body were associated also with hot/dry and cold/wet, respectively. For Galen, femaleness was fundamentally colder and wetter than maleness and so the weaker, wetter, colder left side or chamber of the uterus was the one in which a female foetus was most likely to develop. The writer of *On Regimen* (late fifth or early fourth century BCE), too, explains that femaleness, with its affinity for water, arises from nutrients and regimens that are cold, moist and gentle, while maleness, with an affinity for fire, grows from nutrients and regimens that are dry and warm. So, to beget a girl, both male and female parents must adhere to a watery regimen and to beget a boy they must adhere to a fiery one (*On Regimen* 1.27). It takes the seed from the male parent and female parent coming together in the womb to spark generation, while the sex of the foetus is determined by the fluidity of the sperm and, especially, the uterine environment:

> Now if the fire fall in a dry place, it is set in motion, if it also master the water emitted with it, and therefrom it grows, so that it is not quenched by the onrushing flood, but receives the advancing water and solidifies it on to what is there already. But if it fall into a moist place, immediately from the first it is quenched and dissolves into the lesser rank. On one day in each month it can solidify, and master the advancing parts, and that only if it happen that parts are emitted from both parents together in one place.
>
> *On Regimen* 1.27; trans. Jones 1931

In this model, men with watery bodies and sperm contribute weak semen to generation, semen that can be overpowered by the seed coming from the female parent and the fluidity of the uterine environment. It is the fluidity of the generative matter and in the uterine environment that allows one or the other parent's seed to overpower the other's, with the understanding that male comes from male and female from female. Femaleness arises from what is 'cold, moist, and gentle', while maleness arises from what is 'dry and warm', including both the regimen of the parents and the environment of generation. Here, responsibility for foetal sex lies with both parents; it is the combination of fiery male sperm and a relatively dry uterus that yields the male. The moistness of the uterine environment, though, seems to be the determining factor in femaleness: 'if it [the male semen] fall into a moist place [the uterus], immediately from the first it is quenched and dissolves into the lesser rank [female].'

Generating a male foetus seems to have been a tricky matter. The author of *On Regimen* writes:

> On one day in each month it [the male seed] can solidify and master the advancing parts [the female seed and menstrual fluid?], and that only if it happen that parts are emitted from both parents together in one place.
>
> *On Regimen* 1.27; trans. Jones 1931

The text of *On Superfetation* may offer some help in interpreting this passage. It advises a man to have sex with his wife when her menses are ceasing or have stopped if he wants to beget a male child. If he wishes to beget a female child, they should have intercourse when the menses are 'flowing in their greatest amount' (*On Superfetation* 31; trans. Potter 2010). The 'one day in each month' when the man's semen 'can solidify and master' the woman's seed and create a female foetus is at the end of a woman's menstrual period, whereas if a man ejaculates into a woman's uterus during her period, the fluid from the menstrual flow will dampen the male semen and prevent it from reaching its full completion in a male foetus.

Male seed vs. female seed

The last factor in foetal sex differentiation, according to the ancient Greek medical writers, was the contribution of sexed seed by each procreating parent. In the first place, for successful generation to occur at all, the male and female seed had to be released and meet in the uterus at the right moment. Secondly, each parent contributed seed that mixed with the other parent's seed to create the particular physical attributes of that foetus, including its sex. In *On Generation* (c. 430–420 BCE), the writer explains 'It can be inferred from the visible facts that in both a woman and a man there exist both female and male seed' (*On Generation* 7; trans. Potter 2012). These 'visible facts' are the genetic resemblances between children and their parents, as well as the observation that men and women who procreate with different partners sometimes produce differently sexed offspring with those different partners. The author of *On Generation* earlier explains, 'As the male sex is stronger than the female, it must follow that it is engendered from stronger semen. The matter is like this: if stronger semen comes from both parents, a male is engendered, if weaker, a female' (*On Generation* 6; trans. Potter 2012). He adds that the amount of each type of semen impacts the foetal sex: if there is more strong semen, then a male results, while if there is a greater amount of weaker semen, a female results.[5]

According to the writer of *On Regimen*, also, the male and female partners could each contribute male or female seed; the foetal sex was determined by the strength of each partner's seed relative to the other, as well as by whether the sex of the seed that gained mastery matched the sex of the parent from which it came (i.e. male seed from the male partner was more influencing than male seed from the female partner, and so too female seed from the female partner was more influencing than female seed from the male partner). The writer of *On Regimen*, therefore, imagines six possible combinations of male and female seed and six different sexes that could result from these combinations:

> Now if the bodies secreted from both happen to be male, they grow up to the limit of the available matter, and the babies become men brilliant in soul and strong in body, unless they be harmed by their subsequent diet. If the secretion from the man be male and that of the woman female, should the male gain the mastery, the weaker soul combines with the stronger, since there is nothing more congenial present to which it can go. For the small goes to the greater and the greater to the less, and united they master the available matter. The male body grows, but the female body decreases into another part. And these, while less brilliant than the former, nevertheless, as the male from the man won the mastery, they turn out brave, and have rightly this name. But if male be secreted from the woman but female from the man, and the male get the mastery, it grows just as in the former case, while the female diminishes. These turn out hermaphrodites ('men-women') and are correctly so called. These three kinds of men are born, but the degree of manliness depends upon the blending of the parts of water, upon nourishment, education and habits.
>
> *On Regimen* 1.28; trans. Jones 1931

And the three types of women:

> In like manner the female also is generated. If the secretion of both parents be female, the offspring prove female and fair, both to the highest degree. But if the woman's secretion be female and the man's male, and the female gain the mastery, the girls are bolder than the preceding, but nevertheless they too are modest. But if the man's secretion be female, and the woman's male, and the female gain the mastery, growth takes place after the same fashion, but the girls prove more daring than the preceding, and are named mannish.
>
> On Regimen. 1.29; trans. Jones 1931

Represented another way (Table 9.1):

Table 9.1 Gender possibilities in On Regimen 1.28–9

Man's Seed	Woman's Seed	Dominate Seed	Result
Male	Male	Man (male)	Men brilliant in soul and strong in body
Male	Female	Man (male)	Men less brilliant, but still manly
Female	Male	Man (female)	Men-women (androgunoi)
Female	Female	Woman (female)	Women most womanly and most well-formed
Male	Female	Woman (female)	Women bolder, but still modest
Female	Male	Woman (male)	Mannish women (andreiai)

What we see in this pair of passages is something akin to what we would call gender today, ranging along a spectrum from very masculine men to very feminine women, with four other possible variations in between. The idea of a 'gender' that is independent of sex and socially constructed or performed is a modern concept that originated in the twentieth century (Holmes 2012: 3). But there are many places in the ancient world where we find something like gender, and the way that a particular body presented and performed its sex, being discussed (for sources on sex and gender in antiquity see Holmes 2012).

This passage also seems to be giving a biological explanation for the generation of intersex bodies (e.g. the androgunē) and for varying degrees of maleness/masculinity and femaleness/femininity. However, there does not seem to be room in this model for parents that fall outside of the sex binary. The parents—the ones providing the seed in this generative scenario—are either male or female, and they supply either male or female seed.[6] We do not see mention of an androgunos parent or an andreios parent, for instance, or how their sex would affect the sex of their foetus.

What we do see in the first passage, though, is the idea that regimen and somatic fluidity have a direct impact on 'degree of manliness'. Bodily fluids make for body fluidity, as it were. According to this writer, there are two ways to form a male foetus— when male sperm from the male parent masters male sperm from the female parent and when male sperm from the male parent masters female sperm from the female parent. In both instances, though, the degree of manliness that the adult human can achieve is also determined by 'the blending of the parts of water, upon nourishment, education and habits'. He is able to 'be harmed by his subsequent [post-uterine] diet'.

That is, he can fail to live up to his manly potential if his regimen is too soft, cold, and fluid. In this way, although there is not an implication here that men can slide into being women and vice versa, there is a sense that maleness is unstable. The generative moment, if it is ideal (i.e. male sperm from both parents), supplies the potential for that foetus to grow 'up to the limit of the available matter' and 'become men brilliant in soul and strong in body'. But it is only potential. The male child and adult must adhere to a regimen that keeps his body hot, dry, and strong.

Temperature, fluidity, and foetal sex differentiation in Aristotle

Similarly to the Hippocratic writers, Aristotle wrote about male and female seed meeting in the uterus to form the body of the foetus, through a process of concoction, of mixing and heating. He thought that the male parent contributed generative semen with a vital spark, while the female parent contributed generative and nutritive semen in the form of menstrual fluid.[7] In *Generation of Animals*, Aristotle imagined that foetal sex was determined by how much fluidity and heat were present at the generative moment, from both the male and female semen.[8] If the male semen was able to gain mastery over the female semen and bring it over to itself, the result would be a male foetus. If the male semen, conversely, was mastered by the female semen, then it turned either into its opposite, a female foetus, or was miscarried (*Generation of Animals* 766b14–17). For Aristotle, it was the degree of moistness of the generative materials which determined the capability of the male semen to gain mastery in the womb.

If the male generative substance was present in the uterus in the right quantity and if there was enough dry heat, then the generative material would 'set' or 'concoct' into the form of a male foetus. But if any part of the mixture was too watery, then the result was an imperfect (i.e. incomplete) female foetus. He uses a metaphor of rennet setting milk to make cheese—the proportion of semen to menstrual fluid determines the solidity (sex) of the foetus, just as the ratio of rennet to milk determines the solidity of cheese (*Generation of Animals* 739b22–34; cf. 772a22–5, 729a10–15). Hyperfluidity in generation could be caused by the parents' excessively moist bodies, excessive fluidity of the menstrual fluid, or watery male sperm. He thought that parents with more fluid and feminine bodies were more likely to produce female foetuses and that men with more fluid sperm, due to a deficiency in natural heat, were more likely to produce female foetuses (*Generation of Animals* 766b28–34).[9] Environmental factors, too, could contribute to somatic fluidity and, consequently, to foetal sex: dry winds from the north produced male offspring, while wet winds from the south produced female offspring. The waning moon, associated with the menstrual cycle for women, heralded a colder, more fluid period, ripe for the generation of females (*Generation of Animals* 766b35–767a14).

Femaleness, fluidity, and disability

Although Aristotle is famous for calling the male of the species 'perfect' and the female 'imperfect' and labelling the female a 'deformed' male; nonetheless, he thought that every individual had reached its developmental telos once it became either male or female.[10] While the generation of a female foetus meant that there was something less

150

than complete about the male and female semen—a deficiency of heat or an excess of fluidity—nonetheless, the female was a necessary—and therefore both an ideal and expected—outcome of generation.[11] When Aristotle says that the female has an 'inability' or, literally, 'disability' in opposition to the male 'ability', he is speaking of physical and generative capacity. While Aristotle surely intends us to understand here a medical model of disability, wherein the female lacks a specific physical ability of the male (i.e. the vital spark provided by the male semen that allows generation to take place), he perhaps inadvertently reveals a truth about femaleness in antiquity, understood through a social model of disability. To be female in the ancient world was to be socially inferior and, functionally, disabled.[12]

We should not, however, imagine that Aristotle is not being sexist here. It is his adherence, and the adherence of the other ancient medical writers, to sexist views about femaleness in their explanations of generation that perpetuates the social disability of women in the ancient world—a social disability that is predicated on fundamental beliefs about female physical and mental inferiority. That inferiority is directly tied to the fluidity of female bodies. The works of Aristotle and the other medical and scientific writers from ancient Greece are rooted in cultural and biological notions of the superiority of the hot, dry male physique and intellect, notions that permeated the literature and philosophy of the ancient Greeks. As Ann Carson wrote in 1990, 'Heraklitos told us, the "dry soul is wisest and best"; Homer said, "Zeus has a dry mind"; and Diogenes of Apollonia said, "moisture hinders intelligence"' (Carson 1990: 137–8). Fluid bodies were thought to yield fluid minds which held fluid thoughts. Women were slippery, cold, and leaky, both physically and intellectually.

In his *Gynaecology*, Soranus (first and second centuries CE) does not give an explicit theory of foetal sex differentiation, but his general advice to couples on how to conceive aligns with the advice from the Hippocratic writers of *On Regimen* and *On Superfetation* on how to generate a male child. He advises couples who want to become pregnant to have intercourse when the menses are ending, as this will give the seed the greatest chance of catching onto the uterine walls (*Gynaecology* 1.36, translation Temkin 1956). He also refers to the dangers of becoming pregnant when the body or uterus is in a state of excess fluidity (*Gynaecology.* 1.39). Women, he advises, are fit to conceive when they are between the ages of 15 and 40, 'if they are not mannish, compact, and over-sturdy, or too flabby and very moist' (*Gynaecology* 1.34). He does not say that the result of this suboptimal generation will be a female child, but he does caution that a foetus conceived in too wet an environment will be ill or deformed, drawing an analogy with a farmer who avoids sowing his seed on a field that is too soggy (*Gynaecology.* 1.39).

Galen was more explicit than Aristotle and Soranus in espousing the idea that a female was an imperfect male due to her wet and cold nature. Rather than see her as a necessary telos of generation, for Galen the female was a failed male.[13] He compared her to a blind mole, analogising the incomplete female genitals to the mole's incomplete eyes (*On Seed* 2.5.60; *Usefulness of the Parts* 14.6; cf. King 2013: 36). Because of the weakness of heat in her uterus, her genitals were not able to pop to the outside of her body, as they would with a male foetus. He does clarify that this incompleteness was deliberate on the part of the divine creator, for the advantage of the whole human race—by virtue of being colder and wetter than men, women are able to supply nourishment to the developing foetus (*On Seed* 2.5.69).

Flabby flesh and foetal formation

According to the ancient Greek medical writers, then, the flabbiness and fluidity of the parents had a direct influence on the generative outcome, including the sex of the foetus. Apart from the direct actions that the parents could take to influence foetal sex (e.g. tying up one testicle or the other, as in *On Superfetation*), the parents could also influence the generative outcome through changes in their regimen or the time at which they had intercourse. All of these choices and changes were based on bodily fluidity. The tying off of the testicles had to do with the relative temperature and fluidity of the sperm in each testicle—the left testicle was thought to house colder, more fluid sperm, while the right testicle was thought to hold hotter, more concentrated and concocted sperm. The time of intercourse had to do with how moist the uterine environment was at different points in the month. It would be driest at the end of the menstrual period, and thus suitable for generating a male foetus, while close to or during the time of menstruation it would be more fluid and suitable for generating a female foetus. Finally, there are indications in the texts that we have looked at so far that not only did bodily fleshiness and fluidity have an impact on foetal sex differentiation, but that it was possible for both parents to change their somatic fluidity through regimen. In the texts from the second century CE—those of Galen and Soranus—the female body had the most influence on the developing foetus. Since women were able to change the fluidity of their bodies through their diet and exercise, it appears that women were culpable if a foetus came out deformed or female.

In *On the Usefulness of the Parts of the Body*, Galen situates primary responsibility for the development of the foetus, including foetal sex, with the female parent:

> Hence, since there are two principles for the generation of males, the right side of the uterus in the female and the right testis in the males, and since generally the uterus is the better able of the two to make the foetus like unto itself because it is associated with it for a longer time, there is good reason for the fact that for the most part the male foetuses are found there and the female foetuses in the left (side of the) uterus.
> Galen, *On the Usefulness of the Parts of the Body* 14.7; trans. May 1968

Although both the male testes and the uterus have an influence on the sex of the foetus, the uterus has the most influence because it is in contact with the developing foetus for a longer time.[14] The mother digests (concocts) the nutritive matter that forms the flesh of the developing foetus, and so the same faculties of concoction that form her flesh also form the foetal flesh. If she herself is flabby and fluid, then the foetal matter will be as well, increasing the likelihood of it turning out female.

In some of the Hippocratic writers and in Soranus we see expressions of the idea that men and women had some control over their bodily composition. These changes in regimen were not strong enough to alter the sex of either parent, but within each sex there was the possibility of some somatic variation. According to the writer of *On the Nature of the Child* (c. 430–420 BCE), women could have either dense or loose-textured flesh, the difference being in degrees if not kind (Hippocrates *On the Nature of the Child* 10, edition Potter 2012). In *On Diseases of Women* 1 (late fifth to early fourth century BCE), women who have never given birth have more solid and denser bodies than women who have given birth (*On Diseases of Women* 1.1, edition Potter 2018).

Central to this understanding of somatic variation is a particular notion about the difference between male and female bodies. In the Hippocratic gynaecological texts, the difference between male and female was not thought to reside solely in their genitalia and secondary sex characteristics but in the very nature of their flesh. Women were thought to have loose, spongy flesh, corresponding with their wet and cold natures, while men were thought to have dense, dry flesh, corresponding with their hot and dry natures; men were like woven cloth, women like fleece (King 1998: 29). With their fleece-like flesh, women were capable of taking in and holding more fluid than men.

But even with these fundamental differences between male and female bodies, there was still the notion that ancient men and women could control how fleshy their bodies were. The coldness and wetness of the sperm at the moment of generation, and then the temperature and fluidity of the uterine environment, influenced a person's sex. Because ancient women were seen as fleshier, wetter, colder, and spongier than men, it was thought that they initially came from a wetter, colder, uterine environment. Finally, ancient people had control over their bodily form and composition, and to some extent what we might term their 'gender', i.e. the degree to which they presented as masculine or feminine, through their regimens. The looser, spongier bodies of women were also explained by their (supposed) inactive and frivolous regimens. A woman could present differently along a gendered spectrum based on her daily regimen. If she exercised excessively, she could make herself more manly and compact; if she was idle and ate many rich foods, she could veer too far into a feminine direction and make herself overly flabby and fleshy.

Soranus tells us that women who are idle and inactive become fat and ill-proportioned and bring about their own menstrual difficulties. He attributes amenorrhea to causes that are either pathological (diseases of the uterus, the whole body, or both), or physiological (age, pregnancy, natural barrenness, or athleticism), and he advises physicians to medically treat women who are barren due to gymnastic training or vigorous vocal exercise—activities that can make a body more mannish. In these cases, the doctor should endeavour to make the women restrict their physical activity in order to make their bodies more feminine, and thus more capable of bearing children (*Gynaecology* 3.9). While overly mannish female bodies could prohibit conception altogether, overly fleshy and fluid bodies could yield unsatisfactory generative outcomes: poor conception and foetal development, and a more difficult time in labour (cf. *On Superfetation* 21: 'It is stated that a woman who is unnaturally plump will not conceive'). He reports that Diocles (fourth century BCE), in his gynaecology, had attributed difficult labour in part to women who are very moist and warm, while Cleophantus (third century BCE) attributed difficult labour in part to women who were high strung or lazy and idle, since idleness causes difficult labour while exercise promotes easy labour and enhances the well-being of the foetus (*Gynaecology* 4.1).

It was thought that women in the ancient world could control the composition of their bodies—their fleshiness and fluidity—through their regimens. The ancient physician was supposed to counsel a woman to order her regimen such that she had the correct bodily composition to bring about the best generative outcome. Because a woman would have an easier time in pregnancy and delivery with a male foetus, this advice included efforts to ensure that she produced a male foetus. Pregnant people in the ancient world were also responsible for the harm that they could cause to their developing foetuses from the food and drink that they ingested, the physical objects they gazed at while pregnant or while having sex, and even the images that they conjured up in their own minds, because all of these elements were thought to affect the developing foetus.[15] In addition, they were responsible for the composition and moisture content of their bodies and the impact that they could have on their own fertility and on the foetal development. Depending on the result of their pregnancies, it was entirely possible that pregnant parents could be held responsible for diet and regimen choices that left them flabby, moist, and 'incapable,' and even for producing a female foetus.

The notion that excessive wetness during conception led to suboptimal generation, including the production of malformed or female foetuses, may have also had implications for female sexual pleasure in the ancient world. Female bodies generally produce a vaginal lubricant when they are sexually aroused. Within a heterosexual framework, this lubricant allows for the smooth penetration of the penis into the vagina and guides the male sperm along its path to the female ovum. But if in the ancient world it was thought that excessive bodily fluidity and excess moisture during sexual intercourse were detrimental to healthy conception (and inimical to the generation of a male foetus), then it is possible that the normal responses of the aroused female body were seen as incompatible with optimum generation. Perhaps a drier vagina and vulva were seen as necessary elements for the production of a male foetus. This is conjecture on my part. Figuring out how female sexual pleasure functioned in the ancient world—and how much anyone cared about it—is not an easy task.[16] What we do know is that the ancient Greek medical writers believed that a woman could not become pregnant if she did not experience pleasure during sex. But it is likely that this idea was used more often to show that a woman *had* experienced pleasure if she became pregnant after the fact, rather than influencing anyone to concentrate on facilitating a woman's sexual pleasure so that she would become pregnant (Gellar-Goad 2014). Further, a woman's pleasure during sex was thought to have the same duration as her male partners; her pleasure, we are told, ended when he orgasmed.[17]

Conclusion

According to ancient medical theories, the economy of fluids in the body was able to influence sex and gender at a few points in the ancient life cycle. At the moment of generation, the fluidity of the parents' bodies was a factor in both the quality (and even 'sex') of the sperm that they secreted, and thus was determinative of the foetal sex. The fluidity of these generative substances, including the male and female sperm and, if it was seen as a separate substance, the menstrual fluid, also influenced the foetal sex, with wetter, colder fluids lacking the ability to concoct or set firmly, yielding a female foetus. The temperature and fluidity of the uterine environment could also impact foetal sex—the colder, wetter left side of the womb was thought to be conducive to the development of female foetuses, so sperm deposited there was bound to come out female. Finally, throughout an individual's lifetime, the regimen that they followed could determine the degree to which they were able to achieve the teleological goals of their sex—something akin to what we call gender today. Ultimately, flabby flesh was fluid flesh; it was flesh that had retained excess moisture from its environs and regimen and thus had less-concocted, watery material to supply

154

to the generative process. What we have seen in this chapter is that if sex is not quite fluid in the ancient world, something like gender was. The regimens kept by people in the ancient world could determine the relative masculinity or femininity of their bodies and it could also influence their ability to procreate successfully. Parents with flabby, fleshy, fluid bodies might produce foetuses that were malformed, disabled, and female, since the generative material they supplied was likewise cold, watery, and even impotent.

These ancient Greek medical theories about foetal sex differentiation and about the impact of the parents' body fluidity on foetal sex continued well into the Medieval period. In the Medieval *Trotula*, a woman has the most generative success if she is physically balanced and moderately moist—in such a state she can be fertile up to age 60 or 65. If she is too thin, her fertility will only last until age 50, and if she is on the fatter side, she will only menstruate until age 35 (*Trotula* 4, translation Green 2002). Similarly, women can experience menstrual and generative dysregulation if they eat foods that are too rich or if they exercise excessively (*Trotula* 6), which is reminiscent of Soranus' precautions. In other chapters on reception in this volume, Anastasia Stylianou, Caroline Spearing, and Helen King (Chapters 22, 23, and 24, respectively) show how many of the ideas about fluids and fluidity from the ancient world are ubiquitous in the Medieval and Early Modern periods.

Today, we believe that foetal sex is determined by chromosomes in the male sperm (though even this 'fact' is becoming doubtful under scrutiny), but we attribute most other facets of foetal development to the contact between mother and baby while the foetus is in the uterus—so much so that, while we have studies on the effects of maternal drug and alcohol consumption on foetal pathology, we do not know much about the effects of paternal substance consumption and foetal formation. Like the ancient Greek medical writers before us, we tend to assume that the female parent, due to her more extended contact with the foetus, has a much greater impact on its development and formation.

Notes

1 Dates for the Hippocratic texts are taken from Craik 2015.
2 *Epidemics* 2.6.15–16; in *Epidemics* 6.4.21 men can predict whether they will produce male or female children based on which testicle dropped first at puberty, cf. Dean-Jones 1994: 28. On the weakness of females and the strength of males, see Flemming (Chapter 10) in this volume.
3 Aristotle, *Progression of Animals* 706a20–1. On Aristotle's linking of the male and the right side to the good and the hot and the female and the left side to the bad and the cold, see *Metaphysics* 986a23-b5 and *Parts of Animals* 670b17–22; cf. Dean-Jones 1994: 44–5, including some dispute among the Presocratics about whether women were hotter or colder than men, based on the fact that they menstruated.
4 On Galen's theories of foetal sex differentiation, see Flemming 2000: 205–6. She points out that for Galen, the inability of a woman to conceive could be attributed to 'coldness, fattiness, or another malformation', as could her generation of a malformed foetus. cf. Flemming 2000: 309–10.
5 On the strong and weak seed in *On Generation*, see Flemming (Chapter 10) in this volume.
6 In her discussion of Phaëthusa and Nanno, two female patients in the *Epidemics* ([Hippocrates] *Epidemics* 6.8.32) who exhibit signs of masculinity (hairiness, including the growth of facial hair, amenorrhea), Brooke Holmes points out that, while sex and sexed bodies seem to be fluid in this story, this slide into manliness exhibited by the two women is more

of a pathology than a change of sex. They remain women; the role of the doctor is to make them function as women again (2012: 14–17).

7 In this sense, Aristotle differed from many of the writers of the Hippocratic Corpus, who seem to have envisioned three separate contributions to generation: male semen, female semen, and female menstrual fluid. Via Connell's analysis (2016: 99), it was important for Aristotle that the male semen provide this vital spark; otherwise, there would be no purpose to the male contribution to generation and the female parent would be able to procreate alone.

8 The standard interpretation, which Connell argues against, is that the male parent contributed active semen, while the female parent contributed inert matter to generation. Connell 2016, conversely, argues that according to Aristotle's theory of generation, the female matter was also active semen that had a direct impact on the development of the foetal body and soul—in regard to these ideas see especially 122, 127–32, 141, 146–7. In her chapter in this volume (Chapter 10), Flemming also picks up on this idea that Aristotle's female seed is more than just inert matter. It is not so much that Aristotle has a one (male) seed theory, she argues, but that the two seeds (male and female) are different, to some degree, in kind and function. What we must do, she suggests, is revisit and re-evaluate how we, and the ancient writers, define 'seed'.

9 See also Flemming in this volume (Chapter 10) on how watery regimen leads to the generation of female offspring in Aristotle's embryology.

10 *Generation of Animals* 737a24–8. Mayhew argues that Aristotle's' description of the female as a deformed (or 'mutilated') male is connected to his observations of similarities between women and male eunuchs (2004: 54–68).

11 Connell 2016: 118: 'For Galen, the female animal is literally unfinished—*a-teles* (*On Seed* II.5, 60). Aristotle, in contrast, very seldom uses this term to describe a female animal and, in most instances, considers an animal to be complete (*teleiothen, teleion*) once it has become either sex (*GA* 715a21, 737b11–12)'.

12 Connell 2016: 118; see also Aristotle, *Generation of Animals* 775a12–21 on the female state as a deformity according to Aristotle. The social model of disability holds that people are abled or disabled relative to their physical and social environment and not inherently in their bodies. On this concept see Shakespeare 2006; Adams et al. 2015: 8–9.

13 Connell 2016: 119, 'Both Aristotle and Galen assume that the female produces a weaker contribution to generation and that the female body suffers a related infirmity'.

14 On the role of the uterus in foetal development see Flemming, this volume, Chapter 10.

15 On the diet and regimen of the pregnant woman, see Soranus, *Gynaecology* 1.36.2 and 1.39.3. On the effect of sights and visualisations on the developing foetus ('ideoplasty'), see Soranus, *Gynaecology* 1.39.1; also Pliny, *Natural History* 7.52; Plotinus, *Enneads* 2.3 (52). 14.27–32; Porphyry, *To Gaurum*, 14.27–32; Heliodorus, *Aethiopica* 4.119; cf. Wilberding 2008: 421 for further examples of ideoplasty in ancient literature.

16 On this topic, see Halperin et al. 1990, especially essay by Carson 1990: 135–270; Hubbard 2014; Masterson et al. 2015. James (2012: 81–3) shows how in the Cologne Fragment from Archilochus, the speaker expresses that excessive sexual activity would produce defective children. For sexual pleasure in the Hippocratic *On the Nature of the Child*, see Flemming's Chapter 10, in this volume.

17 In the Hippocratic text *On the Nature of the Child* 4, a woman who is eager for intercourse might experience pleasure before her male partner, but one who is less eager for intercourse will have her sexual pleasure end when her male partner's does. In *On Generation of Animals*, female sexual pleasures lasts the same length of time as male sexual pleasure. Aristotle, conversely, did not think that female sexual pleasure was necessary for generation. See Connell (2016: 112–14) for a discussion of these theories. While some scholars have thought that Aristotle is more sexist than the Hippocratic writer because he dismisses female pleasure as a necessary component to generation, Connell points out that the theory given in *On Generation of Animals* shows disregard for female sexual pleasure because the writer analogises it to male sexual pleasure.

Bibliography

Primary sources

Green, M. H. (2002) *The Trotula: An English Translation of the Medieval Compendium of Women's Medicine*. Philadelphia, PA: University of Pennsylvania Press.

Jones, W. H. S. (1931) *Hippocrates*, vol. IV. Cambridge, MA: Harvard University Press.

May, M. T. (1968) *Galen on the Usefulness of the Parts of the Body*. Ithaca: Cornell University Press.

Potter, P. (1995) *Hippocrates*, vol. VIII. Cambridge, MA: Harvard University Press.

Potter, P. (2010) *Hippocrates,* vol. IX. Cambridge, MA: Harvard University Press.

———. (2012) *Hippocrates,* vol. X. Cambridge, MA: Harvard University Press.

———. (2018) *Hippocrates,* vol. XI. Cambridge, MA: Harvard University Press.

Smith, W. D. (1994) *Hippocrates,* vol. VII. Cambridge, MA: Harvard University Press.

Temkin, O. (1956) *Soranus' Gynecology*. Baltimore, MD: Johns Hopkins University Press.

Secondary literature

Adams, R., B. Reiss, and D. Serlin (eds) (2015) *Keywords for Disability Studies*. New York: New York University Press.

Carson, A. (1990) 'Putting her in her place: Women, dirt, and desire', in D. M. Halperin et al. (eds) *Before Sexuality: The Construction of the Erotic Experience in the Ancient Greek World*. Princeton, NJ: Princeton University Press, 135–70.

Connell, S. M. (2016) *Aristotle on Female Animals*. Cambridge: Cambridge University Press.

Craik, E. M. (2015) *The 'Hippocratic' Corpus: Content and Context*. London: Routledge.

Dean-Jones, L. (1994) *Women's Bodies in Classical Greek Science*. Oxford: Oxford University Press.

Flemming, R. (2000) *Medicine and the Making of Roman Women: Gender, Nature, and Authority from Celsus to Galen*. Oxford: Oxford University Press.

Gellar-Goad, T. (2014) Todd Akin, the Greek doctor Soranus, and "legitimate rape" [Blog] *Society for Classical Studies*. Available at: https://classicalstudies.org/scs-blog/tedgellargoad/todd-akin-greek-doctor-soranus-and-legitimate-rape (Accessed February 2021).

Halperin, D. M., J. J. Winkler, and F. I. Zeitlin (1990) *Before Sexuality: The Construction of Erotic Experience in the Ancient Greek World*. Princeton, NJ: Princeton University Press.

Holmes, B. (2012) *Gender: Antiquity and its Legacy*. London: I.B. Tauris.

Hubbard, T. K. (ed) (2014) *A Companion to Greek and Roman Sexualities*. West Sussex: Wiley-Blackwell.

James, S. L. (2012) 'Case study II: Sex and the single girl: The Cologne fragment of Archilochus', in S. L. James and S. Dillon (eds) *A Companion to Women in the Ancient World*. West Sussex: Wiley-Blackwell, 81–3.

King, H. (1998) *Hippocrates' Woman: Reading the Female Body in Ancient Greece*. London: Routledge.

———. (2013) *The One-Sex Body on Trial: The Classical and Early Modern Evidence*. Farnham: Ashgate.

Masterson, M., N. Sorkin Rabinowitz, and J. Robson (eds) (2015) *Sex in Antiquity: Exploring Gender and Sexuality in the Ancient World*. London: Routledge.

Mayhew, R. (2004) *The Female in Aristotle's Biology: Reason or Rationalization*. Chicago: University of Chicago Press.

Shakespeare, T. (2006) 'The social model of disability', in L. J. Davis (ed.) *The Disability Studies Reader*. New York: Routledge, 197–204.

Wilberding, J. (2008) 'Porphyry and Plotinus on the seed', *Phronesis* 53(4/5), 406–32.

10

ONE-SEED, TWO-SEED, THREE-SEED? REASSESSING THE FLUID ECONOMY OF ANCIENT GENERATION*

Rebecca Flemming

Seed—*sperma* in Greek (or *semen* in Latin)—was a crucial bodily fluid in the ancient world, generally agreed to be the basic stuff of generation, the substance through which new human beings were produced. The word was shared with agriculture (Totelin 2018). From the beginning it designated both plant seeds and a wider realm of small things with the power to initiate larger processes—like the sparks of fires and ideas—and it always ran alongside nouns derived from the verb *gennaō*, 'to create/ engender', such as *gonē* and *gonos*, which could signify both the procreative materials and what was generated, that is offspring, within a wider semantic field of begetting and descent. It was also the subject of lively debate from the time the first sustained investigations into the whole business of 'coming-to-be'—*genesis*—itself were launched by the Presocratic philosophers in sixth-century BCE Ionia onwards. There were differences over the origin, substance, nature, and number of the seed. Did it come from the brain and marrow, the blood, or all of the parts of the body, for instance? What was its physical, causal, and conceptual relationship to the offspring produced? And did 'females emit seed too?'[1]

It was the first two questions, concerning origin and substance (broadly speaking), that dominated Erna Lesky's pioneering study of ancient embryology—*Die Zeugungs- und Vererbungslehren der Antike und ihr Nachwirken* (1951)—but since then the final query has taken centre stage. The division between what are now usually called 'one-seed' and 'two-seed' models, between those classical philosophers and physicians who held that only men produced seed and those who held that both men and women contributed seed to the offspring, has become more prominent.[2] It functions as a convenient organising tool, a way of classifying theories in the doxographic mode, but perhaps also marks an ideological difference; it has been suggested that it speaks to wider issues of gender hierarchy (see e.g. McLaren 1984: 17; Lloyd 1983: 86–111). While this latter idea has been repeatedly called into question and many other complications have emerged with this simple categorisation, both in its classical forms and as it has been applied to or sought out in other historical situations (see e.g. Kessler 2009: 89–126; Flemming 2018), the basic structure has so far remained intact. The notion of a fundamental split between the 'one-seed' and 'two-seed' camps appears to be entrenched in current scholarship.

The argument here, building on these prior objections, is that the one-seed/two-seed division is misleading in itself. It distorts the ancient debate and directs our

DOI: 10.4324/9780429438974-14

attention away from what was most important within it. Key issues of the nature and role of the substances implicated in generation have been overlooked. The ancient enquiry into whether females also emit seed needs to be followed up with questions about whether that seed is the same as emitted by males, in terms of its production and constitution and, perhaps most crucially, its contribution to the offspring. These are the points which really matter in understanding the theories of generation articulated and contested in antiquity: simple counting resolves nothing. Indeed, it might be claimed that if women and men's seed were to be identical in all these respects, then there would be only one seed involved in procreation after all, just provided by both parents.

Sophia Connell has recently raised the same point in her detailed study of *Aristotle on Female Animals*. She distinguishes between what she calls 'the parallel seed theory', found in the Hippocratic medical writings of the Classical Greek world and in those of the great physician of the Roman Empire, Galen of Pergamum, and 'the differentiated seed theory' of the fourth-century BCE philosopher Aristotle, which is the focus of her analysis (Connell 2016: 95). All, she argues, were 'two-seed theories', but the former version was characterised by the view that the female emits seed at the same time and of the same sort as the male, while the latter was committed to the contrary view, that female seed diverges from that of the male on these scores. This is, of course, to challenge the more traditional location of Aristotle in the 'one-seed' camp—his identification, indeed, as the leader of this theoretical faction—as well as to adopt an alternative analytical approach overall. The two do not need to go together; the distinction between 'parallel' and 'differentiated' seed theories may be a valid and useful one even if Aristotle were to remain outside the two-seed fold. It may indeed be even more helpful to allow for a range of possible differentiations between parental contributions, in which both may provide more or less the same seed, different kinds of seed, or different kinds of contribution all together, with only one being seed.

This chapter will make such an allowance, and it will focus on questions of the nature and role of the substances—fluids—involved in human generation rather than the numbers. Two pairs of reasonably well-known case studies, two sets of texts conventionally cited in this context, will be investigated in this way. The first come from the Classical Greek world of the fifth and fourth centuries BCE, involving Hippocratic medical writings on the one hand and the philosophical works of Aristotle on the other. Aristotle does not engage with Hippocratic notions by name, but these were all clearly participants in the same debate at roughly the same time and indeed were understood as such by later contributors to the ongoing discussion about human procreation in the ancient world. These subsequent engagements included the second pair of authors, located this time in the Roman Empire of the second century CE: that is, Soranus of Ephesus and Galen. Both were born in the Greek East, Soranus perhaps 60 or 70 years before Galen, and spent most of their careers as medical practitioners, writers, and teachers, in the imperial metropolis of Rome. Galen did refer to Soranus in his massive surviving oeuvre, though not specifically in relation to seeds and generation; still, their views on this topic can clearly be seen to be in dialogue nonetheless, and both also call on the conceptual resources and authority of Aristotle and Hippocrates in this and other instances.

Hippocratic and Aristotelian seeds

Many of the works collected into the Hippocratic Corpus—that is, associated with the name of Hippocrates of Cos, the legendary founding father of Greek learned medicine—treat practical issues of fertility and its disruption (see Flemming 2013; King 2018). Female health, as it aligned with procreation, was a key concern. The most sustained discussion comes in *On Generation/On the Nature of the Child* (edition Potter 2012), a pair of treatises which circulated separately and together in antiquity (Hanson 2013) and take the story from the production of seed, through sexual intercourse, conception, foetal formation and growth, and pregnancy, to birth, with various excursuses and additions along the way. The aim was to explain everything, how the whole process works, how all the possible outcomes occur, in a coherent and convincing manner.

The key sentence is the opening line of *On the Nature of the Child*:

> If the seed (*gonē*) from both remains in the womb of the woman, first it is mixed together, since the woman is not motionless, and it collects and thickens as it is heated.[3]

> Hippocratic Corpus, *On Generation/On the*
> *Nature of the Child* 12

This is 'conception', the first stage in generation.[4] The womb has received the seeds, its mouth has contracted in response to the moisture, and closing, holds the two seeds within itself so that 'what came from the man and what came from the woman' mix together (*On Generation/On the Nature of the Child* 5). That is the beginning of the process of foetal formation, which will be followed by growth, both phases which require additional resources, mainly blood, from the mother's body. They are nourished by the material which would otherwise have been evacuated in menstruation.

These seeds, from the man and woman, seem identical in this formulation. They are both essential to and play the same role in conception as it moves towards foetal formation. The comparison was explored a bit more in *On Generation*, in the sequences that build up to this moment. The statement about the manufacture of seed is generic:

> I say that seed (*gonē*) is separated from all of the body, from the solid (parts) and the soft (parts), and from all the liquid.[5] There are four kinds of liquid—blood, bile, water and phlegm—for such are the substances a human being innately has within them.

> Hippocratic Corpus, *On Generation/On the*
> *Nature of the Child* 3

However, men and women experience heterosexual intercourse differently, in part because of a divergence in the pace and place of seed production and discharge (*On Generation/On the Nature of the Child* 4; and see Dean-Jones 1992). The first and presumably the second is more sudden in men, so their sexual pleasure is shorter and greater than women's, in any case he always ejaculates externally, whereas she may ejaculate into her own womb or outside, if the mouth of the uterus is too open. His seed adds to her pleasure before bringing it to a close. It is like wine being poured on to a flame: there is an initial increase of heat and incandescence, then the flame is extinguished.

None of these distinctions are about the seed itself, which is essentially the same in all these circumstances, as is further emphasised by the next part of the productive sequence.

> Sometimes what is ejaculated from a woman is stronger, and sometimes it is weaker, and the same for a man. And in the man there is both female seed (*sperma*) and male, and the same in the woman. The male is stronger than the female, and so is necessarily produced from the stronger seed (*sperma*). It works like this: if stronger seed comes from both parents, it becomes male; if weaker, female. Whichever dominates in respect to quantity, that is what it becomes: for if there is much more weaker seed than stronger, the strong is mastered (*kratein*), and mixing with the weak, turned female. But if there is more of the strong than the weak, it masters the weak and turns it male.
>
> Hippocratic Corpus, *On Generation/On the Nature of the Child* 6 (see also Mulder, this volume Chapter 9, p. 148)

The seminal equivalence of the sexes is explicit. Both partners produce seed across the same spectrum of strength and weakness, the uterine contest between them is free and fair, but male is strong and female is weak, weak is female and strong is male. Initially the idea that there is male seed and female seed looks distinct from the assertion that seed can be stronger or weaker, but the terms then collapse into each other.

This equivalence continues as the author moved on to explain parental resemblance, mobilising the same aetiological and narrative resources as before (*On Generation/On Nature of the Child* 8). The seed of both the woman and the man has come from all of their bodies—weak from the weak parts and strong from the strong parts—and the interplay between these strengths and weaknesses will shape the appearance of the offspring. In respect to some parts, the father's seed will dominate, and in respect to others the mother's will, with likeness following domination. Every child will, therefore, share features with both parents, with the precise pattern and balance determined by the seeds as they have been produced by the parental bodies and mixed together in this particular instance. The mechanism is roughly the same as has already decided the sex of the offspring, but the processes are distinct. Overall quantity makes males and females, and then more local seminal interactions within that regulate partial resemblances, a package which has the benefit of both causal economy and of allowing girls to look more like their fathers and boys to look more like their mothers. The fact that this does happen, indeed, that couples can have both male and female children who resemble either parent more or less closely, provides important support for the whole account, the author claimed.

Attention then shifted to non-seminal aspects of generation, especially problems caused by a badly shaped or otherwise faulty womb (*On Generation/On the Nature of the Child* 9–10). This may create weak or malformed offspring, and foetal injuries can also be the result of blows and falls. However, the question of whether the seed itself can be responsible for a less-than-ideal procreative outcome also needs to be addressed (*On Generation/On the Nature of the Child* 11). Although, generally, mutilated or deformed (*pepērōmenos*) adults produce complete seed and whole and healthy children, disease can affect the four fundamental kinds of liquid—blood, bile, water, and phlegm—leading to flawed seed and similarly flawed offspring. A seminal deficiency of this type in respect to the parental mutilation or deformity will engender that same deformity in the child.

The point is again a gender-neutral one. The seed of either parent could have this effect, and the story after the seeds have been held, mixed, and heated is also one in which the contribution of the partners is entirely indistinguishable. Whether that mixture itself is weak and female or strong and male matters, however, for the former sets and articulates more slowly than the latter, taking forty-two days for all its parts to form rather than thirty (On Generation/On the Nature of the Child 18). Maternal blood and breath have provided essential resources for this process—both nourishment and crafting—and they continue to do so throughout pregnancy. Supplies from the mother enable the foetus to grow bigger, become more solid where necessary, and more precisely formed, right through to the somatic extremities—to the fingers and toes. When the hairs and nails have taken root, then the child begins to move, and this occurs at three months for a male and four months for a female (On Generation/On the Nature of the Child 21). The male moves earlier because it is stronger than the female, is made of stronger and thicker seed.

So it is the seed which does the real generative work here, which makes a new human life, of either the male or female variety, and in the likeness of the parents, a compound likeness of both parents. Both womb and maternal material play their part too, but in support of the actual engendering, to complete the process of foetal formation and then growth. They are indispensable but secondary in their actions. This is seed from the man and the woman; there is a positive answer to the ancient question of whether the female emits seed too, but it is not at all clear that there are 'two seeds', that this is a 'two-seed theory'. Both male and female produce seed, but it is the same seed: it is formed in the same way in both bodies and it contributes identically to the offspring. Moreover, seminal equivalence does not in any sense equate with gender equality. Weak seed is what makes females, slowly, and strong seed makes males, rather quicker, because females are weak and males strong.

Other Hippocratic texts mention generative seed, usually in passing and mostly in relation to the male contribution.[6] Female seed may be omitted from the discussion, but its existence was never challenged; there was no debate on the matter. Hippocratic authors held divergent views on many issues—about the number and character of the elemental constituents of the human being, for example, or whether consciousness, the reflective and decision making part of the soul, was located in the head or the heart—but they all seem to have shared the assumption that women emit seed. Certainly the only other surviving treatise in which a theory of generation was expounded reasonably systematically—that is On Regimen (edition Joly 1967; English translation Jones 1931)—shares and builds on that assumption, constructing a model of generation which is structurally very similar to that articulated in On Generation/On the Nature of the Child, even if some of the content is very different.[7]

Turning to Aristotle, however, reveals a greater contrast between male and female contributions to offspring, whether or not that contrast is considered to be contained within the domain of the seminal or not. His Generation of Animals (edition Drossart Lulofs 1961), composed around the middle of the fourth century BCE, is, needless to say, much more extensive and systematically elaborated than any Hippocratic work, and it fits into an overall philosophical programme with claims to completeness and coherence. The name of Hippocrates is absent from the text—only the arguments of Presocratic philosophers are explicitly debated—but, as will be clear, Hippocratic ideas are present, refuted in the course of the discussion along with all the others. Aristotle's position has also been subject to much greater scholarly scrutiny than his Hippocratic antecedents, as already noted, and this account takes its cue from the more revisionist recent engagements with Aristotle's theory of generation, as exemplified by Connell (see also e.g. Mayhew 2004; Falcon and Lefebvre 2018).

For Aristotle, 'conception' occurs when the seed (*gonē*) from the male and the menstrual fluid (*katamēnia*) from the female are held and mixed together in the woman's womb (*Generation of Animals* 739a26–b20; 730a32–b2). He referred to the product of this initial phase of generation as the 'first mixture' (*prōton migma*), 'conception' (*kuēma*), or 'seed' (*sperma*) (*Generation of Animals* 728b32–4). As Ignacio de Ribera-Martín (2019) explains, Aristotle used this last term in order to emphasise the encompassing nature of his system, that this was a stage common to plants and all animals, whether they (like humans) have separate sexes or not. He also sometimes labelled other fluids involved as '*sperma*' or '*spermatikos*' (seminal), though in a different sense of seed, and the combination of the distinct notions with overlapping terminology has added to the confusion surrounding Aristotle's theories in modern scholarship. Returning, for the moment, to the first phase of human generation: the man's seed is discharged externally, into the space in front of the mouth of the uterus, and drawn into the womb if it is in a suitable condition, made hot by the collection of menstrual fluid within. The woman may also have emitted fluid during intercourse, but this is not 'seminal' (*spermatikos*) and does not contribute to the offspring, though it may assist with the passage of male seed into the womb (*Generation of Animals* 727b33–728a35; 739a21–5). Once inside the womb, the seed (*gonē*) from the male and the female menstrual fluid move and mix together. The *gonē* is divided up and begins to act on the *katamēnia* like rennet on milk, 'setting it together' (*sunistēsi*). The specific heat of the seed solidifies the purer part of the menstrual fluid, separating it from the more watery part and forming membranes around it, keeping the liquid at bay, while it itself 'dissolves and evaporates' (*Generation of Animals* 739b21–33; 767a1).

Aristotle was explicit that 'in respect to generation the female contributes to setting together along different lines to the male: the male contributes the principle (*archē*) of motion and the female the material (*hulē*)' (*Generation of Animals* 730a25–7). His formulations could be even more programmatic: 'the male provides the form (*eidos*) and the principle of motion, the female provides the body and the matter' (*Generation of Animals* 729a9–11). There is no cross-over between the two. There is nothing material in the male contribution, while the female contribution is solely material: she provides the stuff on which the heat of the male seed acts and gets moving before vanishing from the scene. 'Female, as female, is passive (*pathētika*), male, as male is active (*poētika*)', Aristotle fundamentally asserted (*Generation of Animals* 729b12–13). Such clear and absolute divisions were hard to maintain across his explanatory narrative as a whole, however, at least on the female side. In accounting for all generative outcomes, menstrual fluid turned out to be a quite particular sort of matter that certainly does more than just be acted upon, though it never challenged the male priority and superiority which Aristotle's theory so openly enacts. Still, the female contribution does possess its own potentialities and movements, to fit with those of the male: it is the right kind of material for 'setting together' and foetal formation.

The more active aspects of the female contribution are found, unsurprisingly, in Aristotle's explanations of sex determination and appearance. The mixing of seed and menses involves 'movements' of both, and if the male movements gain mastery (*kratein*), then the offspring will be male, and if they are mastered, female (*Generation of Animals* 766b15–766b28). The young and the old tend to produce female children, for example, for they are lacking in heat in comparison to those in their prime, and other factors may make either fluid more watery and colder than is optimal (*Generation of Animals* 766b28–767a1). There is also a second—individual—dimension to this contest. Usually, if the male movements gain mastery, they do so as both generically male and individually paternal, so that a male child who resembles his father results, and similarly if they are mastered, that usually creates a female child in the maternal mould (*Generation of Animals* 768a21–768b15). The two determinations can separate, however, and, indeed, the movements can relapse, producing boys who look more like their mothers, girls in the image of their fathers, and children who successively resemble their grandparents, distant ancestors, unrelated, generic, human beings, and, eventually, not even that. These outcomes were all decided by the state of the mixture between *gonē* and *katamēnia*, its balance and dynamism (*Generation of Animals* 767a14–35).

The male movements were always the subject of Aristotle's formulations—they gain mastery or are mastered; the female movements never actually win out, as such—rather the male failure allows their potentialities to come into play. Still, this eventuality, and the possibility of maternal resemblance itself, clearly indicates that there was more to the female contribution than just matter. At a certain level, an equivalence between the male and female role in 'setting together' has emerged, and men's seed and women's menses also converged in the mode and mechanism of their somatic production. Aristotle spent considerable energy refuting the notion that seed is drawn from all the parts of the body, arguing instead that it (*sperma*) is a 'useful residue' (*perittōma chresimon*) derived from nourishment in its final form (*Generation of Animals* 725a1–20; 726a25–8). For blooded animals that is, precisely, blood—in bloodless animals its analogue. So blood, already the most processed stage of the food taken into the body, ready to be directly distributed to the different somatic parts, is then further 'concocted' (*pettomenon*) to produce seed (*sperma*) (*Generation of Animals* 726b2–5; 728a17–25). These transformations are all driven by heat, by the innate heat of the living being. Since women are, by definition, less hot than men—they are the 'weaker' animal—they are unable to propel this process as far (*Generation of Animals* 726b30–727a2). Still, '*katamēnia* is seed (*sperma*) that is not pure but needs working on' (*Generation of Animals* 728a26–7).

Males are able to transform blood into pure seed (*sperma*), a frothy compound of water and *pneuma*, that is 'hot air' (*aer thermos*), as Aristotle says, which has become integrated into the somatic economy (*Generation of Animals* 735b37–736a2; 736a18–21). The hot part is most crucial, for that is where the actual generative power resides. What makes the seed *gonimos* ('engendering') is the 'so-called hot' (*kaloumenon thermon*), intrinsic to the *pneuma*, the 'soul-heat' (*thermotēta psuchikēn*), analogous to the stuff of the stars (*Generation of Animals* 736b29–737a1; 762a18–22). Females are defined by their inability in this last respect. Colder and weaker, they produce impure seed (*sperma*)—*katamēnia*—which lacks this one vital constituent, the 'principle (*archē*) of soul'; thus the female can be considered 'a deformed (*pepērōmenos*) male' (*Generation of Animals* 737a27–30).[8]

The tension and interplay between identity and difference in respect to the male and female contributions to generation in Aristotle's theory are thus very clear, along with the basic hierarchy at work. Both *gonē* and *katamēnia* were produced from blood through the same process and both can be called 'seed' (*sperma*) in the sense of being useful residues with essential generative roles; the former was simply the purer, more concocted, superior, version of the latter. The work of the two fluids in the 'first mixture' then diverged: one provided form and the other matter, with this division articulated as total—the *gonē*, evaporated in the mixing, initiated the movements and setting together and left without material trace, while the *katamēnia* provided only body, possessed no 'principle' (*archē*) of movement or soul. However, as the process of generation proceeded through further stages, it became apparent that the female contribution did have movements and potentialities of its own, was a source of the offspring's nutritive soul, and helped determine sex and resemblance. It converged on the male in terms of nature and operation, though male priority and superiority was only reinforced as a result. This tension is inherent in the hylomorphism central to Aristotle's philosophical system as a whole, and which always requires both identity and difference between the form and the matter that constitute all things. The general point has been emphasised in relation to Aristotle's 'reproductive hylomorphism' by David Lefebvre as part of wider scholarly discussions around the problems involved in that pairing (Lefebvre 2016, building on Henry 2006; see also many of the essays in Falcon and Lefebvre 2018).

The question here, however, is how to seminally characterise Aristotle's theory: was his a one-seed or 'differentiated' two-seed model, or perhaps neither, or indeed both? Such an approach does seem to miss the point, to obscure what was important to his account, as it involved both male and female contributions as the same and different. Connell's distinction between 'parallel' and 'differentiated' seed theories certainly captures part of the contrast between Hippocratic ideas and those of Aristotle, but all were committed to parallel notions of seminal substance and origins, for example, despite the diversity of views about those origins themselves. All also assumed and reinforced male superiority in their work and shared a wider set of concepts and terminology involved in explaining generation.

Soranus and Galen on seeds

The next period from which a rich array of classical medical texts survive, including treatises which discuss and debate human generation, falls from the late first to early third century CE. This was the world of the Roman Empire, which had absorbed and integrated Greek medicine as it had conquered and incorporated Greek lands. Both of the authors who will be discussed here—Soranus of Ephesus and Galen of Pergamum—originated in the Greek East, from major cities of the province of Asia.[9] They composed their treatises in Greek, situated themselves in the Greek learned medical tradition, but spent most of their careers at Rome, in the imperial metropolis. That is where they practised, taught, and wrote, because it was the capital of the Empire, the centre of wealth and power, knowledge and authority (Flemming 2007).

On their ways to Rome, both Soranus and Galen passed through Alexandria, still a key centre of medical education under Roman rule, though no longer the preeminent site it had been in the Hellenistic period. It was in the time of the first Ptolemies, in the early third century BCE, that the physicians Herophilus and Erasistratus had practised systematic human dissection and vivisection in the city, with royal support, and set out newly authoritative descriptions of the human body, both male and female, as a result (von Staden 1989, 1992; Flemming 2003). These texts themselves do not survive, but much of their contents was absorbed into later works, explicitly and implicitly, and at least some of the ideas about generation expressed in them can be found in Soranus and Galen, as will become clear. This is more surprising for the former, since as a Methodic physician, an adherent of the method in medicine, Soranus was fundamentally committed to the notion that the knowledge necessary for the medical art was restricted to knowledge of the manifest generalities—of 'stricture', 'flux', and 'mixture'—which a physician needed to be able to recognise in any sick individual and to take as, in themselves, indicative of their treatment (see Tecusan 2004). Things learnt 'from dissection' are 'useless' (*achrēstos*), he stated in the introductory sequence of his *Gynaecology* (chapter 1.5), but contribute to learning nonetheless, so he will provide an account of the female parts which includes information from this source in order to support a more practical discussion of the workings of women's bodies as they relate to health and procreation in the rest of Book One.

Soranus' account starts with the womb, its rich nomenclature, its shape, situation, flexibility, and composition. This composition is complex, with nerves, veins, arteries, and flesh all implicated, both in the two layers of the uterus itself and in its connections to the global somatic systems. The nerves originate from the spinal cord, while the veins come from the 'hollow vein', and the arteries from the 'thick artery', passing through the kidneys, until four vessels—two veins and two arteries—implant themselves into the womb. 'From these also, one artery and one vein grow into each of the *didumoï*', literally, 'twins', and the term used by Herophilus for the *orcheis*, the 'testicles' in men and women (*Gynaecology* 1.11, edition Ilberg 1927; Galen, *On the Usefulness of the Parts* 14.11, edition Helmreich 1907–9; and see von Staden 1989: 165–9). A further description of this anatomical formation follows which continued to draw on the Alexandrian anatomist:

> Furthermore, the *didumoi* are attached to the outside of the uterus, near its throat, one on each side. They are of loose texture, and like glands are covered by a particular membrane. Their shape is not longish as in males; rather they are slightly flattened, rounded and a little broadened at the base. The seminal duct (*spermatikos poros*) runs from the uterus through each *didumos* and extending along the sides of the uterus as far as the bladder is implanted in its neck. Therefore it seems that the seed (*sperma*) of the female does not contribute to generation (*zōogonia*) since it is discharged towards the outside, a subject we have discussed in the treatise *On Seed*.
>
> Soranus, *Gynaecology* 1.12

Unfortunately, the treatise *On Seed* is lost, as also his books *On Generation*, mentioned a little later in the sequence (*Gynaecology* 1.14).[10] The positive content of Soranus' theories on these subjects is, therefore, largely unknown. It seems, however, that he was committed to the existence of male and female seed, produced in the same anatomical constructions if not through the same process in both cases. Indeed, his parallelism extends to the place where the seminal duct empties, with divergent effects on the what the fluids can then do in relation to procreation. In men, external emission is essential; in women, it entails that their seed makes no contribution to generation. The uterus, in particular its base, is where the (male) seed adheres, however, and 'it itself brings the seed to completion' *Gynaecology* 1.15; see also 1.13 and 33). The womb is also the main site of menstruation, which, though not an action beneficial to health, is necessary for child-production (*paidopoiia*) (*Gynaecology* 1.27–9). This necessity appears to be cleansing and nourishing rather than formative, even in an Aristotelian mode. These are the functions Soranus mentioned: *katharsis*, that is 'purging' or 'purifying' the womb, and '*trophē*', that is food for the embryo (*embruon*) (*Gynaecology* 1.19). He also described the substance involved as 'pure blood' in most women, though sometimes a 'bloody liquid' or 'ichor', as in 'non-rational animals' (*aloga zōa*), without indicating that any kind of further processing was implicated (*Gynaecology* 1.19 and 28).

There is a maternal contribution to the offspring which goes beyond nourishment, Soranus stressed. 'What is generated bears some resemblance to the mother, not only in body but also in soul' (*Gynaecology* 1.39). The effect seems to be quite a direct one, however, not mediated by seed or menstrual fluid. The state of the soul alters 'the mould' (*tupos*) of what is conceived, as demonstrated by women looking at monkeys or beautiful statues during procreative intercourse and producing simian or well-proportioned children as a result (*Gynaecology* 1.39; and see Reeve 1989).

Overall, therefore, much remains uncertain about Soranus' theory of generation. Clearly, both male and female produced and 'emitted' seed, but only the former contributed to generation. That is two seeds, then one seed. The maternal role was, however, not limited to that of carrying and nourishing; she did more than provide a suitable space and sustenance for the embryo. Her soul had some impact on the formation of her offspring as a whole. One possibility would be that Soranus subscribed to something like the later Neoplatonic understanding of generation, in which the male seed passes from the control of the father's soul to that of the mother, attaching itself to the womb in a process analogous to grafting. That is, in a one-seed model, there is a single, male seed, but as James Wilberding (2017: 1) says, 'many Neoplatonists identify the female rather than the male as the immediate active cause of reproduction.' This is speculation, however, and while some of Soranus' language points in this direction, his conception of the soul was certainly not Neoplatonic in character.

There is no such uncertainty regarding Galen's theory of generation, expounded a little over half a century after Soranus' endeavours in the field. His treatise *On Seed* (edition De Lacy 1992) survives and is supported by a host of other extant texts covering the anatomy and physiology of the generative parts, as well as *The Formation of the Foetus*. There is not total consistency across all discussions—emphasis and details shift—but it is straightforward enough to summarise Galen's views on key topics concerning human procreation, views which he explicitly positions in relation to those already outlined, especially those of Hippocrates and Aristotle (Flemming 2018).

For Galen, 'conception' (*sullēpsis*) occurs when male and female seed meet and remain within the woman's womb. He often cited the formulation from *On the Nature of the Child* to this effect, but his understanding of the process involved diverged from that of the 'divine Hippocrates' in various respects (see e.g. Galen, *On Seed* 2.1.12–13 and 31; *On the Usefulness of the Parts* 14.11). Galen's two seeds were not completely parallel or identical in their roles, though their divisions were more of detail, a detail born out of the fine-grained narratives enabled and encouraged by Hellenistic anatomy, than of principle.

The female seed was discharged internally into the 'horns' (*kerata*) of the uterus, one on each side, and coated them as it passed into the body of the womb and there met the male seed, which had also formed membranes where it touched the uterus (*On Seed* 1.7.1–20 and 2.4.16–22). The seeds 'mix' (*misgesthai*) and the membranes 'entwine' (*epiplekesthai*), the female seed, being 'thinner' (*leptoteron*) and 'colder' (*psuchroteron*) than the male, provided the latter with nourishment, and there was a division of labour between the membranes too. That from the female seed (the *allantois*) linked into those—the *chorion* and *amnion*—which enclose the whole seed (*gonē*) and, anchored to the horns, attached itself to other parts of the womb too, allowing further structures to be formed through which foetal nutrition and excretion are organised. Still, Galen was emphatic that both seeds contain two principles—the material and the active—and not just one. Indeed, in *On Seed*, Galen asserted that the menstrual fluid also needs to be considered active in its contribution to the offspring. It is 'mostly material with very weak power', but its power combines with that in the female seed, since they are '*oikeios*' ('suited' or 'congenial') and together, over nine months, the two can outweigh the initial force of the male seed in certain respects (*On Seed* 2.2.19–24). This only mattered for animal hybrids, however, though there is another kind of third-party involvement in determining the sex of the foetus, one which again complicates the seminal aspects of Galen's system.[11]

Seed is made from blood, not 'a melting' (*apotēxeōs*) from all the parts (*On Seed* 2.2.16 and 2.5.3). On this point Galen agreed with Aristotle, while passing over any Hippocratic connections to the latter view. In both male and female bodies, arteries and veins descend from the region of the kidneys towards the generative organs, coiling increasingly as they approach the *orcheis* (testicles) (*Usefulness of the Parts* 14.9–10). In these coils blood and *pneuma* are brought together, and the fluid becomes whiter and more concocted as it goes, a process of elaboration that is completed within the testicles themselves: completed perfectly in the larger and warmer male *orcheis*, less well (*ellipesteron*) in the smaller and colder female versions. Seminal ducts then pass from the testicles to the neck of the bladder in the male and into the 'horns' of the womb in the female—Herophilus was manifestly in error when claiming otherwise— to deliver the seed of both, as necessary for conception (*On Seed* 2.1.15–26). Less female seed is delivered than male, and of poorer quality, through shorter, narrower vessels (*Usefulness of the Parts* 14.10). Still, the woman's *sperma* does more than just contribute to generation; it also incites her to sexual activity and opens the mouth of the uterus in intercourse with a man (*Usefulness of the Parts* 14.11; *On Seed* 2.1.32). The male seed acts similarly in relation to his sexual desire, if nothing more (see Ahonen 2017).

Galen's seminal hierarchy is manifest, but so far it has supported an essential equivalence between the male and female contributions to generation. The female seed vanishes, however, from his explanations of sex determination, a process described as an interaction between the womb and the male seed (*Usefulness of the Parts* 14.7; see also *On Seed* 2.5.35–8; Flemming 2000: 303–29; Mulder, this volume). A basic somatic asymmetry meant that the right womb and the right testicles were hotter than the left and therefore more likely to produce a hotter, therefore male, embryo. The uterus had the upper hand, being 'generally better able to make what was conceived like itself', so that if colder seed from the left testicle fell into the right uterus, it would generally end up male, and vice versa, though this was not inevitable (*Usefulness of the Parts* 14.7). Still, as the Hippocratic aphorism stated: 'male embryos mostly on the right, females on the left' (*Aphorisms* 5.48, edition Jones 1931, repeated in another Hippocratic text, *Epidemics* 6 2.46, edition Smith 1994). The forces at work here—hot and cold—are shared between men and women, as also the anatomical patterns that produce the lateral differences, but the matching of womb and testicle, rather than seeds that one expels the other receives and holds, seems to pull the male and female roles apart. In explicating parental resemblance, Galen returned to the seminal encounter, to unevenness in both seeds which resulted in the male dominating (*kratein*) in some parts and the female in others, so that the offspring was always like both father and mother (*On Seed* 2.5.1–6 and 75). He was, however, less interested in this than in general issues of sexual differentiation, in what that difference consists in, and how it is manifest in all aspects of foetal formation and growth, with the quicker development of the male just one way in which superiority is shown (Flemming 2000).

Despite all the complexities, perhaps Galen was operating with a two-seed model of generation. Men and women emitted seed which contributed to generation in distinct ways. The female seed, in particular, was vital in inciting sexual desire and in providing nourishment for the male seed, as well as forming the *allantoic* membrane, while the male seed (mostly) formed the *amnion* and *chorion* and drove the seminal motions. Galen also explicitly argued against those who held that the female did not produce *sperma*, or at least not 'generative' (*gonimos*) *sperma*, in which category he located Aristotle, Athenaeus (of Attaleia), the founder of the pneumatic medical lineage, and Herophilus: Soranus was not mentioned but must be counted here too (Galen, *On Seed* 2.1.66 and 2.1.15–26).[12] He understood Aristotle's *katamēnia* as essentially non-seminal, though he did allow for a blood/menstrual fluid/seed spectrum. However, Galen clearly misconstrued some aspects of *Generation of Animals*, so his interpretations should not be accepted without question, and he himself could be said to be as much of a 'parallel' as a 'differentiated' seed theorist. What male and female seeds share in his system—the active and the material principles—was fundamental: they were essentially better and worse versions of the same thing. Moreover, given his view of the active role of menstrual fluid, it might be said that there were three seeds in his system, and that is without considering the formative role of the womb, at least in respect to sexual differentiation.

Conclusions

There is much more that could be said about the theories of generation expounded by all of these authors, and this is only a sample of a larger field. The ideas of the Neoplatonists have been briefly alluded to, for example, and would take the discussion in a different direction. Even from this summary survey, however, the problem with the one-seed/two-seed classification should be clear. What counts as seed; how seminal does a fluid need to be? Does a two-seed model simply require that the female emits seed too? Or does that emission have to be *gonimos*, 'generative', and contribute to the offspring? Do the two fluids have to make distinct contributions, play different roles in generation, for there really to be two seeds? In many ways it is this last issue which is the most significant: do male and female make the same or divergent contributions to the process of conception and foetal formation, that is, the crucially creative part of making a new being? Satisfactory answers to this question can really only be provided within the framework of the individual theories, treated on their own terms.

For the cases discussed here, the Hippocratic author of *On Generation/On the Nature of the Child* was the most thoroughly parallel in his approach—there was complete equivalence between male and female contributions—while Soranus appears the most differentiated, despite his two seeds. Between them, Galen's mix of similarity and difference in female and male roles perhaps put the stress on the former, while Aristotle emphasised the latter, including in his vocabulary. Both, crucially, elaborated complex and compound visions of the generative process, however, and while some kind of female contribution was fundamental to all these authors, the assumption of male superiority was even more basic.

Notes

* Research for this paper was undertaken within the framework of the Cambridge University Generation to Reproduction Project, supported by a strategic award from the Wellcome Trust (Grant no. WT 088708).

1 That is one of the regular questions in standard philosophical doxographies, such as transmitted under the name of Plutarch: Pseudo-Plutarch, *Placita* 5.5. It follows questions about the 'substance' (*ousia*) and corporeality (*sōma*) of seed (*sperma*). On these issues, see also Laskaris and Fallas in this volume, Chapters 6 and 7, respectively.

2 Beginning with Preus 1979; Boylan 1984; see also e.g. Laqueur 1990: 38–42; Wilberding 2017: especially 13–32.

3 All translations are my own.

4 'Conception' obviously has a particular modern meaning, but I am going to use it here to label this first stage, whatever the content.

5 What exactly is meant by 'parts' of the body here is unclear; presumably the solid parts would include bones and some organs, and the soft parts flesh and other organs. It is external parts of the body—limbs, hair and eyes, for example—that are alluded to later in the treatise, but without textural qualifications.

6 Discussions of generative failure focus on the womb and its ability to receive and retain male seed, for example; see e.g. Hippocratic Corpus, *On the Diseases of Women* 1, 5, 8, 10, 11, 17, 31 and 32.

7 On seed in *On Regimen*, see Bartoš (2009); and for a wider discussion of the text and its key themes Bartoš (2015). Mulder, this volume Chapter 9, also discusses the theory of sexual differentiation presented in it.

8 I translate 'deformed' to draw attention to the shared vocabulary between Aristotle and the Hippocratic authors (and, indeed, Galen). Aristotle means something particular here, however; see e.g. Gelber (2018); Witt (2012).

9 On Soranus, see Hanson and Green 1994; on Galen, Hankinson 2008; Mattern 2013. On their gynaecology, see Flemming 2000.

10 Ilberg suggested they were the same text; see Hanson and Green 1994: 1031–3.

11 That is it explains why such hybrids—like mules and hinnies—resemble the mother more than the father in terms of their species: *On Seed* 2.1.43–6.
12 On Athenaeus and the *pneumatikoi*, see Flemming 2012: 75–7.

Bibliography

Primary sources

De Lacy, P. (1992) *Galeni De Semine*. Berlin: Akademie Verlag.
Drossart Lulofs, H. J. (1961) *Aristotelis De Generatione Animalium*. Oxford: Clarendon Press.
Helmreich, G. (1907–9) *Galeni De Usu Partium Libri XVII*. 2 vols. Leipzig: Teubner.
Ilberg, J. (1927) *Sorani Gynaeciorum libri IV*. Leipzig: Teubner.
Joly, R. (1967) *Hippocrate Du Régime*. Paris: Les Belles Lettres.
Jones, W. H. S. (1931) *Hippocrates*, vol. IV. Cambridge, MA: Harvard University Press.
Potter, P. (2012) *Hippocrates*, vol. X. Cambridge, MA: Harvard University Press.
Smith, W. D. (1994) *Hippocrates*, vol. VII. Cambridge, MA: Harvard University Press.

Secondary literature

Ahonen, M. (2017) 'Galen on sexual desire and sexual regulation', *Apeiron* 50, 449–81.
Bartoš, H. (2009) 'Soul, seed and *Palingenesis* in the Hippocratic *De victu*', *Apeiron* 42, 17–47.
———. (2015) *Philosophy and Dietetics in the Hippocratic On Regimen: A Delicate Balance of Health*. Leiden: Brill.
Boylan, M. (1984) 'The Galenic and Hippocratic challenges to Aristotle's conception theory', *Journal of the History of Biology* 17, 83–112.
Connell, S. M. (2016) *Aristotle on Female Animals: A Study of the Generation of Animals*. Cambridge: Cambridge University Press.
Dean-Jones, L. (1992) 'The politics of pleasure: Female sexual appetite in the Hippocratic corpus', *Helios* 19, 72–91.
Falcon, A. and D. Lefebvre (eds) (2018) *Aristotle's Generation of Animals: A Critical Guide*. Cambridge: Cambridge University Press.
Flemming, R. (2000) *Medicine and the Making of Roman Women: Gender, Nature, and Authority from Celsus to Galen*. Oxford: Oxford University Press.
———. (2003) 'Empires of knowledge: Medicine and health in the Hellenistic world', in A. Erskine (ed.) *The Blackwell Companion to the Hellenistic World*. Oxford: Blackwell, 449–63.
———. (2007) 'Galen's imperial order of knowledge', in J. König and T. Whitmarsh (eds) *Ordering Knowledge in the Roman Empire*. Cambridge: Cambridge University Press, 241–77.
———. (2012) 'Antiochus and Asclepiades: Medical and philosophical sectarianism at the end of the Hellenistic era', in D. Sedley (ed.) *The Philosophy of Antiochus*. Cambridge: Cambridge University Press, 55–79.
———. (2013) 'The invention of infertility in the Classical Greek world: Medicine, divinity, and gender', *Bulletin of the History of Medicine* 87, 565–90.
———. (2018) 'Galen's generations of seed', in N. Hopwood, R. Flemming, and L. Kassell (eds) *Reproduction: Antiquity to the Present Day*. Cambridge: Cambridge University Press, 95–108.
Gelber, J. (2018) 'Females in Aristotle's embryology', in A. Falcon and D. Lefebvre (eds) *Aristotle's Generation of Animals: A Critical Guide*. Cambridge: Cambridge University Press, 171–87.
Hankinson, R. J. (2008) *The Cambridge Companion to Galen*. Cambridge: Cambridge University Press.

Hanson, A. E. (2013) 'A famous handbook and its relevance for science and medicine: Addendum', *Journal of Roman Archaeology* 26, 738–40.

Hanson, A. E. and M. H. Green (1994) '*Methodicorum princeps*: Soranus of Ephesus', *Aufstieg und Niedergang der römischen Welt* 37(2), 1834–55.

Henry, D. (2006) 'Understanding Aristotle's reproductive hylomorphism', *Apeiron* 39, 269–99.

Kessler, G. (2009) *Conceiving Israel: The Foetus in Rabbinic Narratives*. Philadelphia, PA: University of Pennsylvania Press.

King, H. (2018) 'Women and doctors in ancient Greece', in N. Hopwood, R. Flemming, and L. Kassell (eds) *Reproduction: Antiquity to the Present Day*. Cambridge: Cambridge University Press, 39–52.

Laqueur, T. (1990) *Making Sex: Body and Gender from the Greeks to Freud*. Cambridge, MA: Harvard University Press.

Lefebvre, D. (2016) 'Le *sperma*: forme, matière ou les deux? Aristote, critique de la double semence', *Philosophie Antique* 16, 31–62.

Lesky, E. (1951) *Die Zeugungs- und Vererbungslehren der Antike und ihr Nachwirken*. Mainz: Akad. der Wiss. und der Literatur.

Lloyd, G. E. R. (1983) *Science, Folklore and Ideology*. Cambridge: Cambridge University Press.

Mattern, S. (2013) *The Prince of Medicine: Galen in the Roman Empire*. Oxford: Oxford University Press.

Mayhew, R. (2004) *The Female in Aristotle's Biology: Reason or Rationalization*. Chicago: University of Chicago Press.

McLaren, A. (1984) *Reproductive Rituals: The Perception of Fertility in England from the Sixteenth to the Eighteenth Century*. London: Methuen.

Preus, A. (1979) 'Galen's criticism of Aristotle's conception theory', *Journal of the History of Biology* 10, 65–85.

Reeve, M. D. (1989) 'Conceptions', *Proceedings of the Cambridge Philological Society* 35, 81–112.

Ribera-Martín, I. (2019) 'Seed (*sperma*) and *kuêma* in Aristotle's *Generation of Animals*', *Journal of the History of Biology* 52, 87–124.

Tecusan, M. (2004) *The Fragments of the Methodists, Vol. I: Text and Translation*. Leiden: Brill.

Totelin, L. (2018) 'Animal and plant generation in classical antiquity', in N. Hopwood, R. Flemming, and L. Kassell (eds) *Reproduction: Antiquity to the Present Day*. Cambridge: Cambridge University Press, 53–66.

von Staden, H. (1989) *Herophilus: The Art of Medicine in Early Alexandria*. Cambridge: Cambridge University Press.

———. (1992) 'The discovery of the body: Human dissection and its cultural contexts in ancient Greece', *The Yale Journal of Biology and Medicine* 65, 223–41.

Wilberding, J. (2017) *Forms, Souls and Embryos: Neoplatonists on Human Reproduction*. London: Routledge.

Witt, C. (2012) 'Aristotle on deformed animal kinds', *Oxford Studies in Ancient Philosophy* 43, 83–106.

11

PHALLI FIGHTING WITH FLUIDS

Approaching images of ejaculating phalli in the Roman world*

Adam Parker

The image of the phallus was widespread throughout the Roman world, occurring in a multitude of forms, variations, and media. They were worn as pendants on necklaces, adorned the finger rings of children, hung around the home as windchimes, depicted on mosaics, and were carved or incised into building stones. The use of the phallus as a magical or religious symbol was undoubtedly, at least in part, something inherited from the Classical and Hellenistic Greek worlds (Johns 1982: 9–11) and already bound up in its own mythical heritage. During the expansion of the Roman Empire, particularly in the first century CE, it is primarily the Legions which were the vehicle for transporting this image across the Empire (Turnbull 1978: 199), and they did so with great success. Soldiers were not, of course, the sole users of this imagery for supernatural purposes, merely one of the main conduits for its spread across the Roman world.

Much modern scholarship has been expended on the use of the phallus as an apotropaic symbol:[1] the clearest contemporary evidence we have for presenting a direct relationship between the phallus and this function is the examples in which it may be seen attacking or fighting the Evil Eye (Turnbull 1978: 199–200). The Eye is the Roman embodiment of bad luck and was both feared and respected (Plutarch, *Table Talk* 5.7, edition Babbitt 1936); any object, image, act or ritual designed to deflect or nullify its effects, to promote good luck, or good health (as the antithesis to the negative effects of the Eye) may thus be called apotropaic (a concept falling under the supernatural functions of 'magic').

Apotropaic protection was usually implicit, but this subcategory of phallic images which directly attacked an Evil Eye took a more literal approach to presenting their supernatural function to the ancient viewer; in these scenes, the phalli may be regarded as aggressive, proactive, and direct in attacking the Eye, and this artistic narrative often incorporates ejaculation. Using the bodily fluid from an avatar of protection to physically attack the personification of enmity from a short distance is a powerful image and one which resonates with the magical function of disembodied phalli to distract, disengage, or nullify the negative power of the Evil Eye. This chapter intends to explore the somewhat varied artistic forms that this narrative took across the Roman world and to contextualise them within their original spatial limits, as well as considering material and sensory implications of the objects in these original spaces.

DOI: 10.4324/9780429438974-15

While this is not the place to enter into a wide-ranging semantic and conceptual justification of how I intend to use the term 'magic', suffice to say that it is a contested term (Otto 2013) which has multiple, differing modern definitions.[2] I follow a largely functionalist approach to magic, inspired by Henk Versnel (1991), and it is my belief that there were characteristics of 'magic' which allow for its differentiation from religion, its conceptual ally (Parker 2016: 109–10). In any case, the semantics are not used a great deal in this chapter, but note that I use this approach implicitly in the following. Regardless of an individual's theoretical position on the nature of the relationship magic–religion, it is clear that the exact boundaries between magic and religion (and medicine) and other supernatural elements of the Roman world remain wonderfully vague. In the specific example of the phallic image as an apotropaic device, I argue that its broad function is, quite simply, protection—of both people and places—and that a manifestation of this supernatural quality is visible, to a certain extent, in the archaeological record.

The dialectic between the phallus and the Evil Eye may be as simple as 'good versus evil', but Catherine Johns points out that it is not entirely clear in the artistic narrative whether the Eye itself is fulfilling some sort of apotropaic role (1982: 66).[3] While there are several classical references to this explicit functionality (Aristophanes, *Acharnians* 241, edition Olsen 2004; Varro, *On the Latin Language* 7.97, edition Kent 1938; Pliny the Elder, *Natural History* 28.6.30, edition Jones 1963), it is the material evidence which provides an opportunity to observe this battle on the ground. Materiality involves the study of things, of objects in their own spaces and contexts. The interdisciplinary approach of materiality may provide unique insights into the nature of the relationship between ejaculating phalli and the ways and spaces in which they were use. Unhelpfully, the 'material turn' in archaeology (see Hicks 2010) has only recently reached the study of magic (e.g. Gordon and Marco Simón 2010; Manning 2014; Boschung and Bremmer 2015; Houlbrook and Armitage 2015) and, by extension, the study of apotropaic objects in their original contexts. Previous works have also argued for this broad protective function of the phallic image in the Roman world (Turnbull 1978; Johns 1982; Plouviez 2005), but a spate of recent scholarship has attemped to address the lack of contextualised studies into the use of phallic charms: wearing a phallic pendant (Whitmore 2017); phallic carvings on buildings (Parker 2017); phallic carvings on Hadrian's Wall (Collins 2020). These contextual studies have shown that there is a nuanced use of this apotropaic symbol in the spaces in which it has been studied. As a specific example, Alissa Whitmore (2017) reconstructed, through experimental analyses, a type of copper-alloy phallic pendant with globular testes and used it to show that the shape of the pendant was designed in such a way to keep it pointing outwards (so that it stayed erect) while being worn and, furthermore, that this ithyphallism (in a state of erection) was its default state, despite the regular jostling of physical activity it was exposed to. To my knowledge there have been no archaeological studies which have explicitly addressed examples of phalli which appear to be ejaculating, despite several examples surviving from multiple contexts and materials. At the outset it needs to be pointed out that previous scholarship on phalli has generally focused almost exclusively on the organ itself rather than any associated fluids. The following chapter intends to apply the same contextualised and materialistic approach of the above studies onto this amorphous group in order to establish the material and sensory implications of these objects in the Roman world.

As mentioned earlier, the phallus is clearly depicted as one of the main enemies of the Evil Eye. There are two slightly differing representations of the Eye in this capacity, and in both cases, it is under attack from its 'enemies' (otherwise describable as 'apotropaic symbols'). The most common depiction of the Eye and phalli together is when it is under attack as part of the 'all suffering eye' scene, for example on a mosaic from the so-called 'House of the Evil Eye' at Antioch, a carved stone relief from Leptis Magna (Johns 1982: fig. 77) (see Figure 11.1), and a gold earring from Norfolk (Worrell and Pearce 2014: no. 20, fig.20)—all these examples include a phallus as one of the enemies of the Evil Eye. The second type of representation shows only the phallus attacking an Evil Eye in a one-on-one fight, and often in a very biological way, involving the phallus ejaculating over or towards an Eye. In these specific examples there is variance in the number and type of the associated apotropaic symbols. On the Antioch mosaic, the Eye is surrounded by, clockwise from top: a crow, a trident, a sword, a scorpion, a snake, a leaping dog, a centipede(?), and a leopard. Slightly away from these characters is a macrophallic dwarf, facing left away from the Eye, with his phallus backwards between his legs, directed towards the eye. The Norfolk earring has much in common with the Antioch mosaic (see Figure 11.2).

Figure 11.1 The Evil Eye surrounded by its enemies, mosaic from Antioch, House of the Evil Eye, second century CE. Now in the Hatay Museum, Antioch, inventory number 1024.

Source: © Wolfgang Rieger, public Domain via Wikimedia Commons.

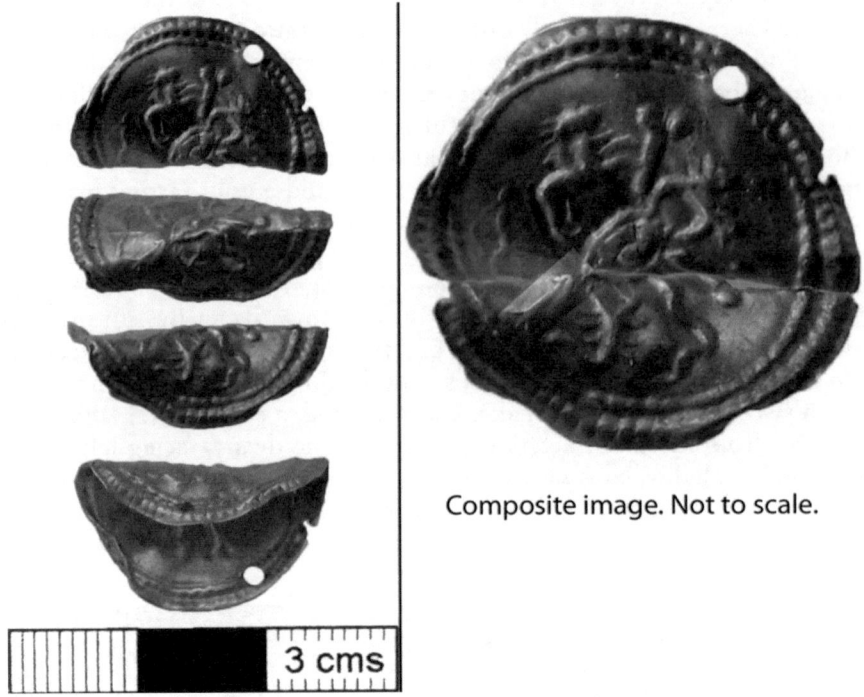

Figure 11.2 The Evil Eye surrounded by its enemies, gold disk from Norfolk, UK, first–fourth centuries CE. Now in the Norwich Castle Museum, PAS: NMS-B9A004.

Source: © Norfolk County Council, via Portable Antiquities Scheme.

It has a central Eye surrounded by, clockwise from top: a phallus, a crab, an arrow, another phallus, a snake, a scorpion, an arrow, a bow and arrow, and a lion or winged horse. Phalli feature prominently in these examples, and its allies fall into two groups: animals and weapons—one natural, one man-made. Indeed, the *materia magica* (the substances, components, and ingredients) of Roman magical rituals often incorporated natural materials such as plants, animals, or body parts in addition to household or mundane objects repurposed for a ritual use (Wilburn 2012: 83–93). However, amongst the images surrounding the Eye, it is only the disembodied phalli (and the macrophallic dwarf) that may be regarded as supernatural in their appearance. Note also the concordance that the animals which walk or crawl attack from beneath, whereas those which fly, and the objects which may be thrown or propelled, do so from the side or from above.

These examples are disparate, separated geographically, temporally, and in physical function, but serve to show knowledge and understanding of this iconography and its associated religious, magical, medicinal, or ritual significance. The Eye and its enemies were known across the entire length of the Roman empire during the early centuries CE, and the phallus featured prominently as one of them.

Let us compare this iconography to what is probably the best-known group of phalli fighting the Evil Eye: the phallic carvings from Leptis Magna (see Figure 11.3). In the most famous scene, a zoomorphic phallus with animal rear legs and hooves takes up the full width of the ansate frame. A small Eye, in the lower left corner of the panel, is being ejaculated upon by a secondary phallus emanating from between the legs of the primary phallus. The Eye also has some manner of arthropod (a crab?) sitting atop it and attacking. Another carving still *in situ*, bearing remarkable similarities in the form of a medium-relief ansate panelled border, depicts a right-facing phallic creature, this time in the form of a bird; it is winged, with avian legs and a tail (Johns 1982: 150, fig. 123; Dumas 2017: 33). A secondary phallus also extends between the legs in a typically mammalian fashion. A small Eye is depicted on the right of the frame, capped above and below by other symbols, though these are not clear. The third and final carving depicts a left-facing centaur (human torso on equine body) which has a macrophallic phallus between the legs as well as a comically enlarged nose (clearly an argument could be made that this too is phallic) (Johns 1982: 94, fig. 77). He is thrusting a spear into an Eye situated at the left of the frame, again attacked above by a crab and below by a scorpion, with a snake coiled at the extreme left. What this group of images is intended to demonstrate is that attacking the Eye is a literal concern—the attackers, human, supernatural, and animal, are often touching the Eye—and that there was a regular team of allies called upon to perform this function.

Figure 11.3 Left-facing zoomorphic phallus attacking an Evil Eye, stone carving from Leptis Magna, Roman Imperial. Archaeological Site of Leptis Magna, Libya.

Source: © Sascha Coachman, via Wikimedia Commons.

177

Amongst these three carvings the first is certainly ejaculating, the second probably is, and the third is thrusting a spear. Thus the three phallic elements are all aggressive. The ejaculate was aggressive in this instance because it intended to make contact with and, presumably, blind the Evil Eye—the ammunition fired by the weapon. The urethra is a bodily orifice and thus existed in the state of potency and vulnerability discussed in the Introduction (see p. 1). In the case of ejaculate towards the Eye, it was demonstrating its potency and its ability to react biologically and meaningfully towards a threat (albeit in a very non-human way), The use of ejaculate is important here as a natural bodily fluid associated with the ithyphallic state of the phallus; its *raison d'être* (see Fallas, this volume, Chapter 7, for ancient connections between semen and erections). In such images the phallus is attacking, deflecting, nullifying, or disengaging the negative effects of the Evil Eye by using its own bodily fluids. This is a supernatural idea depicted in Roman art, and a trope which is visible across the Roman world.

We might consider also the nature of the ejaculated bodily fluid here. It was an erotic or generative fluid—potentially life bringing and also associated with the point of orgasm. These are, arguably, in a semantic sense both positive things, and the use of the ejaculate in an apotropaic or aggressive manner might seem wasteful. What modern scholarship does not embrace is the fact that such images of phalli, albeit supernaturally enhanced, were depicted at this point of orgasm and thus allude to one of the most positive biological experiences of the human male—this concept might have been relevant to dispelling the malignant effects of the Evil Eye. Continuing this idea further, we might suggest that the loss of the generative fluid is somewhat wasteful (in purely biological terms); perhaps there was also an allusion to a sacrifice of this potential life for the greater good? Or that its generative powers lay in producing an apotropaic or beneficial effect rather than anything biological?

The same 'weaponised' use of the phallus is particularly apparent when the phallus and Evil Eye are depicted in combat one-on-one. Three carvings from forts on the northern Hadrianic Frontier in Britain are important here: two from Chesters and one from Maryport. A phallus is joined to a low-relief circular disc on the first carving from Chesters (Coulston and Phillips 1988: no. 407) by a diamond-shaped object which is argued by the author to depict ejaculate (Parker 2017: no. 18) (see Figure 11.4). While the 'Eye' is not as clearly represented as those from Leptis Magna, the sunken relief panel is unlikely to represent anything else. Perhaps in this case the pupil and iris were not carved but painted? The exact provenance of this example within the fort is lost to history. The same cannot be said of the second carving from Chesters, which remains *in situ* on a bridge abutment over the river North Tyne (Coulston and Phillips 1988: no. 404). While the feature the phallus points towards is weathered, the ovoid shape may be interpreted as Eye-like (Parker 2017: 118, no. 15). The location of this carving offers an excellent opportunity to discuss the appropriateness of its position in a bridge over a river. Rivers flood, bridges collapse, people and animals may drown, drought may cause hardship: rivers represent a significant number of uncontrollable and potentially dangerous variables and were thus exactly the kind of liminal spaces enjoyed by the Evil Eye and protected by carved stone phalli in the Roman world. Assuming that the ovoid figure is an Eye, perhaps this scene represents the victory of the apotropaic phallus over the Eye and the dangers of the river itself.

Figure 11.4 A phallus ejaculating towards a stylised Eye, stone carving from Chesters Fort on Hadrian's Wall, second–fourth centuries CE. Now in the Clayton Museum, Chesters Roman Fort, Hexham, CH292.

Source: © English Heritage, photo by Adam Parker.

A conceptual link between a powerful, flowing river and masculine bodily fluids is not beyond the realms of possibility. Roman personifications of the rivers usually depicted them as male figures. For example, a statue of Tiber from a temple to Serapis and Isis in Rome (in the Louvre: MA 593), the male Danube throwing down the female Dacia on a Trajanic sestertius (Ostrowski 1990: 313), and on an earlier coin, the Emperor Domitian standing with his foot on a reclining Rhine (*ibid*). In these examples the rivers are personified as male deities, linking the image of a disembodied male sexual organ and its fluids, to an actual fluid embodied as a male figure. As the Introduction to this volume states on p. 1: 'The human body is full of fluids', and so perhaps this connection is not as tenuous as it might first seem?

Purely speculation, but as a side note it is highly likely that the soldiers who built the bridge themselves urinated into the river, thus linking the reality of their mundane, human experiences with the supernatural world as depicted through their art as well as explicitly connecting their personal fluids with those of the river; one could become the other.

The third example of an ejaculating phallus on the Hadrianic frontier comes from Maryport (Collingwood and Wright 1965: no. 872; Parker 2017: 126, no. 26) (see Figure 11.5). The simply incised phallus points left, towards an oval figure. They are joined by a single line; ejaculate. Interestingly for this carving, it is accompanied by an inscription: 'VER/PAM/SEPT . . .'—translated as 'the phallus of Marcus Septimius' by Collingwood and Wright. *Verpa* is, however, a much more vulgar term than 'phallus'

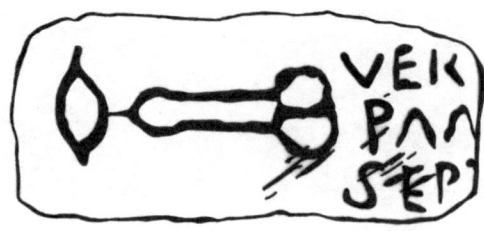

Figure 11.5 A phallus ejaculating towards an ovoid figure, stone carving from Maryport Fort on Hadrian's Wall, second–fourth centuries CE. Now in the Senhouse Roman Museum, Maryport, MAYSM: 1993.55.

Source: © Image redrawn by Adam Parker from the original (see Collingwood and Wright 1965, no. 872).

might suggest. J. N. Adams' discussion of the term is illuminating: '*verpa* was not a neutral technical term, but an emotive and highly offensive word' (1982: 12–14). 'The dick of Marcus Septimius' is a much more realistic translation of this phrase. Marcus was most likely a soldier based at the fort of Maryport; the connection between a phallic carving and a person is interesting as the idea of possession comes into play. Was the phallus simply a crude drawing? Perhaps. Given the supernatural associations given earlier, it could be an attempt of personalising the apotropaic effects of the phallic image towards the man himself.

Unfortunately, for this group of ejaculating phalli there are no nuanced contextual details surviving, as all have come to light from antiquarian investigations. Other than coming from the forts themselves and thus having a strong association to both men and the Roman military, there is little information that can be extrapolated from these. Taking the corpus of carved phalli from the entire frontier zone, however, we learn that there are specific liminal places that such objects frequently feature: boundary places such as defensive curtain walls, building walls, doorways, bridges, and windows (Parker 2017: 118–20). Perhaps of greater interest to the broad study of phalli is the suggestion that the majority of the catalogued carvings, from this study, point to the left and thus may be associated with the ephemeral linguistic link in which left (*sinister*) is inauspicious (Parker 2017: 122).

In all cases of the phalli appearing with the Evil Eye, it is perhaps curious that the fluids are related to the apotropaic phallus rather than the Eye. Biologically speaking, eyes also require fluids in order to function—eyes are moist and are capable of maintaining this state through the unconscious action of blinking. Through crying, eyes are also capable of producing excessive moisture and for this to be spread beyond its immediate environment and away from the eye itself. While there are clearly social or emotional factors which may cause tears, they also result from physical external stimuli touching the eye—this may be from direct force or something insubstantial like dust; in such circumstances, the production of tears may itself be beneficial to the individual by cleansing the eye's surface and ensuring the continuation of the important sense of sight. The depictions of phalli ejaculating towards an Evil Eye in these examples appear to capture the moment of ejaculation or (to continue the analogy started previously) the point of attack rather than the moment of impact of fluid against Eye. Assuming that the supernatural Eye operates in a similarly biological way to the human eye (given that an ejaculating phallus is doing just that, we may expect the same of the Eye), tears may be expected as a result of this action. We may even speculate that the depiction of a crying Eye could be synonymous with defeat or temporary incapacitation, and so it is interesting to observe that this is not evident in the Roman material culture. The Evil Eye is not defeated in such examples; it is *about* to be defeated. Within the supernatural narrative, I argue that this identifies the ejaculating phallus as in a permanent state of guardianship rather than one of bragging victor.

Continuing this theme of the biological representation on such carvings, we should consider also the testes. The disembodied phalli are carved with their accompanying testes. The knowledge that the testes were vital in the production of sperm is discussed at length by Galen (*On Seed* 1.12.5, 1.14, edition De Lacy 1992). At several points in book 1 of his treatise on semen, he decries what he describes as the Aristotelian belief that veins and blood vessels were more important in its production than these organs. An understanding of this biological process is important in connecting what we may crudely describe as the ammunition to the weapon; the two are required for the production of semen. Artistically speaking, a phallus divorced from the remainder of the body may still be recognisably phallic without the addition of testicles, yet here they are. Perhaps, in the context of phalli fighting the Evil Eye, this may incorporate the idea presented by Galen that the testicles were a source of physical strength (*On Seed* 1.15.37) and that men without them are much reduced (on this point see Fallas, this volume, Chapter 7).

The recent find from Catterick, which may also have come from a bridge abutment, provides evidence of a different sort of imagery, as it is prominently ejaculating, but this is not clearly directed towards anything (Parker and Ross 2016). The carving is depicted in medium-relief, with a long stream of bodily fluid projecting and curving away from the glans. The curvature is natural, showing the effect of gravity on the fluid. The concept of ejaculating into the air rather than towards an Evil Eye, is fundamentally different from those given previously because the phallus is no longer attacking anything specific. The Evil Eye can be nullified; in the same way that it may be fought with the teeth and claws of animals or by the sharp edge of the swords and arrows, the phallus fights using its natural bodily fluid. It fights also from a slight range, not requiring direct contact with its enemy. Unless there was a corresponding carving depicting something that the phallus was designed to ejaculate towards, the Catterick carving may instead be spreading its apotropaic effects generally and biologically. Unfortunately, the carving had been reused in a fourth-century road, but its original position within a bridge may have presented the fluid towards the people crossing the bridge. Or perhaps towards the river itself?

The narrative of a phallus fighting an Eye with ejaculate is relatively clear on the carved stone examples mentioned earlier. While the form of the images is broadly similar, the nature of the bodily fluids associated with at least two mosaics are worthy of note here. The first, a second-century mosaic from Sousse (Tunisia) depicts what is described as a fish-shaped phallic image between a pair of equally disembodied vulvas (Dumas 2017: 29) (see Figure 11.6). Both are depicted in flesh-toned *tesserae*—red, pink, and brown. The pointed tip of the fish/phallus appears also to be urinating; a single course of yellow *tesserae* curves outwards and bends towards the left of the frame. Given the orientations of the vulvas and a large, central inscription in black tiles, this apparent disregard for the effects of gravity is somewhat abstract. Both

Figure 11.6 Phallus urinating towards a pair of vulvas, mosaic from Sousse, Tunisia, second–third centuries CE. Now in the Archaeological Museum of Sousse.

Source: © Ad Meskens, via Wikimedia Commons.

vulvas are double-bordered by yellow tiles, the same colour as the urea—is the allusion that both are somehow covered in or physically bound by the fluid of the phallus? The multiplication of the disembodied sexual images certainly has much in common with the apotropaic trope of polyphallism elsewhere visible in Roman artistic media. On the nature of these fluids, clearly it is difficult to differentiate urine and semen in a very simple way. I argue that the depiction of a single stream of fluid is, on balance, more likely to represent urine, whereas a broken stream or individual blobs/spurts is more likely to represent semen. The biological nature of the penis, whether erect or flaccid, in each instance is important in identifying the depicted fluid. The human male has urethral sphincters which work only during a state of erection to prevent the influx of semen into the urinary tract—the internal urinary sphincter is not under voluntary control (it cannot be turned on or off through thought or physical action) and is the reason that men may find it difficult or impossible to urinate while erect.

It is, perhaps, unhelpful to assume, in cases where fluids are shown with phalli, that all erect phalli are ejaculating and all flaccid phalli are urinating, when either may be technically true; biology suggests that the interpretation be weighted in this way, but we should be mindful that the supernaturally disembodied phalli do not necessary have to conform to all the physical laws of their anatomical cousins. In the case of the Sousse mosaic, the yellow coloured tiles of a single, continuous stream load the interpretation towards urine.

This Tunisian mosaic has visual characteristics in common with one from Timgad (Algeria) which has been argued to fulfil an apotropaic function. The latter depicts a standing Black African male holding a fire shovel over one shoulder; the *furnacator* also holds his macrophallus in his left hand and is either urinating or ejaculating. Katherine Dunbabin (1989: 42–4) argued for the latter and that this image was, thus, designed to serve a supernaturally beneficent function for the bathhouse in which it was located. Clarke (1996: 193) suggests that the 'exoticism' of the bathhouse attendant, in that he is outside of social norms, may have added to this apotropaic functionality. This idea can be linked to the supernatural, disembodied phalli and the macrophallic dwarf in the Antioch mosaic. The need to cradle the penis in the left hand suggests that, rather than being erect and capable of physically maintaining its lateral position, the penis is in a flaccid state and simply particularly large; this idea fits well with the anatomical relationship between flaccidity/urea and erection/semen. There is a single line of black tiles surrounding the fish-shaped phallus mentioned earlier—was this example also a macrophallic depiction of an African male? If this is true, and it is comparable with the *furnacator* mosaic in being large but flaccid, this may account for the presence of the urea in the previous mosaic.

In his chapter 'The displays of erotica and the erotics of display in public buildings', John Clarke (1998: ch. 7) argues that the construction of sexual scenes in a suburban Pompeiian bathhouse are designed as form of entertainment, highlighting nonconformist, deviant, or unusual sexual scenes (1998: 239–40). In an earlier paper focusing on the depictions of African men in bathhouse mosaics, Clarke argues that the frequently macrophallic and otherwise nonconformist body shapes (from the perspective of the white, Italic patrician writers) of the black slaves depicted in these spaces also serve an apotropaic function (Clarke 1996: 195–6). The idea that strange or unusual images or substances may be actively sought out as a key element of magical practice is also discussed by Wilburn (2012: 87–8). According to Clarke, it is the construction of an amusing picture to elicit mirth (or the sound of laughter) and appropriate apotropaic imagery, which are used in combination to provide supernatural protection to physical spaces.

In either case, the presentation of all of these mosaics is very public. They appear in domestic homes and public bathhouses, places in which they will be interacted with. A mosaic is, of course, a floor, and floors are designed to be walked over and covered in furniture and domestic decoration. In such a scenario, the apotropaic function of these phallic mosaics is implicit and static. Importantly, however, they may be included in liminal places in need of additional protection. Doorways and open public rooms were in need of supernatural protection in the Roman world. Priapic statues or paintings may be found at the entranceways of houses, such as the House of the Vettii at Pompeii, in order to serve an apotropaic function (Swift 2009: 41). What better substitute as an avatar for Priapus than a disembodied phallus on a floor mosaic?

Protection was not just given to doors; the common practice of hanging curtains in doorways prompts a consideration of whether fabric designs may also have included apotropaic motifs that could protect a space adorned by a curtain (Swift 2009: 42). Elsewhere I have argued that phallic windchimes (*tintinnabula*) may have been particularly well suited to this particular function because of their physical mobility and noise-making characteristics; by opening a door or curtain and unwittingly moving the windchimes, an entrant alerts the occupants of a building that someone has entered as well as highlighting to the entrant that their presence has been identified (Parker 2018). Unlike *tintinnabula*, phallic mosaics do not require any physical interaction to 'work' in an apotropaic sense. That said, they are designed to be physically interacted with through people's feet. The examples in a bathhouse may have a secondary sensory link to the apotropaic function of these bodily fluids if they are themselves wet. We have no indications whether the floor designs could be deliberately targeted or avoided by walkers for supernatural reasons unless the mosaic was somehow differently elevated or positioned in the floor to require an active human choice. As an example of this choice being made available: an *in situ* ithyphallic carving raised several centimetres above the ground on a circular dais in the *principia* of Chesters Fort on Hadrian's Wall would have represented a significant trip hazard to the unwary (Parker 2017: 119). For those aware of its presence, a conscious choice to walk over, on, or otherwise around the carving was necessary.

Not all the relationships between phallic iconography and fluids are quite as obvious as those carved into stone or pictured in mosaics; more subtle, conceptual links between these subjects can be suggested. The *tintinnabula* (see Figure 11.7) mentioned earlier, for example, may have had their musical functions activated by natural forces like the wind; exposure to wind may also provide exposure to rain. Indeed, the use of falling water onto the metallic *tintinnabula* could, we might speculate, produce a different auditory experience to its normal function. If activated by hand or by the wind, it is the movement of the internal clapper against the bell wall that produces the sound. With rain, the whole metal object itself produces sound, ringing off the central figure as well as the bells, and the sound is an important facet of the apotropaic function of these objects—rain may have had transitive and transformative effects on this sound. There are, thankfully, slightly clearer links than this one when it comes to *tintinnabula* because of the existence of such examples with oil lamps suspended beneath them (for example, in the Museo Archaeologico Nazionale di Napoli). Oil lamps are, of course, purpose made for containing fluids; the oil is required to be poured in by hand and then slowly burned away to produce light, heat, and smoke. One example includes a macrophallic dwarf (again a humorous character, like the slaves on the mosaics) with four bells suspended beneath him. Suspended further below is a dual sided lamp, also made from copper alloy. The fluid oil is physically connected to the *tintinnabulum* by chains, the light and heat of which interacted with the figure; the interplay of temperature on the figure (no doubt it would have heated up), the flickering light of the lamp against the shadow of the bright bronze figure, and the movement and sound of the bells created a complex sensory experience. While not associated with bells, a macrophallic faun in ceramic (Johns 1982: pl. 34) was designed as a hanging oil lamp with his urethra used as the spout—the flames emanating from the tip of his penis. In this case the fluids were hidden internally in the void of his body. No doubt the shocked expression of the faun, with both hands aloft, was intended to be entertaining

Figure 11.7 Zoomorphic phallus, copper-alloy *tintinnabulum*, first century CE. Now in the British Museum, 1856, 1226.1086.

Source: © Trustees of the British Museum.

while his phallus was on fire. Other sexual and erotic scenes are depicted on oil lamps and may involve couples or multiple people mid-coitus (Dumas 2012: 16–18; Fedele and Labate 2013) as well as more supernatural scenes, such as Leda and the swan or a woman and a horse (Johns 1982: 108–9, figs. 89 and 90).

Oil lamps are not the only containers which provide conceptual links between phalli and fluids. The Nene-valley colour-coated ware ceramic industry in Roman Britain lasted from the mid-second to the end of the fourth century CE; the ware is characterised by a fine, white clay fabric coated in a dark brown colour coat and often decorated with applied barbotine. The decoration facilitated the depiction of people, plants, animals, and geometric designs in medium-relief on the exterior of its vessels. The vast majority of vessels of this ware are beakers, wide-rimmed, with pedestal bases. A number of examples have been found to have been decorated with phallic or erotic scenes. As a case in point, take a beaker from Horsey Toll, Cambridgeshire (Johns 1982: 95, fig. 80). Clearly incorporating human players, this cup depicts a female, facing right and bent over. In her left hand she holds a huge, disembodied ithyphallic penis (a sex toy?). She turns her head over one shoulder and with her other hand clearly gesticulates, with a pointed finger, to her rear end. Behind her, a naked, macrophallic man is striding or running towards her, but separated by some distance. He is ejaculating as he is travelling. Whether this is an image encouraging sexual penetration of the man into the woman or she is suggesting that the phallus in-hand might be used to replace that of the man is not clear. As Catherine Johns suggests, this image is almost certainly amusing as well as erotic, given the failure of the man to engage in this sexual act despite his best efforts and the encouragement of his partner. No doubt that this is, itself, a suitable function of the imagery, but the supernatural world of Roman experience was a nuanced one; it is entirely possible that the source of amusement acted as a positive apotropaic device—a 'magical sound'. Thus the ejaculating phallus could be part of a scene that was subtle in its function if not in its artistic decoration. There is, again, a sensory link between the function of the object depicting ejaculating phalli and real, physical fluids. In this case the beaker is a vessel designed to hold liquids. Conversely, in this case, the liquid is going into the person using the vessel rather than physically leaving them; perhaps if it contained an alcoholic beverage it could have hastened the need to urinate, as alcohol is a proven diuretic (Hobson and Maughan 2010).

Perhaps the clearest link between a phallic image and a real, physical fluid is a glass *unguentarium* from Roman period Hawara, Egypt (Walker and Bierbrier 1997: 206). The glass vessel is formed in the shape of a moulded phallus and testes; the tip of the glans is missing, from where the sealed vessel was broken to access its contents. When full, the liquid inside would be visible through the glass, though its unusual shape prevented the vessel from free-standing in any obvious way, and if the liquid was a different colour, or was particularly viscous, it may have provided a visual experience before the physical act of breaking released the contents. Given the nature of *unguentaria*, these contents are likely to have released a strong smell, and there is no simple way of resealing or storing this vessel once it was opened. It should also be noted that as a vessel, this object mimics the biological model—fluids leave from the tip despite the narrowed collar between shaft and testes probably being a natural break-point in the glass; care and attention has been given to breaking it at the end.

It formed part of a grave group and, perhaps, fulfilled a performative ritual function during the funerary rites. Recent chemical analyses have demonstrated the presence of resinous plant extracts (mastic, frankincense, and other fragrant lipids) in late Roman mortuary contexts in Britain (Brettell et al. 2015), and the contents of this vial may have fit into this category. The aromatic smell acted as a deodoriser for the smell of decay; the phallic *unguentarium* may take this a step further by linking an apotropaic function of the fluid container to this part of the mortuary practice. It might be a step too far to link the smell of the unguent here to a specific ritual action, but there is a clear conceptual blending between the container, its fluid content, and the function.

186

The 'sensory turn' in Roman archaeology attempts to recapture or reconstruct some of the multisensory experiences that have been difficult to access archaeologically, through phenomenological, embodied, or qualitative approaches (Betts 2017: 7–12). Considering the embodied nature of this unusual unguent bottle—as a hard, portable, fluid- and colour-filled vial, capable of movement and spillage, seems to place fluidity as a primary concern—it was, much like the biological purpose of a penis, designed to harness specific fluids until such a time as they can be released, never to be reclaimed. That said, the potential haptic, olfactory, and kinaesthetic experiences of this vial are all quite removed from the biological analogy; perhaps this difference was meaningful?

As an addendum to this example, a copper-alloy phallic pendant from Meaux, France, is of great importance to the discussion. The pendant is approximately 35 mm in length, ithyphallic, and slightly upcurving, with stylised testes and a large sub-oval suspension loop on its back (Trombetta 2000: 101, fig. 33). The interest comes from the inclusion of an opaque white glass applique on the tip of the glans, interpreted as ejaculate. If correctly identified, this may be the only portable object from the Roman world which depicts an ejaculating phallus. Perhaps in contravention to the other depictions, it is not spurting from the phallus but resting on the tip in a state of permanent ejaculation or, perhaps, it represents the biological state before or immediately after ejaculation. The flat back, behind the testes, in combination with the position of the suspension loop is likely to have resulted in a pendant that rested against the body and projected away from it in a manner wholly comparable to the reconstructed example given by Whitmore (2017). In Whitmore's example, the replica phallic pendant faced forwards, heading towards the direction of travel and towards the supernatural dangers that its wearer may have approached.

In reaching towards a conclusion for this broad discussion regarding the imagery of ejaculating phalli within the Roman world, clearly there were multiple uses for phallic imagery in the Roman world, and not all examples can be linked to ejaculate, urea, or other fluids in the physical or conceptual sense. Such links do, at least, exist, whether it be through the overt depiction of the victory of the phallus over its enemy or through conceptual allusions between images of the phalli and the functions of the objects they are depicted on, and these are of interest as an amorphous group of objects. Beyond its part in mortal life-creation, the phallus (and by extension its fluids) was regarded as a powerful symbol of protection and one part of an apotropaic, supernatural tool-kit which can be efficacious over time and space and, indeed, was used differently in different times and different spaces for carefully considered, explicit purposes. Ejaculate was used to depict the literal spreading of good luck or apotropaic protection away and beyond the reach and scope of the phallus itself. In its artistic forms it depicts a supernatural practice that is inspired by a something that is, in its origins, visceral, organic, and intrinsically human.

Stone carvings on walls show this supernatural duel between phallus and Eye quite clearly and served a relatively static, supernatural function. We should not forget that in all cases, the walls in which they were placed were likely to be adorned with graffiti or covered by market stalls, buildings, trees, or climbing plants and obscured by mud, moss, or grime over time—they need not be interacted with to fulfil their supernatural function. We should give the ancient world credit for being an incredibly complicated place, and this is particularly true of magical practice. Ephemeral things such as words, gestures, sounds, and smells may have fulfilled the exact same functions as these images of ejaculating phalli and are lost to us. Through this chapter I have promoted one possible ephemeral connection which may have been lost. The extent to which the conceptual links between phalli and fluids in the *tintinnabula*, oil lamps, and glass and ceramic vessels were explicit in the Roman world is entirely speculative. It is hoped that these are seen not as disparate strands brought together by an author clutching at straws to prove a point but as one way to consider an embodied experience of the relationship between an apotropaic icon and real people in the Roman world.

Notes

* I would like to thank Victoria Leonard specifically for inviting this contribution following a chance discussion at the Classical Association Conference 2017 and the editors as a team for commenting on and improving this chapter. Thanks are certainly due to Alissa Whitmore for her correspondence and references to relevant materials which have vastly improved the scope of this chapter, and to Helen King, one of my PhD supervisors, for reading and commenting valuably on an earlier draft of this work.
1 The essential work on the apotropaic use of the phallus in the ancient world is Catherine Johns' (1982) *Sex or Symbol? Erotic Images of Greece and Rome*. See also: Turnbull 1978; Jelski 1984; Blasquez 1985; Clarke 1996, 1998; Deschler-Erb and Božič 2002; Dasen 2003, 2015: 185–8; Johns and Wise 2003; Kovač and Koščević 2003; Plouviez 2005; Peña 2008; Parker 2015, 2017, 2018; Whitmore 2017, 2018, 2020; Faraone 2018: 75–8.
2 A selection of archaeological and anthropological definitions: Merrifield 1987: 6; Tambiah 1990: 7; Campbell 1998; Faraone 2001: 16; Stein and Stein 2008: 140; Manning 2014: 1; Chadwick 2015: 37.
3 See Julie Laskaris' contribution in this volume, Chapter 6, particularly on the potentially harmful effects of sight by menstruating women (pp. 113–14) and the dangerous potential of double-pupils (p. 115).

Bibliography

Primary sources

Babbitt, F. C. (1936) *Plutarch. Morals*. London: William Heinemann.
Coulston, J. N. C. and E. J. Phillips (1988) *Corpus Signorum Imperii Romani* (Vol. I, Fascicule VI. Hadrian's Wall West of the North Tyne, and Carlisle). London: British Academy.
De Lacy, P. (1992) *Galen. On Semen*. Text edited with a translation and commentary by P. De Lacy. Berlin: Academie Verlag.
Jones, W. H. S. (1963) *Pliny. Natural History. With an English Translation. Vol. VIII: Books XXVIII–XXXII*. Cambridge, MA: Harvard University Press.
Kent, R. G. (1938) *Varro. On the Latin Language, Vol. I: Books 5–7*. Cambridge, MA: Harvard University Press.
Olsen, S. D. (2004) *Aristophanes. Acharnians*. Oxford: Oxford University Press.

Secondary literature

Adams, J. N. (1982) *The Latin Sexual Vocabulary*. London: Duckworth.
Betts, E. (2017) 'Introduction: Senses of empire', in E. Betts (ed.) *Senses of the Empire: Multisensory Approaches to Roman Culture*. Abingdon: Routledge, 1–12.
Blasquez, J. M. (1985) 'Tintinnabula de Mérida y de Sasamón (Burgos)', *Zephyrus* 38, 331–5.
Boschung, D. and J. N. Bremmer (eds) (2015) *The Materiality of Magic*. Paderborn: Wilhelm Fink.

Brettell, R. C., E. M. J. Schotsmans, P. Walton Rogers, N. Reifarth, R. C. Redfearn, B. Stern, and C. P. Heron (2015) '"*Choicest unguents*": Molecular evidence for the use of resinous plant extracts in late Roman mortuary rites in Britain', *Journal of Archaeological Science* 53, 639–48.

Campbell, C. (1998) 'Magic', in W. H. Swatos (ed.) *Encyclopaedia of Religion and Society*. Walnut Creek: Altamira Press.

Chadwick, A. M. (2015) 'Doorways, ditches and dead dogs: Excavating and recording material manifestations of practical magic amongst later prehistoric and Romano-British communities', in C. Houlbrook and N. Armitage (eds) *The Materiality of Magic: An Artifactual Study Investigation into Religious Practices and Popular Belief*. Oxford: Oxbow, 37–64.

Clarke, J. R. (1996) 'Hypersexual black men in Augustan baths: Ideal somatotypes and apotropaic magic', in N. Kampen (ed.) *Sexuality in Ancient Art*. Cambridge: Cambridge University Press, 184–98.

———. (1998) *Looking at Lovemaking: Constructions of Sexuality in Roman Art, 100 BC-AD 250*. Berkeley: University of California Press.

Collingwood, R. G. and R. P. Wright (1965) *The Roman Inscriptions of Britain, I: Inscriptions on Stone*. Oxford: Clarendon Press.

Collins, R. (2020) 'The phallus and the frontier: The physical and metaphysical barrier of Hadrian's wall', in T. Ivleva and R. Collins (eds) *Un-Roman Sex: Gender, Sexuality and Lovemaking in the Roman Provinces and Frontiers*. Abingdon: Routledge, 274–309.

Dasen, V. (2003) 'Les amulettes d'enfants dans le monde gréco-romain', *Latomus* 62, 275–89.

———. (2015) 'Probaskania: Amulets and magic in antiquity', in D. Boschung and J. N. Bremmer (eds) *The Materiality of Magic*. Paderborn: Wilhelm Fink, 177–203.

Deschler-Erb, E. and D. Božič (2002) 'A late Republican bone pendant from the Münsterhügel in Basel (CH)', *Instrumentum Bulletin* 15, 39–40.

Dumas, C. (2012) 'L'art érotique de la mythologie au spiritual', *Sexe à Rome: Au-delà des idées Reçues. Dossier d'Archéologie* 22, 14–19.

———. (2017) 'Rome, sexe dans la cité', *La Marche de l'Histoire* 19, 26–36.

Dunbabin, K. (1989) '*Baiaum grata voluptas*: Pleasures and dangers of the baths', *Papers of the British School at Rome* 57, 6–46.

Faraone, C. A. (2001) *Ancient Greek Love Magic*. Cambridge, MA: Harvard University Press.

———. (2018) *The Transformation of Greek Amulets in Roman Imperial Times*. Philadelphia, PA: University of Pennsylvania Press.

Fedele, A. and D. Labate (2013) 'Una rara lucerna con scena erotica e iscrizione da *Forum Popilii*', *Forlimpopoli: Documenti e Studi* 24, 65–78.

Gordon, R. L. and F. Marco Simón (2010) *Magical Practice in the Latin West: Papers from the International Conference held at the University of Zaragoza 30th Sept.-1 Oct 2005*. Leiden: Brill.

Hicks, D. (2010) 'The material-cultural turn: Event and effect', in D. Hicks and M. C. Beaudry (eds) *The Oxford Handbook of Material Culture Studies*. Oxford: Oxford University Press, 25–99.

Hobson, R. M. and R. J. Maughan (2010) 'Hydration status and the diuretic action of a small dose of alcohol', *Alcohol and Alcoholism* 45(4), 366–73.

Houlbrook, C. and N. Armitage (eds) (2015) *The Materiality of Magic: An Artifactual Study Investigation into Religious Practices and Popular Belief*. Oxford: Oxbow.

Jelski, G. (1984) 'Pendentifs phalliques, clochettes, et peltae dans les tombes d'enfants de Gaule Belgique. Une découverte à Arras', *Revue du Nord* 66(260), 261–80.

Johns, C. (1982) *Sex or Symbol? Erotic Images of Greece and Rome*. London: British Museum Press.

Johns, C. and P. J. Wise (2003) 'A Roman gold phallic pendant from Braintree, Essex', *Britannia* 34, 274–6.

Kovač, D. and R. Koščević (2003) *The Phallus vs The Curse: The Archaeological Collection of Dr. Damir Kovač*. Zagreb: Muzei Grada Zagreba

Manning, M. C. (2014) 'Magic, religion, and ritual in historical archaeology', *Historical Archaeology* 48(3), 1–9.

Merrifield, R. (1987) *The Archaeology of Ritual and Magic*. London: B.T. Batsford.

Ostrowski, J. A. (1990) 'Personifications of rivers as an element of Roman political propaganda', *Études et Travaux* 15, 309–16.

Otto, B.-C. (2013) 'Towards "historicizing" "magic" in antiquity', *Numen* 60, 308–47.

Parker, A. (2017) 'Protecting the troops? Phallic carvings in the north of Roman Britain', in A. Parker (ed.) *Ad Vallum: Papers on the Roman Army and Frontiers in Celebration of Dr Brian Dobson*. Oxford: British Archaeological Reports Publishing, 117–30.

———. (2018) 'The bells! The bells! Approaching *tintinnabula* in Roman Britain and beyond', in A. Parker and S. McKie (eds) *Material Approaches to Roman Magic: Occult Objects and Supernatural*. Oxford: Oxbow, 57–68.

Parker, A. and C. Ross (2016) 'A new phallic carving from Roman Catterick', *Britannia* 47, 271–9.

———. (2015) 'The fist-and-phallus pendants from Roman Catterick', *Britannia* 46, 135–49.

———. (2016) 'Staring at death: The jet Gorgoneia of Roman Britain', in S. Hoss and A. Whitmore (eds) *Small Finds and Ancient Social Practices in the North-West Provinces of the Roman Empire*. Oxford: Oxbow, 98–113.

Peña, A. G. (2008) 'Amuletico fálico Romano hallado en la Puebla del Rio (Sevilla)', *SPAL Revista de Prehistoria y Arquelogia de la Universidad de Sevilla* 17, 329–34.

Plouviez, J. (2005) 'Whose good luck? Roman phallic ornaments from Suffolk', in N. Crummy (ed.) *Image, Craft and the Classical World: Essays in Honour of Donald Bailey and Catherine Johns*. Montagnac: Éditions Monique Mergoil, 157–64.

Stein, R. L. and P. L. Stein (2008) *The Anthropology of Religion, Magic and Witchcraft*. Boston: Routledge.

Swift, E. (2009) *Style and Function in Roman Decoration: Living with Objects and Interiors*. Farnham: Ashgate.

Tambiah, S. J. (1990) *Magic, Science, Religion, and the Scope of Rationality*. Cambridge: Cambridge University Press.

Trombetta, J. P. (2000) 'La guerre des Gaules: Phallus d'Ile-de-France et d'ailleurs', *Actes des Journées Archéologiques d'Ile-de-France*, 83–110.

Turnbull, P. (1978) 'The phallus in the art of Roman Britain', *Bulletin of the Institute of Archaeology, University of London* 15, 199–206.

Versnel, H. S. (1991) 'Some reflections on the relationship magic–religion', *Numen* 38(2), 177–95.

Walker, S. and M. Bierbrier (1997) *Ancient Faces: Mummy portraits from Roman Egypt*. London: British Museum Press.

Whitmore, A. (2017) 'Fascinating *fascina*: Apotropaic magic and how to wear a penis', in M. Cifarelli and L. Gawlinksi (eds) *What Shall I Say of Clothes? Theoretical and Methodological Approaches to the Study of Dress in Antiquity*. Boston: Archaeological Institute of America, 47–65.

———. (2018) 'Phallic magic: A cross cultural approach to Roman phallic small finds', in A. Parker and S. McKie (eds) *Material Approaches to Roman Magic: Occult Objects and Supernatural Substances*. Oxford: Oxbow, 17–31.

———. (2020) 'Egyptian faience flaccid phallus pendants in the Mediterranean, Near East, and Black Sea regions', in T. Ivleva and R. Collins (eds) *Un-Roman Sex: Gender, Sexuality and Lovemaking in the Roman Provinces and Frontiers*. London: Routledge, 310–45.

Wilburn, A. T. (2012) *Materia Magica: The Archaeology of Magic in Roman Egypt, Cyprus, and Spain*. Ann Arbor: University of Michigan Press.

Worrell, S. and J. Pearce (2014) 'II: Finds reported under the portable antiquities scheme', *Britannia* 45, 397–429.

Part IV

NUTRITIVE AND HEALTHY FLUIDS

12

A NATURAL SYMBOL? THE (UN)IMPORTANCE OF BLOOD IN EARLY GREEK LITERARY AND RELIGIOUS CONTEXTS*

Emily Kearns

Blood, in our culture, is a substance with profound symbolic and metaphorical import. It is also, obviously, supra-cultural in that it is a universal and universally observed substance within ourselves, and biological factors might suggest that at a very basic level there would be shared cultural understandings of blood; visible blood (thus lost blood) is most often a sign of something wrong,[1] and therefore retained blood is good and positive, necessary for life. Something of this sort is easily discerned in a number of different societies. Proto-Indo-European, for instance, had separate words for blood within the body and spilt blood (Mallory and Adams 2006: 187), and the thought-patterns and practices of numerous peoples attest to a view of blood outside the body as a powerful and sometimes dangerous substance.[2] To that extent, blood might seem to qualify as a truly natural symbol, supposing such a thing to exist.[3]

Yet it is also clear that when we look closer, the 'meanings' of blood do, in fact, vary considerably from culture to culture. As Mary Douglas (1973) put it, each symbol has meaning only in connexion with other symbols in a pattern, and the pattern is different in different groups and populations. Her work attempts to move beyond this problem, and yet the problem itself clearly has some validity. To take an obvious example, the central Christian structures of belief, involving the salvific role of Christ's body and blood, dramatically complicated the picture for European societies, especially in dialogue and conflict with Jewish ideas (see especially Biale 2007; Bynum 2007). Further, the very important nexus of ideas relating blood to kinship, so familiar to us, is far from universal. Notably, it is lacking in the Hebrew scriptures and in traditional Jewish thought generally (Anidjar 2014: 170 and nn. 58–9). Despite such cultural differences, when it comes to ancient Greece, although excellent work has been done on medical aspects and on menstrual blood, relatively little attention has been paid to establishing the significance of blood more generally.[4] Superficially, Greek ways of thinking and speaking about blood might seem similar to our own: aside from the literal use, the two main meanings given for αἷμα (*haima*) in Liddell-Scott-Jones's Greek lexicon are 'bloodshed, murder' and 'blood relationship, kin'—even the definitions are phrased in such a way as to highlight the similarities in the usage of the two languages. Modern writers have not always been alert to possible differences in this respect between the two cultures, nor always careful to distinguish between the literal and figurative senses of blood (as highlighted in Anidjar 2014: 167–9). This confusion

DOI: 10.4324/9780429438974-17

between two senses and the partial coincidence of thought-patterns between ourselves and those we are studying seems to have restricted genuine inquiry to a much smaller body of work than one might expect; the subject has not been seen as important or interesting, perhaps has not even been seen as a subject at all. Gil Anidjar, author of an important yet (from a classicist's point of view) flawed study centred on the disastrous symbiosis between Jewish and Christian concepts of the substance, asks, with some evident irony: 'Who would think of writing a book on Greek blood?' (Anidjar 2014: 174).[5]

The present investigation is a little more modest. I shall examine the presence, or absence, of blood in three areas: early epic (chiefly the *Iliad*), sacrificial procedure, and Aeschylean tragedy. Of course there is no one-to-one equivalence between these three areas. Two are composed of literary texts, one set of which has strongly influenced the other; the third is a real-life phenomenon, the evidence for which we piece together from very diverse sources and the significance of which is strongly debated. Nonetheless, all three are expressions of early Greek culture, and similarities and differences between them may help us to see deeper patterns in the meanings of blood in Greek culture, especially as we take into account possible correlations of varying presentation with the aims and functions of the three fields we are looking at.

I begin then with the *Iliad*, a work—according to a famous quotation from M. I. Finley—that is 'saturated in blood' (Finley 1972: 138). Finley, of course, was using blood in a metaphorical, or better, a metonymic sense.[6] Blood, in our way of thinking and speaking, functions as an expected or even a necessary correlate of violent death. But if we approach the question more literally, how bloody in fact is the *Iliad*? The concordance shows a number of relevant terms are used in the poem—*phonos*, *brotos*, and *luthron* (though not the old word *ear/eiar*[7]) all appear in this sense. But *haima* is by far the most frequent term for blood, and this word and its correlates appear roughly ninety-eight times in the poem.[8] This certainly seems quite a lot, especially when we see that only four of these occurrences refer to kinship. The result appears to bear out what we might expect of a war epic—after all, in realistic terms visible blood is surely a feature of the vast majority of fatal and non-fatal injuries alike. But the blood is not necessarily where we might expect it. Out of something over two hundred deaths described in the *Iliad*,[9] only ten mention blood, with two further cases (Sarpedon and Hector) where blood is described in the aftermath of death. Among the frequently used formulas describing the death of a warrior, one might reasonably expect to find a reference to blood. There is in fact one occasionally used phrase which might seem to fit the bill, but as we shall see shortly it is used only in specialised circumstances. If we look for something occurring more regularly, perhaps 'his blood flowed to the ground', 'the black earth drank his blood', or some such, we shall be disappointed. Only in one place we have 'black blood flowed out and wetted the earth' (13.655: ἐκ δ' αἷμα μέλαν ῥέε, δεῦε δὲ γαῖαν). Even in the Hesiodic *Aspis* (*Shield*), where we find the 'pseudo-formula' αἷμ' ἀπελείβετ' ἔραζε ('blood flowed out to the ground'), it is applied quite differently in the two places (174, 268) where it appears, once to an animal fight and once in the description of the personified *Achlus* (Mist or perhaps Sorrow) (see Mason 2015: 97–8). There is no blood at all in the poem's description of the actual conflict, the killing of Kyknos.

On the other hand, in the *Iliad* blood is certainly a feature of the generalised battle-field, and it is conspicuous too in the majority of accounts of non-fatal injury, where, as Tamara Neal has shown, it frequently serves as a badge of honour to indicate a hero's vigour, courage, and endurance.[10] In the case of the first wound in the narrative, the shooting of Menelaus which ends the truce, there is also an aesthetic element: the blood staining Menelaus' comely thighs is compared to the work done by a crafts-woman staining ivory with crimson or purple dye (4.148–52). Blood also appears in post-mortem descriptions, where it may be something which like dust disfigures the body of a previously beautiful and strong fighter; thus in the case of Sarpedon (16.638): 'Even a perceptive man would not have known godlike Sarpedon, since he was covered in missiles and blood and dust from his head right down to his feet.'[11] When corpses are recovered and tended by their own side, the blood is washed off them (6.424–6; 18.345; 23.41, etc.). That Homer and his characters did in fact associate blood with death is shown by Agamemnon's and Menelaus' reaction to the blood spilling from the latter's wound; they shudder, until Menelaus sees that the barbs remain outside his skin (4.148–52). Most tellingly of all, blood is used metonymically to express the slaughter and carnage of fighting: 'Zeus brought Hector away from the missiles, from the dust, from slaughter of men, from blood, and from din' (11.163–4),[12] and metaphorically in the thrice-repeated phrase αἵματος ἆσαι Ἄρηα, 'to glut Ares with blood'.[13] On the level of visual symbolism, it is the striking image of water reddening with blood that introduces the episode of Achilles' fight in and with the river (21.21).

But in individual deaths, where any detail is given, the poet focuses on body parts, frequently the internal organs. Neck tendons are severed, brains are smashed within helmets, hearts and intestines are pierced, bones break, eyes and livers fall out. When blood does appear, it often heightens the gruesome effect: the eyes that fall out in the dust by their owner's feet are bloody (*haimatoenta*), and the brain that gushes out from the spear wound when Ajax kills Hippothoos is likewise *haimatoeis*;[14] the blood is an optional add-on in the virtuoso variety of disgusting deaths. We have seen that blood appears in ten cases out of well over two hundred deaths. Three of these ten cases use what seems to be a formula (5.79–8; 16.330–4; 20.474–7), suggesting a specialised kind of death in which blood is the central observable feature.[15] In one, the victim's forearm is cut off and falls, bloody, to the ground; the second has the warrior's neck struck with the sword; and finally, the defeated man's head is split. In the last two cases, 'the whole sword was warmed with blood' (πᾶν δ' ὑπεθερμάνθη ξίφος αἵματι), and then in all three, 'over the eyes, crimson death took him, and strong fate' (τὸν δὲ κατ' ὄσσε/ἔλλαβε πορφύρεος θάνατος καὶ μοῖρα κραταίη). At least, that is the way the phrase is usually understood, and it makes a very powerful image. It seems very likely that later Greeks read it this way as well (see p. 199). The sense of *porphureos* in the *Iliad* is not, however, entirely clear, and it may not be a colour term at all but refer to the rapid movement of liquid, as when it is applied to the sea or a wave.[16] In that case, it would be *flooding* death that would take hold of the victim, with reference to an unstoppable flow of blood. Either way, with a transferred epithet, these deaths are given one of the conspicuous qualities of blood and characterised as one among many varieties of violent killing. The exception proves the rule—here, blood may be centre stage, but in most deaths it is simply not mentioned.

Most often, the location of an injury and perhaps a detail of the effect is mentioned, but not the accompanying blood. In a few cases, the plausible blood is actually suppressed in favour of an unlikely or impossible alternative. Thus, at 20.478–83 a decapitation is followed by *muelos* (marrow? cerebrospinal fluid?) spilling out of the vertebrae. Kenneth Saunders informs us that this is impossible—what we would see instead would be only venous and arterial blood (see his appendix to Friedrich 2003: 149–50). It may be relevant that this death immediately follows one of the *porphureos thanatos* episodes, so the poet is seeking variety—but he has still neglected realistic blood in favour of fantastic, more 'internal' fluid. Most oddly of all, perhaps, in an image which anticipates the actual violence of the battlefield, the episode of the oath in book 3 compares the flowing wine of the libations not to the *blood* of those who transgress the oath, as might seem obvious (there are plenty of later Greek examples suggesting a wine/blood analogy[17]), but to *brains* flowing and spattered on the ground.[18] Why should this be so—what explains the preference for body parts over blood? It is hard to avoid the conclusion that Homer, or the poetic tradition, is setting out to shock and titillate. When Achilles kills the supplicating Tros with a sword slash through his belly, the liver (impossibly) slips out, and Saunders is moved to comment that the 'use of the special word ἐξολισθάνω is disgracefully onomatopoeic' and that 'Homer enjoys this sort of thing, I believe, however unpleasant it may seem to some modern readers.'[19]

Visible body parts are not only more specific—they supply greater variety—they are less dignified than blood, a substance which may shock but, unlike many bodily fluids, tends not to inspire disgust. The most important heroes, for whom we have been encouraged to feel as individuals, do not die with mangled limbs or spilled guts. They have been defeated, they are no longer able to act as heroes, but their corpses disfigured with blood can evoke admiration and pity, somewhat as blood flow marks the endurance of the hero who survives his injury. Indeed, visible blood may be a feature of either a fatal or non-fatal injury, whereas the appearance of the inner organs will almost always be fatal—the warrior is reduced to a machine which no longer functions.

So blood has an important significance in the *Iliad*, but the poem seems to have moved away from the out-and-out identification of blood and violent death, which it nonetheless attests in some of its more general descriptions. One might suppose that this was just another example of the curious stylisation of the Homeric world, and so in a sense it is—but it may not be purely a literary phenomenon. If we move from a text to a thematic investigation, we can see that there is some similarity in the treatment of blood in sacrifice, both in what we can gather of the reality and in the manner of its portrayal. We regularly talk about 'blood sacrifice' as a synonym for animal sacrifice, but while 'bloodless sacrifice', *anaimaktos thusia*, is a phrase found in the Greek of the Imperial period (although not before), its opposite seems not to be an ancient locution.[20] Of course, blood was necessarily present in animal sacrifice through the slitting of the victim's throat. But in visual depictions, as is well known, the moment of the kill is seldom depicted (Van Straten 1995: especially 186–8). Instead, we see the procession, the animal at the altar, and other elements which stress the sacral nature of the act and divine-human relations; and we see scenes from after the slaughter—the inspection of the liver, the preparation of the *splanchna*, and so on. Blood was indeed splashed on the altar as a sign or a part of the offering made, but as Gunnel Ekroth (2005) has shown, huge amounts were not used, and most of the blood caught in the special basin for the purpose was likely kept for human consumption in black puddings, sausages, and so on. Although at the point of death a sacrificial victim's blood must have been conspicuous, unlike that of a typical warrior in Greek epic, in the post-kill phase, which itself receives more emphasis in art, it is once again body parts which are foregrounded—the liver, so crucial for human-divine communication, the state of which will indicate whether the sacrifice was acceptable or not—and the entrails, consumption of which marks out the priest and a small inner circle as the most honoured and important participants.

So there is perhaps some common ground in this coincidence of sacrificial practice and Homeric narrative. Visible, separated body parts, as we have seen, are undignified, while the same is not necessarily true of blood, which can be spilt and still allow the body to retain a degree of integrity. Sacrificial animals have little dignity on their own account in the Greek world, and Homeric warriors generally wish to ensure that the same is true of their defeated enemies. There are, however, a few important exceptions to the general sacrificial rule: first, the *sphagion*, the 'slaughter sacrifice' performed immediately before joining battle, in which the victim's throat was often pierced with a sword rather than a knife and was then bled directly onto the ground. Although the manner in which the victim bled could have been used for divinatory purposes, the emphasis was clearly on killing, not on elaborate preludes and postludes. Michael Jameson characterises the thought process here as 'we kill—may we kill!' The action is mimetic of what is desired to happen in the battle that follows. Such vase depictions of sacrificial killing as we do see appear to show this type of sacrifice (Jameson 1991). A few heroic sacrifices were of a kind not dissimilar to this, with the victim's throat cut and bled out onto the ground, where, presumably, the blood was received by the dead hero.[21] In oath sacrifice, too, though the offering is made to the deities by whom the oath is sworn, the violence is often emphasised with the dismemberment of the victim, over which the oath itself is made, and sometimes with the participants placing their hand or weapons in the collected blood.[22] Finally we might consider literary and visual depictions of human sacrifice, in which the procedure is based on the *sphagion* rite, with the victim dispatched by sword rather than sacrificial knife and, obviously, no extispicy and no sacrificial meal. Consequently, here too the emphasis is on the kill, and in some descriptions, on blood—although as imagined rather than real events, such descriptions are obviously in a different category and aim among other things to evoke pity for the victim.

Another animal-related ritual—perhaps not strictly sacrifice—where blood is conspicuous is purification. As Heraclitus famously says 'they . . . purify themselves [from blood?] with blood, as if someone who had stepped in mud were to wash it off with mud.'[23] In fact rites of purification take many forms, but in many cases, notably but not exclusively those of purification from homicide, the killing of a piglet was a central part of the procedure, with its blood being made to pour over the polluted person, as in the well-known South Italian vase showing the purification of Orestes.[24] Whether, as is generally assumed, the pollution is then removed along with the blood by various purifying substances, or whether—as Stella Georgoudi suggests—the blood of young animals is particularly pure, sufficient in itself to remove the impurities from the polluted person, it is clear that where an animal is killed in such a ritual, its blood is the central signifier in the ritual and indicates a powerful process.[25] This has some implications for the depiction of blood in tragedy, a genre to which I now turn.

Despite sharing with war epic an interest in violent death, tragedy is very different in its approach to blood. This can be illustrated almost exclusively from Aeschylus, since for our purposes the later two dramatists may be seen as largely following his lead. Although Aeschylus supposedly claimed that his plays were 'slices from Homer's banquet' (Athenaeus 8.347e), several centuries and much intervening literature, as well as generic difference, separate him from the earlier poet. A full explanation of the difference would have to take into account the ways blood is used—or not used—in Hesiod, the Epic Cycle, and the lyric poets. But the problems here are obvious. Either the subject matter of these poems is not much concerned with violence—at least among humans—or the texts are fragmentary, or both. Since Stesichorus is so often and so reasonably thought to be a crucial forerunner of Aeschylean tragedy, it would be particularly interesting to know what role was played by blood in his lyric narratives on heroic themes. Unfortunately, not enough survives for us to make generalisations, but we do have an interesting fragment of his famous *Geryoneis*, the story of Heracles' defeat of the multi-headed (multi-bodied?) monster Geryon:

διὰ δ' ἔσχισε σάρκα [καὶ] ὀ[στ]έα δαίμονος αἴσαι
διὰ δ' ἀντικρὺ σχέθεν οἰ[σ]τὸς ἐπ' ἀκροτάταν κορυφάν,
ἐμίαινε δ' ἄρ' αἵματι πορφ [υρέωι]
θώρακά τε καὶ βροτόεντ[α μέλεα.

The arrow severed flesh and bone through the fate of a *daimon*, and went right through to the top of the head, and stained his breastplate and gory (limbs?) with crimson[26] blood.

Fr. 19 edition Davies and Finglass 2014: 38–43;
translations are my own unless otherwise specified

The hero's wounding of Geryon's first head owes quite a lot to Homer, as the Iliadic simile of the drooping poppy which follows (fr. 19.44–7, cf. *Iliad*, 8.306–8) makes especially clear. And the description of the wound at first seems quite Homeric in its specificity (the site of the injury is described along with the effect of the weapon) but is actually a rather generalised presentation of the encounter of missile and body. Flesh, bone, and (notably) blood: this is rather more how we might, from first principles, expect an Iliadic injury to look like than the *Iliad* itself gives us. Obviously, we cannot say whether this is typical. If so, it already marks a move away from the Homeric pattern, which would not be entirely surprising given the different narrative emphases of lyric.

So while the first tragedies may well have available models of violent death which differ considerably from the *Iliad*, we must look at them on their own terms. *The Persians* uses blood rather sparingly, which is interesting in a play whose subject is warfare, indeed a war which was often seen as comparable to the war at Troy. But it uses body parts even more sparingly—in fact not at all[27]—so it is not a simple case of Aeschylus adopting patterns of Iliadic death. Still, there are some similarities. The messenger speech describing Salamis does not mention blood, but the following stasimon, responding to the report, ends with a reference to the island of Ajax αἱμαχθεῖσα δ' ἀρούραν, 'bloodied in its earth'. This follows what we have seen in the *Iliad*, where blood is used in general, broad-brush contexts to indicate fighting and death but much less so in individual death scenes. But one death among several in an earlier messenger speech stands out as characterised by blood:

Χρυσεὺς Μάταλλος μυριόνταρχος θανὼν
πυρσὴν ζαπληθῆ δάσκιον γενειάδα
ἔτεγγ᾽ ἀμείβων χρῶτα πορφυρᾶι βαφῆι.

Matallos of Chryse, leader of ten thousand, in death soaked his sandy-coloured, full and shady beard, changing his complexion with a crimson dye.[28]

Persians 315–17

The image is a striking one which owes something to the *porphureos thanatos* passages—and the reference to a dye-bath makes it quite clear that Aeschylus understood *porphuros* as a colour—but also to the description of Menelaus' wound and the associated simile in *Iliad* 4.134–47. So Aeschylus uses Homeric description, but uses it selectively to foreground blood in the imagery of death—the other deaths in the speech are far less vivid and individualised.

The Suppliant Women, for whatever reason, also uses blood sparingly and perhaps less interestingly.[29] We shall find more of interest in the *Oresteia* trilogy and the *Seven against Thebes*. In all four plays, blood of every kind is present: literal, metonymic, and metaphorical. Blood-based images are central, often overt, like the famous oath sacrifice in the *Seven* where a bull's blood is collected in a black-bound shield (42; more on this later) or Clytemnestra's account of her murder of Agamemnon, in which the blood signals the welcome death; sometimes figurative, like the visual metaphor of the purple cloth. Of course, imagery of all sorts has been very extensively studied in the *Oresteia* since the groundbreaking work of Anne Lebeck, and my approach owes much to Aeschylean scholars too numerous to mention.[30] More recently, Richard Seaford has explored some of the connexions between the flow of liquids in sacrifice and in tragedy, and in this volume Goran Vidović (Chapter 20) investigates the significance and interrelation of various body fluids in the *Oresteia*.[31]

Agamemnon, with its canvas of events from the quarrel of Atreus and Thyestes to the murder of the titular character, associates blood with all the key episodes, which it presents as linked in a chain of doing and suffering, cause and effect. The two main deaths, those of Iphigeneia and Agamemnon, are both indicated by blood—Iphigeneia's in prospect, as Agamemnon muses over his terrible choice (206–10):

βαρεῖα μὲν κὴρ τὸ μὴ πιθέσθαι,
βαρεῖα δ᾽ εἰ τέκνον δαΐξω, δόμων ἄγαλμα,
μιαίνων παρθενοσφάγοισιν
ῥείθροις πατρώιους χέρας πέλας βω-
μοῦ.

It is a heavy fate not to yield, but heavy if I kill my child, the delight of my house, fouling my fatherly hands with streams of maiden slaughter at the altar.

Agamemnon 206–10

—and Agamemnon's in retrospect, as Clytemnestra relates her action focusing on bloodspill, including the simile in which blood is confused with both rain and, at a remove, semen:

κἀκφυσιῶν ὀξεῖαν αἵματος σφαγὴν
βάλλει μ' ἐρεμνῆι ψακάδι φοινίας δρόσου
χαίρουσαν οὐδὲν ἧσσον ἢ διοσδότωι
γάνει σπορητὸς κάλυκος ἐν λοχεύμασιν.

He coughed up a sharp spurt of blood and hit me with a black shower of gory dew—at which I rejoiced no less than the growing corn rejoices in the liquid blessing granted by Zeus when the sheathed ears swell to birth.

Agamemnon 1389–92; trans. Sommerstein

The literal blood in these descriptions of unseen events is paralleled by the metaphorical blood which the audience actually sees—the purple cloth on which Agamemnon walks into the house (910ff)—and reinforced by numerous verbal references in which blood itself is used in varying shades and combinations of metonym and metaphor. In the first two-thirds of the play, blood is associated with Helen's destructiveness— notably in the parable of the lion cub—with the fall of Troy, and with the fate of slaves who resist (*Ag.* 698, 710–15, 730–4, 827–8, 1065–7). Later, Cassandra hopes for her death to come 'with easy flow of blood' (*Ag.* 1293–4). In the last third, the references accelerate with Cassandra's visions and Agamemnon's death: blood gives a smell to the house, it appears in a murderer's eyes, it falls like heavy rain, it is associated with Ares, it is licked and drunk by the Erinyes and by a nameless δαίμων—the examples are not exhaustive.[32] By contrast, inner body parts are not especially conspicuous in Aeschylus generally,[33] but they do appear once, strikingly, in this play: in Cassandra's vision of Thyestes' children holding their own flesh and inner organs in their hands (*Ag.* 1219–22). This draws attention to the particularly revolting nature of the earlier crime. It indicates how the children have been deliberately dehumanised and stripped of their personal identity, ending up as stew eaten by their father.

The second and third plays of the trilogy both use 'blood' frequently as metonymic for murder, often in the idea of blood*shed* most familiar to us. *Libation Bearers* (*Choephoroe*) is full of the imagery of liquids dripping or poured onto the ground, as blood does when someone is killed. 'What atonement is there for blood which has fallen to the earth?' ask the chorus early on in the play (48: τί γὰρ λύτρον πεσόντος αἵματος πέδοι;). Blood is consistently 'confused' (literally, 'poured together') with other liquids. The Erinys will drink unmixed blood in three draughts (ἄκρατον αἷμα πίεται τρίτην πόσιν), as mortals drink wine (*Libation Bearers*, 577–8). Aegisthus' sword (wielded by Clytemnestra) is said to have dyed the cloth which Agamemnon put on as he stepped out of the bath (1010–13), comparably to the blood-as-dyestuff metaphor we saw in *The Persians*. Other body fluids also come into play: as in *Agamemnon* blood was compared to semen through the intermediary of fertilising rain, so here it is triangulated with tears through the likeness of the libations to the dead.[34] And of course the blood dripping from the eyes of the Erinyes (1058), as Orestes sees them in the final scene, can be read as a ghastly parody of tears from these deities without pity.[35] Above all there is the dream image experienced by Clytemnestra, in which her milk is mixed with her blood, drawn by the teeth of the serpent she is suckling (531–3, 546; see also Vidović, Chapter 20, this volume, p. 329). The effect of all this is to concentrate our minds on the irrevocability of spilt blood (a point made explicitly in the other two plays of the trilogy) and the inevitability of further 'pourings', 'flows', even 'drinkings'.[36] Already the parodos and first episode, with the titular libations ('pourings', χοαί), establish a strong visual impression, setting the mood and materials of the play. The liquid offerings sent by Clytemnestra to placate her murdered husband mirror the blood that has been shed and are intended as compensation for it, an idea rejected by Orestes:[37]

τὰ πάντα γάρ τις ἐκχέας ἀνθ' αἵματος
ἑνός, μάτην ὁ μόχθος.

If you poured out everything in requital for one act of bloodshed, the labour would be in vain.

Libation Bearers 520–1

The only recompense is more blood; the bloodless liquids offered by the queen must be converted into her own blood. Blood as offering, whether to the gods or to the dead, is then another very important image here which was destined for a long career; its use in *Libation Bearers* is anticipated in *Persians*, where the ghost of Darius, himself conjured up by the sort of libations that are given to the dead, predicts the Persian defeat at Plataea with reference to a *pelanos haimatosphagēs*, an 'offering of blood-slaughter' (*Persians* 816). *Pelanos* is a word used chiefly in religious contexts to indicate a liquid or more often a gloopy sort of offering compounded of different materials; in this case it is made from the blood of slaughter, creating a horrible image of battlefield gore.

In *Eumenides*, the counterpart of the metaphorical spilt blood and the actual *choai* is the blood or poison which drips from the Erinyes; they feed on the blood of kin-murderers, taking it into their own systems and draining their victim dry. This dripping of blood makes them resemble the figure of Achlus in the Hesiodic *Shield*; it seems to be a useful way of picturing personified abstractions and nameless deities, but in this case suggests a sort of weird circulation of blood between murder victim, murderer, and avenger deity.[38] The play works out two further ideas: one is the blood of pollution and purification, which makes a first appearance at the end of *Libation Bearers* (1034–8, 1055). Blood clings to the murderer's hand but may be washed off by cathartic blood; the language that Orestes uses recalls Heraclitus' scornful remark about washing off blood with blood:

βρίζει γὰρ αἷμα καὶ μαραίνεται χερός,
μητροκτόνον μίασμα δ' ἔκπλυτον πέλει·
ποταίνιον γὰρ ὂν πρὸς ἑστίαι θεοῦ
Φοίβου καθαρμοῖς ἠλάθη χοιροκτόνοις.

The blood on my hand grows drowsy and fades away, the matricidal pollution becomes washable; for when it was fresh it was driven off by pig-killing purifications at the hearth of the god Phoibos.

Libation Bearers 280–3

But this claim is disputed—Orestes reiterates to Athena that he is clean after purification, but the Pythia has still seen blood on his hands (*Eumenides*, 448–50, 41–2), and of course the chorus of Erinyes are unconvinced until the final resolution. Still more importantly, there is blood as kinship marker, something implicit throughout the trilogy but which comes into the foreground of discussion here. If Orestes does not 'share blood' with his mother, if he is not ὅμαιμος /ἐν αἵματι with her (*Eumenides* 211, 604–8, 653), the spilling of her blood is of less consequence. It is this link between kin blood and spilt blood which drives the particular conflict of *Eumenides*, leading to the claim of the defence in the trial scene that Orestes did not, in fact, commit kin murder. Apollo's biology lesson, in which he maintains that only the father is the true parent, does not use blood terminology, but the question has earlier been posed, by the Erinyes and Orestes himself, in terms of blood.[39] Notably, Orestes asks 'Am I "in blood" with my mother?' (ἐγὼ δὲ μητρὸς τῆς ἐμῆς ἐν αἵματι; *Eumenides* 606). The solution proposed is to limit the application of the idea of kindred blood, as it is normally understood, so that Orestes' dilemma—whether to avenge his father by killing his mother—is no longer a real one. He must avenge his real parent, whoever the murderer was. However, it is far from clear that we should accept Apollo's version uncritically—even Athena seems to incline to it only on the basis of her rather unusual personal experience (*Eumenides* 737–40)—and threatening, sinister blood does not leave the play after the trial. The Erinyes wish to retain their blood 'circulation', and if Orestes has somehow escaped, they will turn instead upon those responsible for letting him go. Images of blood are conjured up in the last part in order to be negated—Athena asks the Furies to let go of their bloody poison, the Furies when pacified wish for Athens that the dust will not drink the dark blood of citizens (*Eumenides* 802; cf. 858, 980)—and finally converted into something positive in the sacrifices which the placated Erinyes will receive as Semnai Theai and the crimson-dyed cloaks worn in the sacrificial procession (1007–28).

Effectively, the resolution of the Oresteia trilogy is done by sleight of hand, and when viewed logically the problem of the shedding of kindred blood does not go away. In the earlier *Seven against Thebes*, the last and only surviving play of Aeschylus's Theban trilogy, the matter is central. Thebes is saved from the onslaught of its attackers, but only at the cost of the mutual killing of the brothers Eteocles and Polynices, first anticipated and then reported. As the chorus say, if a Theban kills an Argive:

> αἷμα γὰρ καθάρσιον.
> ἀνδροῖν δ' ὁμαίμοιν θάνατος ὧδ' αὐτοκτόνος
> οὐκ ἔστι γῆρας τοῦδε τοῦ μιάσματος.

> the blood is purifiable, but for the killing between two men who share blood, the pollution never grows old.
>
> *Seven against Thebes* 680–2

Essentially this is the position of the Erinyes in *Eumenides*—it is of no matter how many purifications Orestes undergoes, none of them has the power to remove the blood of close kin. In the *Seven*, the image of blood is established early on with the oath sacrifice, in which the attackers swear to take Thebes or die in the process:

ἄνδρες γὰρ ἑπτά, θούριοι λοχαγέται,
ταυροσφαγοῦντες ἐς μελανδέτον σάκος,
καὶ θιγγάνοντες χερσὶ ταυρείου φόνου
Ἄρην Ἐνυὼ καὶ φιλαίματον Φόβον
ὠρκωμότησαν ἢ πόλει κατασκαφὰς
θέντες λαπάξειν ἄστυ Καδμείων βίαι,
ἢ γῆν θανόντες τήνδε φυράσειν φόνωι.

Seven men, bold leaders of companies, slaughtered a bull, let its blood run into a black-rimmed shield, and touching the bull's blood with their hands swore an oath by Ares, Enyo, and blood-loving Terror, that they would either bring destruction on the city, sacking the town of the Cadmeans by force, or perish and mix their blood into the soil of this land.

Seven against Thebes 42–8; trans. Sommerstein

This action links oath sacrifice with *sphagion* sacrifice; the participants, as well as the poet, intend the blood as homologous with the blood which will be shed in battle, whether their opponents' or their own. But the substance soon takes on an additional significance. As we might expect, given that the central action is that of mutual fratricide, the play makes persistent use of the juxtaposition of the blood of kinship with the blood of slaughter, shared blood, and spilled blood, culminating after the event with the chorus's grim (and hard to translate) observation about the two brothers:

ἐν δὲ γαίαι
ζόα φονορύτωι
μέμεικται· κάρτα δ' εἴσ' ὅμαιμοι.

Their life is mingled in the earth that runs with blood—they are of one blood indeed.

Seven against Thebes 937–40

To us that might read as a conversion of the metaphorical blood of kinship to actual, literal blood; to the Greeks, perhaps, kinship blood was less metaphorical and the contrast is rather that of living (but separate) blood with spilled, death-signifying (but mingled) blood. In any case, the word ὅμαιμος ('same-blooded', 'together-blooded', often used of siblings) is given a new meaning, underlining the horror of the event. In their mutual slaughter, the quarrelling brothers have shown the impossibility of escaping from their relationship.

I have discussed Aeschylus, whose use of imagery is the most intense of the three tragedians, but Sophocles and Euripides use blood in many of the same ways. For them too it may signify individual death or sometimes wounding and mutilation, often as a detail in a narrative passage evoking a horrific picture; it has a close connexion with the pollution of homicide, often standing for it metonymically, sometimes slipping over into metaphor, and beyond to a literalised vision; it relates to sacrifice, human and other; and it indicates kinship. In all three authors, blood is a powerful, frequently recurring image marking the violence in which so much of tragedy deals. So why the difference from Homer? I have suggested that the *Iliad* does in fact suggest a substratum in which blood is the primary marker of violent death, so perhaps tragedy is just reverting to type—blood is a natural symbol too good to lose. It would also be difficult to imagine tragic heroes dying or suffering violence in the manner of minor Homeric fighters. Oedipus stabs his eyes and blinds himself; the description is graphic and horrible:

> φοίνιαι δ' ὁμοῦ
> γλῆναι γένει' ἔτεγγον, οὐδ' ἀνίεσαν
> φόνου μυδώσας σταγόνας, αλλ' ὁμοῦ μέλας
> ὄμβρος †χαλάζης αἵματος† ἐτέγγετο

At each blow the bloody eye-balls bedewed his beard, and sent forth not slug-gish drops of gore, but all at once a dark shower of blood came down like hail.

Sophocles, *Oedipus Tyrannus* 276–9; edition
Lloyd-Jones and Wilson 1990, trans. Jebb

But it stops just short of the really gruesome—we do not hear of bloody eyeballs fall-ing at his feet.[40] Body parts are dehumanising, comic even in some contexts, while blood allows tragic dignity and decorum.

Tragedy also perceives and exploits a link between sacrifice and violence in other spheres. This idea is not entirely absent from the *Iliad*; as we saw earlier, the mimetic connexion between the oath sacrifice in book 3 and the following violence—both that which is wished on the oath's transgressors and that which actually follows imme-diately the truce is broken. But tragedy goes further, dealing both literally in human sacrifice and also frequently using sacrificial metaphor and imagery. The kind of sac-rifice tragedy evokes, however, is not the convivial festival sort, ending in a nice meal against the backdrop of presumed divine good will, but rather the grimmer rites such as pre-battle sacrifice—in which category human sacrifice also belongs. Purification ritual is also relevant to some tragedies. As we have seen, these are the forms of sacri-fice in which blood is most conspicuous, creating a fruitful nexus for tragic explora-tion (cf. Seaford 2017).

There is another important advantage to blood as a tragic symbol. In much of tragedy, 'violence' and 'kin' ideas are prominent and prominently conjoined, and although this chapter has been concerned mainly with the blood of violence, in fact the two ideas share roughly equal prominence as the chief figurative implications of blood among the Greeks. If the deaths of Eteocles and Polynices had been described in Homeric terms, with the fracturing of their bodies and mutilation of their innards, we would still have plenty of horror surrounding their deaths, but we would not have an expression of the additional horror which comes from their being brothers, with an equal and balanced aim to kill each other. Shared blood and spilled blood; there is an obvious resonance with the stories of the house of Atreus and of Oedipus, but also with many other tragic plots where parents kill children, or children parents, or come close to doing so. Tragedy loves to mix the literal with the figurative and to imbue concrete images with abstract significance; unsurprisingly, perhaps, it never gives us a simple description of blood but always one that resonates with something outside the purely physical.

But can there in fact be a simple, neutral description of the shedding of blood? Looking beyond tragedy, it seems doubtful, even in medical contexts. While different cultures may attach importantly different symbolic values to blood, these are additional to its intrinsic 'natural' significance as a substance indicating both life and death, a substance whose crossing the boundary of the body is fraught with meaning. Blood is then either a natural symbol or one so deeply embedded in Greek culture and our culture that there can be no mere description of blood without some sort of emotional charge, however attenuated; like Menelaus, we all shudder a little in response to the evocation of spilled blood. It is for their own distinct purposes that epic and some sacrificial practice distract from the centrality of blood when they describe or perform mutilation of a body. Tragedy, with a different agenda and only partially overlapping preoccupations, seizes and exploits pre-existing meanings of the substance to the full.

Notes

* I should like to thank Victoria Leonard and Laurence Totelin for organising the conference at which this paper originated, and Mark Bradley, their fellow volume editor. My thanks are also due to Linda McNulty for much stimulating blood-centred discussion.

1 Menstruation is an obvious exception to this rule, and yet the negative value attributed to the phenomenon in many cultures must derive at least in part from the fact that it involves blood emerging from the body.

2 Examples in A. Gottlieb, 'Blood', in Barfield 1997: 41–2.

3 The 'intuitive' existence of such a category is defended by Mary Douglas (1973: 11–12), although *Natural Symbols* as a whole is more interested in different types of social structures in which holistic systems of natural symbols, particularly conceptions of the body, provide an analogy to the functioning of society. Closer to my use of the term is Erich Neumann (one does not have to be a Jungian to find his explanation useful): 'every object in the natural world is at the same time a symbolic reality to us . . . the experience of an "object" itself is always already symbolic experience. The star or tree in us is no less real and no less symbolic than it is in outward experience' (Neumann 1994 [1953]: 212–13). Or, as Gottlieb puts it (Barfield 1997: 41): 'blood is never "just blood".'

4 Among much important work, see Harris 1973; Duminil 1983; on medical writers see Dean-Jones 1994; on women's blood see King 1998. The views of Boylan 2015 are more controversial.

5 The rhetorical question follows some penetrating real questions about the understanding of blood in relation to kinship in the Greek world.

6 Perhaps the classic account of metonymy and its relation to metaphor is Lakoff and Johnson 2003: 35–40. But the influential approach of Jakobson had earlier suggested that metaphor and metonymy occupy opposite poles, the first relying on difference ('two domains'), the second on some type of similarity or contiguity ('one domain'). Jakobson's paper and much related discussion, some of it dealing with interaction between the two types of figure, can be found in Dirven and Pörings (2002). Metaphoric-metonymic interaction within the text seems often to characterise the use of blood in tragedy, as we shall see.

7 Unless we follow the suggestion of the bT scholiast on 19.87 (edition Erbse 1969–88) that the vulgate ἠεροφοῖτις (*ēerophoitis*) 'who walks in the dark', applied to the Erinys, should read εἰαροπῶτις (*eiaropōtis*), 'who drinks blood'.

8 I may have missed a few places where αἷμα (*haima*) is the second part of a compound.

9 References to the *Iliad* are to the Oxford Classical Text of Munro and Allan (1963). Saunders 2004 gives criteria for considering a wound as fatal or otherwise, but his data set is comprised of wounds, not deaths. My own rough figure includes catalogue-killings, with no details given beyond the name of killer and victim.

10 Battlefield: e.g. 4.451; 8.65; 15.715; 17.360; 20.494. Badge of honour: Neal 2006: especially 13–62.

11 No doubt inspired by this passage, but also in partial contrast, the Euphronios krater (*Add.²* 396.404–5, Carpenter et al. 1990) shows blood still gushing from the dead Sarpedon's wounds. Vase-painting in general tends to emphasise blood more than its epic models, perhaps because of the colour possibilities it offers, perhaps because of differing fifth-century sensibilities.

12 Cf. also 9. 326 ('bloody days', ἤματα αἱματόεντα).

13 The same line is repeated verbatim: 5.289 = 20.78 = 22.257. Cf. also 7.329–30 ('whose dark blood Ares spilled'), combining metonymy with metaphor: Ares as agent is metaphorical, but the spillage of blood is a metonym for death.

14 13.627; 17.927–8. But eyes fall out with no accompanying blood mentioned at 16.741–2.

15 The final line is also in the *Little Iliad*, fr. 29, edition West 2003.

16 Thus Liddell-Scott-Jones s.v. πορφύρεος, I; Beekes, *Etymological Dictionary* s.v. πορφύρω ('surge, boil').

17 Most famously the scene in Aristophanes, *Women of the Thesmophoria* 733–64 (Wilson 2007), where Mika's wine-skin becomes her baby and, at 755, the wine spurting out is its blood, in a parody of sacrifice.

18 On the scene and the violence it enacts, see Kitts 2003. I. 225 n. 18 (from the *Proverbia Coisliniana*) attests a similar Molossian rite in which wine is poured out with a prayer that the *blood* of oath-breakers may be analogously spilt. Of course the description could have been inspired by the Iliadic passage.

19 Saunders in Friedrich 2003: 149. The phrase is ἐκ δὲ οἱ ἧπαρ ὄλισθεν (20.470); as Saunders mentions, the verb is used otherwise in the *Iliad* only in the undeniably comic episode when the Lokrian Aias slips in the dung of the sacrificed cattle. The death scene does contain blood for good measure: the liver slips out, the abdomen (probable meaning of κόλπος here) contrastingly fills with black blood.

20 Similarly, in other modern languages, *Blutopfer*, *sacrifices sanglants*, and *sacrificio cruento* have become standard terms. Naiden 2013: 6, traces the first usage to W. Smith's *Dictionary of Greek and Roman Antiquities* (first edn., 1842), although the phrase used there is actually 'bloody sacrifice'. Eckhardt 2014 finds the earliest usage of ἀναίμακτος θυσία ('bloodless sacrifice') in Plutarch, *Numa* 16.1 (edition Ziegler 2014), and considers that the word has been transferred from an earlier context referring to lack of bloodshed between humans.

21 Some heroic sacrifices, often signalled by the word ἐντέμνειν: Ekroth 2002: 257–68. For oath sacrifices, see Kitts 2003.

22 Aeschylus, *Seven* 42–8 (p. 203), but also Xenophon, *Anabasis* 2.2.9 (edition Marchant 1963), recording actual rather than imagined usage.

23 Heraclitus Diels-Kranz B5: καθαίρονται δ' ἄλλως <αἷμα> αἵματι μιαινόμενοι οἶον εἴ τις εἰς πηλὸν ἐμβὰς πηλῶι ἀπονίζοιτο.

24 *RVAp.*, Volume I: 97, 229.

25 Georgoudi 2017: especially 128–32. I am grateful to Irene Salvo for drawing my attention to this chapter in advance of publication.

26 πορφυρέωι, which seems likely to be the correct reading, probably did mean 'crimson' to Stesichorus.

27 The closest approach is 463: παίουσι κρεοκοποῦσι δυστήνων μέλη ('they strike and chop the flesh of the unfortunate men's limbs'). This is both generalised and very much less graphic than the type of Iliadic passage I have been discussing, although the analogy of butchery is shocking.

28 The image is discussed in Garvie 2009: 166–7.

29 It is present largely in the threats of the arriving Egyptians or their herald to inflict violence on their cousins and unwilling brides the daughters of Danaus, and perhaps contrasts with the earlier threat of the Danaids to turn bloodless violence on themselves by hanging, if their

supplication is not granted. It would be interesting to know whether there was more blood in the other two plays of the trilogy. All references to Aeschylus are to the edition by Page (1972).

30 Lebeck 1971: 80–91 is particularly concerned with blood.
31 Seaford 2017 (seen by me only after this chapter was substantially completed).
32 References in the play's last section: 1092, 1121–3, 1188–90, 1277–8, 1292–4, 1309, 1338–42, 1427–8, 1459–60, 1475–80, 1509–12, 1533–4, 158–9, 1656.
33 Even in *Prometheus Bound* (whether or not it is of Aeschylean authorship), the traditional liver eating is not presented particularly graphically (1021–5).
34 Aeschylus, *Libation Bearers* 48, 66–7, establishes an analogy between blood and offerings to the dead, both poured or spilled onto the earth; then as Elektra pours the offerings, the chorus sing of the shedding of tears (152–4). See also Seaford 2017: 229–31.
35 Aeschylus, *Libation Bearers* 1058: κἀξ ὀμμάτων στάζουσιν αἷμα δυσφιλές. (Burges' emendation στάζουσι νᾶμα, although accepted by some editors, is not persuasive.)
36 Aeschylus, *Agamemnon* 1019–21: τὸ δ' ἐπὶ γᾶν πεσὸν ἅπαξ θανάσιμον / πρόπαρ ἀνδρὸς μέλαν αἷμα τίς ἂν / πάλιν ἀγκαλέσαιτ' ἐπαείδων; *Eum.* 261–3: αἷμα μητρῷον χαμαὶ / δυσαγκόμιστον, παπαῖ, / τὸ διερὸν πέδοι χύμενον οἴχεται.
37 Compare Vidović, Chapter 20, this volume, pp. 325–6.
38 Compare the slightly different analysis in Vidović, Chapter 20, this volume, pp. 326–8.
39 Apollo: Aeschylus, *Eumenides* 657–74. Erinyes: 211, 653. Erinyes and Orestes: 604–8.
40 M. L. West proposes the deletion of 1278–9, which would tone down the description without quite eliminating the horror.

Abbreviations of standard corpora and reference works

Add²: Carpenter, T. H., T. Mannack, and M. Mendonça (1990) *Beazley Addenda*. 2nd edn. Oxford: Oxford University Press.

Beekes, *Etymological Dictionary*: Beekes, R. S. P. with L. Van Beek (2010) *Etymological Dictionary of Greek*. Leiden: Brill.

Diels-Kranz: Diels, H. and W. Kranz (eds) (1952) *Die Fragmente der Vorsokratiker: griechisch und deutsch*. Zurich: Weidmann.

Liddell-Scott-Jones: Liddell, H. G. and R. Scott (1968) *Greek-English Lexicon*. 9th edn. Revised by H. Stuart Jones, with a supplement by E. A. Barber and others. Oxford: Oxford University Press.

Paroem. Gr.: Leutsch, E. L. and P. G. Schneidewin (eds) (1839) *Corpus Paroemiographorum Graecorum*. Göttingen: Vandenhoeck & Ruprecht.

RVAp.: Trendall, A. D. and A. Cambitoglou (1978–82) *The Red-figured Vases of Apulia*. Oxford: Oxford University Press.

Bibliography

Primary sources

Davies, M. and P. J. Finglass (2014) *Stesichorus: The Poems*. Edited with introduction, translation and commentary by M. Davies and P. J. Finglass. Cambridge: Harvard University Press.

Olson, S. D. (2012) *Athenaeus Naucratites: Deipnosophistae*. Berlin: Walter de Gruyter.

Erbse, H. (1969–88) *Scholia graeca in Homeri Iliadem (scholia vetera). Recensuit H. Erbse*. Berlin: Walter de Gruyter.

Lloyd-Jones, H. and N. G. Wilson (1990) *Sophoclis fabulae. Recognoverunt brevique adnotatione critica instruxerunt H. Lloyd-Jones et N. G. Wilson*. Oxford: Clarendon Press.

Marchant, E. C. (1963) *Xenophontis opera omnia. Recognovit brevique adnotatione critica instruxit E. C. Marchant, Vol. III. Expeditio Cyri*. Oxford: Clarendon Press.

Munro, D. B. and T. W. Allan (1963) *Homeri opera. Recognoverunt brevique adnotatione critica instruxerunt D. B. Munro et T. W. Allan.* Oxford: Clarendon Press.

Page, D. (1972) *Aeschyli quae supersunt tragoedias. Edidit D. Page.* Oxford: Clarendon Press.

West, M. L. (2003) *Greek Epic Fragments: From the Seventh to the Fifth Centuries BC.* Edited and translated by M. L. West. Cambridge, MA: Harvard University Press.

Wilson, N. G. (2007) *Aristophanis fabulae. Recognovit brevique adnotatione critica instruxit N. G. Wilson.* Oxford: Clarendon Press.

Ziegler, K. (2014) *Plutarchus. Vitae parallelae, Vol. III. fasc. 2.* Berlin: Walter de Gruyter.

Secondary literature

Anidjar, G. (2014) *Blood: A Critique of Christianity.* New York: Columbia University Press.

Barfield, T. (1997) *The Dictionary of Anthropology.* Oxford: Blackwell.

Biale, D. (2007) *Blood and Belief: The Circulation of a Symbol between Jews and Christians.* Berkeley: University of California Press.

Boylan, M. (2015) *The Origins of Ancient Greek Science: Blood, a Philosophical Study.* New York: Routledge.

Bynum, C. W. (2007) *Wonderful Blood: Theology and Practice in Late Medieval Germany and Beyond.* Philadelphia, PA: University of Pennsylvania Press.

Dean-Jones, L. (1994) *Women's Bodies in Classical Greek Science.* Oxford: Clarendon Press.

Dirven, R. and R. Pörings (2002) *Metaphor and Metonymy in Comparison and Contrast.* Berlin: Mouton de Gruyter.

Douglas, M. (1973 [1970]) *Natural Symbols.* London: Barrie & Jenkins.

Duminil, M.-P. (1983) *Le sang, les vaisseaux, le cœur dans la collection hippocratique: anatomie et physiologie.* Paris: Les Belles Lettres.

Eckhardt, B. (2014) '"Bloodless sacrifice": A note on Greek cultic language in the imperial era', *Greek, Roman and Byzantine Studies* 54, 255–73.

Ekroth, G. (2002) *The Sacrificial Rituals of Greek Hero Cults.* Liège: Presses Universitaires de Liège.

———. (2005) 'Blood on the altars? On the treatment of blood at Greek sacrifices and the iconographical evidence', *Antike Kunst* 48, 9–29.

Finley, M. I. (1972 [1954]) *The World of Odysseus.* Harmondsworth: Penguin Books.

Friedrich, W.-H. (2003 [1956]) *Wounding and Death in the Iliad: Homeric Techniques of Description.* London: Duckworth.

Garvie, A. F. (2009) *Aeschylus, Persae: With Introduction and Commentary.* Oxford: Oxford University Press.

Georgoudi, S. (2017) 'Reflections on sacrifice and purification in the Greek world', in S. Hitch and I. Rutherford (eds) *Animal Sacrifice in the Ancient Greek World.* Cambridge: Cambridge University Press, 105–35.

Harris, C. R. S. (1973) *The Heart and the Vascular System in Ancient Greek Medicine from Alcmaeon to Galen.* Oxford: Clarendon Press.

Jakobson, R. (2002) 'The metaphoric and metonymic poles', in R. Dirven and R. Pörings (eds) *Metaphor and Metonymy in Comparison and Contrast.* Berlin: Mouton de Gruyter, 41–7.

Jameson, M. H. (1991) 'Sacrifice before battle', in V. D. Hanson (ed.) *Hoplites: The Greek Battle Experience.* London: Routledge, 197–227.

King, H. (1998) *Hippocrates' Woman: Reading the Female Body in Ancient Greece.* London: Routledge.

Kitts, M. (2003) 'Not barren is the blood of lambs: Homeric oath-sacrifice as metaphorical transformation', *Kernos* 16, 17–34.

Lakoff, G. and M. Johnson (2003 [1980]) *Metaphors We Live By.* Chicago: University of Chicago Press.

Lebeck, A. (1971) *The Oresteia: A Study in Language and Structure*. Cambridge, MA: Harvard University Press.

Mallory, J. P. and D. Q. Adams (2006) *The Oxford Introduction to Proto-Indo-European and the Proto-Indo-European World*. Oxford: Oxford University Press.

Mason, H. C. (2015) *The Hesiodic Aspis: Introduction and Commentary on vv. 139–237*. https://ora.ox.ac.uk/objects/uuid:05a4c022-03d0-4508-800c-9e68e8429999 (Accessed February 2021).

Naiden, F. S. (2013) *Smoke Signals for the Gods: Ancient Greek Sacrifice from the Archaic through Roman Periods*. New York: Oxford University Press.

Neal, T. (2006) *The Wounded Hero: Non-Fatal Injury in Homer's Iliad*. Bern: Peter Lang.

Neumann, E. (1994) *The Fear of the Feminine and Other Essays on Feminine Psychology*. Princeton, NJ: Princeton University Library.

Saunders, K. B. (2004) 'Frölich's table of Homeric wounds', *The Classical Quarterly* 54, 1–17.

Seaford, R. (2017) 'Sacrifice in drama: The flow of liquids', in S. Hitch and I. Rutherford (eds) *Animal Sacrifice in the Ancient Greek World*. Cambridge: Cambridge University Press, 223–35.

Van Straten, F. (1995) *Hiera Kala: Images of Animal Sacrifice in Archaic and Classical Greece*. Leiden: Brill.

13

TASTE AND THE SENSES

Galen's humours clarified

John Wilkins

Background

The bodily fluids traditionally translated as 'humours'[1] are fundamental to many branches of ancient medicine and are vital to our understanding of ancient physiology, which in the influential Hippocratic tradition consisted in balancing blood, phlegm, yellow bile, and black bile in the vessels of the body. This was how the body maintained health and well-being. Galen is the most prominent inheritor and interpreter of the Hippocratic tradition, for good or ill, and the most powerful proponent of the four-humour model. Beyond the Hippocratic authors who composed their treatises in the fifth and fourth centuries BCE and later, the bodily fluids figured largely in Homer, where much blood is spilt, and prominently in tragedy, where protagonists suffer bloody death and fevered sweating in such plays as Aeschylus' *Agamemnon*, Euripides' *Heracles*, and Sophocles' *Women of Trachis*. It was the medical authors, however, who best theorised the humours and their workings. Galen and his predecessors make very clear that the senses are key to understanding the body: touching the skin to determine temperature, seeing the colour and texture of foods and bodily products, as well as tasting and smelling them, together with listening to the patients and the sounds he/she makes are key to diagnosis and prognosis in a medical system lacking chemical and electronic means of measurement. The discussion of bodily fluids in the Introduction to this volume is rather different from Galen's own: sweating, for example, is an essential bodily regulatory process, whose study is as unproblematic as that of the body's other waste products.

In the standard handbook on ancient medicine, Vivian Nutton observes, of writers on the humours, 'the suffocating friendship of Galen has tended to subsume all who agreed with him under the banner of Hippocrates' (Nutton 2013: 207) and 'Galen tried to relate the qualities of foods to the four humours and their various constituent qualities' (247). This chapter attempts to clarify Galen's thought and locates his humoral teaching in his special reading of the Hippocratic *On the Nature of Man* and subsequent thought and isolates certain key features of Galen's humours that set him apart from all the accretions of later centuries that underlie humoral theories of the Medieval and Early Modern periods, which are discussed elsewhere in this volume (see Stylianou, Spearing, and King, Chapters 22, 23, and 24, respectively).[2] I address Galen's thought on these vital bodily fluids from the particular aspect of sensation (*aisthēsis*) in order not to revisit the complexity of humoral theory in general.[3] Galen's humours were generated from food digested by the body's heat, in an ideal case into

DOI: 10.4324/9780429438974-18

the complex form of blood which embraced the four necessary humours. Less than ideal heat or less than ideal food might generate too much of one of the humours (blood, phlegm, yellow, and black bile) and deposit it as a 'residue' (*perittōma*) in an inappropriate part of the body, thereby generating potential imbalance leading to disease if the body could not expel the residue, either diuretically or through the pores of the skin. In Galenic thought, the humours function as a background to other aspects of physiological activity, principally 'mixtures' (*krasis*) of the biological 'qualities' (*poiotētes*) hot, cold, wet, and dry. Galen will often speak in terms of mixtures rather than humours, in particular in the fundamental treatise dedicated to the topic, *On Mixtures*.[4]

Mixtures and humours

Galenic mixtures have hot, cold, wet, and dry qualities and are a biological analogue of the four elements, earth, air, fire, and water. Individual bodies each have a particular 'nature' or 'constitution', that is, a particular combination of 'humours' and a particular mixture. Any food digested might support that nature or unbalance it. The fluids of the body will ideally be in balance with each other and with the vital organs but often are not. The doctor's task is to notice individual peculiarities and adjust the patient's lifestyle accordingly. The foods have profiles, too. Apples and grapes have juices, beets have stored energy,[5] onions are pungent, and meat contains blood. They have taste and texture, smell, and a visual appearance. They make a perceptual impact on the eater before being digested. All of these things indicate to Galen and other doctors on whom he draws how the food is likely to impact on the mixtures and humours once digested. An onion with a pungent taste and smell will cut through thick humours, which are likely to be generated by the most nutritious but also most thick-humoured foods, pork and beef. Pork and onions are therefore in principle to be recommended as a meal if a good balance is to be maintained. Again, wheat is the most nutritious cereal, but unless the eater is a soldier, athlete, or manual worker burning a great deal of energy, wheat will produce thick juices unless properly cooked and processed and eaten in suitable meals. When reviewing nutrition in *On the Capacities of Foods*, Galen devotes the second of three books to plants other than cereals and legumes, plants which he concedes have little 'nutrition' but a great deal of texture that might slow down food in the bowel, or add a piquant juice, such as basil, or some heat. All of these things can be touched, tasted, smelt, and seen. We can see the cleansing quality of onion on the skin and therefore predict its impact on the fluids of the body, just as blood, sweat, urine, faeces, and vomit indicate the state of the fluids internally.

Galen is usually clear on whether a *chumos* is a humour or a juice which produces a humour in the human consumer. When plant juices (*chuloi*) and animal juices (*chumoi*, preferably cooked, ground up, or chopped) are swallowed and taken into the body through the digestive system (*pepsis*, conceptualised as a second 'cooking' in the heat of the body), changes in the body can be comprehended by drawing inferences and analogies from the properties of the food materials before they are eaten. Taste and smell are key.[6] Taste is discussed by Laurence Totelin (2017), with reference to Galen's valuable work *Against Lycus* in which he sets out a classification of tastes that are widely deployed in Galen's works on pharmacology and nutrition. She highlights

(65) a passage from *On the Capacities of Foods* 2.59, where taste and smell indicate qualities of plants and the impact on the body of the human eater:

> It is safer to consider each of the parts of plants in and of itself, first by tasting and smelling and then trying it by eating it. For smell and taste teach us about what sort of *chumos* and vapour each part of the plant possesses, and instantly demonstrates with this its overall mixture. By experiment, if you do it with appropriate distinctions, the capacity[7] is accurately discovered, with the structure of the plant helping to show this, along with its juice (*chulos*). For some have a moist and watery *chumos*, others a thick and viscous *chumos*, each of which, again, should be tasted separately. For some of them are pungent or sharp or bitter, some salty and briny, just as others are harsh, others sour, and others watery and sweet.
>
> Galen, *On the Capacities of Foods* 2.59,
> 166.21–167.11 Wilkins, 6.647–8 Kühn (1821–33)
> (Translations are my own unless otherwise indicated)

Here we see the vocabulary of flavour, taste, *chumos*, and mixture all brought together in the context of plants, which Galen derives from Theophrastus, as Totelin explains. We might have supposed that Galen thought of a *chumos* as a 'humour' in the human body—as indeed he does—but he also seems to see it as inherent in the sap of a plant. Indeed, in *Simple Medicines* he explains what he means:

> But it is not possible to re-teach such people. For the sake of everyone else who is capable of being helped, we must take up and define accurately the account of the objects of sensation that are particular to the tongue. Now the tasting capacity is called by those linked with Theophrastus, Aristotle and Mnesitheus the doctor[8] *chumos*, with the second syllable starting with 'm'. The combination of wet and dry[9] when cooked by heat is called *chulos*, with the second syllable starting with an 'l'. However, among older sources—not only Attic but also Ionian—each is written with an 'm'. For the sake of clear instruction in the current work, I am saying that the capacity of taste, or the quality, or whatever term it might be, is *chumos* with an 'm', while the term for thickened liquid in animals and plants is *chulos*, with the second syllable starting with an 'l'. So we need to know that the *chumoi* in each plant and animal, and indeed those found in the earth are multiple and not easily counted in their species.
>
> Galen, *Simple Medicines*
> 1.38, 11.449–50 Kühn[10]

Galen here speaks of the flavour (*chumos*) of a juice (*chulos*) which has been cooked or ripened by the sun and will bring these properties to bear on the humoral balance of the human body. *Chumos* in plants means principally 'flavour' but can come to mean 'flavour in a moist or viscous fluid', as seems to be meant in the *On the Capacities of Foods* passage cited earlier.

Keith Stewart (2017) surveys the vital bodily fluids in the Hippocratic Corpus and Galen's development of them, showing that the fluids, often termed *hugra*, and variants in the corpus may number two, three, or four, culminating in four only in *On the Nature of Man*, which Galen took as his blueprint for the 'humours', as had done others before him. It is striking, though, that once Galen sets out his four-humour theory and his firm allegiance to Hippocrates, as he interpreted him, that humours are not the primary agents for change in the body. This needs to be stressed, as Véronique Boudon-Millot (2011) and Philip van der Eijk (2015) have argued. Rather, Galen's primary agents are those biological qualities, hot, cold, wet, and dry, as he sets out in *On Mixtures* and many other treatises.[11] Boudon-Millot and van der Eijk go too far, however. The qualities are indeed the primary agents of change, but the humours have an important role also, as Galen makes clear when not foregrounding mixtures. In this chapter I study the role of humours in Galen's dietary and pharmacological works, that is *De sanitate* (*On Maintaining Good Health*), *De alimentorum facultatibus* (*On the Capacities of Foods*), and *Simple Medicines*, showing that at a secondary level the humours effect change in the body; they are closely tied to mixtures; and they perform vital functions through their qualities as made manifest in taste (*chumos*) and smell.

This is not just a question of digestion. Galen's diagnosis and therapy depend on isolating causes of imbalance, which can only be achieved by testing the inputs into the body, the state of the body, and the characteristics of the waste products, identified by *aisthēsis*. What is the taste and smell of the sweat, urine, and faeces? The texture, and appearance to the eye? Taste, smell, texture, and vision are not just avenues to the pleasures of wine and food at dinner. They are the routes to the understanding and establishment of physiological well-being. They enable for each individual balancing of the vital fluids—the humours—and a good mixture.

I am building on Claudia Mirrione's (2017) demonstration of how the 'humours' work in Galen's science to show how taste is key in action and detection. She shows that 'humours' are the link between the ingestion of foods, drinks, and medicines and changes produced in bodily tissue and other parts: Galen in effect fills in a gap in Aristotelian biology that the philosopher does not investigate in detail. If the ingested material is replacing lost energy and lost bodily material, then it is 'food', as he explains at the beginning of *Simple Medicines*:

> Now we call a drug whatever produces change in our nature, just as, I think, food is whatever increases our substance. For both of them are defined by their purpose.
>
> Galen, *Simple Medicines* 1.1, 11.380 Kühn

The distinction is not as watertight as this distinction claims, as we shall see. Foods contain 'capacities' (*dunameis*) which may unsettle the body, as Galen explains in a discussion of energy early in *On Maintaining Good Health* (1.3):

> Since from all animals a large part of their substance flows away every day because of their innate heat, and we need foods and drinks, breathing and pulse to maintain its due proportion (*summetria*), of necessity there will follow the production of residues (*perittōmata*). For if we were able to produce something exactly the same as what was lost, this would be the finest and most healthy option. But since what flows away from each of the parts is the same in nature as the part itself, and none of the things that we eat and drink are exactly the

same as this, it is necessary for the body's nature to modify the food and predigest it and in a word to prepare it to be as similar as possible to the body being nourished. And in this process, what is not exactly digested and assimilated does not grow naturally into the body and moves as a residue through the internal passages. . . . So since eating and drinking are essential for animals, and the production of residues follows from this, nature has prepared organs to remove them and given the organs the capacities (*dunameis*) to attract some of them as they move through the body, to forward others to the right place and to remove others. And it is of course essential that the organs remain unobstructed and in no way weakened in the activities of keeping the body pure and free from residues. Our argument has led us to these two aims for a healthy way of life, one replacing what has been taken out, the other the elimination of residues. And the third is that the animal should not age prematurely.

Galen, *On Maintaining Good Health* 1.3, 5.35–6.22
Corpus Medicorum Graecorum

Foods are not the same as bodily tissue, though a few, such as pork, may have qualities very similar to human tissue.[12] Consequently, foods may, for example, heat or cool, they may be composed of light particles like ground pepper (*Simple Medicines* 1.11, 11.398–404 Kühn) or of thick particles like cheese, they may be 'cutting' or 'thickening' of the vital fluids, and they will nearly all have a taste profile.

Drugs, to go back to Galen's distinction, do produce change, though often not until activated by the body's heat (*Simple Medicines* 1.11, 11.400 Kühn), and may be taken from the same plants as foods or from more toxic plants, or minerals or parts of animals. Drugs will need to be taken if the body is unable to restore its equilibrium by gentler means at mealtimes or in bathing or taking exercise. In a move that many modern readers will find surprising, Galen declares that taste is important in understanding a drug as well as a food (*Simple Medicines* 4.5, 11.632 Kühn). A doctor has

also to work carefully on the impact of the sense of taste on the humours/ flavours,[13] starting from those having any one very clear quality. Now, if you want to identify pungency clearly, then it is garlic and onion and plants similar to them that you should taste many times and chew at length and then with precise testing of what you have remembered compare the sense of the experience you have received. And if it is astringency (*stupsis*), then with castor oil (*kikis*), sumac and similar plants; if it is bitterness [*pikrotēs*] then with soda and bile; and if with sweetness, then with grape juice and honey. And then in addition to these, whether you want to suppose that it is without quality or median in its taste quality.

Galen, *Simple Medicines* 4.5, 11.633 Kühn

Galen in his pharmacology is at pains to stress that drugs and their actions will only be understood if doctors reason from the evidence of their senses rather than from abstract notions:

These things are determined with three senses, taste, sight, and smell. To the taste it is pure if it takes on no quality, but appears to be exactly without

214

quality; pure to the sight when it is pure and exactly translucent; and to the smell likewise no quality must appear in it, such that it takes on some of the extremely bad smells, sharpness, pungency, saltiness, putridity, or some other unspeakable nastiness.

Galen, *Simple Medicines* 1.5, 11.390 Kühn

Foods and taste

I am going to show first how Galen believes foods work on the body, as far as taste and texture are concerned, and then how certain drugs work. Taste (*geusis*) is my criterion since it embraces the idea of 'flavour', which in Greek is *chumos*, the same word as 'humour', as we have seen. The two are intimately linked in Greek thought. In addition, in Aristotle's discussion of how the senses work in *On the Soul* Book 2, taste is identified as a subset of touch and is thus linked to the discussion of touch in van der Eijk's discussion of mixtures in Galen (2015: 676 n. 2). I begin with what Galen considers the two most nutritious foods, wheat and pork, the foods best able to replace the energy lost in heating the body and using muscles, as set out previously in *On Maintaining Good Health* 1.3. These are not distinguished for their taste but rather their energy-producing and textural qualities. As we shall see, the more tasty foods tend to be fruits and vegetables, such as the sourness of a raw apple and the sweetness of ripe grapes. All plants and animal products, though, contain moisture which has an impact on the human body, even if it is 'without quality', *apoios*.[14]

> WHEATS. The latter [loose-textured, porous, and white grains of wheat] have the most bran and, when milled, if one sifts out the very fine meal and makes what are called bran loaves from the remainder, trial will show that while they are poorly nutritive they produce much residue in the stomach and consequently it is passed easily. At the same time, because the bran has a cleansing quality, elimination of the residues, as you would expect, takes place quickly since the bowel is stimulated to excretion. The loaves that are the opposite of these are extremely pure, bringing the greatest weight to the smallest bulk, but of all the breads they pass through most slowly. Indeed you will observe that their dough is quite glutinous, since, when it is drawn out to the greatest extent, it is not torn apart . . . and so these naturally need more yeast and more thorough kneading.
>
> Galen, *On the Capacities of Foods* 1.2, 25–6 Wilkins 2013,
> 6.481–2 Kühn, trans. Powell 2003 (adapted)

The factors that Galen is weighing up here are the colour and texture of the wheat grains (the denser, yellower ones are the best); the sifting, milling, and baking of the flour to make it into bread;[15] the nutritional energy produced, along with any by-products in the form of residues; the speed of passage through the intestines; the cleaning qualities of the material being digested, which will lead to excretion; and the sticky, glutinous quality of the finer wheats, whose gluten (to use a modern term) is activated by kneading and expansion by yeast. These cleansing and glutinous qualities, I suggest, are features of bodily liquids ('humours') rather than of hot, cold, wet, and dry, or at least features of these qualities acting together. We shall return to this issue shortly.

Let us now turn our attention to pork:

> The flesh of pigs is the most nutritious of all foods. . . . Pork is as more gluti-
> nous than beef as beef, in its whole substance, is thicker than pork. But pork
> is much better for concoction[16]—that of mature animals for people in their
> prime who are strong and hardworking; but for other people, that from ani-
> mals that are still growing is better.
>
> Galen, *On the Capacities of Foods* 3.1, 180 Wilkins,
> 6.661–2 Kühn, trans. Powell (adapted)

Here again, Galen is concerned with the energy-giving nutrition and with texture, stickiness, and thickness. Thickness in fluids is determined by cooling and heat-ing: again, qualities and 'humours' acting closely together. The bodily fluids are the medium on which the qualities are acting. Galen is certainly thinking about 'humours' at this point, since he has just mentioned the dangers posed by beef to those who are of a 'more black-biled [*melancholikōteros*] nature in their mixture'.[17] Such natures may find that this bodily fluid causes the spleen to swell, leading to 'bad condition' (*kachexia*) and gout.

With green plants, flavours become prominent in the discussion of bodily fluids, and nutrition in Galen's strict sense of 'energy-supply' is less present. This category tends to undermine Galen's distinction between foods and drugs which I quoted earlier, since these plants, as he often notes, contain 'pharmacological' qualities. First, lettuce:

> I take up again the . . . cultivated lettuce that is customarily eaten by every-
> one . . . and shall say, in summary, that it has moist, cold juice but that nev-
> ertheless it is not of a bad humour.[18] Because of this, neither is it unconcocted
> like other vegetables nor does it inhibit bowel action (just as neither does it
> promote it). It has this effect with good reason since it has neither harshness
> nor sourness, which for the most part are what restrain the bowel; just as it
> is stimulated to excretion by salty and bitter substances and, generally speak-
> ing, by things with some purging quality—none of which exists in the lettuce.
>
> Galen, *On the Capacities of Foods* 2.40, 148–9 Wilkins,
> 6.627–8 Kühn, trans. Powell (adapted)

Galen considers foods in a social context: if everyone eats lettuce, then it is important for the diet, as opposed to red mullet livers eaten only by the extravagant rich. This discussion follows comments on wild lettuce, which has different qualities. Moisture and coldness suggest biological qualities, but bad humour, harshness and sourness, and saltiness and bitterness suggest fluids and flavours: Galen combines the two, since he is interested in the actions of excreting and purging within the body. He is also interested in the contribution of lettuce to blood-making, which is the most important bodily fluid/humour of all. Lettuce he says (*On the Capacities of Foods* 2.40, 147.3–6 Wilkins, 6.625 Kühn), 'of all green plants which generate minimal blood and of a bad "humour", that from lettuce is not great, but not of a bad "humour"'—though it is

216

not completely of a good 'humour' either. Bodily fluids/humours are very much in his mind.

We come now to the celery family:

> These are all diuretic. The celeries are the most commonly used because they are more pleasant and good for the cardia.[19] . . . The Cretan alexander is not uncommon;[20] at any rate a good deal of it is sold in Rome. It is more pungent[21] and far warmer than celery, and also is somewhat aromatic.[22] It is more diuretic than celery, alexander and water-parsnip and in women brings on menstruation. In spring it makes a stalk which can be eaten raw. . . .
>
> All such are food flavourings, just like onions, garlic, leeks, wild leeks, and, in general, all pungent vegetables. Also among this class are rue, hyssop, oregano, fennel, and coriander, regarding which there has been much discussion in the compilations on cookery which in a way are common to both physicians and cooks, though they have their own specific aim and purpose. For we physicians aim at benefits from foods, not at pleasure. But since the unpleasantness of some foods contributes largely to poor concoction, in this regard it is better that they are moderately tasty. But for cooks, pleasantness for the most part normally makes use of harmful seasonings, so that poor rather than good concoction accompanies them.
>
> Galen, *On the Capacities of Foods* 2.51, 157–9 Wilkins,
> 6.637–40 Kühn, trans. Powell (adapted)

The celery family shows nicely that social considerations play a large part; that a Platonic objection to pleasure is not sustainable, since 'unpleasantness leads to poor concoction',[23] and that flavour needs to be thought about seriously in assessing food value.

The complex considerations of pleasure are prominent in my next category, fish.[24] I take the example of the red mullet, which

> is one of the pelagic fish and has been prized by men as superior to the rest in pleasantness in eating. It has flesh that is firmer than just about all of them and is quite friable, which is the same as saying that there is nothing either sticky or oily in it. At any rate, when concocted it is more nourishing than other fish. . . . Red mullet is pleasant, being proper to the nature of humans but, although it is firmer than other fish, nonetheless it can be consumed on a daily basis because it is friable, non-oily and also has some pungency.
>
> Galen *On the Capacities of Foods* 3.26, 226–7 Wilkins,
> 6.715–16 Kühn, trans. Powell (adapted)

These fish, like pigs, have flesh that is *oikeios*, 'proper', in the sense of being related or similar to human flesh. It is more easily digested than many foods and less likely to generate the unwelcome residues discussed earlier.[25] No matter that red mullet are highly prized by rich people: these fish have a good texture for digestion, friable, with no sticky or glutinous difficulties. It also has some pungency, like celery and onions.

Galen does not mention its blood-making qualities, as he did for the lettuce, but he does mention this for the sea bass 'and the other fish':

> Nourishment from this fish and the other fish generates a blood lighter in consistency than that from footed animals, with the result that it nourishes less richly and is rapidly distributed.
>
> Galen, *On the Capacities of Foods* 3.25, 226.3–6
> Wilkins, 6.714 Kühn

Lightness in the consistency of the most important of the 'humours' indicates how important the bodily fluids are in Galen's nutrition and how flavour and consistency must be added to heating, cooling, moistening, and drying. The 'humours' and 'qualities' work closely together, as Mirrione (2017) shows, with the latter taking the leading role, as Galen repeatedly states. The qualities (*poiotētes*) are primarily the quartet hot, cold, wet, and dry, which make up the mixtures of the body (*kraseis*).

Drugs and taste

Let us turn now from 'foods' to drugs in the consideration of bodily fluids and flavours, beginning with the least food-like drugs, the minerals, and moving to plant analogues in Galen's method for identifying astringency. This should be a matter of personal testing, not of dogma, he declares:

> Thus 'copper ore',[26] and 'Cypriot copper ore',[27] and 'flowers of copper'[28] and copper scale,[29] and *sōru*[30] which are astringent and biting, act according to both qualities on each of the parts of the body with which they come into contact, but they cannot teach us clearly whether the burning action is caused by astringency or pungency (*drimutēs*). It is best then to leave these aside and to taste the many other things in turn, considering pure astringency in its particular form and in itself as can best be done, and when you have found it, then you can judge the drug according to the methods you have heard about many times before. . . . Let the test, as has been said many times, be upon bodies that are completely healthy and in the best constitution, and in single conditions (*pathē*), whether hot or cold or dry or wet. For if the healthy body seems in itself to be clearly cooling and the drug fits in with hot conditions, and the drug brings to bear a clearly cold quality to the person who is suffering, then one can be emboldened to reveal something about its astringency, that it is cooling. If, conversely, it is found to have, both with healthy bodies that they are heated by it and hot diseases are made acute by it, and cold ones are helped not to cool then such a drug must be considered to be heating.
>
> Galen, *Simple Medicines* 4.7, 11.640–3 Kühn

When considering sourness and astringency, Galen uses also 'biting', *daknon*, a further regular 'taste' word. 'Burning' is a further effect, along with 'pungency',[31] actions which Galen is at pains to distinguish from astringency. Furthermore, Galen calls astringency and biting *poiotētes*, 'qualities', a term we normally link with the biological qualities of heating, cooling, moistening, and drying. I repeat, a taste, *chumos*, is a quality too, producing change within bodily fluids (*chumoi*) and tissues. Pure astringency for Galen is found in certain plants such as sumac juice (*chulos*). It must be tested on healthy bodies, and if on unhealthy, then the category of disease is crucial, again classified according to hot, cold, wet, and dry. Galen generally links astringency with cooling, matching a taste with a biological quality. One of his examples of astringency is an unripe apple, which as it matures and is heated by the sun becomes less cool, and eventually sweet and good to eat.

By contrast, in Galen's consideration of bitterness and pungency, to which I will only refer briefly, we can see clearly that these flavours and the quality of heat are closely related, as are astringency and cooling:

> On the capacity of bitter flavours (*chumoi*), when distinguishing them and declaring them to be cutting and thinning and cleansing, and clearly hot to the extent that they do not burn, let us go on in the argument to the pungent. First let us say that to be precise they are hot, and then that they corrode and burn and form scabs and melt down. All such items can be placed on the skin; and taken internally, those that are most dangerous to animals in their whole being are all rotting and destructive of those animals. Those that are so only because of the imbalance of heat, if they are composed of thick particles and are earthy cause ulcers internally; while if they are composed of fine particles they are diuretic and productive of sweat—in a word cutting and dispersive.
>
> Galen, *Simple Medicines* 4.19, 11.684 Kühn

Again, we see matching of tastes, fluids, and qualities, and contrasting actions according to whether applied to the skin or taken internally, and to whether fine or thick in composition.

I come, finally, to sweet flavours, as shown in another category of substance, the rich and oily. In the first two of the first five books of *Simple Medicines*, the theoretical books, Galen approaches the qualities and mixtures of drugs first through the medium of fluids, beginning with water and vinegar, and going on to olive oil and rose oil. These are more useful for his argument, it seems, than are solids. They are familiar to all human beings and have important impacts on the body, water most notably for its sustaining of life, as Galen says, and for having 'no quality' if it is pure and free from dissolved minerals such as salts. These fluids all impact on the bodily fluids, with their heating, cooling, and flavour qualities. He addresses these fluids as complex substances: water is changed when salts are dissolved in it; vinegar does not have one single quality; olive oil is complex, depending on where the olives are grown, how it is produced, and how it is applied. In the following passage Galen discusses oiliness (*to liparon*) in general:

> And of course the oily is itself sweet, and all nourishment is in the class of sweet things. This quite reasonably belongs to them in addition through the substance, such as it can be, of food. For nourishment is the filling of what has been emptied. What was emptied out was appropriate to it, so that the nourishment must also be appropriate. And if it is appropriate it must of necessity be pleasant and friendly, and immediately with a balanced warmth

as towards what is being nourished. . . . If . . . they are hotter than they should be, or if they become colder either in the body as a whole or in parts around the tongue or the stomach, they will need not only things which will nourish but also things that will cool and warm. . . . This drug indeed is connected with foods when they are eaten by bodies that are not only emptied but also changed in their quality, to please indeed in two ways, as foods and as drugs. But as certain drugs they will bring benefits and usefulness as agents that cool, warm, dry and moisten, while as foods they will benefit only as they are related and appropriate to the whole range of substances in the bodies being nourished.

Galen, *Simple Medicines* 4.10, 11.651–2 Kühn

Here we see nutrition closely linked with sweetness. We see again Galen's distinction between nutrition and drug-action and their intersection; the notion of nutrition as replacing lost energy; the notion that food needs to be as appropriate as possible to the body and its tissues, as we saw in the cases of pork and red mullet; and considerations of pleasure and the pleasant, which a doctor considers slightly differently from Plato's views, for example, in *Gorgias*, where Plato puts the pleasant and the useful in different categories.

Conclusions

In this chapter I have demonstrated how Galen believed that the qualities and capacities of the fluids of plants and animals and of the compounds of some minerals impacted on the fluids of the human body. In an effort to get away from the complex history of humours and its elaboration after Galen, I have followed Singer and van der Eijk in replacing 'humour' with the term 'bodily fluid'. They and others have insisted on the primacy of heating, cooling, moistening, and drying over humoral pathology in Galen's medicine, quite rightly. But I have shown that other 'qualities' must be added to these four active agents, including the thickness and lightness of particles and the impact of tastes such as sourness and bitterness. The *physical* qualities of heat, cold, moist, and dry act on *fluids* with the important characteristics under discussion. I have shown that these are integral to Galen's pharmacology, where he discusses drugs in close association with foods. Fluids such as olive oil and vinegar are key 'drugs' for Galen as well as foods. Taste in foods is fundamental: it indicates the impact the food will have inside the body in pathology and in the building of tissue and organs. It is clear too, that while Galen distinguishes a drug which triggers change from a food that maintains body energy and tissues, in practice many foods have pharmacological qualities, particularly those green plants with strong tastes. Tastes then belong with qualities as agents of change and must be considered in maintaining health and in diagnosing disorders that may need drug remedies. They belong with other differentiae that Galen deploys to clarify how a simple drug acts on the bodily fluid and tissue of a healthy person, of a person who is healthy but has an imperfect mixture, and on a sick person. Furthermore, bodily fluids will vary from individual to individual, as is amply illustrated in the chapters on breast milk which follow this.

Notes

1 The Cambridge Galen Translation Series has adopted the term 'bodily fluid' for 'humour' (*chumos*) as the most comprehensible approach for the twenty-first century: see Singer and van der Eijk 2018.

2 For example, Galen's understanding of black bile is very distant from the dozens of types of melancholy identified by Robert Burton in *The Anatomy of Melancholy* (1621).

3 See the discussion in Schöner 1964.

4 Translated in Singer and van der Eijk 2018.

5 Consequently 'they distribute in the body a *chumos* thicker than the equilibrium . . . the so called raw *chumos* is collected' (*On the Capacities of Foods* 167.23–168.4 Wilkins, 6.649 Kühn). Raw *chumos* and bad *chumos* are serious challenges to well-being in Galen's physiology. A wet-nurse who had eaten wild greens in spring had developed bad humour and passed it on to the nurseling through her milk, *On the Capacities of Foods* 3.14, 203.5–18 Wilkins, 6.685–6 Kühn. See further Wilkins 2015 and following chapters in this volume.

6 The understanding of the sense of taste in antiquity has advanced greatly with Rudolph 2017a, in which she and Totelin 2017 (among others) show how taste was understood in philosophical and medical thought. There was a broadly agreed typology of eight or more tastes from Plato and Aristotle onwards. Galen is not the first to have a greater number of tastes in the sour range than in sweet, salty, or bitter. Galen's category of sour includes what we know as astringent, sour, and sharp, within which range he places, for example, vinegar and citrus flavours (he only knew the citron itself). Galen bases his discussion of taste in *Simple Medicines* I on Plato's *Timaeus*.

7 *Dunamis* in Greek. In this chapter, I use the term 'capacities' as a translation of dunameis. Powell (2003) translates *dunameis* as 'properties'.

8 Fr. 14 Bertier; see also her comments (1972: 30–2) on the 'saveurs' of Mnesitheus.

9 On wet and dry together, cf. *On Mixtures* 1.566–7 Kühn, on the perfect body of Polyclitus.

10 Translations of *Simple Medicines* are taken from my forthcoming *Galen: Simple Medicines I-V* in the Cambridge Galen Translation Series mentioned in note 1. The standard Greek text of Galen is Kühn 1821–33.

11 Note the additional complication that Galen might call a flavour a *poiotēs* or 'quality' of a substance. *Poiotēs* is not limited to the biological qualities, nor indeed are mixtures, as van der Eijk 2015 noted. I believe that Galen thought flavours were active substances, secondary to heat, cold, wet, and dry, but acting with them, in ways that he was able to describe in *Simple Medicines*, as I discuss later.

12 See *On the Capacities of Foods* 3.1, 182.1–6 Wilkins, 6.663 Kühn.

13 Note primacy of taste for this purpose. Elsewhere taste is third or fourth for Galen: see Nutton 1993 on the five senses.

14 See *Simple Medicines* 1.5, 11.390 Kühn on water and *On the Capacities of Foods* 2.3, on gourds.

15 A particular concern with bakers is improper cooking which leads to a burnt crust and undercooked crumb—very bad for the digestive process, in Galen's view. Note that in discussing foods, and also drugs, in fact, the way in which the natural product is prepared and/ or cooked may be decisive in the impact on the body. Foods normally need cooking since raw plants in particular contain powerful moisture which may overcome the body heat in digestion and lead to dangerous 'raw humour'.

16 *Pepsis*. The body 'cooks' food in the liver and portal vein when converting digestive material into complex blood (which contains due proportion of all four 'humours' or bodily fluids).

17 Van der Eijk 2015: 676 n. 2, briefly notes this passage.

18 *Chulos . . . kakochumos*: cf. *Simple Medicines* 1.38, 11.449–50 Kühn, quoted above on *chuloi*, 'juices' and *chumoi* 'humours', and whether distinctions are to be made between them or not.

19 The upper part of the stomach, where it is joined to the oesophagus. The cardia is toned up by sour flavours.

20 Galen is most concerned with what is normally eaten, as noted on lettuce earlier. There is nothing rare or exclusive about Galen's foods: they are readily available, in contrast to a number of less-familiar drugs.

21 *Drimus*. Powell translates the term 'bitter', as does Grant 2000. I translate *pikros* to cover that class of flavours in Greek (though the Liddell-Scott-Jones lexicon is ambivalent). In support, I would point to the allium family, which Galen says have the same flavour. They are, I believe, 'pungent' with many sulphurous compounds, not bitter: cf. Davidson and Jaine 2006: 558. Davidson and Jaine, however, believe celery is bitter, and rue, also mentioned by Galen, is certainly bitter. Fennel, meanwhile, also mentioned by Galen in this category, is bitter in its seed (Davidson and Jaine 2006: 295). As Galen would be the first to note (*On the Capacities of Foods* 2.59, 164–7 Wilkins, 6.645–9 Kühn), seeds and stalks, as also varieties, may vary in their qualities. Until this difficulty is resolved, I retain 'pungent' for *drimus*. Flavours in the ancient world can be difficult for us to identify precisely, as, notoriously, are colours.

22 A common feature of heating plants in Galen, again linking a biological quality with scent and flavour.

23 Galen here has in mind a discussion in Plato's *Gorgias* where the cook offers a meretricious diet in contrast with the doctor's useful and healthy diet. Galen follows the Platonic lead about cooks but is well aware that the class of flavours identified as *glukus* 'sour' and *hedus* 'pleasant' are fundamental to life, along with salty, sweet, and bitter.

24 On fish and desire in Greek thought, see Davidson 1997. Sweetness (*glukutēs*) in *Simple Medicines* is normally linked with nourishment.

25 Bodily fluids change with age, so that a young pig produces more moist residues than mature pigs, and old people should eat less 'nourishing' food, replacing pork and beef with the lighter flesh of fish and birds (*On Maintaining Good Health* 5.4, 144.3–12 CMG, 6.333–4 Kühn).

26 *Chalkitis*: cf. *Simple Medicines* 10.35, 12.241–2 Kühn; also Dioscorides, *De materia medica* 5.99 (see Beck 2017 for a translation).

27 *Misu*, mined in Cyprus from seams of ore alternating with those of *chalkitis* and *sōru* (melanterite), according to Galen, *Simple Medicines* 9.3.21, 12.226–9 Kühn. *Misu* is composed of finer particles than *chalkitis*, 228: on the importance of fineness of particles in Galen's pharmacology, see Debru 1997.

28 *Chalkanthos*: cf. *Simple Medicines* 10.34, 12.238–41 Kühn.

29 *Lepis*: also at *On the Capacities of Foods* 1.1, 20, Wilkins and *Simple Medicines* 10.16, 12.223–4 Kühn.

30 *Sōru*: a copper ore mined with *chalkitis* and *misu*: see n. 27.

31 If a person is prone to too much yellow bile, their body heat is likely to produce biting and 'smoky' residues. Bathing to open the pores and more exercise are the remedy: *On Maintaining Good Health* 5.11, 161–2 *Corpus Medicorum Graecorum*, 6.365–9 Kühn.

Bibliography

Primary sources

Beck, L. (2017) *Pedanius Dioscorides of Anazarbus. De materia medica*. 3rd edn. Translated by L. Beck. Hildesheim: Olms-Weidmann.

Bertier, J. (1972) *Mnésithée et Dieuchès*. Leiden: Brill.

Grant, M. (2000) *Galen on Food and Diet*. Translated by M. Grant. London: Routledge.

Kühn, K. G. (1821–33) *Claudii Galeni Opera Omnia*. 20 vols. Berlin and Leipzig: Libraria Car. Cnoblochii.

Powell, O. (2003) *Galen. On the Properties of Foodstuffs*. Translated by O. Powell. Cambridge: Cambridge University Press.

Singer, P. and P. van der Eijk (2018) *Galen. Works on Human Nature*. Translated by P. Singer and P. van der Eijk. Cambridge: Cambridge University Press.

Wilkins, J. (2013) *Galien. Sur les facultés des aliments*. Paris: Les Belles Lettres.

Secondary literature

Burton, R. (1621) *The Anatomy of Melancholy*. Oxford: John Litchfield and James Short for Harry Cripps.

Davidson, A. and T. Jaine (2006) *The Oxford Companion to Food*. Oxford: Oxford University Press.

Davidson, J. (1997) *Courtesans and Fishcakes: The Consuming Passions of Classical Athens*. London: Harper Collins.

Debru, A. (1997) 'Philosophie et pharmacologie: La dynamique des substances leptomères chez Galien', in A. Debru (ed.) *Galen on Pharmacology: Philosophy, History and Medicine*. Leiden: Brill, 85–102.

Mirrione, C. (2017) 'Theory and terminology of mixture in Galen: The concepts of *krasis* and *mixis* in Galen's thought', PhD thesis, Humboldt University, Berlin.

Nutton, V. (1993) 'Humoralism', in W. F. Bynum and R. Porter (eds) *Companion Encyclopaedia of the History of Medicine*, vol. I. London: Routledge, 281–91.

———. (2013) *Ancient Medicine*. London: Routledge.

Rudolph, K. (2017a) 'Tastes of reality: Epistemology and the senses in ancient philosophy', in K. Rudolph (ed.) *Taste and the Ancient Senses*. London: Routledge, 45–59.

———. (ed.) (2017b) *Taste and the Ancient Senses*. London: Routledge.

Schöner, E. (1964) *Das Viererschema in der antiken Humoralpathologie*. Wiesbaden: Steiner.

Stewart, K. (2017) *Galen's Theory of Black Bile: Hippocratic Tradition, Manipulation, Innovation*. Leiden: Brill.

Totelin, L. (2017) 'Tastes in ancient botany, medicine and science: Bitter herbs and sweet honey', in K. Rudolph (ed.) *Taste and the Ancient Senses*. London: Routledge, 60–71.

van der Eijk, P. (2015) 'Galen on the assessment of bodily mixtures', in B. Holmes and K.-D. Fischer (eds) *The Frontiers of Ancient Science: Essays in Honor of Heinrich von Staden*. Berlin: Walter de Gruyter, 675–98.

Wilkins, J. (2015) 'Galien et le lait', *Food and History* 13, 273–81.

14

BREASTMILK, BREASTFEEDING, AND THE FEMALE BODY IN EARLY IMPERIAL ROME

Thea Lawrence

Throughout Roman literature, milk is a substance with strong positive associations with the nourishment, growth, and well-being of Rome, its lands, and its citizens. It could evoke the wholesome, fertile Italian countryside, as with animal milk in Virgil's *Georgics* (3.308, 2.519, edition Rushton Fairclough 1916). The 'milky richness' (*lactea ubertas*) of Livy's writing makes him, in Quintilian's eyes, ideal literary food for nourishing the minds of Rome's young men.[1] The milk of the she-wolf in Livy saves Romulus and Remus from death by exposure and is responsible for ensuring and fundamentally shaping the growth and development of Rome itself (*History of Rome* 1.4.6–7, edition Foster 1919). Human maternal milk was also nutritive and wholesome, the acme of bodily fluids. However, unlike symbolic 'milkiness' or the swollen udders of livestock, human milk could not escape its intimate and, as this chapter will explore, often unsettling connection with that most problematic of entities in Roman thought: the female body. This chapter will explore the ways in which the unstable fluidity, mutability, and transgressive tendencies of the female body and mind served to undermine the virtues of milk and the ways in which this fluid posed a potential threat to the growth, stability, and even morality of the Roman state and its citizens.

Breastfeeding in antiquity has received substantial scholarly attention, the vast majority of which has focused on discussing the demographics of and motivations for the use of wet-nurses.[2] Wet-nursing appears ubiquitous in Roman society, although as Claude-Emmanuelle Centlivres Challet notes in this volume and elsewhere, it may have largely been the preserve off the elite (Chapter 8, p. 136; 2017: 902). Scholars have debated how widespread this practice was and what factors shaped the decision to employ a wet-nurse—whether as a result of high maternal death rates, to boost one's fertility, to display one's wealth and status, to emotionally cushion oneself from the painful reality of high infant mortality, and so on.[3] Another strand of scholarship, particularly work by Larissa Bonfante and Patricia Salzman-Mitchell, has discussed the anxieties and ambivalence surrounding Greco-Roman depictions of breastfeeding, noting the disparity between the seemingly positive associations of breastfeeding and the relative paucity of nursing scenes, both literary and artistic (Bonfante 1997; Salzman-Mitchell 2012). Breastfeeding, nursing, or nursing-related scenes were pervaded by complex undertones of unease and danger, particularly in the tension between the female breast as maternal and nutritive and the female breast as sexual (Salzman-Mitchell 2012: 142).[4]

DOI: 10.4324/9780429438974-19

The rich potential of milk as a focus for discussions of the female body is evinced by the many times it appears in this volume alone. The work of Claude-Emmanuelle Centlivres Challet has explored the connection between ancient breastfeeding practices and contemporary beliefs about the quality of breastmilk, and her chapter here highlights the ways in which its presence or absence could be utilised as a metric for the social and moral evaluation of the female body (Centlivres Challet 2017a, 2017b; this volume, Chapter 8, pp. 136–7). Laurence Totelin's chapter (Chapter 15) emphasises the ways in which the reluctance of ancient texts to engage with the confluence of milk and blood immediately following childbirth reflected social anxieties about the female body.[5] Elsewhere, Giulia Pedrucci's work on breastfeeding in ancient Greece considers the connection between menstrual blood and milk and argues that deep-seated prejudices against the former tainted Greek attitudes towards milk (Pedrucci 2013a, 2013b, 2013c). This chapter will build upon Pedrucci's work, examining the ways in which such prejudices manifested during the Roman period.

This chapter will begin by considering the attitudes that can be discerned in early Imperial Roman literature towards menstrual blood and milk, focusing on the account of both fluids in the *Natural History* of Pliny the Elder (23–79 CE). It will explore the ways in which the link between these two fluids in the ancient understanding of female physiology and procreation, in particular the taboos surrounding menstrual blood, served to complicate the otherwise positive value assigned to milk. The chapter will then look at discussions of milk and breastfeeding in Soranus' *Gynaecology* (98–130 CE) and Aulus Gellius' *Attic Nights* (c.123–69 CE) in order to consider the ways in which both medical and non-medical texts from the early Empire located the blame for the perceived instability of milk upon the female bodies that produced it. This discourse was informed by and contributed to broader misogynistic social narratives that reinforced male control over these bodies. Despite its many medicinal and nutritive qualities and its associations with bucolic harmony and maternal care, milk possessed the potential to be physically and morally corrupting. Its inescapable connection to the unstable, fluid, and untrustworthy female body further served to compound the potential threat it posed to the vulnerable bodies it nourished and to the future of Roman society.

Milk and menstruation

Pliny the Elder presented his *Natural History* (hereafter *NH*) as an attempt to gather together the totality of contemporary 'knowledge' about the natural world (*NH*, preface 14; Wallace-Hadrill 1990: 82; Beagon 2013: 84). Firmly situated in the political and intellectual fabric of the early Principate but also making use of a variety of earlier Greek and Roman sources, Pliny's account of breast milk and other bodily fluids reflects much of the information an educated member of the elite might have had access to or been aware of. As with much of the rest of the *NH*, the account seems to reflect or engage with contemporary and enduring cultural attitudes towards gender and the body.[6]

Milk, while vital for infant survival, is not generally presented by Pliny as an appropriate substance for adult nutrition but rather as a medicine. Galen and Dioscorides, similarly, recommended the consumption of breast milk by adults for its curative powers, even suckled directly from the breast (Dioscorides, *De Materia Medica* 2.70.6, translation Beck 2017; Galen, *Method of Medicine* 7.6, edition Johnston and Horsley 2011). Adult breastfeeding appears in antiquity in non-medical contexts, most notably in the story of Pero and Micon, where it occupies an unsteady position between commendably familial virtue and uncomfortable sexual overtone (Deonna 1956: 497–506; Bonfante 1997: 181–2; Valladares 2011). Tara Mulder argues against an incestuous or sexual element both to this and to other stories, including one example from the *NH*. Here, a woman nurses her starving, imprisoned mother, a feat of filial piety Pliny says is beyond compare (*NH* 7.121; Mulder 2017: 228). Despite not recommending this practice in general, Pliny nevertheless presents breastmilk as a worthy and nutritive substance.

Above all, Pliny is concerned with the pharmacological qualities of milk. In book 28, Pliny chronicles the many conditions for which it had great efficacy, including fevers, coeliac disease, and nausea. When combined with frankincense, milk can be used to treat abscesses on the breasts, to cure jaundice, to ease bruising, to tackle lung infections, to ease menstrual cramps, to facilitate healthy bowel movements, and as a protection against or antidote to poisons (*NH* 28.72–3, edition Jones 1963). Combined with other *aromata* such as honey, narcissus, and frankincense, it relieved bloodshot or leaking eyes (cf. Dioscorides, *De Materia Medica* 2.78). For ear infections, it could be mixed with oil or combined with honey soaked in wool and inserted into the ears to eradicate offensive odours. Indeed, milk appears to possess immunological and antimicrobial properties and is seemingly particularly effective in the treatment of eye and ear infections.[7] Regardless of the actual effectiveness of such remedies, however, their broad range demonstrates the considerable power of milk. It makes for such a potent medicine that Pliny claims lifelong immunity from any eye trouble can be achieved by rubbing in the milk of two generations of women, a mother and her daughter (*NH* 28.73). While the modern scientific evidence for this is so far absent, it indicates a perception of milk as remarkably powerful, and suggests that the bodies milk came from mattered.

While different kinds of milk could be beneficial, Pliny makes clear that not all milk was equally valued. Human milk was preferable to animal, with goat, camel, and donkey milk being particularly efficacious. The milk of human mothers was according to Pliny by far the best, being 'the sweetest and most delicate', a description mirrored in ancient medical texts (*NH* 28.72 *dulcissimum esse mollissimumque*; *NH* 28.123; cf. Dioscorides, *De Materia Medica* 2.70.6, Beck translation). Other factors could influence the value of human milk, above all the sex of the child for whom it has been produced. Pliny writes that 'for all uses the milk of a woman who has given birth to a boy is more efficacious' (*NH* 28.72), with milk for male twins even more potent. The milk of a woman who has borne a female child is somewhat less impressive, being beneficial only for cosmetic purposes, specifically for curing spots on the face (*NH* 28.75). The idea that milk produced to feed male offspring was particularly beneficial appears persistent and widespread throughout much of ancient medical literature.[8]

Such hierarchies of milk indicate that the efficacy and nutritional quality of the fluid was determined by the nature of the body within which it was concocted—animal or human—and by the specific context in which it was produced, such as the kind of child it was intended to nourish. Pliny further emphasizes the impact of the female body upon the quality of milk in his stipulation that even the 'best' milk, that of the mother of twin boys, is only at its most efficacious if the mother also abstains from drinking wine and from eating pungent foods (Pliny, *NH* 28.72–3). Mulder (this volume, Chapter 9, p. 152) raises an important point that several authorities, including Aristotle, Soranus, and Galen, place the bulk of responsibility for foetal sex determination on the mother. It was therefore at least preferable, perhaps essential, for a woman's body, regimen, and activity to be carefully controlled in order to produce high-quality milk, suitable for both pharmacological and nutritive purposes.

Pliny's characterisation of another fluid central to female physiology, menstrual blood, could hardly be more different. Menstrual blood was perceived as polluting, unclean, and malodorous in Greco-Roman antiquity, a perception which seems to have been more acute in the Roman period.[9] According to the prevailing theories of Greco-Roman medicine, it was intimately linked with breast milk. Menstruation was generally seen as a side-effect of the spongy wetness of the female body; it was a monthly excretion of the inevitable build-up of excess fluid precipitated by woman's physical deficiency.[10] When a woman became pregnant, this same blood was reappropriated by the body in order to feed and form the foetus, and, when the pregnancy reached its end, redirected to the breasts and concocted into milk.[11] While medical texts often displayed a more ambivalent or even positive depiction of menstruation, Pliny's account reflects the hostility and superstition with which much of contemporary society seems to have regarded menstrual blood:

> Truly fearful and unspeakable things have been raved about the menstrual discharge itself (*dira et infanda vaticinantur*), the other magic (*monstrificus*) of which we have indicated in its proper place. Of these things, one might mention without shame that if the female power [menstruation] coincides with an eclipse of the moon or sun, its effects are incurable, no less also when the moon is not shining; then sexual intercourse is fatal and noxious to men (*coitusque tum maribus exitiales esse atque pestiferos*). At this time also purple is polluted by the woman's touch (*purpuram quoque eo tempore ab his pollui*): so much greater is her power.
>
> Pliny, *Natural History* 28.77–8[12]

Menstrual blood had the power to drive away storms, ruin purple cloth and linen, blunt razors, tarnish mirrors, and to cause bronze and iron to tarnish and take on a foul smell (*NH* 28.78–9). The blood was repulsive to animals and plants, devastating vines and medicinal herbs, driving bees from their hives, killing other insects, and sending dogs into madness.[13] The mere sight of a woman who was in the midst of her first period, or the touch of any menstruating woman, could cause horses to spontaneously miscarry. Humans, too, were vulnerable to the harmful effects of menstrual blood: women, as Pliny says, were not safe from this plague of their sex (*NH* 28.80: *feminis malo*), suffering miscarriage even from proximity to the substance. A man foolish enough to engage in sexual relations with a menstruating woman was risking his life. The language Pliny uses plays on anxieties about pollution, infection, and disgust, suggesting the destructive and even deadly potential of menstruation. Menstrual blood was magical, monstrous, or strange (*monstrificus*). Much like a disease, it polluted on contact (*polluere*) and altered the state of everything it touched (*mutare*), rendering things infected, stained, or spoiled (*infectae*). Its effects were destructive or deadly (*exitialis*), pestilential, baleful, or noxious (*pestifer*), and incurable (*inremediabilis*). That this pollution could be transmitted invisibly beyond the bodily boundaries of the menstruating women compounded the sense of threat, with her gaze capable of tarnishing a mirror (*NH* 7.64; 28.82). This phenomenon, also reported by Aristotle, points to the invasive multisensory nature of the threat, which could pollute even without touch (Aristotle, *On Dreams* 450b–60a, edition Hett 1957; Pedrucci 2013b: 266).

Pliny, determined as ever to catalogue the entirety of natural phenomena, records beneficial and even medicinal uses of menstrual blood—it could, for example, calm storms and kills pests (*NH* 28.71, 82; Lennon 2010: 80). Citing two women, Lais and Salpe, he also mentions the use of menstrual blood to cure fevers, as well as the bite of a rabid dog (Flemming 2007: 271). The midwife Sotira suggests the application of menstrual blood to the feet in order to treat fever and to revive one from an epileptic fit (preferably, for reasons Pliny does not go into, without the knowledge of the patient) (*NH* 28.83). It is notable both that female authorities dominate the pharmacological uses of menstrual blood and that Pliny prefaces the remedies with some scepticism, suggesting that these women might equally be midwives or prostitutes (*NH* 28.70; Flemming 2007: 273–4). It is possible that we are seeing in this instance a disjuncture between revulsion of menstrual blood displayed by Pliny and likely to be representative of elite male attitudes and the more positive attitude particularly of female practitioners, who were more familiar with and less superstitious about the substance. Despite these voices to the contrary, the reader is left with little doubt that menstrual blood was dangerous and polluting. Indeed, Pliny chooses to end his survey of the beneficial applications of the substance with a statement emphasising the taboo nature of the fluid: 'This is all which it would be right (*fas*) to repeat, and much of it I do not say without shame. The rest is detestable and unspeakable (*reliqua intestabilia, infanda*)' (*NH* 28.87).

Appropriately for their physiological connection, milk and menstrual blood emerge in the *NH* as two sides of the same coin. The effects of menstrual blood—disgusting, invasive, and even lethal—aligned closely with the worst deficiencies of the female body, which leaked fluid and odour and constantly threatened to pollute and weaken the bodies and others, above all the dry, rigid bodies of men. Milk reflected instead the benefits of women's liquid physiology, which served as a locus for the growth, nutrition, and well-being of these same men. That milk was the concocted form of blood provides the opportunity for those who shared Pliny's attitude to regard milk as suspect and unreliable as well (Pedrucci 2013b: 287). Also emerging from Pliny's account is the idea that the nature and condition of the female body substantially impacts the quality and effect of the fluids it produces. Not only do some bodies in some contexts naturally produce higher quality milk, there are also hints that the women in possession of these bodies had some control over and therefore responsibility for the milk they produced. It is this feature of milk that will be explored further in Soranus' *Gynaecology* and Aulus Gellius' *Attic Nights*.

Soranus' milk

While Pliny's portrayal of milk emphasises its therapeutic properties, the early second century physician Soranus focuses on its role in infant nutrition and the connection between milk and the body by which it was produced. Soranus' *Gynaecology*, which discusses the physiology and treatment of the female body, is the only extant work of its kind from the Principate, and a significant portion of its second book provides a uniquely detailed account of the correct way to feed and care for a newborn.[14] Informed by his practical experience as a physician, Soranus in general promotes the hierarchy found in Pliny and elsewhere that positioned mothers' milk above all other sources. He asserts that 'it is better to feed the child with maternal milk', as that milk is 'more suited' and 'more natural' than other sources.[15] However, Soranus acknowledges that there were occasions in which the use of a wet-nurse might be acceptable, or even preferred—indeed, he seems to assume that the employment of a wet-nurse is quite likely, regardless of whether it is the best course of action. The *Gynaecology* does not demonstrate much hostility towards the use of a wet-nurse where the mother is concerned for her health, and the desire to avert premature ageing is accepted as a legitimate, if less than ideal, reason that a mother might choose not to breastfeed. This leniency sets it apart from the more censorious attitudes of some other, non-medical discussions of the subject.[16] For the first three days, Soranus recommends against using maternal milk out of a belief that the thicker milk produced during that time, known in modern medicine as colostrum, was dangerous to the baby. This recommendation would necessitate the use of another's milk as a replacement.[17] Overall, while the use of milk from another woman is not depicted as an inherently bad thing, it is less ideal for the physical, moral, and social development of the child.

Soranus' liberal attitude concerning maternal or mercenary breastfeeding has its limits. While the use of wet-nurses is not inherently worthy of criticism, the nature of these sources of alien milk needed to be tightly controlled. Many of his recommendations concern the ways in which the physiology of the specific female body involved could adversely affect the quality of the milk itself. Age played a role, with the ideal wet-nurse being between 20 and 40 years old, as this would ensure that her mind was not too childish, but her body was not too old and inclined to produce watery milk (*Gynaecology* 2.19). Weight, body type, and constitution were also important—the milk of a physically weak woman could become spoiled due to exhaustion brought on by hard work, and so wet-nurses should be 'fleshy and strong' (*Gynaecology* 2.19.29–31). Perhaps unsurprisingly, breast size was deemed significant; excessive milk production, a symptom of which (according to Soranus) was large breast size, carried the risk of milk being retained in the breast after nursing and consequently going off. The infant might then be at risk of drinking this spoiled milk at the next feed, thereby getting sick. Small breasts, however, might not produce enough milk, and so one should only select wet-nurses with medium-sized breasts (*Gynaecology* 2.19.33–45).

In addition to and intersecting with such physiological factors, many of Soranus' recommendations for wet-nurses emphasise the importance of controlling their nature, consumption, and other behaviours. She should be Greek, in order that the child might learn the language, and affectionate, avoiding the perennial trope of the neglectful nurse.[18] The appropriate diet for a wet-nurse required that she 'forgo food that has bad juices, that which is not nourishing and which is hard to digest' (*Gynaecology* 2.25.41–4), a recommendation reminiscent of Pliny's claim that acrid food should not be consumed by female bodies that provided milk. Whereas Pliny's explanation for this is somewhat vague, Soranus makes the reasoning behind the recommendation clear: 'the qualities of the food partaken are conveyed also to the milk' (*Gynaecology* 2.26.83–4). When discussing how to solve infantile constipation or diarrhoea, Soranus expands upon this idea of the transfer of qualities, claiming that any effect the food produces upon the nursemaid will be even more powerfully felt by the child.[19]

While many different elements of female regimen required controlling, two of the most significant factors that Soranus identifies as being detrimental to the quality of a nurse's milk were sexual activity and consumption of alcohol. Soranus draws a link between the two, grouping them both under the heading of 'self-control':

> And the wet-nurse should be 'self-controlled' (*sōphrōn*) so as to abstain from coitus, drinking, lewdness, and any other such pleasure and incontinence. For coitus cools the affection towards \<the\> nursling by the diversion of sexual pleasure and moreover spoils and diminishes the milk (*phtheirō to gala kai meiousin*) or suppresses it entirely by stimulating menstrual catharsis through the uterus or by bringing about conception. In regard to drinking, first the wet-nurse is harmed in soul (*psuksēi*) as well as in body (*sōmati*) and for this reason the milk also is spoiled (*gala diaphtheiretai*) . . . too much wine passes its quality to the milk and therefore the nursling becomes sluggish and coma-tose and sometimes even afflicted with tremor, apoplexy, and convulsions.
>
> Soranus, *Gynaecology* 2.19.65–78

Coitus, as well as threatening to reduce a nurse's affection for her charge, could spoil, reduce, or entirely curtail her milk. The risk of pregnancy, with a subsequent reduc-tion or cessation of milk production, appears to have been a central motivation for recommendations of abstinence on the part of wet-nurses (Bradley 1991b: 215). Cel-ibacy is mentioned alongside sobriety in the recommendations of both Galen and Rufus of Ephesus (Galen, *Hygiene* 9.29, edition Johnston 2018; Rufus in Oribasius, *Uncertain Books* 13–14, edition Raeder 1933) and makes regular appearances in the wet-nursing contracts which survive from Roman Egypt (Bradley 1980: 321–5; Mas-ciadri and Montevecchi 1984: 23–4). Soranus' reference to the risk of 'spoiling' the milk points, on the other hand, to the idea that it was unnatural or unhealthy for both menstrual fluid and breastmilk to flow simultaneously. As discussed by Totelin, Chapter 15 in this volume (pp. 240–1), this distaste for the emission of blood and milk also manifests in the ancient advice not to allow a newborn to drink colostrum; milk consumed while the post-partum lochia were still flowing was a dangerous aberration, a double-flowing of two fluids that were one and the same in humoral thought. We might also see here a pale reflection of the horror with which menstrual blood was elsewhere regarded—Pliny's account makes clear the corrupting power of this fluid when out of its proper context and the danger it posed to unborn children. No won-der, then, that the presence of menstrual blood in such close quarters with milk might render it useless or harmful.

While coitus negatively impacts the quality and quantity of the milk, alcohol con-sumption poses an even greater threat. Soranus claims that drinking harms the wet-nurse 'in soul as well as body'. As with sexual intercourse, sobriety during the nursing period seems to have been a common recommendation for nursing, although some ancient writers, such as Rufus of Ephesus (late first century CE), allowed some alcohol in moderation (Rufus in Oribasius, *Uncertain Books* 13; Masciadri and Montevecchi 1984: 23–4). Some of the risks of alcohol consumption outlined by Soranus have little to do with the effect of the alcohol on the milk itself; a drunken nurse might leave the child unattended while off drinking or even fall into a drunken stupor on top of her charge and crush it to death (*Gynaecology* 2.19.70–3). The stereotype of the drunken, inattentive nurse was enduring and pervasive in both Greek and Roman sources, as was the broader stereotype of female dipsomania (Joshel 1986: 9–10; Burguière et al. 1990: 96; Laes 2011: 62). The quoted passage makes clear that the deleterious effect of wine on the bodily fluid itself was also of grave concern. Wine was considered to be a haematopoietic substance, producing excess blood and disturbing the delicate balance of humours at play (Jouanna 2012: 184). However, dilution does not seem to be Soranus' concern. Milk contaminated with wine causes sluggishness, unconsciousness, tremors, and even more extreme health issues. Re-emphasising his claim that a baby's body is especially vulnerable to negative qualities passed on through milk, Soranus warns those who might consider drinking in moderation to be acceptable:

> That the wet-nurse is not injured by wine now is, therefore, no reason to believe that the infant will not be badly affected either. Rather, one should argue that the wine is stronger than the physique of the newborn, and so the majority of those who are fed carelessly are seized by epileptic convulsions.
>
> Soranus, *Gynaecology* 2.27.110–17

The claim that even a nurse who drank in moderation and who did not drunkenly smother her charge with her breast nor leave it to starve while she slept off her hangover was still dangerous to the baby is a clear indication that it is the milk itself that posed a threat. Soranus characterises the infant body as weak and vulnerable to external influence and presents milk as a substance that could be corrupted and transformed into an invasive, potent, and harmful bodily fluid. It is revealing that even a text that demonstrates a remarkably sympathetic and unsuperstitious attitude towards the female body and character should display such anxiety. Not only were elite mothers themselves unreliable sources of good milk, either through exhaustion or disinclination, their replacements were even more so; the effects of their inferior physiology and behaviour posing a serious threat to the health of socially superior children, a threat embodied in the spoiled milk they produced. It is unsurprising then that intervention and guidance by reliable male physicians like Soranus was required in order to take control of a situation otherwise left in the hands of the very women who could easily cause harm. Furthermore, Soranus' idea that the failings of the nurse could harm infant in soul (or mind, *psuchē*) as well as body suggests that the dangers of milk extend beyond its direct physiological harms. The final source that this chapter will explore further demonstrates that the potential social harms of corrupt milk were, for some Romans thinkers, even greater cause for concern.

Aulus Gellius' milk

The misogynistic mistrust of the female body underpinning Soranus and other ancient writers' desire to oversee the nursing process, reminiscent of Pliny's more extreme depiction of the fluid female body as a destructive force, is brought more explicitly into the foreground in Aulus Gellius' *Attic Nights* (hereafter *NA*, edition Rolfe 1927). Written in the late second century CE, this work contains a vivid account of the discourse of the philosopher Favorinus, in which he attempts to dissuade the mother of an elite woman from employing a wet-nurse instead of allowing her daughter to breastfeed her own child. Unlike Soranus, Favorinus argues vehemently for maternal breastfeeding. Despite, or perhaps because of the prevalence of wet-nurse use by the elite, many literary sources from the late Republic and the early Principate denounce the practice. Tacitus, for instance, contrasts contemporary wet-nurse employment unfavourably with an idyllic past of breastfeeding mothers and praises the Germans for their preference for maternal breastfeeding.[20]

Favorinus' argument conforms to this distaste for mercenary breastfeeding and the withholding of maternal milk, which is characterised within the passage as 'against nature, an imperfect and half-motherhood' (*NA* 12.6). Much of his argument in the *NA* reveals misogynistic attitudes that conceived of women as vain, frivolous, and unreliable. One of his first objections to wet-nursing is based upon the conviction that women shunned breastfeeding in order to preserve their beauty, of which breasts unmarked by signs of nursing were a central element (*NA* 12.7). He condemns this vanity in no uncertain terms:

> In so doing [avoiding breastfeeding for vanity] they do the same madness (*eadem vecordia*) as those who try through thought-out deceits to abort the foetus itself, conceived in their own bodies. . . . But since it is a deed worthy of public detestation and universal disgust to kill a human being in its beginning . . . how far is it from this to deprive a child, already perfect, already a son, of the nourishment of its own accustomed and kindred blood (*consueti atque cogniti sanguinis alimonia privare*)?
>
> Aulus Gellius, *Attic Nights* 12.7–9

Favorinus equates the withholding of maternal milk with abortion, characterising both as 'worthy of public detestation and universal disgust'. This connection between abortion and wet-nursing reflects the belief found elsewhere in Latin literature that abortion was yet another method by which vain women might avoid losing their figures to motherhood—Juvenal, for example, denounces women who drink abortifacient drugs, supposedly for this very reason (Juvenal, *Satires* 6.592–9, edition Morton Braund 2004; see Dixon 1990: 94). In equating wet-nurse use with abortion, the practice is framed as an extreme form of neglect that harmed the child and irreparably damaged the mother–child relationship. His implication that preservation of beauty was a central motive plays directly into widespread stereotypes; refusal to nurse their own children was proof of the vanity, self-absorption, and immorality of contemporary Roman women. Favorinus ignores the alternative explanation offered by the young woman's mother; her daughter has already suffered greatly during childbirth, and they have chosen to use a wet-nurse to prevent her further exhaustion by the trials of nursing. Maternal fatigue following a difficult birth appears in Soranus and other medical literature as an entirely valid reason for using another source of milk, and indeed the connection Soranus makes between a robust constitution and high-quality milk indicates that the use of milk from another woman in better health might be preferable both for the recovery of the mother and the nurturing of the newborn. But for Favorinus, there is no legitimate excuse not to nurse your own child.

While the demonisation of wet-nursing and women who did not nurse their own children was common in Roman textual discussion, Favorinus' rhetoric is notable for its use of medical theories seen in Soranus and other ancient physicians to emphasise the fundamental difference between the nature and quality of milk provided by a mother and by a nurse. Favorinus demonstrates an awareness of the theory that saw milk as a concocted form of the same blood which had nourished the foetus in the womb (*NA* 12.12) and explicitly uses this to argue that maternal breastfeeding was intended by nature and to support his association between denying maternal milk and abortion. In both cases the child is deprived of the same nourishing, life-giving substance, altered only by its concoction and relocation within the body. Like Soranus, Favorinus demonstrates concern that inferior milk poses a threat to the health of the newborn (*NA* 12.20 *pernicioso contagio*). However, Favorinus emphasises how milk alters not just the body but also the mind and character. Milk from a source other than the mother was inherently 'inferior and alien' (*degenerique alimento lactis alieni*) and could 'corrupt the nobility of mind and body' of the nursling (*NA* 12.17). He once again couches his argument in medical theory, claiming that 'just as the power and nature of the seed are able to form likenesses of body and mind, so the properties of the milk have the same effect' (*NA* 12.14–15). Much as its pharmacological uses in Pliny demonstrate its potency, milk here is a fluid with considerable power, with an equal share in shaping the nature of a child as the generative seed.[21]

While Pliny the Elder and Soranus consider the effects of milk upon the body, Gellius' account places more emphasis on the ability of milk to shape the mind (*animus*). Where Soranus' *Gynaecology* refers to the damage alcohol does to the *psuchē* of the nurse, Gellius takes the idea much further: the milk that an infant ingests shapes both body and mind, and Gellius even suggests that the drinking of alien milk (*lactis alieni*) could explain where a child's appearance and character is incomprehensibly different from its parents (*NA* 12.19).[22] Where Soranus focuses mainly on the health risks involved in the transfer of milk, Favorinus focuses more on the potential moral and social damage. He echoes stereotypical concerns surrounding the use of wet-nurses, singling out alcohol consumption and sex as key factors contributing to inferior and harmful milk (*NA* 12.17 *si inpudica, si tementula est*). Throughout, he makes clear that the milk itself is supremely important. When he warns of the dangers of employing a slave or a nurse from a barbarian nation, it is the consumption of unworthy milk that will corrupt (*corrumpere*) the child rather than exposure to her inferior language or customs (*NA* 12.17).

Given the associations between the term *animus* and ideas of breath, air, and inhaling, Gellius' use of the word *spiritus* (breath, air, exhalation, smell, odour) emphasises the penetrative nature of breastfeeding—just as liquid is drawn into the infant's body to nourish or infect it, the mental and moral disposition of the nurse is simultaneously breathed in to the body of the infant, shaping it intimately for better or worse (*NA* 12.17–18). Again, this concept echoes the anxieties found in Pliny's account of menstrual blood; its influence also penetrates other bodies, sometimes mediated through air in the form of a foul, tarnishing odour. The most *animus*-harming effects of menstrual blood occur when, like milk, it is voluntarily consumed through the mouth, as in the case of *cunnilinctores*, and stains the consumer both physically, through malodour, and socially, through the breaking of the most taboo of sexual mores (Seneca, *On Benefits* 4.21.3, edition Basore 1935; Martial, *Epigrams* 12.85, edition Shackleton Bailey 1993; Lennon 2010: 74–5; Lowe 2013: 347). Both fluids demonstrate the ways in which female fluidity was alarmingly penetrating and invasive, but Favorinus' account also suggests that milk could be, if in a more subtle manner, even more destructive than menstrual blood—its ability to shape the mind of an infant meant that it could deprive that child of a moral character befitting its social standing. Favorinus' invective points not only to broader social concerns about widespread wet-nursing but also to an underlying fear that the transmission of milk from an unreliable female body to another might have the power to fundamentally, negatively, and irreparably alter the appearance, spirit, and character of a child—above all, those male children who will grow up to populate the Roman elite.

Conclusions

As with any bodily fluid, anxieties about pollution and the violation of physical boundaries bubble beneath the surface in discussions of milk. As it is an exclusively female fluid, these concerns are exacerbated by an overarching sense of the otherness and physical inferiority of the female body. This inferiority is defined by excessive wetness, and menstrual blood and milk are symptomatic of the liquidity of female corporeality. Ancient writers credit these substances with the unsettling power over the bodies with which they interact. While menstruation is obscured by superstition and taboo, and menstrual blood is ascribed various destructive properties, breast milk possesses power of a different kind. Milk is central to the health and development of the young, particularly of young Roman males, but its disturbing potential for physical and moral contamination makes it more dangerous and in need of control than menstrual blood. Because of the women who produced it, milk could deceive, corrupt, and destroy those with whom it comes into contact.

The extent to which the character and behaviour of a lactating woman influenced the quality and nature of her milk, as well as her fluidity and her porous, unstable corporeal boundaries, generated much anxiety in the Roman world. Gellius and his peers' endorsement of maternal breastfeeding reflects concerns about the pollution of elite offspring with inferior milk produced by a socially inferior nurse who is morally deficient and physically unreliable. Such concerns demonstrate an unwillingness to trust women with the care of their own children. In Gellius' account, the philosopher-orator takes issue with decisions that have been made by a woman—the child's grandmother—on behalf of the health of the new mother.[23] Soranus indicates that male control was also necessary to mitigate the potential damage of using a wet-nurse.[24] Although in reality women may often have arranged these matters themselves, the male-authored literary sources are deeply uncomfortable with decisions concerning infant feeding that excluded male authority figures; fathers, masters, physicians, and even philosophers. Breastmilk and breastfeeding were thus incorporated into wider misogynistic discourses that revealed anxieties about female power and justified male control over the bodies and behaviour of women (Carson 1999: 77–8; cf. Janssen, Salvo, and Mulder in this volume, Chapters 2, 3, and 9, respectively). These same anxieties underpin the founding myth of Rome, where Romulus and Remus were suckled by the she-wolf (or prostitute) and inherited through her milk the savage characteristics that defined their behaviour and their destiny.[25] Milk therefore, and the body from which it came, possessed the power to shape and to destroy not only individual infants but even Rome itself.

Notes

1 Quintilian, *Institutes of Oratory* 1.1.21; 2.4.6; 10.1.31–2, edition Russell 2002; Hays 1987: 113; Kraus 1994: 17; McAuley 2015: 327.

2 See Bradley 1980, 1991a, 1991b, 1994; Fildes 1986: 17–36; Dixon 1990: 105–61; Sparreboom 2009; Dasen 2010; Laes 2011: 50–106; Hackworth Petersen and Salzman-Mitchell 2012; Parkin 2013: 50–3; Sparreboom 2014.

3 See Bradley 1980, Bradley 1991a, 1991b, 1994; Fildes 1986: 17–36; Dixon 1990: 105–61; Sparreboom 2009; Dasen 2010; Laes 2011: 50–106; Hackworth, Petersen, and Salzman-Mitchell 2012; Parkin 2013: 50–3; Sparreboom 2014; Centlivres Challet 2017a; Cicero, *Brutus* 252; Velleius Paterculus, *Compendium of Roman History* 2.59.2–60.2, edition Shipley 1924; Statius *Thebaid* 6.148–50, 152–5, 160–7, edition Shackleton Bailey 2004; Plutarch, *Life of Caesar* 7, edition Perrin 1919; *Life of Gaius Gracchus* 4.1–3, edition Perrin 1921; *Life of Tiberius Gracchus* 1.4–5, edition Perrin 1921; Pseudo-Plutarch, *Morals: The Education of Children* 3.5, edition Babbitt 1927; Tacitus *Dialogue on Oratory* 28.4–6; 29.1, edition Hutton and Peterson 1914a; Suetonius, *Life of the Deified Augustus* 8; *Life of the Deified Julius* 13, edition Rolfe 1914; Appian *The Civil Wars*, 3.10–11, edition White 1913; Cassius Dio, *Roman History* 45.1.1, edition Cary and Forster 1916.

4 Salzman-Mitchell 2012: 142 notes the enduring nature of this tension, and links it to modern American anxieties of a similar nature; Stears 1999.

5 See also Totelin's discussion of the connection between Nymphs and breastfeeding women in ancient Greece: Totelin 2017b.

6 See Lawrence 2019: 37–41, 106–41.

7 Laskaris 2008: 459; Dasen 2013: 53; Isaacs 2014: 214–15; Mulder 2017: 237.

8 Hippocratic Corpus, *On Barren Women* 3.214; 2.162; Laskaris 2008: 459; Pedrucci 2013b: 280–1.

9 Gourevitch 1984: 95–103; Richlin 1992: 281–2; Dean-Jones 1994: 198; Lennon 2010: 81–8; Ripat 2016: 109–11. See Rosalind Janssen, this volume, Chapter 2, on menstruation in Pharaonic Egypt, which explores the containment strategies for this potentially pollutive bodily emission.

10 On the female body in Greek and Roman medicine, see King 1997, 1998; Flemming 2000. On the cool, spongy wetness of the female body, see Aristotle, *Generation of Animals* 775a14, edition Peck 1942; *Parts of Animals* 650–8, edition Peck and Forster 1937; Flemming 2000: 117. On the inherent inferiority of the female body in relation to the male, see Aristotle, *Generation of Animals* 737a; Galen, *On the Usefulness of the Parts* 14.6–7, edition Helmreich 1907–9.

11 Hippocratic Corpus, *On the Diseases of Women* 1.1, edition Potter 2018, *On Regimen* 1.33, edition Jones 1931; Galen, *Hygiene* 1.7; King 1997: 136; Flemming 2000: 116; Pedrucci 2013b: 263–4.

12 Unless stated otherwise, all translations from the Latin are my own.

13 Giulia Pedrucci has explored in greater detail the connection between women, menstruation, and dogs in ancient thought (2013b: 268–70).

14 See the introduction of Temkin 1991 *Soranus' Gynecology*: xxiii–xlix for a detailed discussion of the life and work of Soranus, such as can be discerned from the little we know about him.

15 Soranus, *Gynaecology* 2.18: 'Other things being equal, it is better to feed the child with maternal milk (*mētrōioi galakti*): for this is more suited to it, and the mothers become more sympathetic towards the offspring, and it is more natural (*phusikōteron*) to be fed from the mother after parturition just as before parturition. But if anything prevents it one must choose the best wet-nurse, lest the mother grow prematurely old, having spent herself through the daily suckling.' This and all translations of Soranus are those of Temkin 1991.

16 Plutarch, *Cato the Elder* 20.3, edition Perrin 1914; Pseudo-Plutarch, *On the Education of Children* 3b–f; Soranus, *Gynaecology* 2.18.97–104; Aulus Gellius, *NA* 12.1.

17 Laurence Totelin 2017a: 7 notes that mistrust of colostrum seems to be a widespread cultural phenomenon. For more on colostrum in Soranus, see Totelin in this volume, Chapter 15, p. 243.

18 Aristophanes, *Knights* 5.716, edition Henderson 1998; Soranus, *Gynaecology* 2.19; Tacitus, *Dialogue on Oratory* 2.28–9; Laes 2011: 71

19 Soranus, *Gynaecology* 2.56 'so by eating astringent food the wet-nurse is but slightly constipated; the infant, however, which is fed by her milk much more so (*mallon de to brephos to ex autēs galakti*).'

20 Juvenal, *Satires* 6.9, 592–7; Plutarch, *Cato the Elder* 20.3; Pseudo-Plutarch, *On the Education of Children* 3b–f; Tacitus, *Dialogue on Oratory* 28.2–29.2; *Germania* 20.1; Centlivres Challet 2017b: 369–70.

21 See Rebecca Flemming, Chapter 10, this volume, for further discussion of the role of female and male fluids in foetal generation.

22 This is echoed in the fifth century in Macrobius *Saturnalia* 5.11.15–18, edition Kaster 2011. This argument could have explained cases where the physical appearance of an elite woman's is markedly difference from her husband, avoiding conflict over issues of paternity.

23 Another source, a papyrus from third-century CE Roman Egypt, may preserve the words of another grandmother who similarly adopts a protective role on her daughter's behalf: P. Lond. 9.351, 2–5. Thanks to Laurence Totelin for bringing this to my attention.

24 Both Florence Dupont and Sandra Joshel have argued that the wet-nurse was not so much a surrogate mother as a 'tool' of the father, and indeed Soranus' extensive regulations on the choice of a wet-nurse and her regimen seem to reflect a desire for male intervention: Joshel 1986: 10–11; Dupont 2002; Laes 2011: 75–6.

25 For this and other instances of animal nursing, see Propertius, *Elegies* 2.20; Ovid, *Fasti* 5.111–12; Virgil, *Aeneid* 11.570–3; McAuley 2015: 5; Centlivres Challet 2017a: 901. The centrality of milk to this narrative is further highlighted by the presence of the Ruminal fig tree as the locale at which the twins were nursed; Varro and Plutarch link this tree with the goddess of lactating mothers, Rumina, and to an archaic word for breast, *rumina/ruma*: Varro, *On the Latin Language* 117; Plutarch, *Roman Questions* 236; Mazzoni 2010: 93–5.

Bibliography

Primary sources

Babbitt, F. C. (1927) *Pseudo-Plutarch. Morals. The Education of Children.* Translated by F. C. Babbitt. Cambridge, MA: Harvard University Press.

Basore, J. W. (1935) *Seneca the Younger. On Benefits.* Translated by J. W. Basore. Cambridge, MA: Harvard University Press.

Beck, L. Y. (2017) *Dioscorides. De Materia Medica.* 3rd edn. Translated by L. Y. Beck. Hildescheim: Olms-Weidmann.

Cary, E. and H. B. Forster (1916) *Cassius Dio. Roman History.* Translated by E. Cary and H. B. Forster. Cambridge, MA: Harvard University Press.

Foster, B. O. (1919) *Livy. History of Rome.* Translated by B. O. Foster. Cambridge, MA: Harvard University Press.

Helmreich, G. (1907–9) *Galen. On the Usefulness of the Parts.* Edited by G. Helmreich. Leipzig: Teubner.

Hendrickson, G. L. and H. M. Hubbell (1939) *Cicero. Brutus.* Translated by G. L. Hendrickson and H. M. Hubbell. Cambridge, MA: Harvard University Press.

Hett, W. S. (1957) *Aristotle. On Dreams.* Translated by W. S. Hett. Cambridge, MA: Harvard University Press.

Hutton, M. and W. Peterson (1914a) *Tacitus. Dialogue on Oratory.* Translated by M. Hutton and W. Peterson. Cambridge, MA: Harvard University Press.

———. (1914b) *Tacitus. Germania.* Translated by M. Hutton and W. Peterson. Cambridge, MA: Harvard University Press.

Johnston, I. (2018) *Galen. Hygiene*. Translated by I. Johnston. Cambridge, MA: Harvard University Press.

Johnston, I. and G. H. R. Horsley (2011) *Galen. Method of Medicine*. Translated by I. Johnston and G. H. R. Horsley. Cambridge, MA: Harvard University Press.

Jones, W. H. S. (1931) *Hippocratic Corpus*. Translated by W. H. S. Jones. Cambridge, MA: Harvard University Press. (This volume includes *Regimen*).

———. (1963) *Pliny. Natural History, Vol. VIII. Books XXVIII–XXXII*. Translated by W. H. S. Jones. Cambridge, MA: Harvard University Press.

Kaster, R. A. (2011) *Macrobius. Saturnalia*. Translated by R. A. Kaster. Cambridge, MA: Harvard University Press.

Morton Braund, S. (2004) *Juvenal. Satires*. Translated by S. Morton Braund. Cambridge, MA: Harvard University Press.

Peck, A. L. (1942) *Aristotle. Generation of Animals*. Translated by A. L. Peck. Cambridge, MA: Harvard University Press.

Peck, A. L. and E. S. Forster (1937) *Aristotle. Parts of Animals*. Translated by A. L. Peck and E. S. Forster. Cambridge, MA: Harvard University Press.

Perrin, B. (1914) *Plutarch. Lives. Cato the Elder*. Translated by B. Perrin. Cambridge, MA: Harvard University Press.

———. (1919) *Plutarch. Lives. Caesar*. Translated by B. Perrin. Cambridge, MA: Harvard University Press.

———. (1921) *Plutarch. Lives. Tiberius and Gaius Gracchus*. Translated by B. Perrin. Cambridge, MA: Harvard University Press.

Potter, P. (2018) *Hippocrates*. Translated by P. Potter. Cambridge, MA: Harvard University Press. (This volume includes *Diseases of Women*).

Raeder, J. (1933) *Oribasii Collectionum medicarum reliquiae, Vol 4. Libri XLIX-L, Libri incerti, eclogae medicamentorum, index. Edidit Ioannes Raeder*. Leipzig and Berlin: Teubner.

Rolfe, J. C. (1914) *Suetonius. The Deified Julius*. Translated by J. C. Rolfe. Cambridge, MA: Harvard University Press.

———. (1927) *Gellius. Attic Nights*. Translated by J. C. Rolfe. Cambridge, MA: Harvard University Press.

Rushton Fairclough, H. (1916) *Virgil. Georgics*. Translated by H. Rushton Fairclough, revised by G. P. Goold. Cambridge, MA: Harvard University Press.

Russell, D. A. (2002) *Quintilian. Institutes of Oratory*. Translated by D. A. Russell. Cambridge, MA: Harvard University Press.

Shackleton Bailey, D. R. (1993) *Martial. Epigrams*. Translated by D. R. Shackleton Bailey. Cambridge, MA: Harvard University Press.

———. (2004) *Statius. Thebaid*. Translated by D. R. Shackleton Bailey. Cambridge, MA: Harvard University Press.

Shipley, F. W. (1924) *Vellius Paterculus. Compendium of Roman History*. Translated by F. W. Shipley. Cambridge, MA: Harvard University Press.

Temkin, O. (1991) *Soranus. Gynaecology*. Translated by O. Temkin. Baltimore, MD: Johns Hopkins University Press.

White, H. (1913) *Appian. Roman History. The Civil Wars*. Translated by H. White. Cambridge, MA: Harvard University Press.

Secondary literature

Beagon, M. (2013) 'Labores pro bono publico: The burdensome mission of Pliny's Natural History', in J. König and G. Woolf (eds) *Encyclopaedism from Antiquity to the Renaissance*. Cambridge: Cambridge University Press, 84–107.

Bonfante, L. (1997) 'Nursing mothers in classical art', in A. O. Koloski-Ostrow and C. L. Lyons (eds) *Naked Truths: Women, Sexuality and Gender in Classical Art and Archaeology*. London: Routledge, 174–96.

Bradley, K. (1980) 'Sexual regulations in wet-nursing contracts from Roman Egypt', *Klio: Beiträge zur alten Geschichte* 62, 321–5.

———. (1991a) *Discovering the Roman Family: Studies in Roman Social History*. Oxford: Oxford University Press.

———. (1991b) 'Wet-nursing at Rome: A study in social relations', in B. Rawson (ed.) *The Family in Ancient Rome: New Perspectives*. London: Routledge, 201–29.

———. (1994) 'The nurse and the child at Rome: Duty, affect, and socialization', *Thamyris* 1(2), 137–56.

Burguière, P., D. Gourevitch, and Y. Malinas (1990) *Traité des maladies des femmes de Soranos d'Éphèse, Livre II*. Paris: Les Belles Lettres.

Carson, A. (1999) 'Dirt and desire: The phenomenology of female pollution in antiquity', in J. Porter (ed) *Constructions of the Classical Body*. Ann Arbor: University of Michigan Press, 77–100.

Centlivres Challet, C.-E. (2017a) 'Feeding the Roman nursling: Maternal milk, its substitutes, and their limitations', *Latomus* 17, 895–909.

———. (2017b) 'Roman breastfeeding: Control and affect', *Arethusa* 50(3), 369–84.

Dasen, V. (2010) 'Des nourrices greques à Rome?', in V. Pache Huber and V. Dasen (eds) *Politics of Childcare in Historical Perspective: From the World of Wet-Nurses to the Networks of the Family, Paedagogica Historica* 46, 699–713.

———. (2013) 'Becoming human: From the embryo to the newborn child', in J. Evans Grubbs and T. Parkin (eds) *Oxford Handbook of Childhood and Education in the Classical World*. Oxford: Oxford University Press, 17–39.

Dean-Jones, L. (1994) 'Medicine: The "proof" of anatomy', in E. Fantham et al. (eds) *Women in the Classical World: Image and Text*. Oxford: Oxford University Press, 183–215.

Deonna, W. (1956) 'Les thèmes symboliques de la légende de Pero et de Micon', *Latomus* 15(4), 489–511.

Dixon, S. (1990) *The Roman Mother*. London: Routledge.

Dupont, F. (2002) 'Le lait du père romain', in P. Moreau (ed) *Corps romain*. Grenoble: Collection Horos Jérôme Millon, 115–38.

Fildes, V. A. (1986) *Breasts, Bottles and Babies*. Edinburgh: Edinburgh University Press.

Flemming, R. (2000) *Medicine and the Making of Roman Women*. Oxford: Oxford University Press.

———. (2007) 'Women, writing, and medicine in the classical world', *The Classical Quarterly* 57, 257–279.

Gourevitch, D. (1984) *Le mal d'être femme: La femme et la médecine dans la Rome antique*. Paris: Les Belles Lettres.

Hackworth Petersen, L. and P. Salzman-Mitchell (2012) *Mothering and Motherhood in Ancient Greece and Rome*. Austin: University of Texas Press.

Hays, D. (1987) '*Lactea ubertas*: What's milky about Livy?', *The Classical Journal* 82(2), 107–16.

Isaacs, D. (2014) *Evidence-Based Neonatal Infections*. Oxford: Oxford University Press.

Joshel, S. (1986) 'Nursing the master's child: Slavery and the Roman child-nurse', *Signs* 12(1), 3–22.

Jouanna, J. (2012) *Greek Medicine from Hippocrates to Galen: Selected Papers*. Leiden: Brill.

King, H. (1997) 'Self-help, self-knowledge: In search of the patient in Hippocratic gynaecology', in R. Hawley and B. Levick (eds) *Women in Antiquity: New Assessments*. London: Routledge, 135–48.

———. (1998) *Hippocrates' Woman: Reading the Female Body in Ancient Greece*. London: Heinemann.

Kraus, C. S. (1994) *Livy Ab Urbe Condita Book VI*. Cambridge: Cambridge University Press.

Laes, C. (2011) *Children in the Roman Empire: Outsiders within*. Cambridge: Cambridge University Press.

Laskaris, J. (2008) 'Nursing mothers in Greek and Roman medicine', *American Journal of Archaeology* 112(3), 459–64.

Lawrence, T. (2019) 'Odour, perfume, and the female body in ancient Rome', PhD thesis, University of Nottingham, Nottingham.

Lennon, J. (2010) 'Menstrual blood in ancient Rome: An unspeakable impurity?', *Classica et Mediaevalia* 61, 71–88.

Lowe, D. (2013) 'Menstruation and Mamercus Scaurus (Sen. *Benef.* 4.31.3)', *Phoenix* 67(3/4), 343–52.

Masciadri, M. M. and O. Montevecchi (1984) *I contratti di baliatico*. Milan: s.n.

Mazzoni, C. (2010) *She-Wolf: The Story of a Roman Icon*. Cambridge: Cambridge University Press.

McAuley, M. (2015) *Reproducing Rome: Motherhood in Virgil, Ovid, Seneca, and Statius*. Oxford: Oxford University Press.

Mulder, T. (2017) 'Adult breastfeeding in ancient Rome', *Illinois Classical Studies* 42(1), 227–43.

Parkin, T. (2013) 'The demography of infancy and early childhood in the ancient world', in J. Evans Grubbs and T. Parkin (eds) *Oxford Handbook of Childhood and Education in the Classical World*. Oxford: Oxford University Press, 40–61.

Pedrucci, G. (2013a) *L'isola delle madri: Una rilettura della documentazione archeologica di donne con bambini in Sicilia*. Rome: Scienze e lettere.

———. (2013b) 'Sangue mestruale e latte materno: Riflessioni e nuove proposte. Intorno al'allattamento nella Grecia antica', *Gesnerus* 70, 260–91.

———. (2013c) *L'allattamento nella Grecia di epoca arcaica e classica*. Rome: Scienze e lettere.

Richlin, A. (1992) *The Garden of Priapus: Sexuality and Aggression in Roman Humour*. Oxford: Oxford University Press.

Ripat, P. (2016) 'Roman women, wise women, and witches', *Phoenix* 70(1/2), 104–28.

Salzman-Mitchell, P. (2012) 'Tenderness and taboo: Images of breastfeeding mothers in Greek and Latin literature', in L. Hackworth Petersen and P. Salzman-Mitchell (eds) *Mothering and Motherhood in Ancient Greece and Rome*. Austin: University of Texas, 141–64.

Sparreboom, A. (2009) 'Wet-nursing in the Roman Empire: Indifference, efficiency and affection', M-Phil thesis, VU University, Amsterdam.

———. (2014) 'Wet-nursing in the Roman empire', in M. Carroll and E.-J. Graham (eds) *Infant Health and Death in Roman Italy and Beyond*. Portsmouth, RI: Journal of Roman Archaeology, 145–58.

Stears, C. (1999) 'Breastfeeding and the good maternal body', *Gender and Society* 13(3), 308–25.

Totelin, L. (2017a) 'Milk: The symbolism and ambivalence of a substance', *Viewpoint: The Magazine of the British Society for the History of Science* 112, 6–7.

———. (2017b) 'Motherhood in flux: Greek nymphs, breastfeeding, and ancient gynaecology', in F. Pasche Guignard, G. Pedrucci, and M. Scapini (eds) *Maternità e Politeismi*. Bologna: Pàtron, 359–70.

Valladares, H. (2011) '*Falax Imago*: Ovid's Narcissus and the seduction of mimesis in Roman wall painting', *Word & Image* 27(4), 378–95.

Wallace-Hadrill, A. (1990) 'Pliny the Elder and man's unnatural history', *Greece and Rome* 1, 80–96.

15

BREASTMILK IN THE CAVE AND ON THE ARENA

Early Christian stories of lactation in context*

Laurence Totelin

Introduction: de-centring Favorinus

In the twelfth book of his *Attic Nights*, Aulus Gellius recounts how he witnessed the orator Favorinus (c. 80–160 CE) give an impassioned defence in favour of maternal breastfeeding and against wet-nursing (see also Lawrence in this volume, Chapter 14, p. 231). Favorinus presented his arguments during a visit to one of his pupils, a man of senatorial rank, whose wife had recently given birth to a son (*Attics Nights* 12.1.1–3; edition Rolfe 1927). The family of the mother told him that the labour had been long and difficult, and because she was 'exhausted by the exertion and the lack of sleep', she was napping. This did not prevent the orator from exclaiming 'I have no doubt that she will feed her son with her own milk' (*Attic Nights* 12.1.4). At this, the grandmother of the newborn (the mother's mother) stated that she planned to fetch wet-nurses 'so that to the pains, which she [her daughter] has endured in giving birth, are not added the heavy and difficult task of breastfeeding' (*Attic Nights* 12.1.5). Favorinus then launched into his long speech against a 'type of half-motherhood, against nature and imperfect' (*Attic Nights* 12.1.6), which deprived a child of his mother's milk.

The story does not tell whether Favorinus was successful in changing the grandmother's mind—to Aulus Gellius, this detail was not worth recording. The experiences and opinions of women were of little consequence to Gellius and Favorinus. In this narrative, the young mother's exhausted, leaking, torn body fades into the background, while the orator pours out words of great 'delightfulness, abundance, and richness' (*Attic Nights* 12.1.24). The flow of her bodily fluids, milk and blood, is secondary to the orator's stream of words. Yet we do learn something significant about the mother's breastfeeding, namely that she has not yet put her son to the breast: her nursing is presented as something that will take place in the future if at all (*lacte suo nutritura sit*). As we will see, some women in the ancient world did not start breastfeeding until a few days after birth.

Here, I examine ancient sources relating to the onset of lactation in the first few days after birth, with a focus on sources dating to the first three centuries of the Common Era. Such sources are rare and consist primarily in passages found in medical or philosophical texts, which I examine in the first section of this chapter. Even physicians displayed reluctance towards describing in detail the physiological processes accompanying childbirth. They skirted around the fact that, during the days that follow birth, a mother's body produces a type of milk (which we now call colostrum) which is different

DOI: 10.4324/9780429438974-20

from later milk, while at the same time bleeding quite abundantly from the vagina. Indeed, physicians and philosophers preferred to stress a neat separation between blood and milk (which was concocted blood); they found it difficult to make the space in their world-view for the messiness of a body that exudes the two fluids simultaneously. When they described the first milk, they showed it to be an imperfect substance that was not fit for consumption—something quite different from Pliny's 'most useful' milk (*Natural History* 28.123; edition Jones 1963; see Lawrence, Chapter 14, this volume).

It is fruitful to read these medical sources together with two exceptional early Christian stories that centre young mothers: the story of Mary, the mother of Jesus, in the *Protoevangelium of James*, and that of the slave Felicity in the *Passion of Perpetua and Felicity*.[1] Scholars such as Jennifer Glancy (2010, for Mary), Julia Kelto Lillis (2016, for Mary), and Stamatia Dova (2017, for Perpetua) have recently offered feminist readings of these narratives, reflecting on the prominence they gave to bodily fluids. In the second and third parts of this paper, building upon this scholarship, I draw out what we learn about early lactation and its ambivalence. I conclude that philosophical and medical insights, on the one hand, and early Christian stories, on the other, may be the reflection of broader societal concerns surrounding the maternal body in the days after childbirth.

Ancient medical approaches to the onset of lactation and the first milk

As noted by Thea Lawrence in the previous chapter, the ancients considered milk to be transformed menstrual blood.[2] According to philosophers and physicians, the two fluids normally did not flow at the same time. Thus, Aristotle, to whom we owe the most thorough ancient account of lactation, noted that 'while the milk comes out, the (menstrual) purgation, for the most part, does not occur, although some women experience their (menstrual) purgation while breastfeeding', and 'according to nature, the (menstrual) purgation does not occur in breastfeeding women, nor do breastfeeding women become pregnant' (*History of Animals* 587b30–1, edition Balme 1991; *Generation of Animals* 777a13–14, edition Peck 1942).[3] Closer in time to the early Christian stories that we will examine later, the Methodist physician Soranus (first–second century CE) asserted that 'when the menses occur, the milk dries up; and when lactation occurs, the menses do not appear any longer' (*Gynaecology* 1.15.2; edition Ilberg 1927).

The theories of Aristotle, Soranus, and others had much influence on early Christian thinkers who were asking questions about the nature of Jesus' divinity and humanity. For instance, Tertullian (second century CE), who was familiar with the works of Soranus, wrote that:

Let midwives, physicians, and philosophers (*obstetrices et medici et physici*) give their opinion concerning the nature of the breasts, whether they are accustomed to flow, except when the womb is experiencing generation, when the veins convey the dregs of the lower blood to the breast, and during this transfer, they concoct (*decoquentibus*) it into the more pleasant material of milk. For that reason, during lactation, the monthly shedding of blood is absent.

Tertullian, *On the Flesh of Christ* 20; edition Evans 1956[4]

For Tertullian, who in *Against Marcion* had compared the womb to a sewer (3.11.7; 4.21.11; translation Holmes 1878), an unclean space filled with dirty fluids, it was especially important to separate menstrual blood from the 'more pleasant material of milk'.

Modern gynaecology refers to the suppression of menstruation during lactation as menstrual amenorrhea. That amenorrhea leads to a period of lactational infertility, the length of which is variable and dependent on numerous factors (McNeilly 1993; see also Wiessinger et al. 2010: 169–70). Once an infant receives solid food (around six months after birth), the contraceptive effect of breastfeeding diminishes, and women may menstruate while lactating.[5] This situation, however, might have been different in the ancient world. As noted by Rosalind Janssen, Chapter 2 in this volume (pp. 46–7, citing Harris and Vitzthum 2013), women in natural fertility populations, where they spend long periods pregnant and lactating, experience relatively few ovarian cycles in their lifetime. Factors such as poorer nutrition (see Dean-Jones 1989: 185) might have contributed to long periods of lactational amenorrhea, leading authors such as Aristotle and Soranus to assert that women *normally* did not menstruate while breastfeeding. I would suggest, however, that there is something that goes further than biological 'facts' here; ancient authors felt abhorrence towards the mixing of blood and milk, towards their simultaneous flowing: these substances should not come into contact with each other.

Ancient philosophers and physicians all agreed that milk started accumulating in the breasts during pregnancy, but they expressed various opinions about the exact starting time of this process (see in this volume, Chapter 6, p. 113; Dean-Jones 1994: 222). They observed that milk did not start flowing abundantly until a little while after childbirth. Thus, Aristotle observed that:

> After birth and the [lochial] purgation, a woman's milk becomes abundant, and in some women, it flows not only from the nipples but from the entire breast, and in some even from the armpits. And for some time afterwards, indurations persist, when fluid is not fully concocted and does not come out but rather accumulates.
>
> Aristotle, *History of Animals* 587b19–24

Aristotle noted that milk was only produced in large quantity after the lochial purgation, which is rather puzzling when we consider that, a little earlier, he had stated that these lasted forty days (*History of Animals* 587b5; on ancient theories of lochial bleeding, see in this volume, Chapter 3, p. 58). It is possible, however, that Aristotle had the heaviest post-partum bleeding in mind when he asserted that it is after the purgation that milk became plentiful. This bleeding, referred to as lochia rubra in modern gynaecology, today lasts between three to five days (see Fletcher et al. 2012) and coincides with the production of the first milk, colostrum, which only occurs in small quantities. On the third or fourth day after birth, the milk 'comes in', that is, the breasts start producing more abundant milk (see Wiessinger et al. 2010: 22–3).

Descriptions of colostrum are extremely rare in the ancient Greek and Roman texts. The clearest is in Soranus' *Gynaecology*, where he stressed the significant loss of blood that accompanied the production of the first milk:

> From the second day after the treatment [following birth], one must give to the child the milk of someone who is well able to breastfeed. For a mother's

milk is, until the third day, in most cases, of poor quality:[6] it is thick, very cheesy, and for that reason difficult to digest, oily, and not thoroughly concocted. Also, it is produced by bodies which are in distress, troubled, and which display such change as we see happening after delivery when, having suffered a great loss of blood, the body is dry, toneless, pale, and in the majority of cases, feverish too.

Soranus, Gynaecology 2.18.1; see also 2.20

To Soranus, the first milk was not fit for consumption because it is produced by a maternal body that is in a state of distress. In the words of Susan Holman (1997: 83), there is an attempt at 'distancing it [the infant] from the socially murky realm of its mother's unstable body'. The baby is safer in the hands of 'someone who is well able to breastfeed (*ek tinos tōn tittheuein kalōs dunamenōn*)', a purposefully ambiguous phrase which could refer either to a hired wet-nurse or to a lactating family member.

Soranus then went on to criticise authorities who recommended maternal breastfeeding straight after birth, such as Damastes and Apollonius Biblas (whose writings are unfortunately lost), and asserted that if a woman 'able to provide milk was not easily available', the baby should be given honey or honey with goat's milk instead of maternal milk for the first three days (*Gynaecology* 2.18.2–3). It is only when the mother was properly rested that she could feed her child. Soranus advised the new mother to get rid of some of her milk (by either having an older child suck it out or by expressing) before suckling her infant, to avoid giving to the child the thick part of the milk, 'since it is hard to suck out and is able to clog up (*epinasthēnai*) in newborns because of the tenderness of their gums' (*Gynaecology* 2.18.3). It should be noted that preventing a newborn mammal from taking the first milk means working against a very strong instinct on the part of that newborn to suckle, an instinct that both Aristotle (*History of Animals* 587a33) and Galen (*Hygiene* 1.7, 6.36 Kühn 1821–33) had described.[7]

Today, colostrum, which is indeed thicker than the milk that comes in on the third or fourth day, is seen as the best milk possible, one that is full of antibodies (Laskaris 2008: 460). Yet to Soranus, it was not advisable as a food, as it was of poor quality. Pliny the Elder (24–79 CE) also described the first human milk, which he called *colostra*, as thick and spongy (*Natural History* 28.123; this is the passage where Pliny referred to maternal milk as 'the most useful'). He also noted that in asses, when they were fed too rich a grass, the first milk could be fatal to the foal, who would die of a disease called *colostratio* (*Natural History* 11.237; edition Rackham 1940).[8] Columella (first century CE) too asserted that, before letting a lamb suckle its mother, the milk 'which shepherds call *colostrum*' should be expressed, as it might be harmful (*On Agriculture* 7.3.17; edition Foster and Heffner 1954). While they are not numerous, these passages do point to some anxieties towards the first milk in the Greco-Roman world.[9] Unease towards colostrum is well-attested anthropologically in numerous societies worldwide, and delayed breastfeeding is or was commonly practised (see e.g. Fildes 1986: 81–6; Dennis et al. 2007: 496–7; Hogan 2008 for the British and Irish context). It is helpful to place our scant Greek and Latin sources on colostrum in this anthropological context and not to consider them as exceptional or anomalous.

To the ancients, then, first milk was not fully concocted and cheesy, and therefore difficult to digest and dangerous. These characteristics could occur in milk at other times during lactation. Soranus mentioned them further in his long description of the perfect wet-nurse and the perfect milk. His observations can help us understand better why, and the ways in which, the first milk could allegedly be harmful. Soranus noted that the quality of milk should be determined by its colour, smell, consistency, how it coagulates and its thickness, taste, and the way it changes with time (*Gynaecology* 2.22.1). His comments about colour, consistency, and coagulation/thickness are most relevant to our understanding of harmful milk.[10]

Soranus stressed that the right colour for milk was white, and that 'yellow-red (*xanthon*) milk was not concocted (*apepton*) and raw (*akatergaston*), and for that reason displays a blood-like (*haimatōdē*) colour' (*Gynaecology* 2.22.2).[11] Through its colour, this 'bad' milk advertised its kinship with menstrual blood, which the body had not fully concocted. Interestingly, colostrum is significantly darker than later milk and can appear red, orange, or dark yellow. Soranus made further allusions to blood in his comments on the consistency of milk. Good milk ought to be 'smooth and homogeneous. For if it is fibrous and contains streaks that are red or flesh like (*sarkoeideis*), it is not concocted (*apepton*)' (*Gynaecology* 2.22.2). That is, this milk contains streaks of unconcocted menstrual blood—the description is very graphic. Soranus then turned to the consistency of bad milk, highlighting the risks associated with the consumption of overly cheesy milk:

> For free-running, thin, and watery milk is not nutritious and is apt to disturb the bowels; whereas thick and cheesy milk is hard to digest, and similarly to food that has been partially chewed, it blocks up the passages, it occupies the principal outlets of the body, and it constitutes a danger to life.
>
> Soranus, *Gynaecology* 2.22.3

The stakes were high: cheesy milk could cause death by choking. Fortunately, there were means available to test that milk had the right thickness. One involved dropping some milk on one's nail or on a leaf of sweet bay and observe whether it ran down too fast or remained too motionless (*Gynaecology* 2.22.4). Another test consisted in dropping milk in water and checking how fast it dissolved. 'Cheesy' milk settled at the bottom of the water; milk that dissolved too fast was too watery; but worse still was milk which was 'reduced to fibrous shreds, like water in which meat has been washed, for this milk is not concocted (*apepton*)' (*Gynaecology* 2.22.5). Again, milk that had not been fully concocted signalled its rawness through its similarity with uncooked meat.

To sum up, maternal milk was a wholesome food that was perfect for an infant's nourishment, but not all maternal milk was good, and first milk according to Soranus was bad milk. Its rawness evoked meat, and its thickness brought cheese to mind. Now, a relatively common analogy for the formation of the embryo in antiquity was that of cheese making: the female menses were like milk that was curdled under the effect of sperm (rennet). The analogy is to be found in several places in Aristotle (*Generation of Animals* 729a13–15, 739b21–7, 772a21–5); in the Hebrew book of Job, which was translated into Greek in the Hellenistic period (10:10); and in Tertullian's *On the Flesh of Christ* 19. In fact, that analogy between generation and cheese making (or butter making) exists in various cultures: sperm is comparable to rennet; woman's blood is like milk; embryos are like cheese or butter (see Belmont 1988).

Is it possible that some fears relating to the consumption of early milk in Greek and Roman society are linked to fears of cannibalism: there are similarities between thick, yellowish-reddish first milk and what may happen at the very beginning of pregnancy, when the menses curdle under the effect of sperm.[12] But while cannibalism might be a little too far-fetched, it remains that some people in antiquity expressed concerns about the first milk and that those concerns hinged on the fact that this milk could be tainted by blood. With these considerations about first milk, the mixing of blood and milk, we can now turn to the stories of Mary and Felicity.

The birth of Jesus

The *Protoevangelium of James* is an apocryphal gospel, composed in Greek probably in the second half of the second century CE, which tells the story of Jesus' birth and childhood (for an introduction, see Elliott 2005: 48–52).[13] It has a strong focus on the figure of Mary and includes details of her own childhood. One of the author's motives for writing was to show not only that Mary had been a virgin when she conceived Jesus but also that she remained a virgin even after birth—a much disputed question at the time (see Lillis 2016).

As in Luke's Gospel, Mary was travelling with Joseph to Bethlehem for the census when she felt the birth pangs (*Protoevangelium* 17.3). Joseph found a cave (Greek: *spēlaion*; contrast with the story of the lodgings, Greek *kataluma*, in Luke 2:7) where he left her to fetch a Hebrew midwife (*Protoevangelium* 18–19). When he returned to the cave with an unnamed woman, they witnessed an extraordinary scene:

> And a blinding cloud overshadowed the cave. . . . And soon the cloud drew back from the cave, and there appeared a great light in the cave, which the eyes could not bear. After a little while, that light drew back, until a baby appeared. And the midwife cried out, and she said: 'Today is a great day for me, for I have seen this extraordinary sight.' And the midwife left the cave, and she met Salome, and said to her: 'Salome, Salome, I have an extraordinary sight to tell you about: a virgin has birthed things for which her genitals do not have space.' And Salome said: 'As the Lord my God lives, unless I put in my finger and inspect her genitals, I will not believe that a virgin has given birth.'
>
> *Protoevangelium of James* 19.2–3; trans. based on Lillis 2016

Salome, whose occupation is not specified by the author (was she a midwife too?), did perform her test, and her hand caught on fire.[14] After she had repented, however, her hand healed (*Protoevangelium* 20.1–2).

Jennifer Glancy (2010: 108–12) and Julia Kelto Lillis (2016) have argued that, in this story, Mary gave birth in a 'clean' manner. Her body did not bleed, either through tearing (which frequently occurs in childbirth) or through the lochial bleed that follows birth.[15] In the absence of blood, it is through her breastfeeding that Mary signalled her motherhood. In the words of Glancy (2010: 109), '[d]enied genital insignia of motherhood, Mary leaks news of her maternity from her virginal breasts.' It should be noted, however, that the text does not disclose whether Mary is lactating and her breasts are leaking. Rather, it states that the baby took to the breast, which, as mentioned earlier, is an instinct in newborns which the ancients had observed.

Later versions of the birth of Jesus added details about Mary's lactation, removing the ambiguity found in the *Protoevangelium of James*. In pseudo-Demetrius of Antioch's *Birth of our Lord and on the Virgin Mary*, a text preserved in Coptic and perhaps dating to the fourth century CE, Mary felt her breasts fill with milk immediately after she was visited by the Angel Gabriel.[16] In the Gospel of Pseudo-Matthew, a Latin text dated to the eighth or ninth century CE (Elliott 2005: 86), Mary was visited in the cave by a midwife called Zelemi (a doublet of Salome), who witnessed the amazing scene:

> When she had entered [the cave], Zelemi said to Mary: 'Allow me to touch you.' And when Mary had allowed her to touch her, the midwife cried out in a loud voice and said: 'Lord, Lord, take mercy on me! To this day I had never heard nor had any inkling that breasts might be filled with milk (*ut mamillae plenae sint lacte*) and that the birth of a son might show his mother to be a virgin. But there has been no blood defilement (*nulla sanguinis pollutio facta est in nascente*) in his birth, no pain in labour. A virgin has conceived, a virgin has given birth, and a virgin she remains.'
>
> *Gospel of Pseudo-Matthew* 13.3; edition Tischendorf 1853

Where the unnamed midwife in the *Protoevangelium of James* had only hinted at the flow of bodily fluids, Zelemi was clearer: Mary's breasts were full of milk, and there was no *pollutio* in her birth. The word *pollutio* here refers to the blood that accompanies childbirth. Birth was seen as polluting in both Greco-Roman and Jewish religion, a pollution that was linked to the shedding of blood (see e.g. Cole 1992: 107–11; Branham 1997).[17] But Mary was not polluted by giving birth to Jesus (see Glancy 2010: 114, who does not make reference to these later versions of the story).

Glancy (2010: 110–14) has argued that Mary's immediate breastfeeding should be contrasted to her mother's nursing in the *Protoevangelium*. Unlike Jesus, Mary was not breastfed immediately after birth by her mother, Anna:

> Anna said to the midwife, 'What have I borne?' And the midwife said, 'A female.' And Anna said, 'My soul is exalted this day.' And she laid her down. And when her days were completed, Anna cleansed herself (*apesmēxato*) and gave her breast to the child and gave her the name 'Mary'.
>
> *Protoevangelium of James* 5.2[18]

The author did not state how many days Anna abstained from breastfeeding but asserted that she cleansed away blood before she gave the breast. Glancy (2010: 113) has noted that, while Jewish women were considered impure after birth, there was no Levitical Law against breastfeeding during that period. Instead of reading the *Protoevangelium* in the context of Levitical Law, one should read it alongside the views about breastmilk preserved in Tertullian and Soranus. Lisa Straus (2013), for her part, has discussed Anna's delayed breastfeeding in relation to one of the Dead Sea Scrolls, which makes provision for the hiring of a wet-nurse while a mother is impure after birth (4Q226.6.ii, to be compared to Leviticus 12:8), and noted the similarities between those provisions and those made by Soranus.

It is beyond the scope of this chapter to discuss the Leviticus and Dead Sea Scroll in detail and to determine whether the author of the *Protoevangelium* was influenced more by Jewish law or by Greek philosophy.[19] I would suggest that it is best to consider ancient concerns about the first milk, whether they be Greek, Roman, or Hebrew, in a broader anthropological context—such concerns are common. Mary in the *Protoevangelium* did not have to worry about her milk because she had miraculously skipped a post-partum stage: that when the body bleeds heavily. Because she was not bleeding, there was no risk that her milk might be unconcocted, tainted with blood, and cheesy. Her milk was fit for immediate consumption. Unlike the exhausted mothers in Soranus' *Gynaecology*, and unlike her own mother, Mary could put a child to her breast straight after birth. Such was the miracle of a virgin birth.

Felicity's bleeding

The *Passion of Perpetua and Felicity* tells the story of the martyrdom of a group of Christians arrested in the town of Thuburdo Minus (near Carthage) in 203 CE, under the rule of Septimius Severus.[20] It is a complex text, with three authorial voices: that of Perpetua herself—a rare female voice from antiquity—who recorded her experience of arrest and imprisonment, as well as visions sent by God; that of an editor, who was 'closely connected to Tertullian'; and that of Saturus, another martyr, who also recounted one of his visions (Bremmer and Formisano 2012: 5).[21]

The *Passion* focuses on two figures: Vibia Perpetua, a noblewoman who was 'of good birth, nobly educated, honourably married', and Felicity, a slave (*Passion* 2.1; see Bremmer 2012: 37). Both women were mothers: at the time of the arrest, Perpetua had a young son 'at the breast' (*Passion* 2.2), and Felicity was pregnant (*Passion* 15, see later).[22]

Breastfeeding plays a crucial role in the *Passion*, as both Perpetua and Felicity are presented as lactating women, albeit at two ends of the nursing cycle—cessation (Perpetua) and onset (Felicity). As Stamatia Dova has recently argued 'Perpetua's struggle to transition from motherhood to martyrdom is inextricably intertwined with her experiences of lactation suppression and cessation' (2017: 245). In her narrative, Perpetua made several references to the feeding of her son. She recounted how, after their arrest, she and her companions were first placed under home surveillance, then put in a dungeon in prison, where she was 'distressed by anxiety for my baby' (*Passion* 3.6). Thanks to the intervention of the deacons Tertius and Pomponius, however, the companions were transferred to a better part of the prison, where Perpetua was able to nurse her son, who by then was 'weakened because of the lack of food' (*Passion* 3.8). A few days later, she was allowed to have the baby stay with her in prison 'and immediately I regained strength, as I was relieved from pain and anxiety for my baby' (*Passion* 3.9). As Dova (2017: 254) has suggested, the pain that Perpetua had experienced might have been partly physical, linked to possible breast engorgement following a break in breastfeeding. The child then remained with her, 'receiving the breast', until the day of the companions' hearing, when they were condemned to fight wild animals in the arena (*Passion* 6.7). At that point, Perpetua's father took away the child, refusing to let the young woman feed him again. God, however, intervened in favour of Perpetua: 'But this was God's will, not only did the child no longer long for my breasts, but also they did not cause me fever. Thus, I was distressed neither by anxiety for my baby nor by pains in my breasts' (*Passion* 6.8). God saved Perpetua from the pains and fevers that accompany sudden lactation cessation (Dova 2017: 257–8).[23] On the day before the combat with the beasts, Perpetua had a final vision. She dreamt of being in the amphitheatre, where she was made to fight against an Egyptian man:

> He was to fight against me. Some beautiful young men approached me, to help and support me. I was stripped naked, and I became masculine, and my assistants started rubbing me down with oil, as they usually do in a contest.
>
> *Passion* 10.6–7

By the grace of God, she had stopped lactating and had become masculinised (for an analysis of Perpetua's gender, see Williams 2012).

By contrast, Felicity remained feminine to the last in the *Passion*. Her martyrdom too is tied to motherhood and its attendant fluids, which never ceased to flow. The narrator introduced her as a pregnant woman:

> Regarding Felicity, in truth, the Lord's grace touched her too in this way. When she was in the eighth month of her pregnancy (indeed she was pregnant when she was arrested), as the day of the spectacle was approaching, she was in a state of great anxiety that it might be delayed on account of her belly (because it is not permitted to exhibit pregnant people in punitive shows) and that she might shed her holy and innocent blood later, together with others who were criminals.
>
> *Passion* 15.1–3[24]

This also worried her companions, who prayed to the Lord 'two [or three] days before the spectacle' (*Passion* 15.4, see later). Their prayers were answered as she immediately went into labour, and she gave birth to a daughter, whom one of her sisters raised as her own (*Passion* 15.5–7).[25]

The timing of the birth of Felicity's daughter is noteworthy is two respects. First, she was born during the eighth month of pregnancy, a time which the Greeks and Romans considered to be particularly dangerous for a delivery. The *Passion* stresses that Felicity's labour was especially painful, as is natural in the eighth month (*Passion* 15.5).[26] Felicity's daughter had defied the odds and survived despite the ominous time of her birth. Second, the baby was born two or three days before the games (*ante tertium diem muneris*), depending on whether one counts days inclusively (as was common in the Roman world) or exclusively.[27] Whether she had given birth two or three days before the spectacle, however, Felicity went to her death during the liminal period when a mother's body bleeds abundantly and produces the first milk.

The *Passion*'s narrator alluded to Felicity's bleeding later, when he described the procession of the future martyrs from the prison to the amphitheatre, where they were to die:

> Perpetua followed with a luminous face and a calm pace, like a wife of Christ or a favourite of God, forcing all to cast down their gaze with the strength of

her eyes. And Felicity too, rejoicing that she had given birth safely so that she might fight against the beasts—from blood to blood, from midwife to gladiator (*a sanguine ad sanguinem, ab obstetrice ad retiarium*)—to be cleansed after birth in a second baptism.

Passion 18.2–3[28]

Jan Bremmer has noted that we here have a very rare allusion to the 'blood drenched clothes from birth' (2012: 46; see also Horace, *Epodes* 17.50–2, edition Rudd 2004). This, however, might also be read as a reference to lochial bleeding, which Felicity would have been experiencing two or three days after birth.

The martyrs were supposed to wear special costumes for the spectacle—outfits of the priests of Saturn for the men, and of the devotees of Ceres and Proserpina for the women—but following Perpetua's protest, they were stripped naked instead (*Passion* 18.4–6). Thus, the women were brought before the crowds to fight a fierce heifer (*Passion* 20.1), a rather unusual animal to appear in ancient games, and perhaps chosen to deride Perpetua and Felicity as mothers (Bremmer 2012: 47; Heffernan 2012: 339). The appearance of the two women created ripples of shock among the audience:

Thus, stripped naked and covered with nets, they were brought forward. The crowd shuddered when it gazed upon them: one a tender girl [Perpetua], the other [Felicity] with her breasts dripping [milk] (*stillantibus mammis*), as she had recently given birth.

Passion 20.2

The scene was so outrageous that the martyrs were called back and dressed in loose clothing before the spectacle resumed, ending inescapably in death by the sword of gladiators. Scholars have rightly stressed that this scene is both highly eroticised and implausible.[29] How could the spectators (even those nearest to the women) have seen the dripping breasts of Felicity? This scene may well be an invention of the narrator, but as argued by Alicia Myers (2017: 144), it served an important function in that it 'reinforce[d] the physicality of the miracles provided by God'. The narrator contrasted the beautiful body of Perpetua, now athletic and masculine, to that of Felicity, bleeding and lactating.[30] Their bodies, exemplifying two very different stages in the maternal cycle, have both been transformed by the grace of God.

Scholars often translate the Latin *stillantibus mammis*, as 'her breasts still dripping milk'.[31] I would suggest, however, that what we witness in this scene is the very moment when Felicity's milk comes in, when it starts to flow abundantly. Her breasts are not 'still leaking'; they leak out for the first time. Felicity has passed the stage in which her body produces only the thick, cheesy milk that is unfit for consumption. Her body would now be able to sustain an infant's life, and it is as a lactating mother that she goes to her death.

Conclusions

Milk and blood do not mix in the neat economy of ancient bodily fluids. Milk, as concocted menstrual blood, ought not to be produced by a menstruating or pregnant body. Ancient philosophers and physicians stressed the separation of those fluids.

249

Occasionally, however, they acknowledged that the reality was more complex and that there were times when a mother's body could exude milk and blood simultaneously: her periods could return a few months after she had given birth (a phenomenon which the ancients believed to be spurred on by sexual intercourse); and her body did experience the lochial bleed for several weeks post-partum. Physicians and natural scientists considered the milk produced by a bleeding body to be dangerous: this raw and cheesy substance could kill nurslings. Their solutions were to delay maternal breastfeeding for a few days, and to regulate—or be seen to regulate—a mother's sexual activity so that she would not become pregnant while lactating.

One could dismiss the concerns expressed by these authors as marginal—after all, Soranus reported that some medical writers were proponents of direct breastfeeding after birth. However, we know that, anthropologically, anxiety towards colostrum is common, and that women were routinely prevented from breastfeeding in the hours (even sometimes days) following birth in the not very distant past in Europe—a practice still occurring in some societies. I would argue that when Soranus and others described their concerns about the first milk and early maternal breastfeeding, they were reflecting broader societal anxieties which went beyond religious boundaries.

It is in this context that we should read the stories of Mary, her mother Anna, and Felicity. The author of the *Protoevangelium* took pains to describe Anna's delayed breastfeeding merely to contrast it to that of Mary, but through this rare depiction we get a glimpse of what might have been fairly typical in the ancient Mediterranean world. Mary, for her part, was able to nurse Jesus directly because she had experienced a virgin birth, in which blood was not shed. Her milk was not tainted by the lochia and was therefore suitable for consumption from the start.

Felicity's milk, on the other hand, was never to be consumed by her daughter; instead, it was shed like her blood on the arena. The narrator of the *Passion* might have taken a little too much pleasure in describing the crowd's shock at the sight of her dripping breasts, which very few in the audience could have seen. But in doing so, he conveyed some of the awe that the body of a new mother would have inspired, especially a new mother about to die in the arena. As noted by Joyce Salisbury (1997: 142), a martyr should produce blood, not milk. The two substances should not mix in the sands of amphitheatres.

And the unnamed mother in Favorinus' tale? She sleeps for eternity, her narrative unresolved. Unwittingly, Favorinus and Aulus Gellius have left her at the stage where her body could not yet produce fully concocted milk and where her breastfeeding was yet to happen. Their words survive; her experience of lactation is forever lost.

Notes

* It has been a privilege to work with Mark Bradley and Victoria Leonard on the publication of this volume. I wish to thank them for their advice and encouragement in writing this chapter. Versions of this paper were presented at a meeting of the Society for the History of Medieval Technology and Science and at a departmental seminar in Classics at the University of Glasgow. I am grateful to the audiences of these seminars for their insightful questions. Unless stated otherwise, all translations from the Greek and the Latin in this chapter are mine.

1 There has been some important work on the importance in early Christianity of breastfeeding as a means to form the soul and transmit Christian virtues: see e.g. Myers 2017: 77–108;

Penniman 2017. In this chapter, I focus on the physicality of breastfeeding rather than on the soul.

2 On ancient theories of lactation as well as the link between milk and blood, see e.g. Dean-Jones 1994: 219–24; Pedrucci 2013.

3 Aristotle's views in this matter strongly influenced Galen, whom I will not discuss much here, as Galen paid relatively little attention to the physiology of breastfeeding. See De Lacy (1992: 251) for a list of passages in which Galen discussed the physiology of lactogenesis.

4 Tertullian cited Soranus in several chapters of *On the Soul* (6, 8, 14, 15, 25, 44); see Waszink 2010: 22–38. It is not certain whether Tertullian had read Soranus' *Gynaecology* or not—Soranus was a prolific author—but his views on the female body bear similarities with those of Soranus; see Glancy 2010: 119; Barr 2017: 169–71.

5 On whether the Greeks and Romans understood the contraceptive effect of lactational amenorrhea, see Centlivres Challet 2017: 376.

6 One of the manuscripts indicates that the milk is of poor quality until the twentieth day. The most recent editors, however, have opted for the reading 'three days': Burguière et al. 1990: 26; *contra*, see Temkin 1956: 89, note 35.

7 The reader will find a modern edition and translation in Johnston 2018. The reference to the traditional Kühn edition is added for ease of reference.

8 Pliny (*Natural History* 28.123) also referred to children who were breastfed after their mother had conceived again as '*colostrati*'. He stated that this was very dangerous for the children because the milk had 'thickened into a sort of cheese'. Physicians recommended that breastfeeding women avoid sexual intercourse, because it would sour the milk by bringing on the menses or pregnancy. See e.g. Galen (*Hygiene* 1.9, 6.46 Kühn); Soranus, *Gynaecology* 2.19.11.

9 While the first milk was considered dangerous for newborns, it did not present the same risk for adults who consumed it. Ancient authors refer to beestings, animal first milk, as a delicacy. See for instance Martial, *Epigrams* 13.38 (edition Shackleton Bailey 1993), where goat's beestings are mentioned.

10 Similar comments are made by Galen, *Hygiene* 1.9, 6.47 Kühn.

11 I would suggest that Soranus had the colour of saffron in mind when he described this colour of milk. On the synaesthetic properties of this colour, see Bradley 2013: 135.

12 Anthropologist Ludwig Feuerbach (1804–72) compared the newborn to a cannibal, see Penniman 2017: 2. One can note that drinking too much curdled milk was considered poisonous in antiquity. See for instance Paul of Aegina, *Medical Compendium* 5.56, edition Heiberg 1924.

13 The edition followed here is that of Tischendorf 1853: 1–50. A more recent edition by Quecke and de Strycker 1961 is available for the earliest copy of the *Protoevangelium*, Papyrus Bodmer 5 (third–fourth century CE).

14 Salome's doubts clearly echo those of Thomas when faced with Jesus' resurrection (John 20:24–9); see Lillis 2016: 6.

15 Lillis further argues that it is the lack of distension of her genitals, rather than the presence of a hymenal membrane (the very existence of which was debated in antiquity; see Sissa 1990), that marked Mary as a virgin. Other ancient authors presented the birth of Jesus as rather messier. Tertullian (*On the Flesh of Christ* 4), for instance, suggested that Mary gave birth in the 'normal' way, but Jesus, as the son of God, cleansed her from the defilement of birth; see e.g. Glancy 2010: 117.

16 Pseudo-Demetrius, *Birth of our Lord and on the Virgin Mary*, translation Budge 1915: 687: 'The virgin herself was marvelling at the salutation [of Gabriel], and she was troubled, saying in her heart "Behold, the sweet odour has reached me through the angel. And behold, his word is fulfilled, for lo, I have conceived, lo, my breasts are full of milk, and lo, my womb is swollen."'

17 *Pollutio* might also refer to Mary's menses. Glancy (2010: 108) has suggested that Mary was not yet menstruating when she conceived.

18 Neither Pseudo-Demetrius' *Birth of our Lord and on the Virgin Mary* nor the *Gospel of Pseudo-Matthew* include any detail of Anna's breastfeeding. The text of Papyrus Bodmer 5 is more specific about Anna cleansing her bleed (*apesmēxato tēs aphedrou*).

19 Glancy (2010: 113) has noted that the author of the *Protoevangelium* had a poor knowledge of Judaism. Vuong (2013: 90, 2019: 60), on the other hand, suggested that Anna's delayed breastfeeding should be read in the context of over-interpreted Levitical Law. Straus (2013) brought all the sources together but had no conclusions as to how they might have influenced each other or not.

20 There is much literature on the martyrdom of Perpetua and Felicity. Heffernan 2012 and Bremmer and Formisano eds 2012 are good starting points. As noted by Jan Bremmer (2012), Felicity's story, however, is far less explored than that of Perpetua.

21 There are two versions of the *Passion of Perpetua and Felicity*: one in Latin (shorter), the other in Greek. There is debate as to which version should take priority (for a brief summary of the debates, see Bremmer and Formisano 2012: 2). Here, I follow the Latin version as edited by Farrell and Williams 2012. There also exists a shorter version of the story of Perpetua and Felicity in Latin, the *Acta*, on which see recently Cobb 2019. This version has no detail of Felicity's lactation.

22 Heffernan (2012: 151) suggested that Perpetua's son was around 18 months at the time of the arrest. While one cannot be so certain, it is clear that the child was no longer a newborn and could have taken supplementary food. See also Dova 2017: 255. On the maternal body in the *Passion*, see Perkins 2007, who argues that the focus on the maternal body in the narrative raises suspicions as to its historicity. The question of historicity is beyond the scope of this chapter.

23 Greek medical texts include some recipes to assist with lactation cessation and prevent inflammation of the breasts. See for instance pseudo-Galen, *Remedies Easily Procured* 15, 14.447–8 Kühn.

24 Heffernan (2012: 20) suggested that Felicity might have been intended as a *nutrix* for Perpetua's children.

25 On the significance of this daughter in the narrative of the *Passion*, see Ronsse 2006: 323–7.

26 See for instance the Hippocratic treatise *On the Eighth Month Infant*, or the pseudo-Galenic *Medical Definitions* 160, 19.454.6–10 Kühn; sources discussed in Hanson 1987; Bremmer 2012: 44. The months in this reckoning are lunar rather than solar.

27 Bremmer and Formisano (2012: 20) translate 'three days before the event'. Heffernan (2012: 309) translates 'two days before the games'. Counting days inclusively was common in the Roman world, and I therefore tend to favour Heffernan's translation.

28 Bremmer (2012: 46) noted that the epigrammatic style of this passage is reminiscent of that Tertullian.

29 See e.g. Frankfurter 2009: 221–4; Heffernan 2012: 340; Bremmer 2012: 48–9.

30 Myers (2017: 144) suggested that Felicity was bleeding lochial blood. To my knowledge, she is the only one to have drawn this conclusion, which seems correct when we consider the timing of the games a few days after the birth of Felicity's daughter. Joyce Salisbury (1997: 142) argued that one possible explanation for the audience's horror at the view of Felicity's dripping breasts is that it reminded them of processions of Isis during which priests poured out milk through a golden breast. She concluded, however, that the main reason would have been abhorrence for the mixing of blood and milk and martyrdom with motherhood. On the figure of Isis lactans, see Tran 1973.

31 See e.g. Perkins 2007: 329, 2008: 167; Heffernan 2012: 339; Solevåg 2016: 320; Gold 2018: 113.

Bibliography

Primary sources

Balme, D. M. (1991) *Aristotle. History of Animals, Vol. III: Books 7–10*. Edited and translated by D. M. Balme. Cambridge, MA: Harvard University Press.

Budge, E. A. W. (1915) *Miscellaneous Coptic Texts in the Dialect of Upper Egypt*. Edited with English translations by E. A. Wallis Budge. London: British Museum.

Burguière, P., D. Gourevitch, and Y. Malinas (1990) *Soranos d'Éphèse. Maladies des femmes. Tome II: Livre II.* Texte établi et traduit par P. Burguière, D. Gourevitch et Y. Malinas. Paris: Les Belles Lettres.

Elliott, J. K. (2005) *The Apocryphal New Testament: A Collection of Apocryphal Christian Literature in an English Translation.* Oxford: Oxford University Press.

Evans, E. (1956) *De carne Christi liber. Tertullian's Treatise on the Incarnation.* Text edited with an introduction, translation and commentary by E. Evans. London: Society for Promoting Christian Knowledge.

Farrell, J. and C. Williams (2012) 'Passio Sanctarum Perpetuae et Felicitatis', in J. N. Bremmer and M. Formisano (eds) *Perpetua's Passions: Multidisciplinary Approaches to the Passio Perpetuae et Felicitatis.* Oxford: Oxford University Press, 24–32.

Forster, E. S. and E. H. Heffner (1954) *Columella. On Agriculture, Vol. II: Books 5–9.* Translated by E. S. Forster and E. H. Heffner. Cambridge, MA: Harvard University Press.

Heffernan, T. J. (2012) *The Passion of Perpetua and Felicity.* Oxford: Oxford University Press.

Heiberg, J. L. (1924) *Paulus Aegineta. Libri V-VII.* Edidit J. L. Heiberg. Leipzig and Berlin: Teubner.

Holmes, P. (1878) *The Five Books of Quintus Sept. Flor. Tertullianus Against Marcion.* Translated by P. Holmes. Edinburgh: T. & T. Clark.

Ilberg, J. (1927) *Sorani Gynaeciorum libri iv, de signis fracturarum, de fasciis, vita Hippocratis secundum Soranum.* Leipzig: Teubner.

Johnston, I. (2018) *Galen. Hygiene, Vol. I: Books 1–4.* Edited and translated by I. Johnston. Cambridge, MA: Harvard University Press.

Jones, W. H. S. (1963) *Pliny. Natural History, Vol. VIII: Books 28–32.* Translated by W. H. S. Jones. Cambridge, MA: Harvard University Press.

Kühn, K. G. (1821–33) *Claudii Galeni Opera Omnia.* 20 vols. Berlin and Leipzig: Libraria Car. Cnoblochii.

Peck, A. L. (1942) *Aristotle. Generation of Animals.* Translated by A. L. Peck. Cambridge, MA: Harvard University Press.

Quecke, H. and E. de Strycker (1961) *La forme la plus ancienne du Protévangile de Jacques.* Brussels: Société des Bollandistes.

Rackham, H. (1940) *Pliny. Natural History, Vol. III: Books 8–11.* Translated by H. Rackham. Cambridge, MA: Harvard University Press.

Rolfe, J. C. (1927) *Gellius. Attic Nights, Vol. II: Books 6–13.* Translated by J. C. Rolfe. Cambridge, MA: Harvard University Press.

Rudd, N. (2004) *Horace. Odes and Epodes.* Edited and translated by N. Rudd. Cambridge, MA: Harvard University Press.

Shackleton Bailey, D. R. (1993) *Martial. Epigrams, Vol. III: Books 11–14.* Edited and translated by D. R. Shackleton Bailey. Cambridge, MA: Harvard University Press.

Temkin, O. (1956) *Soranus' Gynecology.* Translated by O. Temkin. Baltimore, MD: The Johns Hopkins Press.

Tischendorf, C. (1853) *Evangelia apocrypha: adhibitis plurimis codicibus Graeci et Latinis maximam partem nunc primum consultis atque ineditorum copia insignibus collegit atque recensuit Constantinus Tischendorf.* Leipzig: H. Mendelssohn.

Vuong, L. C. (2019) *The Protevangelium of James.* Eugene, OR: Cascade Books.

Secondary literature

Barr, J. (2017) *Tertullian and the Unborn Child: Christian and Pagan Attitudes in Historical Perspective.* London: Routledge.

Belmont, N. (1988) 'L'enfant et le fromage', *L'homme* 105, 13–28.

Bradley, M. (2013) 'Colour as synaesthetic experience in antiquity', in S. Butler and A. Purves (eds) *Synaesthesia and the Ancient Senses*. London: Routledge, 135–48.

Branham, J. R. (1997) 'Blood in flux, sanctity at issue', *RES: Anthropology and Aesthetics* 31(1), 53–70.

Bremmer, J. N. (2012) 'Felicitas: The martyrdom of a young African woman', in J. N. Bremmer and M. Formisano (eds) *Perpetua's Passions: Multidisciplinary Approaches to the Passio Perpetuae et Felicitatis*. Oxford: Oxford University Press, 35–53.

Bremmer, J. N. and M. Formisano (2012) 'Perpetua's passions: A brief introduction', in J. N. Bremmer and M. Formisano (eds) *Perpetua's Passions: Multidisciplinary Approaches to the Passio Perpetuae et Felicitatis*. Oxford: Oxford University Press, 1–13.

Centlivres Challet, C.-E. (2017) 'Roman breastfeeding: Control and affect', *Arethusa* 50(3), 369–84.

Cobb, S. (2019) 'Suicide by gladiator? The acts of Perpetua and Felicitas in its North African context', *Church History* 88(3), 597–628.

Cole, S. G. (1992) '*Gynaiki ou themis*: Gender difference in the Greek *Leges sacrae*', *Helios* 19(1–2), 104–22.

Dean-Jones, L. A. (1989) 'Menstrual bleeding according to the Hippocratics and Aristotle', *Transactions of the American Philological Association* 119, 177–91.

———. (1994) *Women's Bodies in Classical Greek Science*. Oxford: Clarendon Press.

De Lacy, P. (1992) *Galeni De semine, edidit, in linguam Anglicam vertit, commentatus est Ph. De Lacy*. Berlin: Akademie Verlag.

Dennis, C.-L., K. Fung, S. Grigoriadis, G. E. Robinson, S. Romans, and L. Ross (2007) 'Traditional postpartum practices and rituals: A qualitative systematic review', *Women's Health* 3(4), 487–502.

Dova, S. (2017) 'Lactation cessation and the realities of martyrdom in *The Passion of Saint Perpetua*', *Illinois Classical Studies* 42(1), 245–65.

Fildes, V. (1986) *Breasts, Bottles and Babies: A History of Infant Feeding*. Edinburgh: Edinburgh University Press.

Fletcher, S., C. A. Grotegut, and A. H. James (2012) 'Lochia patterns among normal women: A systematic review', *Journal of Women's Health* 21(12), 1290–4.

Frankfurter, D. (2009) 'Martyrology and the prurient gaze', *Journal of Early Christian Studies* 17(2), 215–45.

Glancy, J. A. (2010) *Corporal Knowledge: Early Christian Bodies*. Oxford: Oxford University Press.

Gold, B. K. (2018) *Perpetua: Athlete of God*. Oxford: Oxford University Press.

Hanson, A. E. (1987) 'The eight months' child and the etiquette of birth: *Obsit omen*', *Bulletin of the History of Medicine* 61(4), 589–602.

Harris, A. L. and V. J. Vitzthum (2013) 'Darwin's legacy: An evolutionary view of women's reproductive and sexual functioning', *Journal of Sex Research* 50(3–4), 207–46.

Hogan, S. (2008) 'Breasts and the beestings: Rethinking breast-feeding practices, maternity rituals, and maternal attachment in Britain and Ireland', *Journal of International Women's Studies* 10(2), 141–60.

Holman, S. R. (1997) 'Molded as wax: Formation and feeding of the ancient newborn', *Helios* 24(1), 77–95.

Laskaris, J. (2008) 'Nursing mothers in Greek and Roman medicine', *American Journal of Archaeology* 112(3), 459–64.

Lillis, J. K. (2016) 'Paradox *in partu*: Verifying virginity in the *Protevangelium of James*', *Journal of Early Christian Studies* 24(1), 1–28.

McNeilly, A. S. (1993) 'Lactational amenorrhea', *Endocrinology and Metabolism Clinics of North America* 22(1), 59–73.

Myers, A. D. (2017) *Blessed Among Women? Mothers and Motherhood in the New Testament*. Oxford: Oxford University Press.

Pedrucci, G. (2013) 'Sangue mestruale e latte materno: Riflessioni e nuove proposte. Intorno all'allattamento nella Grecia antica', *Gesnerus* 70(2), 260–91.

Penniman, J. D. (2017) *Raised on Christian Milk: Food and the Formation of the Soul in Early Christianity*. Yale: Yale University Press.

Perkins, J. (2007) 'The rhetoric of the maternal body in the *Passion of Perpetua*', in T. Penner and C. Vander Stichele (eds) *Mapping Gender in Ancient Religious Discourses*. Leiden: Brill, 313–32.

———. (2008) *Roman Imperial Identities in the Early Christian Era*. London: Routledge.

Ronsse, E. (2006) 'Rhetoric of martyrs: Listening to Saints Perpetua and Felicitas', *Journal of Early Christian Studies* 14(3), 283–327.

Salisbury, J. E. (1997) *Perpetua's Passion: The Death and Memory of a Young Roman Woman*. London: Routledge.

Sissa, G. (1990) *Greek Virginity*. Harvard: Harvard University Press.

Solevåg, A. R. (2016) 'Hysterical women? Gender and disability in early Christian narrative', in C. Laes (ed.) *Disability in Antiquity*. London: Routledge, 331–43.

Straus, L. M. (2013) 'Got milk? Lactation and purification among covenanters', *Horizons in Biblical Theology* 35(1), 42–60.

Tran, V. T. T. (1973) *Isis lactans*. Leiden: Brill.

Vuong, L. C. (2013) *Gender and Purity in the Protevangelium of James*. Tübingen: Mohr Siebeck.

Waszink, J. J. H. (2010) *Quinti Septimi Florentis Tertulliani De Anima*. Leiden: Brill.

Wiessinger, D., D. West, and T. Pitman (2010) *The Womanly Art of Breastfeeding*. London: Pinter & Martin.

Williams, C. A. (2012) 'Perpetua's gender: A Latinist reads the *Passio Perpetuae et Felicitatis*', in J. N. Bremmer and M. Formisano (eds) *Perpetua's Passions: Multidisciplinary Approaches to the Passio Perpetuae et Felicitatis*. Oxford: Oxford University Press, 54–77.

Part V

DISSOLVING AND LIQUEFYING BODIES

16

TEARS AND THE LEAKY VESSEL

Permeable and fluid bodies in Ovid and Lucretius

Peter Kelly

In Ovid's *Metamorphoses*, many characters dissolve into fluid bodies. The shedding of tears often anticipates and instigates this form of transformation. On the one hand, tears indicate the porous or permeable nature of the human body and the capacity for the boundaries of the body to be transgressed. Tears, like sweat and other bodily fluids, pass from the interior to the exterior of the body; this presupposes an opposition between inner and outer, and the dualist concept that the body may be considered a container of the mind or soul. On the other hand, the shedding of tears in the *Metamorphoses* is depicted as a reduction of the physical substance of the body as a whole, especially when weeping leads to liquefaction. This suggests a different identity model where there is no opposition between inner and outer, and the mind and the body together constitute an undifferentiated psychosomatic whole. In transforming into a fluid, the interior and exterior elements of the mind–body complex merge together, and the individual's barrier to exterior reality loses definition. Not only do tears signal the permeability and instability of the body, as highlighted in relation to sweat (see the Envoi to this volume), but they actively mediate its boundary and challenge the unitary concept of the body, as defined through the opposition between the inner body and outside world. Identity becomes defined exclusively through its fluidity and lack of stable form. This form of transformation then poses an epistemological dilemma; as the different identity models merge and flow together, it becomes impossible to hold a single or stable image of the body as fluid.

This chapter explores how Ovid develops an image of the fluid body by juxtaposing the ideas of permeability and fluidity and how these concepts may be seen as tied to two different yet interrelated theories of identity. The first may be seen as the Homeric model, what A. A. Long (2015: 37) defines as 'psychosomatic identity'. The second is the concept of mind–body dualism developed by Lucretius, especially in book 3 of *On the Nature of Things*. In weaving together these two distinct theories of identity, Ovid portrays a concept of the body that is itself in flux, while emphasising the fluid boundaries between myth and natural philosophy and their ability to seep into each other. This chapter uses the stories of Leucothoe from book 4 and Cyane from book 5 of the *Metamorphoses* as case studies and draws specific connections with *On the Nature of Things*.[1] It will follow on from recent works on concepts of the body in antiquity, especially the volume on bodies and boundaries by Thorsten Fögen and Mireille Lee (2009), which draws upon Julia Kristeva's theory of 'abjection' (1982).[2] Here the boundaries of the body are seen as permeable and unstable and the body itself as existing in a liminal state between subject and object. The orifices and the

DOI: 10.4324/9780429438974-22

substances that pass through them become indicators of the body's potential disintegration and the inability for it to maintain a physical and conceptual distinction between the inner and outer: 'accordingly, bodily refuse—such as spit, faeces, hair clippings—become ritually charged with special powers that threaten its integrity' (Fögen and Lee 2009: 3). In the *Metamorphoses*, tears frequently prefigure liquefaction; thus, in a similar sense they may be read as indicating and even instigating bodily dissolution and the breakdown of the distinction between the individual and external reality, as metamorphosis materialises the experience of recognising the self as something fundamentally other.

Weeping in Homeric epic is frequently depicted as a loss of vital substance which is linked to psychological turmoil. Michael Clarke (1999: 115) shows that Homer 'does not oppose mental life to the life of the body but takes them as an undifferentiated whole'. In the *Odyssey* (19.204),[3] Penelope's body is described as melting when she listens to a stranger talking about her husband: 'tears flowed from her while she listened and her skin melted'(τῆς δ' ἄρ' ἀκουούσης ῥέε δάκρυα, τήκετο δὲ χρώς). The psychological experience of imagining the travails of her husband causes her to weep, which in turn results in the dissolution of her body. Odysseus is likewise described as melting when he listens to Demodocus' song about the Trojan War (8.521–2): 'But Odysseus melted and a tear, from under his eyelids, wet his cheeks' (αὐτὰρ Ὀδυσσεὺς τήκετο, δάκρυ δ' ἔδευεν ὑπὸ βλεφάροισι παρειάς). In his discussion of tears in Greek myth, Richard Buxton (2009) goes as far as to equate αἰὼν 'life essence' with tears. This parallel is evident in book 5 of the *Odyssey*, when Odysseus' αἰὼν (life essence) ebbs away in the tears he sheds for his lost home (5.151–3): 'Nor were his eyes ever dry of tears, as his sweet life essence flowed down, lamenting for his journey home' (οὐδέ ποτ' ὄσσε δακρυόφιν τέρσοντο, κατείβετο δὲ γλυκὺς αἰὼν νόστον ὀδυρομένῳ).[4] Michael Clarke has argued that tears in this context are almost equivalent to 'the vital moist stuff that characterizes the living body', while their fluidity encapsulates the temporal aspect of the flow of life (Clarke 1999: 114). Although a full discussion of the precise nature of liquid vitality in Homer is beyond the scope of this chapter, these passages illustrate that psychological turmoil in the Homeric body can trigger a physical experience in the form of the shedding of tears, which can in turn result in the melting or dissolution of the individual who weeps. As we shall see, a similar dynamic characterises Ovid's portrayal of liquefaction.

The second identity model which Ovid draws upon is the philosophical or dualist conception of identity, which sees the soul and the body as distinct entities. This concept was developed in Greek philosophy and was most fully realised by Plato (Long 2015). For instance, in the *Timaeus* (44a), the human body is described as 'the whole vessel of the soul' (τὸ τῆς ψυχῆς ἅπαν κύτος), while Cicero in the *Tusculan Disputations* (1.52) likewise depicts the body as a container: 'for the body is like a vessel or receptacle for the soul' (*nam corpus quidem quasi vas est aut aliquod animi receptaculum*).[5] In *On the Nature of Things*, Lucretius compares the human body to a cracked vessel or leaky pot in order to illustrate the permeable boundaries of the body and the danger that the soul might seep through its porous structure.[6] As Woldemar Görler (1997: 207) states, Lucretius' 'body-vessel-analogy is not a metaphor or a simile as it is in Plato and the Stoics, but a physical doctrine'. For an Epicurean the body is a container of the soul in a very literal sense, while in Lucretius' materialism, the soul also has a distinct corporeal form. Unlike in Plato and the Stoics, both the body and the soul experience dissolution when sundered from each other; in life, since the body contains void the soul moves within it like a liquid, while in death it flows away like water from a cracked vessel. Matthew Joncock (2016: 184) has, however, shown that in Lucretius the 'body and soul's relationship is more complex than simply container and contents. Rather, the two are interwoven from first birth like a cloth woven from two types of thread' (e.g. *On the Nature of Things* 3.555–7). Lucretius' body-vessel analogy must be read alongside his depiction of the mind and body as intertwined and interdependent structures. Lucretius' materialist representation of mind–body dualism—and especially his attempt to combine the image of the body's permeability with the interconnected relationship of the mind and the body—exerts a considerable influence on Ovid's depiction of liquefaction.[7]

Lucretius begins his exposition on the composition of human identity by describing how the soul is composed of exceedingly minute and perfectly round bodies. He frequently compares these soul particles to those which constitute water and smoke, as well as poppy seeds, specifically in relation to their ability to flow according to the smallest momentum (3.186–205). Lucretius argues that the soul and body must be linked, as when the body is wounded the soul is affected equally and suffers together with the body (3.168–9). Lucretius describes how the body is a poor container, as it is perforated with *foramina* ('apertures'), which allow the atoms of sense-perception to enter but also provide a means for the soul to seep out.[8] When this occurs, the body dissolves while the soul is dispersed. Lucretius uses the leaky-vessel analogy to illustrate this process:

> nunc igitur quoniam quassatis undique vasis
> diffluere umorem et laticem discedere cernis,
> et nebula ac fumus quoniam discedit in auras,
> crede animam quoque diffundi multoque perire
> ocius et citius dissolvi in corpora prima,
> cum semel ex hominis membris ablata recessit.[9]

> Now as a consequence, since you perceive that liquid, once its vessels are shattered, flows out from every side and moisture is dispersed in different directions, and since mist and smoke disperse into the breezes, you must believe the soul is also poured out and vanishes more swiftly and more quickly dissolves into its original particles, once removed and departed from human limbs.
> Lucretius, *On the Nature of Things* 3.434–9[10]

The soul flows out and disperses into its composite particles once the vessel of the human body becomes too fragile or porous (*rarefactum*) to contain it. The leaky-vessel analogy, by extension, pictures tears, sweat, sperm, and any other liquid emissions as indicators of the porous nature of the human body and the potential for the soul to seep through its structure. In her treatment of tears in Lucretius, Christina Clark (2009: 162) argues that 'to Lucretius, our ability to weep at all depends on the porous condition of our bodies. . . . This permeability of our bodies leads to constant danger of boundary violations; our bodies are like leaky vessels.'

Lucretius describes how the particles of the mind, spirit, and body are woven together and that when they are severed from each other, the body dissolves. He uses a

261

metaphor comparing the body and the soul to plants with common roots to illustrate their joint nature. He then compares the soul and body together to a bush of frankincense, with the soul being the equivalent of the plant's smell:

> quod genus e thuris glaebis evellere odorem
> haud facile est quin intereat natura quoque eius.
> sic animi atque animae naturam corpore toto
> extrahere haut facile est quin omnia dissoluantur.

> Just as it is not easy to pluck the scent from clumps of frankincense without its whole nature also being lost, so it is not easy to extract the nature of mind and spirit from the body without all being dissolved.
>
> Lucretius, *On the Nature of Things* 3.327–30

The odour of the frankincense bush is depicted as integral to the plant, just as the soul and the body are a compound, and if separated from each other, the body dissolves. The metaphor is consistent with imagery elsewhere in book 3 that pictures the decay of the corpse as a form of liquefaction. It extends the imagery associated with the leaky-vessel analogy, with the body not merely being permeable but now dissolving altogether. The frankincense analogy illustrates the close association between Ovid's depiction of liquefaction in the *Metamorphoses* and Lucretius' theory of identity from book 3 of *On the Nature of Things*.

In book 4 of the *Metamorphoses*, Ovid recounts the story of Leucothoe, who is buried alive by her father for having an affair with Phoebus Apollo. When Apollo discovers this he unearths Leucothoe and attempts to revive her, but he is too late, and after he sprinkles her body with heavenly nectar her body dissolves:[11]

> protinus inbutum caelesti nectare corpus
> delicuit terramque suo madefecit *odore*,
> virgaque per *glaebas* sensim *radicibus actis*
> turea surrexit tumulumque cacumine rupit.[12]

> Immediately her body, wet with heavenly nectar, melted away and soaked the earth with its perfume and a sprout of frankincense gradually rose up, through the clumps of earth, driven by its roots, breaking through the mound with its tip.
>
> Ovid, *Metamorphoses* 4.252–5

After Leucothoe's body dissolves, a sprout of frankincense grows from her liquefied body. When Phoebus sprinkles her corpse with nectar, he says to her 'nevertheless you will touch the sky' (*tanges tamen aethera*), as in the form of burning incense she will rise as smoke into the air (4.251).

Ovid's depiction of the transformation of Leucothoe contains a series of close visual and verbal parallels to Lucretius' frankincense analogy. The 'perfume' (*odore*) of Leucothoe's body soaks the earth, and a 'shoot of frankincense' (*uirga turea*) rises up in its place, through the 'clumps of earth' (*glaebas*), 'driven by its roots' (*actis radicibus*). In Lucretius' frankincense analogy, the difficulty of separating 'scent' (*odorem*) from 'clumps of frankincense' (*glaebis thuris*) is stressed, while soul and body are said to have 'common roots' (*communibus radicibus*) (3.325). The image in the *Metamorphoses* of the burning frankincense rising into the air is also comparable to Lucretius' frequent use of the image of smoke to illustrate the soul dispersing when separated from the body (3.425–33, 455–6, 582–3). Not only is Ovid alluding to Lucretius here, but Lucretius' analogy is mythologised and personified.

Lucretius' frankincense analogy is a key image in his representation of the material reality of the spirit, mind, and body and their interdependent relationship, while the analogy also prefigures many of the proofs of the soul's mortality which follow. Lucretius' depiction of the body and soul as plants having the same roots is a striking image that recurs at *On the Nature of Things* 5.554, where it is applied to the earth and the cushion of air which mysteriously supports it from below, a relationship that is explicitly compared with that of the body and soul. In alluding to and adapting the metaphor, Ovid demonstrates a keen awareness and intricate knowledge of Lucretius' theory of identity. In mythologising Lucretius' analogy, however, Ovid removes the surface traces of the philosophical dialogue to which it pertains. In setting out a series of key allegorical touch points with Lucretius' discourse on the relationship between the soul and the body, Ovid encourages a reading of Leucothoe's transformation as representing the decay of the corpse and the seeping out and dispersal of the soul. This link is further emphasised through the metaphoric relationship between the soul and smoke. There is, however, no indication as to whether any significant part of Leucothoe survives her transformation; the smoke which rises from her body is only smoke. Ovid appears to utilise Lucretian imagery to emphasise the insurmountable uncertainty on which the philosophical theory is based. Ovid's mythologising of Lucretius' metaphor reflects back on the philosophical discourses to which it alludes, which has the effect of portraying the philosophical theory as nothing more than a hopeful myth. Lucretius' materialist version of mind–body dualism is as much a myth as that of the story of Leucothoe.

Leucothoe's transformation is juxtaposed with that of her sister Clytie, who was madly jealous of Apollo's affection for Leucothoe. Following Lecuthoe's liquefaction, Clytie wastes away and is transformed into a heliotrope (4.267). She can only gaze longingly at the sun while her sister unites with him once more in the form of burning incense. As Clytie wastes away, she consumes nothing for nine days, resorting only to nourish herself on dew and tears (4.263). Clytie attempts to prevent the loss of the physical substance of her body through the consumption of her own tears. Clytie's drinking of her tears may be read as an attempt to re-internalise her diminishing vital essence and consolidate her corporeal boundaries. In having Clytie wither into a bloodless plant, Ovid contrasts the fates of the two sisters, as Clytie's body dries up instead of dissolving or evaporating. Clytie's transformation can be interpreted in relation to Lucretius' model of identity and specifically the leaky-vessel analogy. In consuming her own tears, Clytie attempts to prevent her vital essence from seeping out from her body. In the *Metamorphoses* excessive weeping is often the result of psychological turmoil. It is Clytie's unrequited love for Apollo which is the ultimate catalyst for her transformation. The juxtaposition of physical and psychological trauma can also be seen in the story of Cyane, from book 5 of the *Metamorphoses*. Cyane, a water nymph, witnesses the abduction of Proserpina by Dis and attempts to prevent him from returning to the underworld with his stolen bride. Stretching out her arms, she uses her body to block his path (5.439). Dis, however, ignores her blockade and plunges via her waters into the underworld, causing Cyane to dissolve:

> at Cyane, raptamque deam contemptaque fontis
> iura sui maerens, inconsolabile vulnus
> mente gerit tacita lacrimisque absumitur omnis
> et, quarum fuerat magnum modo numen, in illas
> extenuatur aquas.

> But Cyane, grieving for the ravished goddess and the disregarded rights of her own spring, sustained an inconsolable wound in her silenced mind; all of her was consumed by tears, and she was reduced to the waters of which she had just been the great divinity.
>
> Ovid, *Metamorphoses* 5.425–9

Andrew Zissos (1999: 100) has argued that the plunging of Dis through Cyane's waters into the underworld has clear undertones of physical assault and rape.[13] But it is the psychological wound which Cyane sustains that instigates her physical transformation. This has the effect of undermining or even inverting what at first appears to be a clear distinction between mind and body, as Ovid describes how Cyane's mind (*mens*) has suffered a physical wound (*vulnus*). Her weeping, a physical symptom of her psychological trauma, results in her liquefaction as she is reduced (*extenuatur*) into the waters over which she had previously ruled. Ovid also plays upon the distinction between water and water divinity, as Cyane no longer has the ability to control her physical form. As she is diminished through weeping, Cyane loses both her psychological and corporeal identity as she is transformed into an undifferentiated, fluid body.

Ovid precisely details Cyane's transformation as follows:

> molliri membra videres,
> ossa pati flexus, ungues posuisse rigorem;
> primaque de tota tenuissima quaeque liquescunt,
> caerulei crines digitique et crura pedesque
> (nam brevis in gelidas membris exilibus undas
> transitus est).

> You could have seen her limbs soften, her bones suffer bending, her fingernails lose their hardness; it was smallest parts of her whole body that first became liquid, her cerulean hair, her fingers, her legs and her feet (for it is a short crossing from slender limbs to icy waters).
>
> Ovid, *Metamorphoses* 5.429–34

Both the interior and exterior elements of Cyane's body soften and dissolve, as the whole of her body is liquefied, from her bones to her hair. This dissolution begins with the smallest parts of her body before spreading to her extremities, her slender limbs being likened to the cold waters into which she is transformed. Her shoulders, back, and breast then vanish into narrow streams (5.434–5). At the end of Cyane's transformation, it is the intangibility of her body that is stressed: 'finally, instead of living blood, water entered her corrupted veins, and nothing remained which you could have grasped' (*denique pro vivo vitiatas sanguine venas | lympha subit, restatque nihil, quod prendere possis*) (5.436–7). This image is different to the gradual melting of her entire body. Cyane loses the ability to maintain her corporeal boundaries, as the water seeps through her now fully permeable skin and enters into her bloodstream. Ovid presents two different yet intertwined images of Cyane's metamorphosis: the first portrays the dissolution of her entire body into an undifferentiated substance, stemming from excessive weeping; the second portrays an increase in her body's permeability until it is no longer able to maintain its distinction from the exterior world. Ovid, again following in the vein of Lucretius, visualises the body as a fragile structure unable to contain its own vital essence, while its fundamental fluidity emphasises the interwoven nature of the mind and the body both physical and conceptually.

The imagery of bodily dissolution is again employed in the transformation of Byblis from book 9 of the *Metamorphoses*. Byblis, a daughter of Miletus, falls in love with her brother Caunus, and when he rejects her, she weeps profusely until Naiads transform her into a spring. Ovid illustrates her liquefaction through a series of similes:

> protinus, ut secto piceae de cortice guttae
> utve tenax gravida manat tellure bitumen,
> utque sub adventum spirantis lene Favoni
> sole remollescit, quae frigore constitit, unda,
> sic lacrimis consumpta suis Phoebeia Byblis
> vertitur in fontem, qui nunc quoque vallibus illis
> nomen habet dominae, nigraque sub ilice manat.

At once, as pitchy drops from severed bark, or clinging bitumen from the swollen earth flow, and as, on the arrival of the gentle breath of the west wind, water, which had been fixed by the cold, melts in the sun, so Phoeban Byblis, consumed by her own tears, was turned into a spring, which even now in those valleys holds the name of its mistress, and flows under a dark holm-oak.

Ovid, Metamorphoses 659–65

These similes present us with a number of different ways to interpret Byblis' transformation. Her tears are compared with pitch, which seeps through the surface of wounded bark as well as bitumen flowing out of the earth. Both of these images portray a liquid moving through a permeable boundary. In contrast, the third simile compares the liquefaction of Byblis' body to the melting of ice, indicating that her transformation is a straightforward transition from a solid to a liquid state. A tension exists in Ovid's presentation of the dissolution of the body in two distinct phases or in two separate ways, which frequently function to undermine and reconceptualise each other.

A comparable dynamic is found in Ovid's description of the metamorphosis of Arethusa from book 5 of the *Metamorphoses*. Arethusa is transformed into a stream to escape the advances of the river god Alpheus. Arethusa's fluid nature becomes a further means of highlighting the dissolution of corporeal boundaries. In this case, it is not so much the distinction between inner and outer which is eroded but the separation of self and other. As bodies of water, Arethusa and Alpheus can flow together. Instead of tears, however, it is the sweat which breaks out upon her body that prefigures her dissolution:

> occupat obsessos *sudor* mihi frigidus *artus*,
> caeruleaeque cadunt toto de *corpore guttae*,
> quaque pedem movi, *manat* locus, eque capillis
> ros cadit, et citius, quam nunc tibi facta renarro,
> in latices mutor.

> A cold sweat seized my limbs and dark blue drops fell from all of my body and where I stepped with my foot, the place grew wet, and dew fell from my hair and more swiftly than I now recall what happened, I was changed into water.
>
> Ovid, Metamorphoses 5.632–6

The beads of sweat and drops which fall from her body suggest that her liquefaction is at least in part the result of an increase in the body's permeability; the seeping of sweat through the barrier of the skin further confounds the corporeal structure and can be seen as a physical indicator of uncontrollable emotion. Ovid's description of Arethusa's transformation again looks towards a metaphor from *On the Nature of Things*, where Lucretius compares the porosity of the body to that of a cave. Lucretius describes how water seeps like sweat through the walls of a cave and uses this to illustrate the permeability of the human frame and its composition of a mixture of matter mixed with void:

> principio fit ut in speluncis saxa superna
> *sudent* umore et *guttis manantibus* stillent.
> *manat* item nobis e toto *corpore sudor*,
> crescit barba pilique per omnia membra, per *artus*.

> First as it usually happens within caves, the rock above sweats moisture and is wet with flowing drops. Likewise sweat flows from all of our body, the beard grows and hairs through every joint and limb.
>
> Lucretius, On the Nature of Things 6.942–5

If we examine this passage from book 6 alongside the earlier one from the *Metamorphoses*, it is clear there are a number of close verbal and visual parallels: Arethusa's body breaks out in drops (*guttae*) and beads of sweat (*sudor*) which flow (*manat*) from her body and wet the ground; likewise Lucretius describes how the cave sweats (*sudent*) and is wet with flowing drops (*guttis manantibus*), before making a direct comparison between the cave and the body. Ovid's description of the dew falling from Arethusa's hair may also recall Lucretius' use of hair in this passage as a further indicator of the void which permeates the body. Sarah Myers (1994: 49) also notes the connection between these two passages and describes Ovid as 'recontextualizing [Lucretius'] scientific theory in this mythological scene'. For Lucretius, one of the conditions that allows for the body to be porous is that the world is constituted of matter and void. It is the void that provides the *foramina* that allow the soul particles to escape from the body as well as the space into which they can dissipate (1.419–40). Void is a crucial aspect then of Lucretius' understanding of both cosmic and corporeal structure and is the central theory behind the sweating cave metaphor. Not only then is Ovid familiar with Lucretius' model of identity, but he is also distinctly aware of how this relates to Lucretius' theory of the makeup of the cosmos at large.

Fluidity and flux are central aspects of the world of the *Metamorphoses*, especially as portrayed in the speech of Pythagoras, while flux is also specifically tied to the experience of material reality. The fluid body presents a fundamental challenge to perception, which leads to a multiplication of shifting perspectives and the potential for knowledge itself to collapse. In book 15 of the *Metamorphoses*, Pythagoras states that 'there is nothing in the whole world which persists. All things flow, and every image formed is fluctuating' (*nihil est toto, quod perstet, in orbe. | cuncta fluunt, omnisque vagans formatur imago*) (15.177–8).[14] Pythagoras compares the flux and change that govern the universe to the continuous flow of the river; the totality of existence becomes in effect a fluid body. The phrase *cuncta fluunt*, 'all things flow' (178), has been read as a direct translation of Heraclitus' πάντα ῥεῖ, while the comparison of time to the river is a likely allusion to Heraclitus' image of the river into which you cannot step twice, which in turn is mediated through Plato's *Theaetetus* (160d8 and 182a1) and *Cratylus* (402a). As Philip Hardie (2015: 509) has shown in this image, 'each form is given shape as something fluctuating', while Michael Goyette's description of flux in Seneca in Chapter 17 of this volume could equally be applied here: 'the fluidity of each individual body in the universe both amplifies and is amplified by the fluidity of other bodies, culminating in a state of perpetual flux throughout the larger body of the universe' (p. 273). The ever-changing nature of material reality, that is, the flux of the world-body, finds expression in the embodied experience of the individual. Ovid encapsulates this in the following image: 'but wave is driven on by wave, the first being pushed by the one coming and itself pushing the one before' (*sed ut unda inpellitur unda | urgeturque prior veniente urgetque priorem*) (15.181–2). Each human life, like a wave rippling on the surface of reality, is but a brief peak that appears in the material substrate of the world and which cannot be separated from the peaks that rose previously and those that will follow. The onomatopoeic quality of the repetition of *unda* captures the image of the rolling waves yet also serves to link this image with the text itself. Each *unda*, like each allusion or each new text, follows from and flows into and yet is different to the previous one.[15]

Lucretius also employs the leaky-vessel analogy to illustrate a more abstract representation of the flow of life and our inability to hold on to what we perceive or experience. Helen King in this volume (Chapter 24, pp. 391–3) draws out further the extended influence of the image of the perforated vessel. Towards the end of book 3 of *On the Nature of Things*, Lucretius criticises a man in old age, who, still unsatisfied with that which he has achieved, fails to withdraw from an active life, with the result that his memories and blessings are poured away (*perfluxere*), as if consigned to a leaky vessel (*pertusum uas*) (935–43). The analogy has shifted from being an illustration of the relationship between the soul and the body to a more abstract representation of the inability to hold onto memories and experiences. Lucretius is here alluding to Plato's use of a similar analogy in the *Gorgias*, where Socrates, referring to people who are easily influenced, compares the part of their soul which contains desire to a 'perforated jar' (τετρημένος πίθος), since it can never be filled up (493a–494c). He then compares the soul of the fool to a 'sieve' (κόσκινον) since he cannot retain anything that he has experienced or learned. In the *Gorgias*, Socrates makes an oblique reference to those uninitiated in the underworld who continually pour water into a leaky vessel out of an equally leaky sieve. Plato is here referring to the myth of the Danaids, the daughters of Danaus who, as punishment for killing their husbands, are condemned for eternity in Tartarus to pour water into a leaking vessel. In his final use of the leaky-vessel analogy in book 3, Lucretius too evokes the myth of the Danaids:

> hoc, ut opinor, id est, aeuo florente puellas
> quod memorant laticem pertusum congerere in uas,
> quod tamen expleri nulla ratione potestur.

> This, as I see it, is the story of the girls in the bloom of their years gathering liquid in a leaky vessel which despite their efforts can in no way be filled full.
> Lucretius, *On the Nature of Things* 3.1008–10

Lucretius uses the leaky-vessel analogy as a physical doctrine: the human frame is not like but simply is a porous container. In referring to the story of the Danaids, however, Lucretius, like Plato, betrays the myth underlying his philosophical doctrine. This demonstrates that already in Lucretius, and indeed Plato, we can observe the instability of the divisions between natural philosophy and mythology. The shifting meanings that Lucretius uses the leaky-vessel analogy to demonstrate—on the one hand the body's inability to retain the soul and on the other the inability of the soul to retain experiences and memories—expresses how the life of an individual cannot be separated from the experience of that life. We have returned to the Homeric model of identity, where the oozing of a liquid expresses not just the seeping out and dissolution of vital essence but the ongoing process of living.

Both Ovid and Lucretius utilise imagery concerned with permeability and fluidity to express a series of interconnected ideas concerned with the nature of personhood, the experience of material reality, and the illustration of these ideas within literature and through language. Ovid responds directly to Lucretius in his representations of liquefaction and in particular the transformations of Leucothoe, Cyane, Byblis, and Arethusa. Ovid models this form of transformation on Lucretius' theory of mind–body dualism, where weeping and liquefaction can be read in terms of the movement of vital substance from the interior to the exterior of the body, stemming from an increase in the body's permeability. In the *Metamorphoses*, however, the image of the body's permeability would seem at least in part to jar with the image of the outright liquefaction which ultimately occurs. One of the central paradoxes of the *Metamorphoses* is that transformation more often than not leads to permanence. In most cases, once an individual undergoes metamorphosis, they become fixed in a moment of transition. Transforming into a fluid body highlights this paradox, as in this case the fixed state in question is itself characterised by flux and mutability. In a world where the only unifying principle is change, there can be no fixed representation of identity or model with which it can be viewed.

Despite this, the body as fluid comes closest to realising a true expression of unembodied embodiment. Ovid's representations of the dissolution of the entire body, both interior and exterior together, appears to be more in keeping with the examples from the *Odyssey* highlighted at the beginning of this chapter, where Odysseus and Penelope are described as physically melting and where weeping is likewise the result of psychological trauma (8.521–2, 19.204). Liquefaction can be interpreted as the ultimate expression of a psychosomatic whole. In actualising Lucretius' metaphors and removing them from their philosophical context, Ovid fully realises the link established in Lucretius between the material structure of the mind and the body and their interdependent dissolution at the moment of separation. In placing this in a mythological context, Ovid demonstrates the ease by which the Homeric body can become the Lucretian and the inability for myth and natural philosophy to maintain their boundaries. Ovid collapses mythological and philosophical representations of personhood, while the unitary nature of fluid and the loss of distinction that transforming into a fluid entails provide a means of interpreting both bodily and textual structures. Finally, these ideas extend outwards from models of identity to the embodied experience of reality. Both Ovid and Lucretius share more than a close affinity with Homer, where the loss of vital essence, the shedding of tears, and the experience of the temporal flow of life itself operate within the same somatic and semantic domains.

Notes

1 Other examples from the *Metamorphoses* where liquefaction results from excessive weeping are as follows: Hyrie dissolves from crying and is transformed into a lake (7.380–1); Byblis weeps profusely before mountain nymphs replace her with a stream (9.1049); while the transformation of Arethusa should also be noted, as drops of sweat pour out from her body before she is transformed (5.632–6). P. M. C. Forbes Irving (1992: 299–307) provides a useful list of characters from Greek myth that transform into bodies of water, including those who transform from weeping.

2 This research also draws on the theoretical approaches of recent studies such as Thorsten Fögen (2009), and Manfred Horstmanshoff et al. (2012), as well as those set out in the Introduction to this volume.

3 All displayed text from the *Odyssey* follow Monro and Allen's Oxford Classical Texts (Allen 1963; Monro 1963). Translations, unless otherwise stated are my own.

4 Following R. B. Onians (1951), Buxton (2009: 217) uses this passage to conclude 'that one's αἰὼν can be imagined as in some sense a liquid, which may flow out of the body in the form of tears'. Clarke (1999: 114), also discussing this passage, argues that it is not the substantial similarity between αἰὼν and tears that should be stressed but rather the process of flow from the body: 'The picture [of αἰὼν] that emerges is of something visible and tangible, the vital moist stuff that characterizes the living body; yet in the temporal sense . . . this same word names something entirely abstract, the duration of a life. The common ground of meaning is not in a particular static thing but in the on-going *process* of living, which can be seen and encapsulated in different contexts by a length of time or by an oozing liquid.'

5 Cicero (*Tusculan Disputations* 1.61) also describes how our memories and experiences are poured into the vessel of the soul: *utrum capacitatem aliquam in animo putamus esse, quo tamquam in aliquod vas ea, quae meminimus, infundantur?* 'What then should we believe, that there is a space for holding within the soul, into which, like into a vessel all that we remembered is poured?' This may be compared to the Homeric model of identity, where the flowing of vital essence is as much concerned with the flowing of lived time.

6 Lucretius frequently compares the human body to a *vas*, a 'vessel' or 'container' for the *anima* (3.434, 3.555, 3.936, 3.1009, and 6.17).

7 How we are to translate *anima* is somewhat problematic. Lucretius' use of it is often in opposition to *animus*. *Animus* is often equated with *mens* and may be read as the rational or thinking part of human identity, something akin to mind, consciousness or psyche, while *anima* may be interpreted as the spirit or the breath of life. In general I will translate *anima* as 'spirit', *animus* as 'mind', and when the two are taken together, as frequently occurs in book 3 of *On the Nature of Things*, the term 'soul' will be used as a somewhat catch-all term.

8 Kelly (2014) argues that Ovid in his description of house of *Fama* (12.44–5) alludes to Lucretius' analogy of the house to describe the porous nature of the body (4.596–601), and specifically how the *foramina* allow voice and sound to enter.

9 All displayed text from the *On the Nature of Things* follow Bailey's (1922) Oxford Classical Text.

10 Displayed translations from book 3 of *On the Nature of Things* are adapted from M. P. Brown's (1997) translation.

11 The sprinkling of Leucothoe's body with nectar is paralleled in the story of Adonis, when Venus also sprinkles perfumed nectar in Adonis' blood (10.732). Ovid uses a strange analogy, as the blood then 'swelled up' (*intumesco*) like a clear bubble rising through yellow mud (733–4). We might also compare Lucretius' description of the embalming of the corpse in honey (*DRN* 3.891).

12 All displayed text from the *Metamorphoses* follow Tarrat's (2004) Oxford Classical Text unless otherwise stated, while Anderson's (1998) Teubner is frequently consulted. All displayed translations from the *Metamorphoses* largely follow Hill's (1985–2000) translation, with some minor adaptations in places.

13 Weiberg (2018) provides a useful discussion on the psychological wound in Sophocles' *Trachiniae*.

14 Hill (2000: 207) points to Diogenes Laertius' summarising of Heraclitean principles (9.8.3–4): γίνεσθαί τε πάντα κατ' ἐναντιότητα καὶ ῥεῖν τὰ ὅλα ποταμοῦ δίκην 'all things come into being from opposition and all things in their entirety flow in the same way as a river.' It is not feasible in the present context to discuss the long-running debate concerning the extent to which Heraclitus actually held a doctrine of flux and to what degree the river fragments were intended to depict this flux. One side of the debate argues that Heraclitus' association with the doctrine of universal flux and the reading of the river fragments as illustrating πάντα ῥεῖ is largely if not entirely a literary construct by Plato and one especially derived from the *Theaetetus* and *Cratylus*.

15 For Plato, this Heraclitean principle presented a serious epistemological and ontological challenge, because if everything is constantly in a state of changing into something else, there can be no fixed or stable identity, and knowledge and discourse become impossible.

Bibliography

Primary sources

Allen, T. W. (1963) *Homer. Opera*, vol. III. Oxford: Oxford Classical Texts.

Anderson, W. S. (1998) *P. Ovidi Nasonis, Metamorphoses*. Berlin: Bibliotheca Scriptorum Graecorum et Romanorum Teubneriana.

Bailey, C. (1922) *Lucretius. De Rerum Natura*. Oxford: Oxford Classical Texts.

Brown, M. P. (1997) *Lucretius' De Rerum Natura 3*. Warminster: Aris and Philips.

Hardie, P. (2015) *Ovidio. Metamorfosi, vol. VI: Libri XIII-XV*. Milan: Fondazione Lorenzo Valla.

Hill, D. E. (1985–2000) *Ovid's Metamorphoses*. Oxford: Aris and Philips.

Monro, D. B. (1963) *Homer. Opera*, vol. 4. Oxford: Oxford Classical Texts.

Tarrat, R. J. (2004) *Ovid. Metamorphoses*. Oxford: Oxford Classical Texts.

Secondary literature

Buxton, R. (2009) *Forms of Astonishment: Greek Myths of Metamorphosis*. Oxford: Oxford University Press.

Clark, C. A. (2009) 'Tears in Lucretius', in T. Fögen (ed.) *Tears in the Graeco-Roman World*, Berlin: Walter de Gruyter, 161–78.

Clarke, M. (1999) *Flesh and Spirit in the Songs of Homer*. Oxford: Oxford University Press.

Fögen, T. (ed.) (2009) *Tears in the Graeco-Roman World*. Berlin: Walter de Gruyter.

Fögen, T. and M. M. Lee (2009) *Bodies and Boundaries in Graeco-Roman Antiquity*. Berlin: Walter de Gruyter.

Forbes Irving, P. M. C. (1992). *Metamorphosis in Greek Myth*. Oxford: Clarendon.

Gorler, W. (1997) 'Storing up past pleasures: The soul-vessel-metaphor in Lucretius and his Greek models', in K. A. Algra, M. H. Koenen, and P. H. Schrijvers (eds) *Lucretius and his Intellectual Background*. Amsterdam: North-Holland, 193–207.

Horstmanshoff, M., H. King, and C. Zittel (eds) (2012) *Blood, Sweat and Tears: The Changing Concepts of Physiology from Antiquity into Early Modern Europe*. Leiden: Brill.

Joncock, M. (2016) 'Metaphor and argumentation in Lucretius', PhD thesis, Royal Holloway, London.

Kelly, P. (2014) 'Voices within Ovid's House of Fama', *Mnemosyne* 67, 65–92.

Kristeva, J. (1982) *Powers of Horror: An Essay on Abjection*. New York: Columbia University Press.

Long, A. A. (2015) *Greek Models of Mind and Self*. Cambridge, MA: Harvard University Press.

Myers, S. (1994) *Ovid's Causes: Cosmogony and Aetiology in the Metamorphoses*. Ann Arbor: University of Michigan Press.

Onians, R. B. (1951) *The Origins of European Thought about the Body, the Mind, the Soul, the World, Time and Fate*. Cambridge: Cambridge University Press.

Weiberg, E. L. (2018) 'The writing on the mind: Deianeira's trauma in Sophocles' *Trachiniae*', *Phoenix* 72, 19–42.

Zissos, A. (1999) 'The rape of Proserpina in Ovid "Met." 5.341–661: Internal audience and narrative distortion', *Phoenix* 53, 97–113.

17

SENECA'S *CORPUS*

A sympathy of fluids and fluctuations[*]

Michael Goyette

To delve into the diverse literary *corpus* of the first-century CE Stoic philosopher and Roman tragedian Seneca the Younger is to immerse oneself in a world of fluidity. This preoccupation is frequently evident in Seneca's representations of bodies—not only human bodies, but also those of animals and other embodied aspects of the natural environment, celestial bodies, and the body of the universe at large. Examining selections from Senecan prose (the *Moral Letters to Lucilius* and the *Natural Questions*) and two of his poetic tragedies (*Oedipus* and *Thyestes*), this chapter demonstrates that Seneca repeatedly employs language and imagery of fluids—especially water and blood—and metaphors of fluidity to highlight the permeability of the human body and the mutability of human emotions, while also interlinking these conditions with a pervasive inclination toward flux in the broader macrocosm of the universe. Furthermore, I contend that these recurring representations of fluids and fluidity underscore the challenges inherent to maintaining physical integrity and health, as well as the difficulties of achieving the mental and emotional stability associated with Stoic ethics. This chapter thus investigates conceptualisations and rhetoric of fluids and fluidity in a range of Senecan texts and their context within broader philosophical and medical discussions in Greco-Roman antiquity.

Across the disparate genres of his writings, Seneca consistently inscribes a universe governed by principles of *sympatheia*. This influential doctrine, promulgated by earlier Greek Stoic philosophers such as Zeno, Chrysippus, and Posidonius, held that all of the universe's components exist in a constant state of symbiotic tension and reciprocal interaction, yielding a state of continuous change, or fluidity.[1] For Stoics, these principles of integration and interdependence applied not only to large-scale components of the universe but also to smaller entities, including the microcosm of the human body (Lapidge 1978: especially 168–80). A similar framework underlies many works in the Hippocratic Corpus,[2] where the human body is persistently characterised as a unity of sympathetically interacting parts (*On the Places in Man* sec. 1). The principles of *sympatheia* bear a particularly close conceptual kinship with humoralism, a holistic theory notably articulated in the Hippocratic text *On the Nature of Man*, which describes illness as a state of imbalance in the body's four constituent fluids, or humours.[3] The sympathetic basis of humoralism is also manifested in the humours' associations with aspects of the natural world that lie beyond the boundaries of the human body, such as the seasons, winds, and physical elements.[4] Stoic Seneca, then, is akin to the Hippocratic humoralist in his attentiveness to corporeal fluctuations and

DOI: 10.4324/9780429438974-23

the human body's dynamic interactions with external forces. This fixation on bodily fluids and the fluidity of bodies invites further enquiry in scholarship interested in ancient medicine and the body, which has previously given little attention to these aspects of Seneca's works.

I begin with selected passages from Senecan prose, where Seneca sometimes addresses the fluidity of the body quite directly. I then turn to passages from Senecan tragedy, demonstrating continuities in language usage and philosophical approach, and then in closing discuss one important difference between these literary *corpora*. Seneca repeatedly emphasises the fluid nature of the human body—and the tenuous quality of human existence (Busch 2009: 255)—in his *Moral Letters* (*Epistulae*, hereafter *Ep.*), a collection of 124 prose letters addressed to Seneca's friend Lucilius. Scholars generally agree that these are not records of actual correspondence but rather an anthology of philosophical essays that are often concerned with how to live a Stoic life amid circumstances of flux. In *Letter 58*, Seneca discusses several ontological and ethical questions that arise from his view of the physical universe, including the obstacles posed by the fluid conditions of human existence. In one section he contrasts the eternal, abstract Forms familiar from works of Plato with the fluctuating and vulnerable nature of the human body:

> We feeble and fluid beings exist among wavering things: let us then direct the mind to those things which are eternal. Let us marvel at the flittering forms of all things on high, and at the god who circulates among them and foresees how he could defend from death that which he was not able to make immortal because substance was preventing it, and how with reason he could overcome the defect of the body.
>
> Seneca, *Moral Letters* 58.27[5]

Seneca characterises human life as doubly imperilled by forces of fluidity: not only is the constitution of the human body 'feeble and fluid' (*inbecilli fluidique*), but human beings also exist 'among wavering things' (*inter vana*). In using the adjective *vana*, Seneca employs a word with a very broad semantic range—other relevant translations include 'empty', 'false', and 'ineffectual' (Glare 2012: 2215–16 s.v. *vanus*)—but 'wavering' is particularly pertinent in this context as it highlights the transience of all physical matter, especially in light of the contrasting reference to 'things which are eternal' (*aeterna*). According to Seneca, the wavering conditions of the universe not only parallel the fluid nature of the human body, they also compound the precarity of embodied existence. Norman Pratt detects similar working of Stoic physics in Senecan tragedy, observing that 'in an interconnected universe the effect of bodily causation in turn operates as a cause of other effects.'[6] Pratt is not addressing Seneca prose here, but he identifies an important logical implication of Stoic *sympatheia* which is also conveyed in *Letter 58*: the fluidity of each individual body in the universe both amplifies and is amplified by the fluidity of other bodies, culminating in a state of perpetual flux throughout the larger body of the universe.[7] The Introduction to this volume (p. 5) hypothesises that bodily fluids mediate the relationship between the inner body and the outside world; while this certainly applies here, in the Senecan context there are also larger forces of fluidity at play beyond the fluids of the body themselves.

Here Seneca asserts, then, that material existence is antithetical to physical continuity (cf. *quae immortalia facere non potuit, quia materia prohibebat*), and he comments on the implications of these conditions for the human body. Noting its susceptibility to both internal and external forces of fluidity, Seneca describes the body as a 'defect' (*vitium*)—a word with strong connotations of physical debility and sickness (Glare 2012: 2292–3 s.v. *vitium*). Such states are, according to Thomas Rosenmeyer (1989: 111), 'an inevitable implication of *sympatheia*'. *Vitium* also possesses moral and ethical connotations (other possible translations include 'fault', 'vice', and 'blemish') which are evident in the passage's appeal to reason as a remedy for minds that may be focused on fleeting bodily concerns instead of that which is abstract, unchanging, and eternal. I will return to the ethical implications associated with Seneca's perspectives on fluidity and embodiment later in this chapter.

Seneca's *Letter 58* accentuates its points about the fluid nature of bodies (human and otherwise) by invoking a famous aphorism attributed to the Presocratic philosopher Heraclitus:[8]

> Our bodies are swept along like rivers. . . . Even as I myself say that these things are changed, I myself am changed. This is what Heraclitus says: 'We go down twice into the same river, and yet it is a different river.' For the name of the river remains the same, but the water has already flowed past. This is more obvious in a river than in a human being, but no less swift a course passes over us. . . . I have spoken about a human being, a substance which is fluid and fleeting and subject to every influence; the universe, though eternal and indestructible, is also changed and never remains the same.
>
> Seneca, *Moral Letters* 58.22–4

This reference to Heraclitus' constantly-changing river presents a vivid image of literal fluidity which is reinforced by language of flux in this passage[9] (multiple instances of *flumen*, along with uses of *aqua*, *amnis*, and *fluvida*) and language of change (multiple instances of the verb *mutare*). Seneca acknowledges that forces of fluidity are more evident in a river, but he emphasises that human beings are subject to those same forces. The parallelism between the nature of rivers and the realities of the human condition is punctuated by the use of *cursus* ('course'), which can denote the way a person leads one's life as well as the path that a river takes.[10] In such ways, Seneca's *Letter 58* professes that all forms of matter and all physical bodies—including the human body—are governed by universal laws that predispose conditions of fluidity. In Chapter 16 in this volume, Peter Kelly reveals a similar representation of bodies (human and non-human) in Lucretius and Ovid. Kelly connects these authors' representation with the instability, permeability, and (especially in Lucretius) porosity of bodily boundaries, but in Seneca's works these fluid conditions of embodiment are ultimately more rooted in principles of *sympatheia*.

Seneca draws further connections between the nature of the human body and rivers in his *Natural Questions* (*NQ*), a prose encyclopaedia of the natural world. In Book 3, Seneca investigates bodies of water, including rivers, springs, and lakes. In one section, he imagines that these sources of water are arranged upon and within the earth in a way that is analogous to the configuration of arteries, veins, and channels for other fluids in the human body:[11]

> But I firmly assert this: the earth is governed by nature, indeed after a model of our own bodies, in which there are both veins and arteries—the former being containers of blood, the latter containers of air. On the earth there are likewise certain channels through which water courses, others through which air courses; nature has fashioned its likeness to that of human bodies to such an extent that even our ancestors spoke of veins of water (i.e. springs). And just as there is not only blood, but also many other types of fluid in our bodies—some indispensable, others tainted and somewhat thicker (the brains in the head, the marrow in the bones, mucus, saliva, tears, and something included in the joints by which they are bent more quickly because of lubri-cation [i.e. synovial fluid])—so, too, in the earth, there are many types of fluid. . . . [15.4] In other respects, as in our bodies, the fluids in the earth often contract defects; either a blast, shaking, the decrepitude of the place, cold, or heat damage the natural quality [of the fluid].
>
> Seneca, *Natural Questions* 3.15.1–2, 4

As in *Letter 58*, Seneca expresses a congruence between the nature of the human body and the workings of larger bodies—in this case, the body of the earth. His portrait of the human body teems with bodily fluids: blood, bone marrow, mucus, saliva, tears, synovial fluid, even brains.[12] One might also include air in this catalogue of fluids.[13] The sheer number of bodily fluids mentioned creates an impression of extensive fluid-ity, which intensifies the potential for imbalance and concomitant illness. Referring to these bodily fluids collectively as *umores*, a word which can also denote various kinds of moisture, including the water present in rivers and springs (Glare 2012: 2303 s.v. *umor*), Seneca underscores the correspondence between human bodily fluids and bod-ies of water in the natural environment. A similar effect is accomplished with the use of *venae* (here translated as 'veins'), which can refer to human blood vessels as well as channels of water inside the earth (Glare 2012: 2232 s.v. *vena*); earlier Romans, according to Seneca, found this semantic duality quite fitting. As he matches part for part and function for function, and in some cases uses overlapping pieces of termi-nology, it becomes apparent that Seneca is presenting a correspondence which is not merely metaphorical in his conceptualisation.

Seneca thus depicts the earth as an embodied repository of waters that possess an intrinsic fluidity, mirroring the fluid and precarious situation of the human body as described in *Letter 58*. Furthermore, he relates that just as bodily fluids tend to become 'tainted' (*corrupti*), the earth's channels of water are inclined to contract 'defects' (*vitia*) from external forces such as blasts, shaking, cold, and heat. Thus, the human body and the earth alike are threatened by internal and external propensities toward instability. The use of *vitia* in reference to the fluids of the earth recalls Sen-eca's characterisation of the human body in *Letter 58*, revealing further parallelism in his conceptualisation of these entities. Seneca's explanation of the earth's physical structure thus calls attention to the frailties of human condition that stem from the inherently fluid world human beings inhabit and from the fluid nature of the human body itself (cf. Inwood 2005: 171; Scott 1999: 63). As Rosenmeyer puts it in his comments on this passage: 'Liquidity is both the setting and instigator of everything that is wrong with the world' (Rosenmeyer 1989: 128).

Images of fluidity also play a central role in Seneca's prognostications about the earth's eventual demise, which he discusses in the closing sections of Book 3 of the *Natural Questions* (3.27–30). Following traditional Stoic cosmogony, Seneca asserts that the earth undergoes periodic cycles of destruction and subsequent renewal that are triggered by flooding. Each deluge, he explains, begins with excessive rain which ruins crops, causes famine, and makes homes collapse and sink into soil that becomes 'soft and fluid' (*molle fluidumque*, 3.27.5). As in *Letter 58*, these fluid conditions produce a situation in which 'nothing is stable' (*nihil stabile est*, 3.27.6). As the rain and flooding accumulate, entire cities are swept away and submerged, while rivers and other bodies of water overflow their banks and eventually envelop the entire earth in a single whirlpool (3.27.9). The source of the earth's destruction arises not from an external source but from the substances and nature of the earth itself, as Seneca stresses that the earth is 'susceptible to changing and to being dissolved into liquid' (*mutabilem et solvi in umorem*, 3.29.4).[14] A few paragraphs later, he goes so far as to suggest that Nature has placed water in all regions of the earth so that it can besiege the world from all sides whenever it chooses (*NQ* 3.30.3; cf. Fedeli 2000: 42). Such descriptions illuminate why Seneca, in agreement with Stoic orthodoxy, considered water to be the most powerful of nature's elements (*NQ* 3.13.1).

In outlining the process by which the deluge cyclically overwhelms the world, Seneca also draws connections between the earth's susceptibility to inundation and the human body's liability to illness. He describes the deluge in language of putrefaction and wasting away (e.g. *NQ* 3.29.6), including *tabes*, a word which the contemporary medical encyclopaedist Celsus applies to a set of diseases that are characterised by symptoms of emaciation (Celsus, *On Medicine* 3.22). The rhetoric of illness is even more explicit in a simile in which Seneca compares the progression of the deluge to an infection that spreads from one part of the body to adjacent areas:

> Just as healthy [bodies] pass over into illness and ulcers infect adjacent parts, in the same way those things which are nearest to the liquefying land will themselves be washed away and dissolved, and then rush away.
>
> Seneca, *Natural Questions* 3.29.7

This analogy is predicated on principles of Stoic *sympatheia*, as the very interconnectedness of body and earth exacerbates the level of instability in each embodied system. For the human body, health remains elusive and illness the normative state because of the ease with which ulcers and other forms of illness proliferate through its interconnected parts; for the body of the earth, flooding spills over from one region to the next as if spreading by contagion until the entire world is plunged into a state in fluidity.[15] While Seneca does not always view floods as negative—he describes, for instance, the Nile's ability to impart fertility through its annual flooding (*NQ* 4A 1.2–2.2)—he nonetheless has a prevailing tendency to associate instances of flooding with the dissemination of impurities, instability, and annihilation. Indeed, Seneca repeatedly notes that these qualities of the Nile's flooding make it unique and exceptional among rivers and bodies of water (e.g. *NQ* 3.1.2; 4.1.2; 4.1.10).

This analysis of passages from Senecan prose can shed light upon conceptualisations of bodies and experiences of corporeality in Senecan tragedy, where representations of fluidity and sympathetic interactions are sometimes less obvious or straightforward. While the relationship between Seneca's works is highly complex and one should not assume congruity between a statement in one of his philosophical essays and a passage from one of his poetic tragedies,[16] I observe that Senecan tragedy also utilises extensive language and imagery of fluidity—as well as literal fluids—in its representations of bodies and its construction of a sympathetic universe. In this way, I follow the trend of the last few decades in which scholars have become more inclined to consider how Seneca's philosophical prose can inform understandings of Senecan tragedy (and vice versa).[17] My analysis builds particularly upon Rosenmeyer, who has persuasively demonstrated that Seneca's tragedies frequently evince precepts of Stoic natural science, including *sympatheia*, and anxieties concerning the fragility of the human body that are characteristic of Senecan prose (Rosenmeyer 1989: especially 99, 116). I expand upon Rosenmeyer's analysis by focusing on Seneca's use of language and imagery of fluidity in two tragedies, *Oedipus* and *Thyestes*, where they play a particularly prominent role.

In Chapter 12 in this volume, Emily Kearns observes that the genre of Greek tragedy (and the plays of Aeschylus' *Oresteia* in particular) frequently fixates on blood in literal or metaphorical ways. Such preoccupations are perhaps even more prominent in Senecan tragedy. Seneca's *Oedipus* and *Thyestes* both brim with images of blood emerging in ways that are surprising, disturbing, and profoundly interlinked with large-scale deviations from stability and orderliness. This is very much reflected in the vocabulary of 'blood': *Oedipus* alone contains sixteen combined uses of the noun *sanguis* and the adjective *sanguineus* and thirteen combined uses of the noun *cruor* and the adjective *cruentus*.[18] Like its Sophoclean counterpart, Seneca's *Oedipus* relates the struggles of the Theban king Oedipus to alleviate the bloodthirsty plague that is devastating his city. For much of the play, Oedipus remains ironically unaware of his acts of incest and patricide and their connection with the causation of the plague. These are familial transgressions, involving the debasement of the royal bloodline (*cruore semper laeta cognato domus*, 627; cf. 1022) and literal bloodshed.

The effects of Oedipus' transgressions ripple far beyond the narrow circle of his family, staining the Theban people and natural environment with a wide-sweeping trail of blood. This is evident in the first choral ode (especially 178–90), which relates the effects of the plague, including several blood-imbued symptoms that afflict the Theban populace: cheeks swollen with blood, cascading nosebleeds, and bursting blood vessels. The startling eruptions of blood are not limited to human bodies, as the local river Dirce is astonishingly said to have 'twice swirled with blood' (178). Considering Seneca's view that bodies of water are inclined to encroach upon nearby land and ultimately submerge it (*NQ* 3.29.7), one might infer that Thebes is in danger of becoming completely engulfed by blood. The fact that blood emerges so strikingly in both human bodies and in a local body of water suggests that it has in a sense already enveloped Thebes, functioning as a polyvalent substance that joins and mediates between parts of the cosmic whole.[19] Indeed, these appearances of blood in seemingly disparate bodies and domains—all of which are ultimately linked to Oedipus' acts of bloodshed and defilement of the royal bloodline—illustrate the interconnectedness of parts both small and large, exactly as one would expect in a sympathetic universe.

277

The result is an impression of inescapable physical and moral contagion,[20] as Oedipus himself gloomily articulates when assessing the devastation wrought by the plague near the beginning of the play: 'no part is immune or free from destruction' (*nec ulla pars immunis exitio vacat*, 52).

Sympathetic interactions involving blood also extend to the world of animals,[21] as is apparent in a scene in which the prophet Tiresias sacrifices a heifer and his daughter Manto examines the heifer's organs (especially 353–80).[22] In a ritual of divinatory *extispicium*, Manto observes myriad examples of bodily disturbance, including several shocking and ominous abnormalities in the heifer's internal organs and bodily fluids. Several of these peculiarities involve blood: the heifer has 'strange blood' (*novus . . . cruor*, 355) that wildly spurts out, its blood vessels have turned black, dark blood stains its entrails, and one of its lungs is suffused with blood and bereft of air. The spurting of blood from the heifer recalls the burst veins suffered by the inhabitants of Thebes in the choral ode (190), and this parallelism is reminiscent of the analogy drawn between the veins of the human body and the 'veins' of water within the earth in the *Natural Questions*.

These instances of gushing blood also prefigure Oedipus' gruesome act of self-blinding in the final act of the play (Boyle 2011: 196; Fitch 2004: 48), in which 'copious blood pours forth from torn blood vessels' (979). Oedipus' act of self-blinding is particularly bloody because he roots his eyeballs out with his own bare hands, rather than poking them out with a hairpin as in Sophocles' comparatively less bloody play. Manto also notices that the heifer's liver is frothed with 'black bile' (*felle nigro*, 358), a bodily fluid often associated with depressive and/or manic mental states in Greek and Roman medical prose,[23] thus connecting the conditions of the heifer's body with the people of Thebes as they suffer in their melancholic state of affairs.[24]

Other aspects of the heifer's body bear more specific parallels with the situation in Thebes. The fact that the liver has two heads (360) instead of the usual one can be interpreted as portents of the struggle between Oedipus' two sons for the kingship of Thebes, and the seven veins on the hostile side serve as harbingers of the 'Seven against Thebes' (Fitch 2004: 49). Furthermore, the heifer's heart, lungs, and other organs are situated in a bizarre arrangement (366–70) which embodies the turmoil of Thebes' civic body and its natural environment, and the unmated heifer's enigmatic pregnancy (371–3) appears to allude to the strange circumstances in which Oedipus unwittingly conceives children with his own mother (Busch 2007: 248–51). Shocked by the disordered state of the heifer's body and the convoluted conditions of the womb, Manto declares that 'nature has been overturned' (*natura versa est*, 371). But as she focuses only on the body immediately in front of her, Manto remains unaware of the extent to which this pronouncement holds true. While the significance of these signs eludes both Manto and Tiresias, an audience aware of the sympathetic nature of Senecan bodies and bodily fluids can grasp the broader implications at play.

Seneca's *Thyestes* is similarly saturated with notions of fluidity and language of blood, which gushes during gory violations of human bodies and triggers disturbances of a more grandiose scale. The play opens with a deeply fluid image, as the shade of Tantalus laments the accursed thirst and hunger he experiences as nearby waters and food perpetually drift away from him (*Thyestes* 1–6). The subsequent scenes portray a horrific episode in the incessant feud between the royal brothers Thyestes and Atreus, who have repeatedly quarrelled over the throne of Mycenae prior to the action of the play. Inviting Thyestes to a feast as a supposed offering of reconciliation, Atreus accomplishes a shocking and sickening measure of revenge by murdering Thyestes' three sons, dismembering them, and surreptitiously inserting parts of their body and blood in Thyestes' food and drink. The brothers' back-and-forth grasping for power and revenge yields a constantly shifting state of affairs (in other mythological sources, this one-upmanship continues beyond the actions of Seneca's play). With such a tumultuous plotline, the play is primed for metaphors of fluidity and the spilling of blood.

Fractured relations within a bloodline and bloodlust drive the plot of the play.[25] Atreus alludes to the multifaceted significance of blood when he deviously offers to Thyestes a *poculum gentile*, or 'family cup' (982–3). Since the cup is not only a family heirloom but also contains wine mixed with blood from Thyestes' sons, this is a morbidly cruel pun epitomising blood's centrality to the play.[26] A messenger who witnessed Atreus murdering and mutilating his nephews relates the extreme bloodiness of the scene, including several grisly details that provide a vivid and disturbing illustration of the permeability of the human body (especially 732–43). This a striking characteristic of Seneca's tragedies in general;[27] indeed, every extant Senecan tragedy features mutilation and/or amputation of human bodies (Most 1992: 395). When Atreus viciously stabs his third victim in this scene, the child's blood spurts onto the palace's sacrificial altars and extinguishes their flames. For scholars such as Richard J. Tarrant, this occurrence has much broader consequences, suggesting that proper ritual observance and religious piety in general have been extinguished (Tarrant 1985: 196–7.). It may also imply that Atreus' actions are violations of hospitality and the household itself, since the extinguished altar fires are situated upon the hearth, a frequent symbol of the home. Atreus' acts of bloodshed are thus not only assaults upon the bodies of his nephews but also injuries upon the royal household itself.

The messenger insinuates such effects when he observes that Atreus drove his sword out 'beyond the other side of the body' (*exegit ultra corpus*, 740) of the third child. In addition to underscoring the cruelty of Atreus' actions and the penetrability of the body, this phrase also suggests that Atreus' actions will have effects beyond the immediate corporeal injuries he is inflicting upon Thyestes' children (Tarrant 1985: 196). Indeed, the play repeatedly shows that Atreus' murders have extremely wide-ranging implications. The connection between his acts of bodily mutilation and larger-scale obliteration is particularly apparent in Tantalus' reference to the 'limbs of the mutilated household' (*lacerae domus . . . artus*, 432–3), a metaphor which grotesquely melds the limbs of the children with parts of the royal house. The phrase foreshadows Atreus' tearing apart of Thyestes' children and household alike, while also reminding the audience about the acts of murder, dismemberment, and familial rupture previously perpetrated by Tantalus himself.

These sympathetic correlations between body parts and parts of the royal house resemble the analogy between veins and rivers in the *Natural Questions*. They also resonate with rhetoric in Seneca's *Letter 95*, which discusses ethical social interaction and specifically denounces the spilling of human blood:

> Here is another inquiry: how people ought to be treated. In what way do we act? What instructions do we give? To spare human blood? What minor a thing it is not to do harm to the one whom you ought to assist! . . . Nevertheless I might bestow upon him this definition of human duty: everything that

you see, in which divine and human things are encompassed, is one: we are limbs of a great body.

Seneca, Moral Letters 95.51–2

Advocating for people to treat each other with kindness and gentleness, Seneca exhorts his readers to avoid violence and bloodshed since human beings are all limbs—or members (*membra*)—of a great body. A sympathetic rationale underlies this championing of beneficence and condemnation of bloodshed, as Seneca asserts that doing harm to another person effectively does harm to the larger body of humanity. The allusive reference to the 'limbs of the mutilated household' in *Thyestes* exhibits a highly similar rationale, and by reading these two Senecan works in tandem we can gain further insight into Seneca's characterisation techniques. If we recognise that Atreus defies the precepts outlined in this section of *Letter 95*, he appears an even more deeply antisocial character, as he violates not only his own family members but humanity itself.

Atreus' acts of bloodlust are driven by turbulent emotional flux. In Stoic thought, every emotion causes a bodily movement and triggers an action, reflecting a perceived sympathy between emotional and physical states. This causal relationship is particularly evident in Stoic discussions of anger (Vogt 2006: 58), which emphasise anger's tendency to galvanise impulsive human actions and which sometimes liken those impulses to aggressive animal behaviours.[28] It is in this manner that the messenger's speech (especially 732–43) compares slaughtering Atreus to a lion with blood-soaked jaws (*cruore rictus madidus*)—an indication that the lion has recently fed. Motivated not by ordinary reasons of hunger, but by *ira* (anger), the lion relentlessly pursues a herd of cattle; Atreus likewise thirsts for even more blood after killing two of Thyestes' children, as his uncontrollable anger swells (*ira tumet*)[29] and impels him to slaughter yet another child. The messenger emphasises the swelling of Atreus' emotions through using the verb *tumere*, which can denote flux of either a physical or emotional nature (Glare 2012: 2189 s.v. *tumere*; see also Segal 1983: 148). This verb, along with the related nouns *tumor* and *tumultus* and the adjective *tumidus*, occur repeatedly in *Thyestes*, establishing a widespread sense of instability in both physical and emotional realms.[30]

Like the bodies of water that overflow their banks and flood nearby stretches of land in the *Natural Questions*, Atreus' wild fluctuations of emotion spill over and bring about drastic emotional and physical shifts in others. This is apparent when Thyestes reports symptoms of distress shortly after consuming the feast Atreus has prepared for him (especially 957–62). It is striking that Thyestes experiences these reactions prior to realising the true nature of the feast, as if the emotional disturbances present in Atreus have spread to Thyestes in a way so subtle that Thyestes cannot detect the cause (cf. *nulla surgens dolor ex causa*, 944). Moreover, Thyestes compares these seemingly inexplicable 'disturbances' (*tumultus*, 961) to seas that swell (*tument*, 960) without any wind. Such language and imagery connect Thyestes' rumblings of indigestion and his vacillating emotions with fluctuations in the wider realm of nature, thereby accentuating the magnitude of his distress.

Much like the transgressions of Oedipus, the emotional and bodily disturbances that take place within Atreus' household forebode turbulence in the natural environment and in the cosmos at large.[31] According to the messenger, when Atreus prepared to slaughter the children on the sacrificial altars, an earthquake shook the grounds of the royal palace, a shooting star inauspiciously darted across the sky, libations of wine turned into blood, and ivory statues wept (696–702). The commotion of the geological, celestial, and supernatural phenomena in this passage is punctuated by language of flux (*fluctuanti, fluunt, flevit*) and language of shaking (*tremescit, succusso, nutavit*). The play makes frequent use of such language: there are seven instances of the verb *fluere* ('to flow'), four instances of the noun *fluctus* ('wave'), and seven instances of the noun *unda* ('wave'). The latter word is repeatedly used in contexts where upheaval in the household is envisaged as turbulence at sea.[32]

The final choral ode relates further astronomical aberrations that occur immediately before the scene in which Atreus reveals to Thyestes the cannibalistic nature of the feast. The chorus observes that the sun has reversed its course and brought about premature darkness, and that numerous constellations are about to flee from their usual positions in the sky. The sense of instability is so extreme that the chorus predicts these constellations will be plunged into the sea and submerged in a whirlpool (867–8). Since these constellations are comprised of fixed stars which should not set below the horizon (Tarrant 1985: 213), such an occurrence would conflate the sphere of the heavens and the earth before drowning them both in a whirlpool. This image of swirling fluidity vividly encapsulates the raging disorder that pervades every level of this tragedy, while also evoking the destructive deluge that Seneca predicts in Book 3 of the *Natural Questions* (Tarrant 1985: 214). Interpreted through this lens, the chorus seems to be anticipating the annihilation of the entire cosmos. One is left with the impression of a universe that is exceptionally unstable and fragile, and the extent of the devastation makes Atreus' machinations appear all the more grandiose.

In such ways these Senecan tragedies present a dynamic continuity between bodily conditions, emotional states, environmental circumstances, and celestial phenomena which collectively echo conceptualisations of embodiment and *sympatheia* in Senecan prose.[33] Across disparate genres, Seneca presents a consistent view of the human body as a permeable vessel containing fluids that are highly susceptible to imbalance, contamination, and displacement due to their constant interactions with erratic internal and external forces. He frequently uses language, imagery, and metaphors of flux to illuminate the physical and emotional instabilities that relentlessly besiege human beings, while similarly pathologising the macrocosm of the universe as it affects and is affected by action in the human sphere. In both his tragedies and prose, Seneca affords special attention to water and blood as fluids with particular capacities to reveal these intrinsic sympathies and incipient instabilities. As observed in the Introduction to this volume and by Helen King in Chapter 24, classical writers often characterise the female body as particularly moist and/or erratic; Seneca, however, emphasises the fundamental fluidity and instability of bodies in general.

This situation seems rather bleak for those striving to sustain physical health or the emotional constancy associated with a Stoic lifestyle,[34] but Seneca does supply the readers of his prose treatises certain remedies to cope with the scourge of incessant flux. Firstly, these Senecan texts fortify their readers with a keen awareness of the vicissitudes that all people face[35] as a consequence of the body's fluidity and its interconnectedness with other fluid bodies. Furthermore, Seneca's philosophical writings repeatedly prescribe the practice of philosophy itself, as in this passage from *Letter 53*:

> The power of philosophy to blunt every blow of fortune is beyond belief. No missile can settle in her body; she is protected and firm. She weakens the force

of some missiles and wards them off with the loose folds of her gown as if they are light; others she dashes aside and casts all the way back to the one who had sent them.

Seneca, *Moral Letters* 53.12

Seneca personifies *philosophia* as an embodied figure who is impervious to the blows of fortune[36] and other external forces. Unlike other bodies in Seneca's writings, the *corpus* of philosophy is firm and sealed off from the tumultuous conditions of the physical universe and the ills of the human condition. Aron Sjöblad's analysis of Seneca's letters similarly observes that 'Seneca ties wisdom to immobility, while a Stoic *proficiens* or disciple still sways or moves slightly' (Sjöblad 2015: 37; cf. *Ep.* 35.4). In Seneca's estimation, the successful cultivation of Stoic philosophy enables a person to transcend ubiquitous conditions of fluidity and thereby achieve solidity, stability, and security (cf. Seneca, *De Constantia Sapientis* especially 3.5).

Seneca similarly celebrates the power of philosophy against forces of flux in *Letter 104*, where he discusses some of his own health issues and questions his physician's ability to provide treatment that addresses the root causes of his maladies (see especially *Ep.* 104.1–2). He goes on to recommend philosophy as a universal panacea:

> But if it is pleasing to live with the Greeks, pass your time with Socrates, with Zeno: the former will teach how to die if it be necessary, the latter how to die before it is necessary. Live with Chrysippus, with Posidonius—they will pass on knowledge of things human and divine . . . they will bid you how to steel your spirit and how to raise yourself up in the face of threats. There is only one harbour for this wavering and turbulent life—to pay no heed to things that might come to pass, and to stand confidently and prepared to receive the missiles of fortune head-on, neither hiding nor holding back.
>
> Seneca, *Moral Letters* 104.21–2

As in *Letter 53*, Seneca stresses philosophy's ability to provide stability and protection against 'the missiles of fortune' (*tela fortunae*). This rhetoric, along with familiar language of flux and turbulence (e.g. *huius vitae fluctuantis et turbidae*), underlines the precarity of human life.[37] He provides even more specific advice than in *Letter 53* by recommending a few specific philosophers to study, including Zeno, Chrysippus, and Posidonius, who seem fitting recommendations because they are all early Stoics who described the universe as a fluid, sympathetic entity. Also apt is the mention of Socrates, considering the emphasis he places on the impermanence of the body in works such as Plato's *Phaedo* (e.g. *Phaedo* 115c–e). Several sections later in *Letter 104*, Seneca states that the practice of philosophy enabled Socrates to overcome the exigencies of corporeality, to abide various other hardships, and to face a death sentence with complete emotional tranquillity (104.27–8). Having called into question the authority of his doctor, Seneca recommends reading the Stoics and emulating Socrates as a more deeply curative remedy in a universe permeated with flux. This optimism about the possibility of attaining emotional tranquillity is an important aspect of Seneca's prose, but it is conspicuously absent from Seneca's tragedies, with their larger-than-life figures who experience and provoke passionate emotional impulses. This variance in outlook concerning emotional flux reveals a key difference between Seneca's literary *corpora*.[38]

Notes

* I would like to thank Victoria Leonard, Laurence Totelin, and Mark Bradley for their work bringing this volume to fruition. I also thank everyone whose work went into organising the *Bodily Fluids/Fluid Bodies in Greek and Roman Antiquity* conference, and the other participants and attendees for their valuable questions and feedback.

1 For an ancient account of *sympatheia* in the Stoic system of thought, see Diogenes Laertius' *Life of Zeno*, especially 7.140. For modern scholarship on *sympatheia* in the Stoic tradition, see Rosenmeyer 1989: 106–9 et passim; Pratt 1983: 46–8.

2 Holmes 2014: especially 123, 128–36; Rosenmeyer 1989: 110–11, 136.

3 *On the Nature of Man*, especially section 4. While principles of *sympatheia* are influential in many Hippocratic texts, this is the only treatise in the Hippocratic Corpus where these four humours are mentioned, and it was Galen who later made them canonical.

4 Lloyd 1966: 252–3; Arikha 2007: 3–6; Nutton 2013: 81–3.

5 When quoting from Seneca's *Moral Letters*, I use the textual edition of Reynolds 1965. I use Reynolds 1977 for Seneca's philosophical treatises, Hine 1996 for the *Natural Questions*, and Zwierlein 1986 for the tragedies. All of the translations from the Latin are my own.

6 Pratt 1983: 48. See also Stough 1978: 205–6; Rosenmeyer 1989: 106–7, 111, 128, 141.

7 cf. Boyle (2011: 118): 'Seneca's *natura* is inextricably entwined with the unimpedable movement of the universe.'

8 Heraclitus frag. 49a Diehls. On links between the thought of Heraclitus and the emphasis on fluids and fluctuations in Hippocratic medicine, see Nutton 2013: 81.

9 In this volume, Kelly (Chapter 16) discusses similar language in the final speech of Pythagoras in Ovid's *Metamorphoses* (see especially 15.177–8) which seems to allude to the image of Heraclitus' ever-changing river.

10 Glare 2012: 523 s.v. *cursus*. The semantics of the English derivative 'course' are highly similar.

11 Seneca also compares the veins of the human body and the 'veins' of the earth in *NQ* 6.14.1–2; cf. the Hippocratic *On Sevens*, section 6.

12 The inclusion of brains in a list of bodily fluids is striking. Seneca may have in mind the cerebrospinal fluid that can be found inside and around the human brain, or the brain's condition shortly after death when its tissues undergo autolysis, or rapid self-destruction, resulting in the liquefaction of brain cells (D'Arceuil and de Crespigny 2007: 64). It is also possible that the spongy nature of the brain may have given the impression that it is a (highly viscous) fluid. For discussion of another possible reference to the brain as a fluid (Homer, *Iliad* 20.482–3), see Kearns' chapter (Chapter 12) in this volume, pp. 195–6.

13 Seneca reiterates that the earth has passageways for air that are akin to the human body's respiratory tract in *NQ* 5.4.1–2 and 6.24.1–4, and he constructs analogies between the function of blood in the human body and the function of air in the earth's atmosphere in *NQ* 2.3.1. (cf. Williams 2005: 154–5).

14 For a discussion of dissolution as a potential consequence of embodiment in Lucretius and Ovid, see the Chapter 16 by Kelly in this volume.

15 Seneca similarly correlates the cyclical flowing and drying up of natural springs and other bodies of water with the periodic effects of gout, quartan fever, and menstruation in the human body. (*NQ* 3.16.1–3; cf. 3.30.4) On connections between the periodicity of these human phenomena and Seneca's descriptions of the seasonal flooding of the Nile and the recurrent deluge of Stoic cosmogony, see Williams 2008: 236–8. On Seneca's analogies between bodily distress and terrestrial deluge, see Inwood 2005: 173.

16 On the complex relationship between Senecan prose and Senecan tragedy, and the complicated roles of Stoic philosophy across the corpus of Senecan literature, see Pratt 1948; Boyle 2006: 198, 201; Wray 2009: 237, 254; Chaumartin 2014, especially 657–60, 668–9; Fischer 2014: 745–8.

17 For example Nussbaum 1994, especially 448–53; Fischer 2014. Earlier scholarship (e.g. Dingel 1974; Curley 1986) more often maintained that Senecan prose and tragedy should be considered separately.

18 Seneca's tragedies typically employ *cruor* and *sanguis* with very little semantic distinction. (Glare 2012: 507: s.v. *cruor*, Glare 2012: 1861–2 s.v. *sanguis*; cf. Mencacci 1986).

19 Kearns (Chapter 12 in this volume) similarly observes that blood plays extensive causal roles in Aeschylus' *Agamemnon*.

20 Seneca also uses language of infection and disease to describe the spread of moral corruption in *Letter* 7.1–2.

21 In this volume (Chapter 12) Kearns similarly discusses the tendency of Greek tragedies to correlate the spilling of blood during animal sacrifice with violence in other (often more metaphorical or abstract) spheres.

22 Busch 2007 analyses the sympathetic relationship between the conditions of the heifer's body and circumstances at Thebes (especially 227, 228, 230, 237, 249–53, 257 n. 60), including the situation of Oedipus and his later act of self-blinding (especially 259–64; see also Segal 1983: 141–2).

23 Arikha 2007: 9, 10, 57, 60, 115–16, 278; Nutton 2013: 84–5.

24 Oedipus describes the miserable state of his people in the play's opening scene (57–70, especially 58–9).

25 cf. Kearns' discussion in this volume (Chapter 12) of blood as a marker of kinship in Greek tragedy (including examples involving the house of Atreus).

26 Tarrant 1985: 227. A similar juxtaposition of wine and blood occurs in Seneca's *Oedipus* (324), where Manto observes a libation of wine ominously transform into blood shortly before she examines the body of the heifer.

27 On Seneca's preoccupation with corporeal integrity and 'primary boundary anxiety', see Segal 1983: 149 et passim.

28 On connections between anger and animality in Seneca's letters, see Sjöblad 2015: 20.

29 In *On Anger* 1.20.1, Seneca describes the onset of *ira* in similar language of swelling and likens this occurrence to disease rampantly spreading through a body.

30 On connections between bodily suffering and mental distress in Seneca's letters, see Sjöblad 2015: 28 et passim.

31 On connections between human transgressions and cosmic disruption in Senecan tragedy, see Volk 2006: especially 90.

32 Seneca uses similar imagery of flux in describing human life as a 'sea of troubles' in *Consolation to Polybius* 9.6.

33 On philosophical connections between the physics and ethics of the *Natural Questions*, see Williams 2012: 2–5, 11, 71.

34 On the tension between Seneca's fixation on bodily suffering and his Stoic ethics, see also Segal 1983: 155; Edwards 1999: 255.

35 Vottero 1989: 39; Scott 1999: especially 63–8; Fedeli 2000: 45.

36 On the importance of overcoming the vicissitudes of *fortuna* in Stoic philosophy, see Sjöblad 2015: 38–41.

37 cf. *Letter* 95.57. Here Seneca asserts that *tranquillitas* eludes most people because they lack sound judgement and waver (*fluctuantur*) in their decision making.

38 On the point that Seneca's tragedies aim to test the limits of Stoic views concerning the passions, see also Nussbaum 1994: 448–53; Wray 2009: especially 254.

Bibliography

Primary sources

Hine, H. M. (1996) *L. Annaei Senecae: Naturalium Quaestionum Libros*. Stuttgart and Leipzig: Teubner.

Reynolds, L. D. (1965) *L. Annaei Senecae: Ad Lucilium Epistulae Morales*. Oxford: Oxford University Press.

———. (1977) *L. Annaei Senecae: Dialogorum Libri Duodecim.* Oxford: Oxford University Press.

Zwierlein, O. (1986) *L. Annaei Senecae: Tragoediae.* Oxford: Oxford University Press.

Secondary literature

Arikha, N. (2007) *Passions and Tempers: A History of the Humours.* New York: Harper Collins.

Boyle, A. J. (2006) *Roman Tragedy.* London: Routledge.

———. (2011) *Seneca: Oedipus: Edited with Introduction, Translation, and Commentary.* Oxford: Oxford University Press.

Busch, A. (2007) 'Versane natura est? Natural and linguistic instability in the extispicium and self-blinding of Seneca's Oedipus', *Classical Journal* 102, 225–67.

———. (2009) 'Dissolution of the self in the Senecan corpus', in S. Bartsch and D. Wray (eds) *Seneca and the Self.* Cambridge: Cambridge University Press, 255–82.

Chaumartin, F. (2014) 'Philosophical tragedy?', in G. Damschen and A. Heil (eds) *Brill's Companion to Seneca: Philosopher and Dramatist.* Leiden: Brill, 653–69.

Curley, T. F. (1986) *The Nature of Senecan Drama.* Rome: Edizioni dell'Ateneo.

D'Arceuil, H. and A. De Crespigny (2007) 'The effects of brain tissue decomposition on diffusion tensor imaging and tractography', *Neuroimage* 36(1), 64–8.

Dingel, J. (1974) *Seneca und die Dichtung.* Heidelberg: C. Winter.

Edwards, C. (1999) 'The suffering body: Philosophy and pain in Seneca's *Letters*', in J. I. Porter (ed.) *Constructions of the Classical Body.* Ann Arbor: University of Michigan Press, 252–68.

Fedeli, P. (2000) 'Seneca e la natura', in P. Parroni (ed.) *Seneca e il suo tempo: Atti del convegno internazionale di Roma-Cassino 11–14 novembre 1998.* Rome: Salerno, 25–45.

Fischer, S. E. (2014) 'Systematic connections between Seneca's philosophical works and tragedies', in G. Damschen and A. Heil (eds) *Brill's Companion to Seneca: Philosopher and Dramatist.* Leiden: Brill, 745–68.

Fitch, J. G. (2004) *Seneca: Oedipus, Agamemnon, Thyestes, Hercules on Oeta, Octavia.* Cambridge, MA: Harvard University Press.

Glare, P. G. W. (2012) *Oxford Latin Dictionary.* Oxford: Oxford University Press.

Holmes, B. (2014) 'Proto-sympathy in the Hippocratic Corpus', in J. Jouanna and M. Zink (eds) *Hippocrate et les hippocratismes: Médecine, religion, société: Actes du XIVe Colloque International Hippocratique.* Paris: Académie des Inscriptions et Belles Lettres, 123–38.

Inwood, B. (2005) *Reading Seneca: Stoic Philosophy at Rome.* Oxford: Oxford University Press.

Lapidge, M. (1978) 'Stoic cosmology', in J. M. Rist (ed.) *The Stoics.* Berkeley: University of California Press, 161–85.

Lloyd, G. E. R. (1966) *Polarity and Analogy: Two Types of Argumentation in Early Greek Thought.* Cambridge: Cambridge University Press.

Mencacci, F. (1986) 'Sanguis/cruor: Designazioni linguistiche e classificazione antropologica del sangue nella cultura romana', *Materiali e discussioni per l'analisi dei testi classici* 17, 25–91.

Most, G. (1992) '*Disiecti membra poetae*: The rhetoric of dismemberment in Neronian poetry', in R. Hexter and D. L. Selden (eds) *Innovations of Antiquity.* New York: Routledge, 391–419.

Nussbaum, M. (1994) *The Therapy of Desire: Theory and Practice in Hellenistic Ethics.* Princeton, NJ: Princeton University Press.

Nutton, V. (2013) *Ancient Medicine.* 2nd edn. London: Routledge.

Pratt, N. T. (1948) 'The Stoic base of Senecan drama', *Transactions and Proceedings of the American Philological Association* 79, 1–11.

———. (1983) *Seneca's Drama.* Chapel Hill: University of North Carolina Press.

Rosenmeyer, T. G. (1989) *Senecan Drama and Stoic Cosmology.* Berkeley: University of California Press.

Scott, J. (1999) 'The ethics of the physics in Seneca's *Natural Questions*', *The Classical Bulletin* 75(1), 55–68.

Segal, C. (1983) 'Boundary violation and the landscape of the self in Senecan tragedy', *Antike und Abendland* 29, 172–87.

Sjöblad, A. (2015) *Metaphorical Coherence: Studies in Seneca's Epistulae Morales*. Lund: Lund Centrefor Languages and Literature, Lund University.

Stough, C. (1978) 'Stoic determinism and moral responsibility', in J. M. Rist (ed.) *The Stoics*. Berkeley: University of California Press, 203–31.

Tarrant, R. J. (1985) *Seneca's Thyestes*. Atlanta: Scholars Press.

Vogt, K. (2006) 'Anger, present injustice and future revenge in Seneca's *De Ira*', in K. Volk and G. Williams (eds) *Seeing Seneca Whole: Perspectives on Philosophy, Poetry and Politics*. Leiden: Brill, 57–74.

Volk, K. (2006) 'Cosmic disruption in Seneca's *Thyestes*: Two ways of looking at an eclipse', in K. Volk and G. Williams (eds) *Seeing Seneca Whole: Perspectives on Philosophy, Poetry and Politics*. Leiden: Brill, 183–200.

Vottero, D. (1989) *Questioni Naturali di Lucio Anneo Seneca*. Torino: Unione tipografico-editrice Torinese.

Williams, G. (2005) 'Interactions: Physics, morality, and narrative in Seneca *Natural Questions* 1', *Classical Philology* 100(2), 142–65.

———. (2008) 'Reading the waters: Seneca on the Nile in *Natural Questions*, book 4A', *The Classical Quarterly* 58(1), 218–42.

———. (2012) *The Cosmic Viewpoint: A Study of Seneca's Natural Questions*. Oxford: Oxford University Press.

Wray, D. (2009) 'Seneca and tragedy's reason', in S. Bartsch and D. Wray (eds) *Seneca and the Self*. Cambridge: Cambridge University Press, 237–54.

18

BODILY FLUIDS, GROTESQUE IMAGERY, AND POETICS IN PERSIUS' *SATIRES*

Andreas Gavrielatos

From its outset, Roman satire prioritises senses of solidity, dryness, and masculinity and values what is firm, in other words, what is not fluid. Images of fluidity are then naturally employed in order to create caricatures, mock, and make references to the degradation and the decay of the targets. In satire's moral criticism, fluidity is associated with luxury, effeminacy, pleasure, and corruption, while contemporary poetic criticism offers images of fluidity that suggest notions of smoothness, either depicted with distortive images of delivery or which are apparent in the satirist's stylistic techniques (Edwards 1993: 174). In both cases, fluids from the body are a common choice in the imagery of fluidity. Images of fluidity not only involve liquids in their literal sense but also figurative expressions of fluidity, frequently expressed with the use of *fluere* ('to flow').[1] The application of the degraded quality of fluidity on the body has the side-effect of initiating or enhancing a grotesque depiction, while the target is ridiculed.[2] Although the effects of the bodily grotesque in Roman satire have been examined and theorised, the exploitation of bodily fluids in this direction have not been studied in their own right before (see also Centlivres Challet, this volume, Chapter 8, on Juvenal). Looking at the literary exploitation of the fluids of the body, this chapter examines their role in satire's use of grotesque imagery, with a focus on Persius' *Satires*.

Persius (34–62 CE) wrote his book of *Satires*, which was prefaced by a fourteen-verse prologue, under Nero. He targets the moral corruption and the poor literary taste of his time and follows the Roman satirical tradition of Horace and Lucilius. The most obvious example of Persius' interaction with fluidity is that the avoidance of smoothness resulted in the obscurity of his style and imagery.[3] Shadi Bartsch (2015: 149–60) argues that Persius' harsh and obscure style and metaphors—his *acris iunctura* (sharp connection)—are essential for his poetics, as they demonstrate the extremes of Persius' thesis about poetics. Images of fluidity are used to ridicule, while images of bodily fluids create caricatures of Persius' targets. This also accords with another quality of Persius' *Satires*: his predilection for rustic simplicity (*rusticitas*). In 1981, Johan C. Zietsman investigated Persius' programmatic self-presentation as half countryman (*semipaganus*) and half rustic poet in the prologue as a demonstration of a departure from the imitation of Greek norms and practices in favour of a genuine Roman style of writing. He concludes that Persius 'urges his contemporaries to rediscover the old traditional values' (Zietsman 1981: 56). *Rusticitas* is the concept that encapsulates

DOI: 10.4324/9780429438974-24

these values. In an article informed by Tibullus' poems (first century BCE), Spyridon Tzounakas (2006: 111–12) interprets Persius' *rusticitas* as containing the Roman ideals of manliness, simplicity, and frugality, in contrast to the luxury and the effeminacy found in urban refinement (*urbanitas*). From a literary perspective, the manliness finds its expression in Roman satire and effeminacy in Callimachean elegy (Tzounakas 2006: 119–23). Consequently, the same dichotomy applies to the poetic themes and style; Persius' *Satire* eschews the erotic themes of his contemporaries—the mellifluous and smooth style, and the liquid sounds—in favour of a *sermo* ('discourse'), which is a realistic and everyday poetry that develops rather than imitates. The metaphorical language that is the backbone of fluid imagery is in line with Persius' style of programmatically avoiding smoothness and effeminacy (Bartsch 2015: 164–5). Finally, this also aligns with Stoic philosophical principles that denounce passions and favour philosophical rusticity instead of smoothness (see also Goyette, Chapter 17, this volume; Bartsch 2015: 152–5; Tzounakas 2006: 125).

With such a multidimensional programme, Persius' grotesque imagery is vivid and often enhanced with a sense of fluidity. When it comes to his moral criticism, images of fluidity are imbued with a sense of distortion and instability and are integral to the construction of the satirist's social criticism. Additionally, Persius uses notions of fluidity in his literary criticism, when he disparages the style and the themes of his contemporary authors and their audience (e.g. the liquefaction of bodies in Ovid; see Kelly, Chapter 16, this volume). These two levels of the notion of fluidity highlight its significance and effectiveness for his satirical purposes. Persius' medium for the creation of this imagery is very often the body, and in most cases it is achieved with the employment of bodily fluids. This chapter presents a comprehensive approach to the exploitation of the imagery of bodily fluids and fluidity in the satires before focusing on the employment of the two most frequent cases: saliva and sweat. The examination of this technique will result in a better understanding of both Persius' unique style and the presence of bodily fluids in Roman literature.

Imagery of fluidity and bodily fluids in Persius' *Satires*

Body imagery is prominent in satire: the bodily grotesque creates caricatures and facilitates mockery or evokes moral criticism. Using Bakhtinian terms, Paul A. Miller (1998: 259–60) defines the grotesque bodies in satire as 'negative creations' of degradation. Often, the body betrays the target's moral decay, passions, and weaknesses. In the series of episodes that Persius includes in his satires, we rarely hear the characters' voices. Instead, we are offered descriptions of their actions through an imagery where the body is the protagonist. Degrading the body by presenting it as corrupted and depraved is not merely a means for mockery but also one of Persius' ways of exposing the truth. In his imagery, his targets are deprived of their elaborate appearances and public images, thus revealing their true characters.[4] Persius explicitly stresses the idea in 4.43–5: *ilia subter | caecum vulnus habes, sed lato balteus auro | praetegit* (below your loins you have a hidden wound, but a belt with its golden band conceals it). In the process of satiric degradation, fluidity participates in two ways. First, there are images of fluids that are applied to the body, implying an external source or influence that has a degradating effect. Second is the fluidity that springs from the body, thus indicating the inherent qualities of the target.

288

Fluidity in Persius' *Satires* is comprised of three main types of fluidity: these are always associated with human actions, feelings, or beliefs, and are organised here in Tables 18.1–3. Fluidity appears either in a literary sense in the form of liquids or in a metaphorical sense as a descriptor for style, and it appears twice with the use of *fluere* (flow).[5] The majority of these references are linked to the body, sometimes directly, as demonstrated in Table 18.2, and at other times indirectly, as liquids 'applied' to the body, set out in Table 18.1.

Before we address the special function of bodily fluids, an examination of the role of external fluidity and how it is associated with the body will help us to establish the

Table 18.1 Liquids linked to the body: external applications

Function	Fluid	Passage	
Artificial intelligence	Nature: Spring (Hippocrene)	*Prol.* 1 (edition Clausen 1992)	Nec **fonte** labra **prolui caballino** (I neither **washed** my lips in the nag-spring)
That intoxicate/distort	Gargle	1.17–18	**liquido** cum **plasmate** guttur I mobile **conlueris** (when you have **rinsed** your lively throat with a **liquid gargle**)
	Wine	1.30–1	inter pocula quaerunt I Romulidae saturi (the sons of Romulus, stuffed, are enquiring **between drinks**)
		3.93	lenia loturo sibi **Surrentina** rogabit (he will ask for light **wine** to drink while bathing)
	Hellebore	1.50–1	non hic est Ilias Atti I **ebria veratro**? (isn't Attius' *Iliad* **soaked with hellebore**?)
	Hellebore	3.63–4	**elleborum** frustra . . . I poscentis videas (you will see that you ask for **hellebore** in vain)
	Hellebore	4.16	**Anticyras** melior **sorbere** meracas (you'd rather **gulp down** undiluted **hellebore**)
	Hellebore	5.100–1	diluis **elleborum** . . . ? (do you mix **hellebore**?)
	Vinegar	4.32	pannosam faecem morientis **sorbet aceti** (he **gulps down** the tattered dregs of expiring **vinegar**)
	Oil	3.44	saepe oculos . . . tangebam parvus **olivo** (when I was little . . . I often used to dab my eyes with **oil**)
That clean	Decoction	1.125	aspice et haec, si forte aliquid **decoctius** audis (have a look at this too, if by chance you are listening to something better **boiled down**)
	Nature: River (Tiber)	2.15–16	**Tiberino** in **gurgite** mergis I mane caput . . . et noctem **flumine** purgas (you immerse your head into the **Tiber's flood** . . . and wash away the night in the **river**)

Table 18.2 Liquids linked to the body: bodily fluids

Function	Fluid	Passage	
Oral fluids and mouth fluidity	Saliva (direct reference)	1.104–5	summa delumbe **saliva** I hoc natat in labris (this spineless stuff swims on foaming **saliva**)
	Saliva (direct reference)	2.32–4	uda labella I infami digito et lustralibus ante **salivis** I expiat (his forehead and wet lips she purifies with her wicked middle finger and magical **spittle**)
	Saliva (direct reference)	5.111–12	inque luto fixum possis transcendere nummum I nec gluttu sorbere **salivam** Mercurialem? (and could you pass over a coin stuck in the mud and not soak up Mercury's **spittle** in one gulp?)
	Throat (see also Table 18.1)	1.17–18	liquido cum plasmate **guttur** I mobile conlueris (when you have rinsed your lively **throat** with a liquid gargle)
	Mouth	1.35	**tenero** subplantat verba **palate** (he stumbles over the words on his **delicate palate**)
	Tongue	1.80–1	quaerisne unde haec sartago loquendi I venerit in **linguas** (do you ask from where this hotchpotch of words has come to their **tongues**?)
	Catarrh	2.57	somnia **pituita** qui purgatissima mittunt (who send us dreams mostly purged from **catarrh**)
Excitement	Sweat	2.53–4	**sudes** et pectore laevo I excutiat guttas laetari praetrepidum cor (you would **sweat** and from your left breast your fluttering heart would let out drops)
		3.47	pater adductis **sudans** audiret amicis (my father listened in a **sweat** with his gathered friends)
Runny-eyed ancestors	Tears/runny eyes	1.79–80	hos pueris monitus patres infundere **lippos** I cum videas (when you see the **runny-eyed** fathers pour this sort of advice into their sons)
Runny-eyed descend-ants		2.71–2	quin damus id superis, . . . quod . . . I non possit magni Messalae **lippa** propago? (why don't we give to the gods that which the **runny-eyed** descendant of great Massala cannot ?)
Irritability	Bile	3.8	turgescit vitrea **bilis** (my greenish **bile** is swelling)
Disrespect	Urine	1.113–14	pueri, sacer est locus, extra I **meiite** (guys, this place is sacred, go **piss** elsewhere!)
Boils/anger	Blood	3.116–17	nunc . . . fervescit **sanguis** et ira I scintillant oculi (now the **blood** boils with passion and the eyes spark with anger)
Ejaculation (figura-tively)	Semen	1.18	**patranti** fractus **ocello** (broken with a **sexually aroused eye**)

satirical significance of fluidity in the way that it distorts and produces caricatures, as reproduced in Table 18.1. The imagery is created with the use of liquids and fluids that (a) are found in nature, (b) are 'inserted' into the body or are absorbed (in the case of the oil) and have a distorting and intoxicating character, and (c) have cleansing and favourable features.

Table 18.3 Poetic fluidity

Notion of Fluidity	Passage	
Liquidity in rhythm/themes	1.33–5	rancidulum quiddam balba de nare locutus ǀ Phyllidas, Hypsipylas, vatum et plorabile siquid, ǀ eliquat (saying something nauseating through his lisping nose, he strains his Phyllises, his Hypsipyles, something dismal of the bards)
	1.63–5	'quis populi sermo est?' quis enim nisi carmina **molli** ǀ nunc demum numero **fluere**, ut per **leve** severos ǀ **effundat** iunctura unguis? ('What do the people say?' What else indeed but that poetry now at last **flows** in **smooth** measure, so that **smoothly glide** over the junctures the rough fingernails)
	1.105	in labris et in **udo** est Maenas et Attis (on the lips and in **wetness** *Maenas* and *Attis* float)
Influence	1.79–80	hos pueris monitus patres **infundere** lippos ǀ cum videas (when you see the **runny-eyed** fathers pour this sort of advice into their sons)
Medium	3.12–14	tum querimur **crassus** calamo quod pendeat **umor**. ǀ nigra sed infusa vanescit **sepia** lympha, ǀ dilutas querimur geminet quod fistula guttas (then we complain because the **ink** is hanging thick from the **nib**. But the black **ink** mixed with water is thinned down, and then we complain because the pipe doubles the diluted drops)

The fluidity's external source in the cases above means that it modifies the body, that its character is a distortive one, and that it has a negative effect. The body is the projection of these effects, and its distorted character is based on the absorption of fluidity. The only exception is the case of the *decoctius aliquid* (1.125).[6] This liquid has a positive character as a metonymy for Persius' work that stresses its superiority over the conventional poetry of his contemporaries. Contrariwise, the liquid gargle that the reciter uses to clean his throat in 1.17 results in his effeminate recitation, making him the caricature of the image. The use of liquids can also indicate prodigal pleasures, and accusations of luxury go hand in hand with those of effeminacy (Edwards 1993: 174–6). Hellebore has a special role in this imagery; there are four references to it throughout Persius' *Satires*, which have recently been discussed collectively by Tzounakas. To summarise Tzounakas' study, the reference in the first satire denounces the poetic composition of his time as the product of artificial inspiration, which is also attacked with the use of Hippocrene (*Prologue* 1). The references in 3.63, 4.16, and 5.100 point to insanity and function within the paradigm of medical imagery, common in Stoic thought, yet their efficacy is doubted by Persius when compared to 'sound philosophical principles' (Tzounakas 2014: 190–1). The reference to hellebore in *Satire* 3.63–4 ('you will see that you ask for hellebore in vain') also appears in relation to a skin swollen by the bodily humours as the result of dissolute practices (see Freudenburg 2018b: 277–8). The thesis offered here, that the external application of liquids (Table 18.1) implies that a new quality has been imposed on the characters, depriving them of their authenticity, is further corroborated by the function of hellebore in *Satire* 1. The metaphorical intoxication of Attius' *Iliad* is at the same time part of Persius' literary criticism and an indication of effeminacy (see Tzounakas 2014: 192–5).

For Roman satirists, effeminacy is a common target, and although treated with exaggeration, it often becomes the means for denoting the degenerated body (Williams 2010: 287–8; Rosen and Keane 2013: 390–2). The discussion of these examples stressed the significance of fluidity's association with effeminacy in satire that offers the framework within which Persius uses fluids for satirical ends. Images of fluidity deviate from the ideal stability and masculinity of *Romanitas* and are best manifested through their application on the body. The exploitation of bodily fluids contributes to the role of fluidity in this paradigm through the creation of caricatures that support Persius' mockery and facilitate his criticism. Just as alimentary and medical images and metaphors place the body into the grotesque regime, so too does fluidity, not least through the use of bodily fluids, since they constitute its manifestations (see Hudson 1989; Gowers 1993: 109–219; Bartsch 2015). In imagery and metaphors of external fluidity, the grotesque is bestowed upon the body and the satirist's attack targets its source, whereas with bodily fluids the grotesque springs from the inside; its cause is the body itself. Table 18.2 lists Persius' references to bodily fluids.

Of the bodily fluids that appear in Persius' *Satires* (Table 18.2), those related to oral fluids and the mouth, with direct references to saliva and its implicit uses, dominate the imagery. The most exquisite sense of fluidity in this regard is found in the description of the poet's preparation for recitation in 1.17–18. I will return to these references in the next section. Overall, the fluidity attached to the otherwise solid mass of a body results in the augmentation of its grotesquerie and stresses its corruption.[7] Images of bile and swollen bodies are enriched with a sense of instability and vice, which requires introspection in order to be resolved. Once again, the imagery of bodies distorted by their own fluids is indicative of their moral disease, of which the bodily fluids are the signs; they have broken out of swollen bodies (Freudenburg 2018b: 267–8). The idea of corruption falls in line with the social critique that Persius shapes in his first satire, where Persius accuses his contemporaries of moral decay and employs charges of effeminacy, luxury, and vanity. Moreover, social and literary criticism go hand in hand, through interwoven images and multi-levelled semantics (see Ferriss-Hill 2012: 379–85). Bodily fluids participate in these techniques and allow the faults to appear, intensifying the degradation of Persius' targets.

At this point, it is worth outlining how the notion of fluidity permeates Persius' literary criticism before we evaluate the application of bodily fluids in the respective imagery. A direct reference to this fluidity occurs in Persius' response to his interlocutor's question about public opinion:

'Quis populi sermo est?' quis enim nisi carmina molli
nunc demum numero **fluere**, ut per leve severos
effundat iunctura unguis.

'What do the people say?' What else indeed but that poetry now at last **flows** in smooth measure, so that the rough fingernails glide smoothly over the junctures.

Persius, Satires 1.63–5

The metaphorical use of the verb *fluere* encapsulates all of Persius' criticism on contemporary poetry. The obvious reference here is to smoothness, a product of poetic artificiality that alters the truth and the real meanings, resulting in a corrupted style. A smoothness in the style can be achieved, among other ways, simply with the use of 'liquid' sounds, based on the letters l, m, n, and r, which Persius systematically associates with the poetry he disdains (Freudenburg 2001: 6). In fact, there are many references to fluidity that underlie writing practices and style; these are collected in Table 18.3.

Yet Persius' attacks on stylistic choices need to be understood on another level. Of the references listed, perhaps the most revealing one is the use of ink in 3.12, as ink is the medium of writing—a material depiction of poetic composition. The verbal depiction of this image also resembles a bodily fluid (*umor*), which is thick and unmanageable (Freudenburg 2018b: 272). This is not a random choice; the importance of bodily fluids in Persius' literary criticism enriches the imagery and supports its links with moral criticism. As Bartsch (2015: 151) stresses with regards to Persius' obscurity,

> Like almost all issues connected to style, *compositio* too was charged with ethical and sexual ramifications; there existed so-called smooth combinations and harsh combinations of both sounds and syllables, smoothness treading a dangerous line in its proximity to effeminate softness and overnicety, harshness in turn the tool of those who wanted to seem stern and rugged, old-fashioned, and averse to the sweetness of euphony.
>
> Bartsch 2015: 151

The framework established so far enables the examination of images of bodily fluids for their participation in the social satirical caricature and their function at a metapoetic level as an indication of fluidity in the author's style. Having set this premise, I will discuss the function and effectiveness of two examples of bodily fluids employed by Persius: the fluidity of the mouth and the images of sweating bodies.

A mouth full of saliva

A pivotal theme in Persius' composition is the social interaction that involved the sending and receiving of a message. His criticism is built on how this social function shapes characters at a moral level, while it affects the literary production. This is articulated through images of poetic recitation, where the poet is the one who sends a message and the audience is the receiver. This image is not merely effective in attacking the poetry itself, but sexual innuendos position the poet as the homosexual penetrator and the audience as the receiver (see Miller 1998: 276; Freudenburg 2001: 171–2). The body is again crucial in this imagery: the poet's mouth (and by extension the throat) is the medium for expression, sending a message, articulating ideas, and reciting poetry; in short, the medium for one's projection of oneself. On the other hand, there is an emphatic use of the ear as the medium for receiving a message and judging it. As a pure, cleansed ear is associated with a clear mind and judgement, so an undistorted mouth is the one that tells the truth, but in satire this ideal communication needs to be reversed, and the notion of fluidity serves this purpose. Whereas the ear requires liquid medicine from an external source to be administered in order to clean it, saliva is the prevalent cause of distortion of the mouth. As shown by Table 18.2, direct references to saliva or oral fluids make up most of the citations of bodily fluids. Lines 1.17–18 are the most indicative example for our purposes: the corruption of the poet is owed to the fluidity that has an external source. The following recitation is ready for the penetration of the audience's ears/loins.

Besides the creation of the grotesque, saliva operates in this imagery at another, more visible level. Caricatures are founded on exaggeration. The over-elaboration and smoothness of style has charged contemporary poetry with copious Graecisms and 'easy gliding liquids', and l, m, n, and r are apparent in verbal expressions as well as in the themes of *Phyllis*, *Hypsipyle* (1.34), and *Callirhoe* (1.134) (See Freudenburg 2001: 177, 183). Emotive, smoothed, mellifluous, and effeminate themes like these would also require the alternation in voice to sound 'more feminine'. The struggle for enunciation results in 'a mouth full of saliva'; the transmission of the corrupted choices is complete, and the fluidity is passed on to—or rather spat at—the audience. At this point, Persius' imagery of mouth fluidity can be informed by Roman oratory and what was considered to be good rhetorical practice. The significance of delivery was stressed in a big part of *Rhetoric for Herennius* (3.11–14), a treatise on the effectiveness of oratory and persuasion, formerly attributed to Cicero. Quintilian (12.10.27–39, edition Winterbottom 1970) comments extensively on the voice and, in particular, on the 'sweetness' of Greek sounds. He particularly stresses the inclusion of Greek names in poetry for the sake of their sound, in contrast to the harsher sounding Latin. James Adams (2003: 108) deduces from Quintilian 12.10.57 that Greek names sometimes maintained their pronunciation in prosody. Adams (2003: 108–10) stresses the prestige of Greek pronunciation, the effect of aspiration in Greek words, and the reception of the 'sweet' sounds of upsilon and zeta, based again on Quintilian (see also Biville 1995). The evidence suggests that Greek words are included in order to increase smoothness in the poet's style, a practice which, in addition to the Greek themes of the poetry, would invoke Persius' criticism. Furthermore, the solidity of the voice that articulates ideas appropriately becomes the paradigm for the approved rhetorical practice. On the contrary, it is the emotions and passions that affect the voice and make it fluid. Although an emotive tone may be appreciated in specific circumstances, a distorted and heavily emotional tone is bereft of solidity and has become emasculated.[8] This notion of effeminacy as the result of distortion in the delivery is what satire has in common with rhetoric, and it serves its purposes. In practice, it results from a luxurious and extravagant tenderness of the voice, and its manifestation finds its place in the image of saliva.[9]

I will further elaborate the correlation between effeminate themes and effete recitations with two extracts from Persius' first satire.

> (a) ecce inter pocula quaerunt
> Romulidae saturi quid dia poemata narrent.
> Hic aliquis, cui circum umeros hyacinthyna laena est,
> rancidulum quiddam balba de nare locutus
> Phyllidas, Hypsipylas, vatum et plorabile siquid,
> eliquat ac tenero subplantat verba palato.
>
> Look! over their cups are enquiring
> the sons of Romulus—stuffed full—what's new from divine poesy.

Then someone, with a hyacinth mantle draped around his shoulders,
lisping something nauseating through his nose, pours out
his Phyllises, his Hypsipyles, and any other gloomy stuff of the bards
and stumbles over the words on his delicate palate.

Persius, Satires 1.30–5

(b) summa delumbe **saliva**
hoc natat, in labris et in udo est Maenas et Attis

This spineless stuff swims on foaming saliva, on the lips and in wetness *Maenas* and *Attis* float.

Persius, Satires 1.104–5

In these two passages, saliva, and mouth fluidity more broadly, has a grotesque effect, distorting the depiction of the messenger and by extension the message itself, that is, the poet and his recited poetry. It is not free from defect because it is an exaggerated struggle to communicate ideas in an artificial style overloaded with pompous expressions that results in an excessive and flamboyant recitation. It is this elaboration of style that distorts the articulation of the message, since it removes the message from its true, relevant, and pure character. The style incorporates smooth sounds and Greek themes as well as Graecisms in its expression (see e.g. Zietsman 1981: 55; Tzounakas 2008a). In a similar vein, mythological themes of smooth and effeminate erotic elegy have the same effect (Miller 2013: 218–19). The smoothness endows the saliva with meta-poetic colouring and is included in Persius' manifestation in favour of a less elaborate and more honest, solid, and masculine style. Effectively, the audience—the sons of Romulus—become intoxicated, and the poet is caricatured; notions of effeminacy are apparent in the scene depicted in the first extract and especially in the description of the bard and his poetry (see Bramble 1974: 101; Bartsch 2015: 38).

Besides the imagery, the disdain for fluidity is further stressed through Persius' stylistic choices. Drawing examples again from the previous extracts, line 35 is perhaps the most indicative in this regard. There is a very careful arrangement of the liquid letters l and r, one in each word in alternate order: *eliquat ac tenero sublantat verba palato*. This comes after the Greek sweetness that is achieved with the two y's in the names of the denounced themes, *Phyllis* and *Hypsipyle*, which are described as *plorabile*, a rare adjective that is not only lugubrious in its connotation but also in its liquid sound (Bramble 1974: 104; Lee and Barr 1987: 72; further on tears, see Kelly, this volume). The fluidity in this line indicates the smoothness of the poetry and emphatically becomes bodily with the use of *tenero palato* at its climax: an appropriately delicate mouth for the soft and effeminate elegy.[10] The second passage is a salient example of poetic criticism embedding the saliva metaphor. The poetry is *delumbe* in its theme, but it is articulated through a medium with a distinctive grotesquerie, which Persius delineates with the use of mouth fluids.

A similar manifestation of fluidity with the use of liquid letters and sounds is to be found in 1.79–82:

hos pueris monitus patres infundere **lippos**
cum videas, quaerisne unde haec sartago **loquendi**

venerit in **linguas**, unde istud dedecus in quo
trossulus exultat tibi per subsellia levis?

when you see the **runny-eyed** fathers pour this sort of advice
into their sons, need you ask where this hash of language
came from into their tongues, or this monstrosity that
stirs the blood of your gentlemen along the benches?

Persius, *Satire* 1.79–82

The runny-eyed fathers affected with bodily fluids are the cause of the fluidity in both the poetry and the style of Persius' contemporaries. The repetition of the liquid l in the keywords of each line (*lippos, loquendi, linguas*) finds its climax in the alteration of l and s in the last line of the extract. At its climax, the effect is emphasised with the use of the soft s.

Direct imagery of saliva occurs at a different level in two more instances, which are more associated with divine powers than with poetic composition and performance.

ecce avia aut metuens divum matertera cunis
exemit puerum frontemque atque uda labella
infami digito et lustralibus ante **salivis**
expiat, urentis oculos inhibere perita

Look! a grandmother or fearful aunt has taken out
the baby boy from his cradle, and his forehead and wet lips
she purifies with her wicked middle finger and magical spittle,
experienced at preventing the Evil Eye.

Persius, *Satires* 2.31–4

inque luto fixum possis transcendere nummum
nec gluttu sorbere **salivam** Mercurialem?

And could you pass over a coin stuck in the mud
and not soak up Mercury's spittle in one gulp?

Persius, *Satires* 111–12

In both extracts, the references to saliva convey Persius' disdain for contemporary society (Bramble 1974: 102). The first comes from a satire that attacks Roman religious practices of praying, which stem from superstition rather than Stoic principles.[11] The use of the middle finger and the saliva were related to the Evil Eye, as they both had 'magical qualities'.[12] The combination of the two is a common practice assigned mainly to elderly females; Petronius (*Satyricon* 131.4) also provides a similar image of the practice: *mox turbatum sputo pulverem medio sustulit digito frontemque repugnantis signavit* (then she took with her middle finger dust mixed with spittle and made a mark on my forehead, although I resisted) (see more in Nelson 2017: 77–8; see also Parker, this volume). The opposition of this practice to Stoic doctrines provokes Persius' attack, and it is within this context that the saliva creates the grotesque that transforms the targets into satirical caricatures. Therefore, it is perhaps not a coincidence that saliva is employed again in *Satire* 5, the most Stoic one. The grotesque here is attached to the greed that is emphasised by the adjective *Mercurialem* (Harvey 1981: 158; Kissel 1990: 682–3). Persius mocks the dependence of humanity on gold and money, which results in the lack of true freedom. This dependence, the feeling of obsession that chains a human to greed, obliges them to pick up a coin from the mud, metaphorically making money through humble or impure actions. This escalates into a grotesque image, the drinking of Mercury's saliva. The grotesque is achieved with the *iunctura acris*, a paradox between the god of wealth and prosperity and his mouth fluid (Harvey 1981: 158). Once again, Persius uses saliva in his social criticism as the bodily fluid upon which his caricatures are constructed.

Nonverbal behaviour: sweat

In recent criticism, there has been an increased interest in nonverbal behaviour in Latin literature, with a focus primarily on gestures (Corbeill 2004; Nelson 2017). At the same time, readings on Roman satire have been greatly informed by theories on the relations between the body and food, as well as on the bodily grotesque (Miller 1998; Bartsch 2015). In 2008, Christina Clark included the category of 'affect displays' as one of the types of nonverbal behaviour, which she defined as 'autonomic responses of the nervous system such as sweating, pallor or trembling, and types of impulsive but not entirely uncontrollable responses, such as weeping' (Clark 2008: 258, after Lateiner 1992: 257). Affect displays are often found in Roman satire, and they result in the creation of caricatures by portraying characters manifesting emotional distress, such as the many cases of trembling. Perhaps the most revealing one is the reaction of the Romans to a mellifluous and pompous recitation:

> tunc neque more probo videas nec voce serena
> ingentis trepidare Titos, cum carmina lumbum
> intrant et tremulo scalpuntur ubi intima versu.

> Then, neither in seemly manner nor with serene voice,
> you can see huge Tituses thrill as poems enter
> their loins and vibrant verses tickle their inmost parts.
>
> Persius, *Satires* 1.19–21

Trembling is another form of fluidity, since the body has lost its solidity and become subject to the fluidity of the message it receives, through cause and effect. The audience is affected by a poet 'whose quivering adds yet another level of fluidity that threatens to undermine the dry, solid virtues of the masculinist, Roman norm' (Miller 1998: 267). It is, however, possible to identify an implied reference to bodily fluids here as well. The *Titi* are found here in a state of ecstatic sexual arousal and, given that the name bears the connotation of a penis, the sexual climax which would lead to ejaculation is conspicuous by its absence in the image. At the same time, the trembling implies effeminacy, and Clark (2008: 267) sees this as the reason why Catullus avoids it as an affect display in his 51. Both these notions become more likely when seen in their respective contexts. The reason for the excitement of the audience is the smooth and effeminate recitation (1.15–18), a most fluid image, as it appears in Table 18.1. Kirk Freudenburg's (2001: 162–6) interpretation of *patranti* as 'ejaculating', however perplexing the image may be, helps us to include semen in the list of bodily fluids by pointing to the ecstatic eye at the pinnacle of excitement, while it also corroborates the idea of full sexual tension in the scene.

Another affect display that Persius uses less implicitly is that of sweating as a manifestation of passionate excitement and excessive effort:

> saepe oculos, memini, tangebam parvus olivo,
> grandia si nollem morituri verba Catonis
> discere non sano multum laudanda magistro,
> quae pater adductis sudans audiret amicis.

> When I was little, I remember, I often used to dab my eyes with oil
> if I didn't want to memorise the lofty speech of dying Cato
> which my insane teacher would much praise,
> while my father listened in a sweat with his gathered friends.
>
> Persius, *Satires* 3.44–7

The critique of the practices in education is directed towards the persona's father and teacher as well as the student's ignorance.[13] The caricature of the father is achieved with the secretion of sweat from his body. His body, and masculine solidity, is distorted by the fluidity it produces as the consequence of his emotions, becoming the affect display that completes the image together with the father's nonverbal behaviour. The ancient scholiast supports an understanding of sweat as the result of the father's anxiety about his ignorant son's recitation in front of his friends (*sollicitus de spe filii recitantis*; see also Harvey 1981: 91). Walter Kissel (1990: 427) is right, however, in allowing an understanding of the affect display as pointing to joy. He is excited to hear the recitation of his son and at the same time is ignorant of the son's unwillingness to learn. The excitement is in line with the traditional usage of sweat as an affect display. At the same time, the setting of this recitation and the audience bear similarities with the earlier poetic recitation and the reaction of the *Titi* in the first satire (1.19–21), but purged of its sexual innuendos. The allusion to the first satire stresses the cause of the sweat, which in this instance is the recitation. In both episodes, the recitation is characterised by a well-rehearsed delivery, praised here by the mad teacher: *non sano multum laudanda magistro* (Lee and Barr 1987: 108).

Clark's discussion of the nonverbal behaviour in amatory poetry points to the absence of sweat from the list of affect displays in Catullus and Horace. Horace, however, favours the use of this affect display for satirical purposes four times in his satires: 1.4.72, 1.9.10, 1.10.28, and 2.2.20–1. These satirical images are mostly used to express the excessive efforts of the characters. Out of these images, it is the sweaty orators in *Satire* 1.10.27–9—*Latine* | *cum Pedius causas exsudet Publicola atque* | *Corvinus* (while Pedius Publicola and Corvinus sweat over their cases in Latin)—who may have influenced Persius' imagery in this extract.[14] At the same time, sweat in old age has been perceived as paradoxical and rare, if it is not associated with bodily effort, according to Theophrastus' treatise (*On Sweat* 18–23, edition Fortenbaugh

2003). In that respect, it is possible to recognise the mocking that is taking place in Persius' image. The characters' condition has resulted from their passion and excitement, which are opposites of the Roman masculine ideal of having control of one's emotions (see also Clark 2008: 270; Williams 2010: 155–6).

Conclusion

The human body has a special place in Roman satire: the bodies of the depicted characters make them visible, projecting them into their social lives. The parallel criticism of poetry and society in Persius' *Satires* fully exploits the possibilities the bodily images offer. In most cases, a smooth and tender body is a cover for the inner corrupted self. Bodily fluids distort this perfect image: they leave the body open, breaking through from the body itself and, as a result, they reveal what really lies inside. The satirist makes this happen by colouring the body with a strong sense of fluidity, and he becomes the one that reveals the truth, which is one of the objectives of the genre.[15] Therefore, Persius' treatment of bodily fluids corroborates what the Introduction of the present volume states as one of its main hypotheses: that 'it is through fluids that the relationship between the inner body and the outside world is mediated' (p. 5).

The grotesque depiction of Persius' targets has first and foremost a moral objective. From a Stoic perspective, bodily fluids are exuded as the result of a moral disease; they come from the inside, and they reveal the true, degraded characters. They expose the passions and the excitement that derive from their effeminate and luxurious lifestyle, and their outbreak provokes Persius' criticism. The moral decay also affects the poetic taste of the time and vice versa, apparent both in the compositions and recitations of poetry and in the audience's applause. A recitation of smooth poetry that flows over a delicate mouth is an indicative example of this imagery, where bodily fluids are employed to stress Persius' disdain for the tear-jerking themes and the mellifluous style. Persius' use of bodily fluids is not arbitrary; the imagery functions as a literary device that yields a sensory tone to the description of the caricatures.

It is not a surprise that Roman satire and Persius in particular have exploited the debasing qualities of bodily fluids in such a way. His choices, however, indicate that this aspect of the human body did not go unnoticed by the critical eye of the satirists, but they were used to denote effort, passions, and excitement. By incorporating bodily fluids into his techniques and satirical media, Persius makes them a literary device to mock and invoke the grotesque.

Notes

1 All translations of Latin texts are my own.
2 This has been previously argued by criticism. It is a recurring theme in Paul A. Miller's (1998: 262) discussion on the bodily grotesque, which is fundamental to the development of this chapter. See also Gold 1998: 375; Clark 2008: 258–9.
3 Persius' obscurity has been the reason his poetry has been dismissed by scholars until fairly recently; see Dessen 1968: 3–5.
4 It is again possible to associate fluidity with smoothness, which is found here in the appearance, covering an entirely 'rotten inside' (Plaza 2006: 196). See also Horace, *Satires* 2.1.64–5.

5 1.64 *carmina molli* | *nunc demum numero fluere*; 3.20 *effluis amens*
6 The cleansing effects of the Tiber are here used sarcastically as the cleansing and purifying power of running water in the morning that washes away the vices of the night before in preparation for prayer. The reference here acts as the moral counterpart to the ironic reference to the other 'fluid in nature' of Table 18.1, i.e. Hippocrene (*Prologue* 1), whose positive literary effects are satirised again. For a different use of rivers and satire's water imagery, see Freudenburg 2018a.
7 Saliva and sweat also appear in perhaps Catullus' most grotesque depictions in 23.16, edition Mynors 1958: *a te sudor abest, abest saliva* (there's no sweat in you, there's no saliva). Here, the grotesque of the bodily fluids is an exaggeration of Furius' frugality that has led to the dryness of his body. However, this dryness is depicted with all sorts of grotesque descriptions, the combination of which results in the mockery of Furius.
8 See Quintilian, *The Orator's Education* 1.11.1: *femineae vocis exilitate* (feminine and weak voice) and the notes of Rose 1924. For public recitation, see Bexley 2015: 783–8. See also Dugan 2001: 420–5; Salm 2015 (especially Section 3) for the 'psychology of the voice' in antiquity, after Pernot 1993.
9 For this note I am indebted to Professor Anthony Corbeill (University of Virginia) for bringing Cicero, *Brutus* 260 to my attention and for his suggestions on the form and the meaning of *sputatilica*. Looking at rhetorical practices, the notion of spittle occurs in the *hapax legomenon sputatilica*, a neologism formed by Sisenna in Cicero, *Brutus* 260, following the example of Cotta, who avoided resemblance to Greek sounds in favour of a rustic style (*Brutus* 259). Sisenna uses the neologism derived from *sputo* with the double suffixation *-ilis* and *-icus*. The *-icus* suggests an implied Graecism, and Alan E. Douglas (1966: 190) suggests that the characterisation of the accusations as 'worthless' (*sputatilica crimina*) is a Latin neologism for the Greek κατάπτυστος, although the Greek adjective seems to be used solely (or mainly) for individuals with denotations of morality, which is not often found in rhetoric. Perhaps, the diminishing adjective was directed to the delivery—rather than the moral implications—of the accusations, indicating that it was based on the Graecism 'spat at'.
10 See Wyke 1994: 119–20 for the association of elegy with effeminacy. Persius also uses *tenerus* for descriptions of verse, e.g. 1.98: *quidnam . . . tenerum et laxa cervice legendum* (something delicate to recite with a flabby neck).
11 See e.g. Cicero, *On Divination* 1.126. See also White 2003: 138–46.
12 Because of its magical qualities, the efficacy of spittle was also exploited in medicine; see Pliny, *Natural History* 28.36. See also Corbeill 2004: 13. The magical qualities of spittle are also seen by the effect they have on prophetic ability, e.g. Polyidus and Glaucus in Apollodorus, *Bibliotheca* 3.3.2; Apollo and Cassandra in Servius, *Commentary on the Aeneid* 2.247.
13 The person can be distinguished from the persona from the autobiographical references to Persius; see Lee and Barr 1987: 107–8. See also Ferriss-Hill 2015: 92–3.
14 For the scene in Horace, see Gowers 2012: 320–1. The passage also influenced Persius in his scene with a Pedius in 1.85–87, for which see Tzounakas 2008b: 126–30.
15 Freudenburg (2018b) also interprets the pot imagery in *Satire* 3 as the symptom of what lies on the inside.

Bibliography

Primary sources

Clausen, W. V. (1992) *A. Persi Flacci et D. Iuni Iuvenalis Saturae*. Oxford: Clarendon Press.
Fortenbaugh, W. W., R. W. Sharples and M. G. Sollenberger (2003) *Theophrastus of Eresus. On Sweat, On Dizziness and On Fatigue*. Leiden: Brill.
Garrod, H. W. (ed.) (1963) *Q. Horati Flacci Opera*. Oxford: Clarendon Press.
Mynors, R. A. B. (1958) *C. Valerii Catulli Carmina*. Oxford: Clarendon Press.
Winterbottom, M. (1970) *M. Fabi Quintiliani Institutionis Oratoriae Libri Duodecim*, vol. I–II. Oxford: Clarendon Press.

Secondary literature

Adams, J. N. (2003) *Bilingualism and the Latin Language*. Cambridge: Cambridge University Press.

Bartsch, S. (2015) *Persius: A Study in Food, Philosophy, and the Figural*. Chicago: University of Chicago Press.

Bexley, E. (2015) 'What is dramatic recitation', *Mnemosyne* 68(5), 774–93.

Biville, F. (1995) *Les emprunts du latin au grec: Approche phonétique, Tome II: Vocalisme et conclusions*. Louvain: Peeters.

Bramble, J. C. (1974) *Persius and the Programmatic Satire: A Study in Form and Imagery*. Cambridge: Cambridge University Press.

Clark, C. A. (2008) 'The poetics of manhood? Nonverbal behavior in Catullus 51', *Classical Philology* 103(3), 257–81.

Corbeill, A. (2004) *Nature Embodied: Gesture in Ancient Rome*. Princeton, NJ: Princeton University Press.

Dessen, C. S. (1968) *The Satires of Persius: Iunctura Callidus Acri*. London: Duckworth.

Douglas, A. E. (1966) *M. Tulli Ciceronis Brutus*. Oxford: Clarendon Press.

Dugan, J. (2001) 'Preventing Ciceronianism: C. Licinius Calvus' regimens for sexual and oratorical self-mastery', *Classical Philology* 96(4), 400–28.

Edwards, C. (1993) *The Politics of Immorality in Ancient Rome*. Cambridge: Cambridge University Press.

Ferriss-Hill, J. L. (2012) '*Talis oratio qualis vita*: Literary judgements as personal critiques in Roman satire', in I. Sluiter and R. M. Rosen (eds) *Aesthetic Value in Classical Antiquity*. Leiden: Brill, 365–91.

———. (2015) *Roman Satire and the Old Comic Tradition*. New York: Cambridge University Press.

Freudenburg, K. (2001) *Satires of Rome: Threatening Poses from Lucilius to Juvenal*. Cambridge: Cambridge University Press.

———. (2018a) 'Satire's censorial waters in Horace and Juvenal', *Journal of Roman Studies* 108, 141–55.

———. (2018b) 'The po(e)ts and pens of Persius' third Satire (the waters of Roman satire, part 2)', in S. Finkmann, A. Behrendt, and A. Walter (eds) *Antike Erzähl- und Deutungsmuster: Zwischen Exemplarität und Transformation*. Berlin: Walter de Gruyter, 267–84.

Gold, B. K. (1998) ' "The house I live in is not my own": Women's bodies in Juvenal's *Satires*', *Arethusa* 31(3), 369–86.

Gowers, E. (1993) *The Loaded Table: Representations of Food in Roman Literature*. Oxford: Clarendon Press.

———. (2012) *Horace Satires I*. Cambridge: Cambridge University Press.

Harvey, R. A. (1981) *A Commentary on Persius*. Leiden: Brill.

Hudson, N. (1989) 'Food in Roman satire', in S. Braund (ed.) *Satire and Society*. Exeter: University of Exeter Press, 69–87.

Kissel, W. (1990) *Aules Persius Flaccus Satiren: Herausgegeben, übersetzt und kommentiert*. Heidelberg: Carl Winter.

Lateiner, D. (1992) 'Affect displays in the epic poetry of Homer, Vergil, and Ovid', in F. Poyatos (ed.) *Advances in Nonverbal Communication: Sociocultural, Clinical, Esthetic, and Literary Perspectives*. Amsterdam: John Benjamin, 255–69.

Lee, G. and W. Barr (1987) *The Satires of Persius*. Liverpool: Francis Cairns.

Miller, P. A. (1998) 'The bodily grotesque in Roman satire: Images of sterility', *Arethusa* 31(3), 257–83.

———. (2013) 'Mythology and the abject in imperial satire', in V. Zajko and E. O'Gorman (eds) *Classical Myth and Psychoanalysis: Ancient and Modern Stories of the Self*. Oxford: Oxford University Press, 213–30.

Nelson, M. (2017) 'Insulting middle-finger gestures among ancient Romans and Greeks', *Phoenix* 71(1), 66–88.

Pernot, L. (1993) *La rhétorique de l'éloge dans le monde gréco-romain*. Paris: Institut des Études Augustiniennes.

Plaza, M. (2006) *The Function of Humour in Roman Verse Satire*. Oxford: Oxford University Press.

Rose, H. J. (1924) 'Some traps in Persius' first Satire', *The Classical Review* 38, 63–4.

Rosen, R. M. and C. C. Keane (2013) 'Greco-Roman satirical poetry', in T. K. Hubbard (ed.) *A Companion to Greek and Roman Sexualities*. Malden: Wiley-Blackwell, 381–97.

Salm, É. (2015) 'Écouter l'orateur dans le monde gréco-romain', *Pallas* 98, 199–213.

Tzounakas, S. (2006) '*Rusticitas* versus *urbanitas* in the literary programmes of Tibullus and Persius', *Mnemosyne* 59(1), 111–28.

———. (2008a) 'The rejection of Graecisms in *Persius* 1, 92–106: From form to culture', *Grazer Beitraege* 26, 1–30.

———. (2008b) 'Persius' re-reading of Horace: The case of some proper names', *Classica et Mediaevalia* 59, 123–37.

———. (2014) 'The hellebore in Persius' *Satires*', *Euphrosyne* 42, 189–95.

White, M. J. (2003) 'Stoic natural philosophy (physics and cosmology)', in B. Inwood (ed.) *The Cambridge Companion to the Stoics*. Cambridge: Cambridge University Press, 124–52.

Williams, C. (2010) *Roman Homosexuality*. 2nd edn. Oxford: Oxford University Press.

Wyke, M. (1994) 'Taking the woman's part: Engendering Roman love elegy', *Ramus* 23(1–2), 110–28.

Zietsman, J. C. (1981) 'Persius: The rustic poet', *Akroterion* 26, 52–7.

Part VI

WOUNDED AND PUTREFYING BODIES

19

'EFFLUX IS MY MANIFESTATION'

Positive conceptions of putrefactive fluids in the ancient Egyptian coffin texts*

Tasha Dobbin-Bennett

As the editors noted in the Introduction to this volume (p. 5), the release and subsequent control of bodily fluids mediates the relationship between the corporeal body and the outer world. The permeability of the body's boundaries, reflected through the release of fluids such as sweat, blood, tears, milk, urine, sperm, and putrefaction, highlights the fluidity of the human form and condition mirrored in concepts of gender, sexuality, social status and identity, and spiritual transformation.

In the ancient world—Egyptian, Greek, or Roman—liquid emanating from living and deceased bodies offered ancient authors analogues through which they could comment on the behaviours, attitudes and the general image of society as Andreas Gavrielatos argued in the preceding chapter on Persius' *Satires* (pp. 287–8). The uncontrolled discharge of fluids permeating through the seemingly perfect and yet porous visage, such as sweat, became a mechanism through which Persius could reflect on internal corruption of both the body and society as a whole. In Part II of this volume, Rosalind Janssen (Chapter 2) also demonstrated how bodily fluids and fluidity of the gendered body represented a microcosm of society. Janssen stressed that menstruation, and therefore the female body, influenced daily life at the ancient Egyptian village of Deir el-Medina. However, she also revealed the ambiguous dual nature of bodily fluids, arguing that menstruation was also a microcosm of the divine—aspects can hold positive and negative connotations and context determines which applies at what time. For example, while menstruation is the antithesis of divine regeneration (the presence of blood negates the possibility of pregnancy and therefore further spiritual rebirth), it also reveals the potential for regeneration—for without menstruation no pregnancy can occur.

Picking up these two threads from Gavrielatos and Janssen, I argue that the fluidity and permeability of the post-mortem body symbolises the concept of divine regeneration in the Middle Kingdom Coffin Texts (c. 2160–1650 BCE), a set of 1,185 religious spells and recitations inscribed on the inside and outside of ancient Egyptian coffins.[1] Here, rather than bodily fluids associated with the vitality and charged emotion of the living, the fluids in question are post-mortem putrefactive liquids released from the body during the natural decomposition processes that occur following death.

Our modern understanding of ancient Egyptian conceptions of putrefactive fluids within the Coffin Texts is inextricably bound together with our understanding of the purpose of mummification: the physical and ritual process of preserving and wrapping

DOI: 10.4324/9780429438974-26

the deceased.[2] Indeed, a large majority of the religious recitations and spells within the Coffin Texts deal with the mitigation of decomposition, along with the resulting putrefactive discharge, as a way to preserve a life-like body.[3] These texts often equate the negation of the decomposition process and the desire to live again in the afterlife with a complete, properly functioning body. Some Coffin Texts absolutely expressed the desire to hinder or halt the decay of the body post-mortem. Nevertheless, what interests me are other Coffin Texts where decomposition and putrefaction are elements that are both welcome and a necessity for the deceased. This seemingly contradictory juxtaposition is not unusual in the wider ancient Egyptian world-view, for example the deity Seth,[4] but why some Coffin Texts record decomposition as necessary and 'good' and others present decomposition as 'bad' needs a little more nuancing.

Rooted in earlier mythology, New Kingdom texts (1550 BCE onwards) considered the efflux of the deceased/corpse Osiris the origin of the Nile inundation (Darnell 2004: 99–100). Consequently, decomposition and putrefaction as forces for regeneration and rejuvenation of Egypt itself were reflected through royal invocation to Osiris: 'I consecrate for you the Inundation, produced through your efflux' (Griffith 1970: 436–7). The Osirian myth also applies this cosmic cycle to the deceased themselves. Mummification transforms the deceased into an Osirian being, who in turn engenders and rejuvenates her/himself through decomposition and putrefaction (Manassa 2007: 65–6). On the other hand, texts such as Book of the Dead Chapter 154 ritually negate carefully delineated decomposition elements.

How do we then resolve these two viewpoints: the body must be whole and preserved, but the body must also function as an Osirian substitute—leaking essential putrefactive fluids, bathing the body in efflux, and thereby returning the body to the primordial waters? The answer might be in careful management of the spiritual and corporeal body, allowing for selective release of putrefactive fluids by representatives of the sphere of embalming. Some decomposition is good—it must occur—but decomposition that leads to a compromised divine vessel is negative. The body must not rupture, for if the external visage of the body is breached or distorted the primary purpose of the body as a post-mortem vessel for the *ba*-soul is also negated. The answer might then lie in the decomposition process itself.

Perhaps the most important point about decomposition is that it is an ongoing process, not a single event. In the first few minutes after death, the human body begins to decompose. While there are, broadly, five phases of decomposition (fresh, early decomposition, advanced decomposition post-bloat, natural mummification, skeletonisation) the first phase is the micro-breakdown of the soft tissues via autolysis (Vass 2001: 190). Deprivation of oxygen to the cells governs autolysis, which results in an increase in the levels of carbon dioxide in the blood. As the carbon dioxide rises, cell pH decreases, thereby increasing the acidity of the cells. Alongside the change in cell structure, cellular enzymes (lipases, proteases, and amylases) begin to dissolve the cell walls from the inside out. The compromised cellular walls then release nutrient-rich fluid into the surrounding tissue (Zhou and Byard 2011: 7).

The development of autolysis progresses much more quickly in tissue environments that have a high content of enzymes, for example the liver, or high water content, such as the brain (Vass 2001: 190). Eventually, all cells in the body will succumb to autolysis. After enough cells have ruptured, releasing fluid, the process of putrefaction commences (Parkinson et al. 2009: 380). The first indicators of autolysis and putrefaction are the presence of fluid-filled blisters on the skin, the start of skin slippage (Vass 2001: 190), marbling and/or greenish discoloration of the skin (Galloway et al. 1989: 610), and the release of putrefactive fluids from the nose, mouth, and anus (Janaway et al. 2009: 326). Broadly speaking, the earliest stages of decomposition visually demonstrate the deceased nature of the body without compromising the integrity of the human vessel.

Based on examinations of ancient Egyptian mummies, we know that decomposition did occur prior to the application of artificial mummification. During their examination of 51 mummies from the Necropolis of Dush in Kharga Oasis, Françoise Dunand and Roger Lichtenberg (2006: 174, figures 229 and 230) noted that some of the mummies displayed an advanced state of decomposition. They suggested that the cause of the variation in the final appearance of the mummies seemed to be a delay 'separating the death of the individual from the beginning of the mummification process' (Dunand and Lichtenberg 2006: 182). So while mummification halts decomposition and can obscure evidence of early putrefaction, advanced decomposition events are still evident in the human form post-mummification. Nevertheless, distortion and perforation of the skin, including unintentional decapitation due to compromised soft tissue, were mitigated by embalmers through the reconstruction of the human form during the embalming process. The process of reconstructing the deceased and accounting for any damage or loss symbolically mimicked Osiris' reconstitution and therefore equated the deceased in mummiform with the mummiform god Osiris (Assmann 1989: 138–9).

The early decomposition stage can begin as early as the first day after death, but these first signs of decay can also begin as late as the fifth day post-mortem, depending on exogenous factors such as climate (Galloway et al. 1989: 610) or endogenous reasons such as obesity, fever, or infections (Zhou and Byard 2011: S86). Because the tissues with the highest content of water decompose the fastest, the cranial region, which houses the brain and eyes, tends to demonstrate the earliest signs of decay. This element may indeed be the reason why most of the Coffin Texts that demonstrate positive associations with decomposition reference the release of putrefactions from the orbital socket.

Using the calculation results from Tasha Dobbin-Bennett (2014: Appendix 6), the onset of the early decomposition stage could occur in the Egyptian climate as early as the first to second day post-mortem depending on the season. Mary Megyesi et al. (2005: 6, table 5) suggest that Total Body Scores (TBS)—a scoring system that gives each stage of decomposition a numerical range to help quantify the post-mortem changes—from 8–13 represent the stage during which the body releases putrefactive fluids in the facial region. The results suggest that the early decomposition stage could have occurred as early as the first to second day after death depending on the season and would have lasted approximately one to two days. At a TBS of 13, putrefactive fluids would have most likely purged from the mouth, nose, and eyes, as well as the anus. The face, genitals, and abdomen would likely demonstrate signs of bloat, although full bloating of the torso might not occur for an additional two days.

Advanced decomposition—which includes the distortion of skin and tissue through bloating, the resulting collapse of the abdomen and loss of bodily fluids, and the substantial loss of biomass through insect colonisation and consumption–may have contributed to the negative associations attached to decomposition. However, the elements that constitute the early decomposition stages, including livor mortis, pallor mortis, algor mortis, and rigor mortis, as well as early putrefaction, appear within religious contexts that seem to equate these physical signs with the symbolic transformation of the deceased into a divine being in the afterlife.

Fluids begin to purge from the post-mortem body within the first few days after death but continue to flow forth during the early and advanced stages of putrefaction. However, there are generally two distinct purging events of fluids from the body. The first occurs during the first few days post-mortem via the mouth, eyes, nose, and anus. The second occurs after the fluids and gases that were produced during the bloating stage are released through weakened points on the body. During this time the body enters advanced decay and releases larger quantities of putrefactive fluid, not only from the rectum but also from breaks in the skin (Vass 2001: 190–1; Janaway et al. 2009).

Table 19.1 shows the estimated days to a Total Body Score of 8 and 20. At a Total Body Score of 8, the deceased would begin to display signs of fluids purging from the cranio-facial region (the first stage of putrefactive purge). When a body reaches a Total Body Score of 20, bloating has most likely been resolved, and the body would be releasing larger amounts of putrefactive fluids (Megyesi et al. 2005: 6, table 5). Table 19.1 displays four data points for each of the Egyptian sites. For each site, the black column references the time in days to reach a TBS of 8, while the grey column represents the time in days to reach a TBS of 20. The low and high data labels present the change in air temperature and humidity across the calendar year (Dobbin-Bennett 2014: Appendix 2 for full environmental data for each site).

Table 19.1 Comparison of estimated days to Total Body Scores 8 and 20 across the four Egyptian sites

According to the results of the calculations using Megyesi et al. (2005) formulae (Dobbin-Bennett 2014: Appendix 4), a post-mortem body in the Egyptian climate would begin showing signs of putrefactive fluids in the cranio-facial region within the first to second day after death. Larger putrefactive fluid release, signifying the end of bloating and the onset of advanced putrefaction, would occur approximately five to seven days post-mortem. At this later stage of decomposition, a post-mortem body undergoes significant putrefaction, often in tandem with insect colonisation (Prahlow 2010: 169–70). It is this distinct difference, both in location on the body and in time elapsed since death, which may suggest why cranio-facial putrefactive fluids formed part of the positive ritual transformation in the Coffin Texts.

Bodily fluids associated with the deceased in the Coffin Texts include *iȝf*,[5] *iw.tyw*,[6] *id.t*,[7] *fd.t*,[8] *ḥwȝȝ.t*,[9] *qis.wt*,[10] *rḏw*.[11] It is not always possible nor appropriate to assign a direct biomedical denotation to an ancient Egyptian term, particularly where there is only a single attestation.[12] Nevertheless, the context, collocation with lexemes depicting body parts and verbs that indicate directionality (i.e. *pri* 'to come forth') suggest that these terms refer to secretions that come forth from the body; in the Coffin Texts the 'body' in question is deceased.

Several scholars have considered lexemes, which refer to bodily fluids from the deceased, in the Coffin Texts including Rune Nyord (2009: 320–3, 2012: 165–84), Ursula Köhler (1975: 369–99), Lauri Pantalacci (1981: 57–66, 1983: 297–311), Jérôme Rizzo (2007: 123–36), Harco Willems (1996: 118–19, 407), Andreas Winkler (2006: 125–39), and Jan Zandee (1966, 1975–6: 1–47). With the exception of Zandee (1966), who assigns wholly negative associations to these lexemes, these scholars have identified contexts where the release of fluids from the deceased's body are connected to successful resurrection.

Winkler (2006: 126) put forward a very interesting approach to the duality inherent in the concept of *rḏw* 'efflux'. He applied a 'two-level system' to the resurrection process in the Pyramid Texts, following Nils Billing's terminology: reconstitution and manifestation. Billing (2002: 25) understood the concept of reconstitution as a generic term for the process of the restoration of the body. In this state, the body returns to an original state away from the disorder and chaos of death and decomposition. The 'manifestation' of the deceased is the reappearance of the reconstituted deceased in his divine form; manifestation is the ultimate sign of life over death. Billing's theological construct is fascinating and closely follows one of the overriding hypotheses of this study. My point of departure from Billing concerns the 'original' state of the body. I, too, argue that reconstitution of the body moves the body away from a death-state, but I contend that this process moves the physical and spiritual body toward disorder and chaos. The mechanism for this transformation remains the same, *rḏw*, but the reconstitution of the body can only occur if the body is moved out of the ordered, structured, known, world and is returned to the moment of creation; creation stems forth from the primordial waters of Nun, of which *rḏw* can be viewed as a liquid filament linking the deceased with the healing and dissolving power of Nun. As Eric Hornung has so eloquently stated:

> In temporal terms, regeneration is possible only outside of the ordered world of creation. In order to be rejuvenated, that is, to reverse the course of time, one must step for a little time outside time and see oneself at the beginning of the temporal world, at the point of creation or even in the world before

creation, which knows no time. Rebirth in the morning is therefore a renewal of creation, and is achieved with the help of the primeval gods, who sent the sun forth from their midst at the beginning of creation. Like the creation of the world, sunrise can be called the first occasion.

Hornung 1983: 161

It is the bathing of the deceased in these creative waters that allows the all-important step outside of time and order. Once the deceased has released the *rḏw*, he has become reconstituted. In touching the primordial waters s/he is spiritually and physically recreated, allowing for the next phase of the process: manifestation. This 'manifestation' is the reappearance of the reconstituted deceased in his/her divine form, here represented by the ritual transformation into the divine mummiform god, Osiris.

A number of Coffin Texts link putrefactive fluids to the resurrection of the deceased in the afterlife. In these texts, putrefactive fluids *iȝf*, *qis.wt*, *ḥ:rḏw*, and *fd.t*, come forth from the creator god so that the deceased can gain nourishment from the fluids or can be recreated anew by bathing in the fluids.

> *i-Rꜥ*[13]
> *wn n=i ḏnḥ=k pw*
> *ink wꜥ m psḏ=k pw kȝ.w*[14]
>   *ꜥnḫ.w m iȝf n ir.ty=k*
>   *m šnw imw ḫtt.t=k*
> *pr ḥkk*[15] *r ḥȝ.t wiȝ*[16]
>   *wḏ=f mdw n ḥw*
>   *wḏ=k wi n ḥw*
> *mk wi wḏ(.w) n=k in dbn-wr*

O Re,
Open these your wings for me!
I am one of those nine *kȝ*-spirits of yours,
            who live on the *iȝf*-fluid of your eyes,
            and in the hair which is in your armpits.
Hkk goes up to the bow of the barque.
Just as he gives orders to Hu,
so shall you commend me to Hu.
For I am one who has been commended to you by the Great Encircler.

Coffin Text Spell 617 (VI 228e–229f)[17]

Coffin Text Spell 617 occurs on three coffins as well as on the Papyrus Gardiner III.[18] The spell opens with a vocative to Re, requesting he open his wings for the deceased, because the deceased has taken on the role as one of the Ennead. The deceased positions himself as 'one who lives on the *iȝf*-fluid of your eyes'.[19] Here, the putrefactive fluid is paralleled with the following prepositional phrase which includes the hair of the armpit (*šnw imy.w ḫtt.t=k*) as a life-giving source. This equation between the *iȝf*-fluid and the hair of the armpit may allude to the ritual act of nwn-mourning, which involved male and female mourners flipping their hair forward over their face and raising their arms over their heads.[20] The ritual act of the nwn-gesture may have simulated the creative waters of Nun, the creative fluid from whence all life originated (Bonneau 1964: 285; Manassa 2007: 30–1). That a deity would perform a mourning gesture, weeping or releasing fluids from the eye while raising arms/wings high enough to reveal armpit hair, is not unusual in the later Underworld Books (Manassa 2007: 31, note 128). Here the ritual act of resurrection by one of the creator deities, Re, may add further efficacy to the recitation. The association between hair, the inundation, and putrefaction is further explained in Coffin Text Spell 168.

ṯs n šnw n ꜣs.t n šnw n nb.ḥw.t
ṯs pẖr
iw ḥwꜣꜣ.t
wšr itrw
ḏmd iḥm.wt[21]
    ꜥm.n n=f Gb mḥw
ḏmd ḏr.ty šsmw ḥr smꜣ nb.ty

The hair of Isis is knotted to the hair of Nephthys,
vice versa.
Putrefaction is boat-less.
Joining the river-banks.
The rivers are dried up,
    because Geb has swallowed the flood-waters for himself.
The two hands of Shesmu are joined over the lungs of the Two Ladies.
                    Coffin Text Spell 168 (III, 28a–29c)

This short spell appears on twelve coffins and may relate to a state of drought, either temporal or spiritual. Spell 168 presents the opposite situation to the free-flowing hair of Isis and the equation to the inundation demonstrated in Spell 617; instead, the tangled divine locks prevent the inundation. The spell clearly delineates the lack of divine water through the nominal *sḏm=f* (*wšr*) and the emphasised perfective *sḏm.n=f* (*ꜥm.n Gb*) that follow. The rivers have dried up because Geb had already swallowed the floodwater. The deceased has neither the necessary fluids on which to live, characterised by the lack of putrefactive fluid indicated here in Spell 168, nor the movement of the divine hair to simulate the re-creative Nun waters.

In Spell 168, the putrefaction (*ḥwꜣꜣ.t*) is boat-less (*iw*). Raymond Faulkner (2004, Volume I: 145, note 3) equates this phrase with the removal and disposal of 'fouler forms of garbage' and suggests that the low Nile may have prevented the removal of the putrefaction. However, while this spell may indeed relate to the temporal sphere, it may also relate to the mortuary sphere, especially as the following spell, Coffin Text Spell 169, which appears on five coffins,[22] directly relates to the inundation as putrefactive fluid that occurs upon the face.

On the foot-section of Coffin B5C, Spells 169, 165, 166, and 167 precede Spell 168. This unit of spells first equates putrefactive fluids from the face with the inundation (Spell 169) and then provides a series of spells that ensure that the deceased has the appropriate offerings and food (Spells 165–6). Spell 167 equates the deceased with the creator deity Re-Atum and invokes the 'mourner' (*ḥꜣ.t*) to *ir smꜣ=ṯ* 'prepare your hair'. The lexeme, *smꜣ*, literally meaning scalp but with the hair determinative ⸗ may have the extended sense of 'hair upon the scalp'. The cycle of spells appears to place Spell 168 in the midst of the mortuary context.

Within this textual milieu, two possible reinterpretations of the 'boat-less putrefaction' spring to mind. First, in Coffin Text Spell 617, the putrefactive fluid, which has come forth from the Eye of Re, is the mechanism by which the deceased lives; the *iȝf*-fluid, together with the hair of the armpit, are precursors to the journey in the solar barque (*wiȝ*). The context of the Spell 617 situates the deceased within the solar realm and places the deceased alongside the deity, Hu, upon the solar barque. The epithet that concludes that spell, *dbn-wr*, the Great Encircler, refers to the daily circuit of the sun-god. Within a similar thematic context, Spell 168 perhaps alludes to a situation prior to the release of the inundation; without the creative waters of Nun, the putrefactive fluids cannot act as the mechanism for the deceased to ascend in the solar barque. The spells that follow Spell 168, across various coffins, remedy this situation, or perhaps proceed in an ordered, expected, fashion: with the presence of putrefactive fluids, the deceased can take up his place on the solar barque.

The 'boat-less' state of putrefaction may also allude to the vessel in which the waters of Osiris were caught and transported. In the context of inundation rituals, Osiris can be considered a creator deity, given his equation with the life-giving properties of the inundation waters. The efflux (*rḏw*) that seeped from the side of Osiris was the inundation waters.[23] Inundation rituals from the Dynastic period, which revolved around the collection of floodwaters within theomorphic water vases, developed into the well-known Greco-Roman phenomenon of 'Osiris-Hydreios' or 'Osiris-Canopus' (Barrett 2011: 134, note 426). These composite manifestations of Osiris are expressed through the form of a water jar. Caitlín Barrett (2011: 134, note 427) has recently contradicted the traditional view of scholars that an Osiris-Canopus style of cult did not exist before the Roman Empire. She highlighted that the textual evidence from as early as the Old Kingdom strongly suggests that the ritual had Dynastic antecedents.

Willems (1996: 118, and notes 479, 481) has linked the depiction of the snw-vase on the foot of the Middle Kingdom coffin of the Heqata with the receptacle for the waters of Osiris. While the ancient Egyptian information on the purpose of the snw-vase is limited, he argues that the context within the object friezes on the footboard of the coffin suggest that the vase may have been used during the rites at the Purification Tent. The association between the place of purification and snw-vase is present in Coffin Text Spell 314.[24] In this spell, the deceased is assimilated with Osiris (*ink Wsir*—'I am Osiris') (De Buck 1935–2006: 94a) and also Thoth (*ink ḏḥwty*—'I am Thoth') (De Buck 1935–2006: 94k) in his capacity as the 'legal counsel' for Horus in his lawsuit against Seth for the right to rule.[25]

> *wn=ʾi ḥnꜥ ḥw hrw ḥbs tštš*
> *wn.t(w) snw.w iꜥ.w[26] wrḏ-ib.w sštȝ.w*
> *hrw imn sštȝ.w m ḏ.t m r-stȝw*

> May I exist together with Hu on the day of clothing the one who is hacked up,
> when the snw-jars of the washings of the mysterious Weary of Heart,
> on the day of hiding the mysteries in Rosetau for eternity.
> Coffin Text Spell 314 (IV, 94t–95b)

Coffin Text Spell 314 places the deceased within the Horus-Osirian constellation and stresses the *perpetuum mobile*—the indefinite repetition of the action—of the cyclical regeneration of the deceased. As an Osirian-being (*ink Wsir*), the deceased can bestow the rejuvenating embalming and mummification rites upon Osiris, who will, in turn, bestow that same treatment upon the deceased. Two references in this section of Spell 314 link the recitation to the embalming of Osiris: *tštš* and the well-known epithet of Osiris—*wrḏ-ib* (Leitz 2002, Volume II: 512–13). The use of the perfective active particle, *tštš*, to describe the state of the one who is receiving treatment is apt. *tštš* has the verbal denotation of 'to hack up, to pound',[27] and occurs as an epithet of Osiris (Leitz 2002, Volume VII: 441). The use of this lexeme must refer to the dismemberment of Osiris by Seth and the ritual reconstitution of his limbs by his siblings: Isis and Nephthys. As in Coffin Text Spell 237, the clothing (*ḥbs*) of the hacked-up body of Osiris (*tštš*) alludes to the wrapping of his body with the funerary linens (De Buck 1935–2006: 312a-e).

The snw-jars in Spell 314 are used, perhaps, either to catch the waters that pour off Osiris, through the ritual purification with *qbḥw* 'cool' waters, or to catch the release of his putrefactive fluids.[28] In the Greco-Roman period papyrus, P. Jumilhac, the snw-vase was said to contain the efflux (*rḏw*) of Osiris and was located within the *wʿb.t*, or place of embalming. The boat-less (*iw*) state of the putrefaction (*ḥwȝȝ.t*) in Coffin Text Spell 168 may indeed reference the need for a vessel into which the life-giving efflux of Osiris must go.

Willems (1996: 119, and notes 486–91) links Coffin Text Spell 235, which occurs in proximity to the depiction of the snw-vase, to the restoration of the deceased through the efflux of Osiris. He argues that the gift of the deceased's efflux (*rḏw*) back to the deceased, from the divine tribunal, has a dual purpose: the capacity for free movement and the return of the life-giving waters that flowed out of the deceased after death.

> *rd.wy=ṯ n=ṯ*
> *ṯs n=ṯ ḥʿ.w=ṯ*
> *inq n=ṯ ʿ.wt=ṯ*
>> *šȝs=ṯ nm.t=t*
>> *r ḏȝḏȝ.t*
> *r bw n.t nṯr.w im*
> *rdi=sn n=k[29] rḏw*
> *pr im=k[30]*
> *ni wrḏ-ib=ṯ ḥr=s*
> *ẖns=ṯ sp-snw*
> *ni wrḏ=ṯ sp-snw*

Your two legs are yours.
Lift up your body for yourself,
Gather together your limbs for yourself,
>> so that you shall tread out the strides to the tribunal
>>> to the place of the gods therein.
They will give you the efflux that came forth from you.
May you not be inert, under it.
May you travel, may you travel.
May you not tire, may you never tire.

Coffin Text Spell 235 (III, 301c–302d)

We can, perhaps, take that argument one step further. The concept of free movement is certainly applicable in this spell and is a common motif throughout the Coffin Texts (Barguet 1971: 15–22). But the presentation (*rdi*) of the efflux (*rdw*) to the deceased is not just a return of the bodily fluids that the deceased has lost during the putrefactive phrase, it is the mechanism by which the deceased is revived. The semantic structure reveals this emphasis: the freedom of movement allows the deceased to move (*šss*) towards the divine tribunal (*dsds.t*), where he will receive his efflux that came forth from his body. The use of the prospective *sdm=f* (*rdi=f*) expresses the indicative future; this is a certain outcome of the freedom of movement. The perfective active participle (pr) following the noun (*rdw*) provides the key adjectival information: the efflux came forth (pr) from the deceased. This refers not to lustration waters (*qbhw*) nor purification waters (*w'b*) but to the putrefactive fluid itself. Finally, the use of the negated prospective *sdm=f* ensures the deceased of his future. He will not remain inert (*wrd-ib*); he will transition past the *wrd-ib* state of the Osirian-deceased, which is the state prior to the rejuvenation and regeneration of the deceased, because he has received the efflux. The same message resonates in Spell 617, where fluid emanating from the Eye of Re is the means by which the deceased shall again live.

The link between the release of putrefactive fluids from the eye, the inundation, rebirth as the newborn solar deity, and the ultimate moment of creation is brought together in Coffin Text Spell 318. Here, the deceased is, at once, the efflux and the inundation that brings life to not only the deceased himself but also the land through the release of the life-giving waters.

> *ink is ȝgb n ir.t 'Itm*[31]
> > *ḥ'i ḫpr rdw*
> > *sȝ wḫ' sȝ m ȝgb wr 'ȝ*
> *ink is ḫpri m ḫpr.w=i wsr.w nb*
> > *ȝḥ.w=i ḫpr.w=i nb*
> *iw=i ḥtp ib.w*
> > *s:wȝd=i s:'nḫ tȝ.wy*
> *nnk ȝḥ.t*[32]
> *nnk šmw*
> *ink ir.(w) pr.t n ns pw tp r=i*
> *ḫpr.w=i pw ȝḥ.t ḥ:rdw*
> *fd.t=i pw prr.t m iwf=i pr.t*

Indeed, I am the one who is the inundation of the Eye of Atum,
> who rejoices when the efflux occurs,
> > the son of the one who loosens the son as the great and large inundation.

Indeed, I am Khepri in all my manifestations and powers
> and all my effective spirits and manifestations.

I come so that the hearts are content,
> and I make green that which nourishes the two lands.

The inundation-season belongs to me.

The summer-season belongs to me.
I am the one who created the going forth of this flame upon my mouth.
The inundation of efflux is my manifestation.
Peret-season is the sweat that goes forth from my flesh.

Coffin Text Spell 318 (IV 140d–142a)

Coffin Text Spell 318 links together the reconstitution and manifestation elements of recreation through the release of putrefactive fluids. In the reconstituted state, the deceased is one with the inundation, the essential life-giving fluids that have come forth from the Eye of the creator god *par excellence*: Atum. Placing the deceased, here described as the floodwaters themselves, within the root of the eye of Atum returns the deceased to a 'pre-created state' that then enables him to reappear in the created universe anew (Koemoth 1993: 120–2). The element of original creation by Atum is clearly displayed in Coffin Text Spell 1130: *s:ḫpr.n=i nṯr.w m fdt=i iw rmt m rmwt n ir.t=i* 'The gods I created from my sweat, mankind is from the tears of my eye.'[33] Further, it is the close association between the facial putrefactive fluids and the creative potential that comes forth from the Eye that lends the positive association to *ḥwꜣꜣ.t*, 'products of putrefaction' in Spell 755: *iw rm.wt nṯr m ḥwꜣꜣ.t ꜥ.t im=i* 'The tears of the god are the putrefaction of the body part therein me.'[34]

The negative associations attached to putrefactive fluids and the process of decomposition are well understood. Less obvious are the reasons why religious texts also include many positive references to elements of the decomposition process. The Coffin Texts examined here help us consider the early decomposition process as a mechanism of transformation, moving the body from one state to the next. Early stages and events of decomposition are necessary to the transformative process, as they assist in moving the deceased away from the ordered, defined world and allow the deceased to rejoin the chaotic waters of creation. Efflux and putrefactive fluids that flow forth from the deceased's facial region may indeed relate to the earliest release of putrefactive fluids during the early decomposition stage. These fluids visually indicate that the person is deceased and is now bathed in efflux, mimicking the waters of Osiris, but the body has not yet experienced any processes that would compromise the integrity of the body. The body has moved physically and spiritually from the realm of the living into the realm of the death, while still functioning as an intact vessel. This recognisable form becomes the basis for the transformation of the deceased into a mummiform divine being.

These selected Coffin Texts demonstrate ritual control over the irrepressible release of putrefactive fluids, which mediate the relationship between the corporeal body and the spiritual world. The permeability of the body's boundaries is necessary, up to a point. Complete fluidity of the body exemplified through advanced putrefaction is undesirable, because the body has become chaotic and formless. Instead, the transitory body is managed, mitigated, and removed from society. Much like Janssen's discussion on the management of menstruation, decomposition is managed through post-mortem physical and ritual attention; the fluidity of the body's transformation acts as a microcosm of the divine; the individual rebirth mimics and replicates the original moment of creation and ongoing cyclical recreation of the cosmos.

315

Notes

* A version of this paper was presented at the Bodily Fluids/Fluid Bodies Conference in Cardiff, Wales (12 July 2016). The author would like to thank the organisers of the conference, as well as the editors of this volume, for the inclusion of her work despite it being outside the temporal parameters of the conference.

1 Numerous scholars have discussed the Coffin Texts. See, for example, Eric Hornung (1999: 8 ff.) for a historiographical summary.

2 Halting putrefaction and decomposition is often cited as the primary reason for the implementation of the mummification process; see, for example, Rosalind David (2000: 373); and Alfred Lucas (1962: 272–326).

3 See, for example, Book of the Dead Chapters 45 and 154, as well as Coffin Text Spells 432, 519, 755, 756, 810, 822.

4 The conception of Seth most commonly demonstrates the balanced dualistic nature of the ancient Egyptian world-view. Seth fulfils both a positive and negative role in ancient Egyptian mythology; see H. te Velde (1967). However, the positive and negative components were also an integral part of the more obscure aspects of ancient Egyptian theology, for example, the headless Osiris. In the form of the Greco-Roman deity *Akephalos*, this headless form of Osiris was a magically powerful, solar deity. On the other hand, headless Osiris could also represent his dismemberment at the hands of Seth. See John Coleman Darnell (2004: 115–17 and note 363) for extensive references therein. For a discussion of the positive aspects of putrefaction with particular association to Osiris, see, for example, Lauri Pantalacci (1983: 306–8); and Colleen Manassa (2007: 47 and note 244).

5 Dimitri Meeks 1998, Volume II: 15, no. 78.0152.

6 Meeks 1998, Volume II: 17, no. 78.0208; J.F. Borghouts 1978: 27; Manassa 2007: 51–2.

7 Adolf Erman and Hermann Grapow 1926–61, Volume I: 152.7–12.

8 Erman and Grapow 1926–61, Volume I, 582.6–11; Hildegard Von Deines and Wolfhart Westendorf 1962, Volume I: 309; Rami van der Molen 2000: 147.

9 Erman and Grapow 1926–61, Volume III: 51.1–5; van der Molen 2000: 323.

10 Meeks 1998, Volume I: 385, no. 77.4368.

11 Erman and Grapow 1926–61, Volume II: 469.5–18; Von Deines and Westendorf 1962, Volume I: 558. rDw is translated in a variety of different ways, including 'Ausfuß' or as 'Flüssigkeit', in the *Wörterbuch*. Drioton (1949: 156) translates the term as 'excrétion'. Pantalacci (1981: 62) uses the term 'viscères en décomposition'. Zandee (1960: 11, 57–8) suggests that the secretion that issues from the body when decaying is a form of *cadaverositas*, which takes the form of foul-smelling liquid that is released from the cells when they die. Kettel (1993: 317–18) argues that the texts, themselves, do not offer an exact denotation for the term but rather connects *rḏw* with an element that comes forth (*pri*) from the limbs of the god: *rḏw pr.w m ḥˁ.w* 'the efflux has come forth from the limbs.' Kettel also adds: 'Or, le terme *rḏw.w* à proprement parler intraduisible, désigne tout liquide qui sort d'un corps divin.'

12 See Katherine Eaton (2018: 44) for a discussion on alternative methodologies to translate ancient Egyptian lexemes and a critique of various approaches, including the methodology employed within this chapter.

13 The text in Coffins B3B0a and B2L clearly reads Ra (De Buck 1935–2006: 228e). However, Raymond Faulkner (2004, Volume II: 201, note 1) has translated the name as Thoth because 'the opening of the wings is an action more appropriate to the ibis Thoth than to the sun-god.' However, this rational could be applied to any of the winged gods, including Horus and Khepri. Furthermore, this could be a reference to the winged sun disk.

14 Following P. Gardiner III.

15 Leitz 2002, Volume V: 561 has only one attestation listed for the deity: this spell.

16 Following Coffin B2L; P. Gardiner III has a large lacuna in these lines.

17 All translations in this chapter are the author's own, unless otherwise noted.

18 Coffins B3B0ᵃ, B3B0ᵇ, B2L, P. Gardiner III (De Buck 1935–2006: 228).

19 The lexeme iAf is rare and is only recorded by Meeks (1998, Volume II: 15, no. 78.0152) as occurring in this spell. It is not listed in the *Wörterbuch*, nor in the *Wörterbuch der Medizinischen Texte*.

20 See Manassa (2007: 30–1, note 122) for a discussion on the gender of mourners, with particular focus on the later Underworld Books.

21 The title only occurs on Coffin M5C.

22 B2B0, B4B0, B1Be, M5C, M46C.

23 Barrett 2011: 133–4; Bommas 2005: 268–9; Centrone 2005: 355–9; Colin 2005: 282–92; Darnell 1995: 63, note 87; Delia 1992: 181–90; Koenig 2005:103; Klotz 2006: 31; Malaise 1985: 126–9; Manassa 2007: 65, 325, 373; Meeks 1971: 24, 68.

24 Coffin Text Spell 314 was an antecedent for the later Book of the Dead Chapter 1, where similar phraseology describes the ritual washing of the one who is hacked up (tštš). The Book of the Dead versions lack the reference to the snw-vase—see Munro 1994: pls. 2–3.

25 For further discussion on this myth, see Allam 1992: 137–45, with references therein.

26 Following the suggestion of Faulkner 2004, Volume I: 237, note 6, who reads ia.w with an erroneous t.

27 Erman and Grapow 1926–61, Volume V: 330. 5–11. The lexeme has just one attestation in the Medical Texts: P. Ebers 504. The word appears in conjunction with the production of a remedy (Von Deines and Westendorf 1962, Volume II: 962).

28 In his seminal essay on decomposition as a transitional rite of passage, Hertz 1960: 33 noted that the custom of catching putrefactive fluids within vessels is not confined to ancient Egypt or Greece. For example, one of the funerary customs of Bali was to keep the deceased in the house for many weeks. During that time, the coffin was pierced at the bottom 'to permit the escape of the liquids, which are gathered in a basin that is emptied every day with great ceremony'. In addition, the Dayak of Borneo collected putrefactive fluids in earthenware vessels and mixed those fluids with rice, which the close relatives would eat during the period of mourning.

29 Coffin T3C does not have the dative, but G1T and A1C both preserve the n=k.

30 Coffin T3C uses the second person feminine suffix pronoun, t, but at this point in the text the scribe has used the masculine pronoun, k. The remainder of the spell uses the feminine form.

31 S1P has ir.t Hr 'the eye of Horus'.

32 Following De Buck 1935–2006, Volume IV: 141, note 3*.

33 De Buck 1935–2006, Volume VII: 464g–465a. This essential difference between humans and gods could be a juxtaposition of two creation myths and is not mentioned elsewhere.

34 De Buck 1935–2006, Volume VI: 385d–e. The spell only occurs on one coffin—B1C.

Bibliography

Primary sources

Billing, N. (2002) *Nut, The Goddess of Life: In Text and Iconography*. Uppsala: Uppsala University.

Borghouts, J. (1978) *Ancient Egyptian Magical Texts*. Leiden: Brill.

De Buck, A. (1935–2006) *The Egyptian Coffin Texts*. Chicago: Chicago University Press.

Faulkner, R. O. (2004) *The Ancient Egyptian Coffin Texts: Spells 1–1185 & Indexes*. Oxford: Aris & Phillips.

Griffith, J. G. (1970) *Plutarch's De Iside et Osiride*. Cardiff: University of Wales Press.

Secondary literature

Allam, S. (1992) 'Legal Aspects in the "Contendings of Horus and Seth"', in A. Lloyd (ed.) *Studies in Pharaonic Religion and Society in Honour of J. Gwyn Griffiths* (London: The Egypt Exploration Society), 137–45.

Assmann, J. (1989) 'Death and initiation in the funerary religion of ancient Egypt', in W. Kelly Simpson (ed.) *Religion and Philosophy in Ancient Egypt*. New Haven: Yale Egyptological Studies III, 135–59.

Barguet, P. (1971) 'Les textes spécifiques des différents panneaux des sarcophages du Moyen Empire', *Revue d'Égyptologie* 23, 15–22.

Barrett, C. E. (2011) *Egyptianizing Figurines from Delos: A Study in Hellenistic Religion*. Leiden: Brill.

Billing, N. (2002) *Nut, The Goddess of Life: In Text and Iconography*. Uppsala: Uppsala University.

Bommas, M. (2005) 'Situlae and the offering of water in the divine funerary cult: A new approach to the ritual of Djeme', in A. Amenta, M. M. Luiselli, and M. N. Sordi (eds) *L'acqua nell'antico Egitto: Vita, rigenerazione, incantesimo, medicamento*. Rome: L'Erma di Bretschneider, 257–72.

Bonneau, D. (1964) *La Crue du Nil, divinité égyptienne, à travers mille ans d'histoire (332 av.-641 ap. J.-C.) d'après les auteurs grecs et latins, et les documents des époques ptolémaïque, romaine et byzantine*. Paris: Librairie C. Klincksieck.

Centrone, M. C. (2005) 'This is the form of [. . .] Osiris of the mysteries, who springs from the returning waters (south wall of the Osiris room at the great temple of Philae)', in A. Amenta, M. M. Luiselli, and M. N. Sordi (eds) *L'acqua nell'antico Egitto: Vita, rigenerazione, incantesimo, medicamento*. Rome: L'Erma di Bretschneider, 355–60.

Colin, M. -E. (2005) 'Presenting water to the deities within the Baroque sanctuaries of Graeco-Roman times', in A. Amenta, M. M. Luiselli, and M. N. Sordi (eds) *L'acqua nell'antico Egitto: Vita, rigenerazione, incantesimo, medicamento*. Rome: L'Erma di Bretschneider, 283–92.

Darnell, J. C. (1995) 'Hathor returns to Medamud', *Studien zur Altägyptischen Kultur* 22, 47–94.

———. (2004) *The Enigmatic Netherworld Books of the Solar-Osirian Unity: Cryptographic Compositions in the Tombs of Tutankhamun, Ramesses VI, and Ramesses IX*. Freiburg: Vandenhoeck & Ruprecht.

David, A. R. (2000) 'Mummification', in P. T. Nicholson and I. Shaw (eds) *Ancient Egyptian Materials and Technology*. Cambridge: Cambridge University Press, 372–89.

Delia, D. (1992) 'The Refreshing Water of Osiris', *Journal of the American Research Center in Egypt* 29, 181–90.

Dobbin-Bennett, T. (2014) 'Rotting in hell: Ancient Egyptian conceptions of decomposition and putrefaction', PhD thesis, Yale University, New Haven.

Drioton, É. (1949) 'Review of De Buck's *The Egyptian Coffin Texts III*', *Bibliotheca Orientalis* 5, 149.

Dunand, F. and R. Lichtenberg (2006) *Mummies and Death in Egypt*. Ithaca: Cornell University Press.

Eaton, K. (2018) 'Ancient Egyptian concepts of bodily decay in the Old Kingdom part 1, PT 684', *Zeitschrift für Ägyptische Sprache und Altertumskunde* 145(1), 43–56.

Erman, A. and H. Grapow (eds) (1926–61) *Wörterbuch der Aegyptischen Sprache*. Berlin: Akademie-Verlag.

Galloway, A., W. H. Birkby, A. M. Jones, T. E. Henry, and B. O. Parks (1989) 'Decay rates of human remains in an arid environment', *Journal of Forensic Science* 34(3), 607–16.

Hertz, R. (1960) *Death and the Right Hand*. Aberdeen: Cohen & West.

Hornung, E. (1983) *Conceptions of God in Ancient Egypt*. Ithaca, NY: Cornell University Press.

———. (1999) *The Ancient Egyptian Books of the Afterlife*. Ithaca, NY: Cornell University Press.

Janaway, R. C., S. L. Percival, and A. S. Wilson (2009) 'Decomposition of human remains', in S. L. Percival (ed.) *Microbiology and Aging: Clinical Manifestations*. New York: Springer, 313–34.

Kettel, J. (1993) 'Canopes, rDw.w d'Osiris et Osiris-Canope', in C. Berger (ed.) *Hommages à Jean Leclant III*. Cairo: IFAO, 315–26.

Klotz, D. (2006) *Adoration of the Ram*. New Haven: Yale Egyptological Seminar.

Koemoth, P. P. (1993) 'La "racine" wAb: du mythe à la métaphore', *Studien zur Altägyptischen Kultur* 20, 109–23.

Koenig, Y. (2005) 'L'eau et la magie', in A. Amenta, M. M. Luiselli, and M. N. Sordi (eds) *L'acqua nell'antico Egitto: Vita, rigenerazione, incantesimo, medicamento*. Rome: L'Erma di Bretschneider, 91–106.

Köhler, U. (1975) *Das Imuit: Untersuchungen zur Darstellung und Bedeutung eines mit Anubis verbundenen religiösen Symbols*. Wiesbaden: Harrassowitz.

Leitz, C. (2002) *Lexikon der ägyptischen Götter und Götterbezeichnung*. 8 vols. Leuven: Peeters.

Lucas, A. (1962) *Ancient Egyptian Materials and Industries*. London: Edward Arnold.

Malaise, M. (1985) 'Ciste et hydrie, symboles isiaques de la puissance et de la présence d'Osiris', in J. Ries (ed.) *Le symbolisme dans le culte des grandes religions*. Louvain-la-Neuve: Centre d'histoire des religions, 125–55.

Manassa, C. (2007) *The Late Egyptian Underworld: Sarcophagi and Related Texts from the Nectanebid Period*. Wiesbaden: Harrassowitz Verlag.

Meeks, D. (1971) 'Génies, anges, démons en Égypte', in D. Meeks (ed.) *Génies, anges et démons: Égypte, Babylone, Israël, Islam etc., Sources Orientales VIII*. Paris: Éditions du Seuil, 19–84.

———. (1998) *Année Lexicographique Égypte Ancienne*. Paris: Cybele.

Megyesi, M. S., S. P. Nawrocki and N. H. Haskell (2005) 'Using accumulated degree-days to estimate the postmortem interval from decomposed human remains', *Journal of Forensic Science* 50(3), 618–26.

Munro, I. (1994) *Die Totenbuch-Handschriften der 18. Dynastie im Ägyptischen Museum Cairo*. Wiesbaden: Harrassowitz Verlag.

Nyord, R. (2009) *Breathing Flesh: Conceptions of the Body in the Ancient Egyptian Coffin Texts*. Copenhagen: Museum Tusculanum Press.

———. (2012) 'On (mis)conceptions of the body in ancient Egypt', *LingAeg* 20, 165–84.

Pantalacci, L. (1981) 'Une conception originale de la survie osirienne d'après les textes de basse époque', *Göttinger Miszellen* 52, 57–66.

———. (1983) '[Ounem-Houaat]. Genèse et carrière d'un génie funéraire', *Bulletin de l'Institut Français d'Archéologie Orientale* 83, 297–311.

Parkinson, R. A., K.-R. Diaz, J. Horswell, P. Greenwood, N. Banning, M. Tibbett, and A. A. Vass (2009) 'Microbial community analysis of human decomposition on soil', in K. Ritz, L. Dawson, and D. Miller (eds) *Criminal and Environmental Soil Forensics*. Heidelberg: Springer, 379–94.

Prahlow, J. (2010) *Forensic Pathology for Police, Death Investigators, Attorneys, and Forensic Scientists*. New York: Humana Press.

Rizzo, J. (2007) 'Le terme Dw comme superlatif de l'impur: L'exemple de abw Dw', *Revue d'Égyptologie* 58, 123–35.

te Velde, H. (1967) *Seth, God of Confusion: A Study of his Role in Egyptian Mythology and Religion*. Leiden: E.J. Brill.

van der Molen, R. (2000) *A Hieroglyphic Dictionary of Egyptian Coffin Texts*. Boston: Brill.

Vass, A. A. (2001) 'Beyond the grave: Understanding human decomposition', *Microbiology Today* 28, 190–2.

Von Deines, H. and W. Westendorf (1962) *Wörterbuch der medizinischen Texte*. 2 vols. Berlin: Akademie Verlag.

———. (1996) *The Coffin of Heqata (Cairo JdE 36418): A Case Study of Egyptian Funerary Culture of the Early Middle Kingdom*. Leuven: Uitgeverij Peeters en Departement Oriëntalistiek.

Winkler, A. (2006) 'The efflux that issued from Osiris: A study on rDw in the pyramid texts', *Göttinger Miszellen* 211, 125–39.

Zandee, J. (1966) *An Ancient Egyptian Crossword Puzzle*. Leiden: Ex Oriente Lux.

———. (1975–6) 'Sargtexte um über Wasser zu verfügen (Coffin Texts V 8–22; Sprüche 356–62)', *Jaarbericht van het Vooraziatisch-Egyptisch Genootschap Ex Oriente Lux* 24, 1–47.

Zandee, Z. (1960) *Death as an Enemy: According to Ancient Egyptian Conceptions*. Leiden: Brill.

Zhou, C. and R. Byard (2011) 'Factors and processes causing accelerated decomposition in human cadavers: An overview', *Journal of Forensic and Legal Medicine* 18, 6–9.

20

THE PHYSIOLOGY OF MATRICIDE

Revenge and metabolism imagery in Aeschylus' *Oresteia**

Goran Vidović

'[T]hose aspects of our animal humanness that get excluded from or bottled up in other genres' are common subjects in ancient satire and Aristophanic comedy. So argues Daniel Hooley, drawing a distinction between 'indecent' and 'decent' literature (the latter represented by 'epic') on an intuitive classification of bodily fluids and emissions. While listing '[s]hit, vomit, pus, gas, semen' to exemplify the domain of comedy and satire, he adds parenthetically that there is 'not much blood, an epic fluid', and that *nobody pisses in epic*' (Hooley 2007: 8, emphasis mine). Indeed, scatology is the endemically comic grammar: 'Farting and excreting . . . are an important component of the comic hero's "arsenal" of self-expression'; the few extended scenes of defecation in Aristophanes are 'invariably for thematic purposes . . . which transcend the merely farcical' (Henderson 1991: 54, 397). For example, it is thematically justified that Aristophanes' protagonist defecates on an arms-dealer corselet and calls personified War the 'lord of shitting down your legs' in the *Peace* (1226–37; ὁ κατὰ τοῖν σκελοῖν, 241; trans. Silk 2002: 154; see Edwards 1991).

A fortiori, explicit references to bodily discharge—other than 'blood, sweat, and tears'—call for special alertness when they occur in 'decent', high-register literature. A scene where someone *does* piss—not in 'epic' but, equally inappropriately by Hooley's standards, in Classical Athenian tragedy—occurs at the critical moment in Aeschylus' *Libation Bearers* (*Choephoroe*), the central play of his *Oresteia* trilogy. Orestes and Pylades, disguised as foreigners, bringing news of Orestes' alleged death, return to avenge the death of Agamemnon. After an unsuspecting Clytemnestra welcomes them and they go into the house, Orestes' childhood nurse Cilissa comes out to lament his reported death, recalling how she took care of him in infancy. Her speech will be analysed in detail (*Libation Bearers* 743–63):

ὦ τάλαιν' ἐγώ,	(743)
ὥς μοι τὰ μὲν παλαιὰ συγκεκραμένα	
ἄλγη δύσοιστα τοῖσδ' ἐν Ἀτρέως δόμοις	(745)
τυχόντ' ἐμὴν ἤλγυνεν ἐν στέρνοις φρένα,	
ἀλλ' οὔ τί πω τοιόνδε πῆμ' ἀνεσχόμην.	
τὰ μὲν γὰρ ἄλλα τλημόνως ἤντλουν κακά·	
φίλον δ' Ὀρέστην, τῆς ἐμῆς ψυχῆς τριβήν,	
ὃν ἐξέθρεψα μητρόθεν δεδεγμένη	(750)
καὶ νυκτιπλάγκτων ὀρθίων κελευμάτων	
< . . . . >	

DOI: 10.4324/9780429438974-27

καὶ πολλὰ καὶ μοχθήρ' ἀνωφέλητ' ἐμοὶ
τλάσῃ· τὸ μὴ φρονοῦν γὰρ ὡσπερεὶ βοτὸν
τρέφειν ἀνάγκη—πῶς γὰρ οὔ;—τροφοῦ φρενί·
οὐ γάρ τι φωνεῖ παῖς ἔτ' ὢν ἐν σπαργάνοις   (755)
εἰ λιμὸς ἢ δίψη τις ἢ λιψουρία
ἔχει· νέα δὲ νηδὺς αὐτάρχης τέκνων.   [MSS. αὐτάρκης]
τούτων πρόμαντις οὖσα, πολλὰ δ' οἴομαι
ψευσθεῖσα, παιδὸς σπαργάνων φαιδρύντρια,
κναφεὺς τροφεύς τε ταὐτὸν εἰχέτην τέλος.   (760)
ἐγὼ διπλᾶς δὲ τάσδε χειρωναξίας
ἔχουσ' Ὀρέστην ἐξεθρεψάμην πατρί·
τεθνηκότος δὲ νῦν τάλαινα πεύθομαι.   (763)

O wretched me!
For I found the old griefs that have happened in this
house of Atreus hard enough to bear, all mixed together as they
were, and they pained my heart within my breast;
but I have never yet had to endure a sorrow like this.
The other troubles I patiently put up with.
But dear Orestes, who wore away my life with toil,
whom I reared after taking him straight out of his mother!
<Over and over again I heard> his shrill, imperative cries, which forced
me to wander around at night <and perform> many disagreeable tasks
which I had to endure and which did me no good.
A child without intelligence must be reared
like an animal—how else?—by the intelligence of his nurse [?];
when he's still an infant in swaddling clothes he can't speak at all
if he's in the grip of hunger or thirst, say, or of an urge to make water—
and the immature bowel of small children self-governing. [MSS.: self-sufficient]
I had to divine these things in advance,—and often, I think, I
was mistaken, and as cleaner of the baby's wrappings—
well, a launderer and a caterer were holding the same post.
Practicing both these two crafts,
I reared up Orestes for his father;
and now, to my misery, I learn that he is dead![1]

Aeschylus, *Libation Bearers* 743–63

What to make of these 'ordinary things that nowhere else find a place in extant Greek tragedy' (Garvie 1986: 243–4), which 'take domestic detail further than any other scene in Greek tragedy' (Gregory 2009: xxiii)? Why are we hearing about infant Orestes' urge to urinate, soiling diapers, and incontinent bowels at all—let alone at such a climactic moment, right before he murders his mother?

One earlier commentator describes this monologue as the 'pithy illiterate babble of the old woman', essentially dismissing the Nurse's perceived disruption of the tragic register as a timely 'comic relief'.[2] Alan Sommerstein reminds us that this murky label, originally invented to account for scenes in Shakespeare 'that offended against what were thought to be fundamental aesthetic canons', has been occasionally applied to the lowborn characters of *Oresteia* (Sommerstein 2002: 152). But his survey of comic language in the trilogy shows that only less than 10 percent of it is assigned to low-status characters, demonstrating instead that linguistic breach of aesthetic decorum coincides with the outburst of violence in the *Oresteia*. From the lexical distribution he persuasively concludes that 'far from being light relief of any sort, comic language is used in the *Oresteia* to heighten the blackness and bleakness of the vicious cycle of retaliatory violence, and disappears at the point where that cycle is broken'; simply put, 'ugly deeds that can only be described in an ugly way' (Sommerstein 2002: 163–4; cf. Seidensticker 1982: 65).

With different methodology but along similar lines, A. F. Garvie observes that the Nurse's speech is more than a sideshow: she provides a rare display of genuine affection in the trilogy, standing out as the mother-figure of Orestes as a foil to Clytemnestra, while the helpless baby creates a 'grim contrast' with the grown murderer (Garvie 1986: 243–4 on *Libation Bearers* 730–82). Specific correspondences are conspicuous: baby Orestes mirrors the newborn snake in Clytemnestra's dream (*Libation Bearers* 523–53; cf. Rousseau 1963; Catenaccio 2011 on dreams in the *Oresteia*); Cilissa describes the baby's cries keeping her 'wandering in the night' (νυκτιπλάγκτων, 751) with an Aeschylean coinage used only in the *Oresteia*, which Garvie sees as a 'deliberate echo' of Clytemnestra haunted by the nightmare (νυκτιπλάγκτων δειμάτων πεπαλμένη, 524; Garvie 1986: 248, on *Libation Bearers* 751–3).[3] Infant Orestes cannot speak and is reared like an animal (*Libation Bearers* 753–5), which Garvie connects with the prominent animal imagery in the trilogy (for which see e.g. Knox 1952; Heath 1999; Saayman 1993). In the *Libation Bearers*, he is 'in swaddling-clothes' (ἐν σπαργάνοις, 755), like the snake (ἐν σπαργάνοισι, 529; cf. perhaps οὖφις †επᾶσα σπαργανηπλείζετο† 544). Moreover, 'Orestes' situation in a sense parallels that of Aegisthus, who was driven into exile while still ἐν σπαργάνοις (*Ag[amemnon]* 1606), only to return later (1607) as the avenger' (Garvie 1986: 188, on *Libation Bearers* 529; more on these passages later).

Cilissa thus addresses in one way or another some major themes of the trilogy, such as parenthood and speech,[4] and Orestes' infancy becomes a miniature figurative re-enactment of the Orestes myth at large. Even the most graphic of the details, the washing of Orestes' diapers (σπαργάνων φαιδρύντρια, *Libation Bearers* 759), transcends the mundane 'pithy illiterate babble' if nothing else by alluding lexically to the bath of Agamemnon (λουτροῖσι φαιδρύνασα, *Agamemnon* 1109);[5] the bathtub allusion is activated by a brutally ironic move in the immediately preceding scene where unsuspecting Clytemnestra welcomes Orestes, offering a warm bath (θερμὰ λουτρὰ, *Libation Bearers* 670).

The question now concerns the thematic valence of Orestes' soiling the diapers in the first place and his incontinence, which makes the whole cycle uncontrollable and frustrating for the Nurse. Such a specific and inappropriate image calls for attention if only because '[t]he single most compelling feature of the artistry of the *Oresteia* is its elaborate network of image and metaphor' (Zeitlin 1965: 463), and every image is 'part of a larger whole: a system of kindred imagery' (Lebeck 1971: 1).[6] Especially developed among these networks of metaphor is the imagery of liquids; blood, for example, is likened to sacrificial wine (Zeitlin 1965) or textile dye (and more: Lebeck 1971: 80–91). Nor are the associations limited to 'epic' fluids: the 'quasi-erotic' overtones of the pleasure Clytemnestra felt on being sprinkled by Agamemnon's blood

(*Agamemnon* 1389–92) had been arguably noticed already by Sophocles (*Antigone* 1238–9; Sommerstein 2002: 154). If, then, Sommerstein is right, that the horrors of the *Oresteia* are intensified by the ugliness and inappropriateness of comic register—and comedy, for its own part, enjoys 'dramatizing [a character's] dependence upon, and frequently his lack of control over, his bodily needs' (Henderson 1991: 54)—then there is something thematically significant about Orestes' metabolism.

In particular, I interpret the image of the physiological cycle of intake and discharge of bodily fluids as a symbolic expression of another kind of cycle, indeed the fundamental one for the trilogy: of crime and retribution. The infant Orestes' incontinence in disposing bodily waste allegorically captures the essence of the vicious cycle of revenge in the Atreid myth: while the drive is instinctive, understandable, and sometimes felt as necessary, it should not be left to go on unrestrained. As a background for the detailed analysis of the symbolic function of the imagery of Orestes' metabolism, I first discuss select references to the quantity and circulation of fluids, both inside and outside the body.

Aeschylus persistently exploits the dichotomy of wetness and dryness, especially when playing with the ambiguity regarding which one is more desirable. The messenger speech in the *Agamemnon*, for example, presumes two contrasting ancient commonplaces: that water is a vital fluid and dangerous as a means of traffic.[7] Agamemnon's messenger reports that after the tumultuous sailing of the Greek fleet, dry land at Troy was even *worse*—but only because the soil was wet and it was constantly raining, whereby their clothes got infested with vermin (*Agamemnon* 558–62). That is, water is both the sailor's nightmare and the source of life after all, only not of the sort of living creatures that the Greek warriors would want. That Agamemnon escaped the 'sea Hades' (Ἅιδην πόντιον, *Agamemnon* 665) pointedly foreshadows his humiliating death in the tub: the domestic, downscaled sea (parallels for this type of murder collected by Bremmer 1986).

Frequently, wetness and dryness logically stand for vitality and perishing, respectively. The leaves of the bone-marrow (μυελὸς) of an old man are already 'withering' (τό θ' ὑπεργήρων φυλλάδος ἤδη | κατακαρφομένης, *Agamemnon* 74–80; on the bone-marrow as the 'vital fluid that is the stuff of life', see Sommerstein 2008: 11, ad loc.). Clytemnestra poses as a withering wife who cried herself dry: 'In my eyes the gushing fountains of tears have dried up, there's not a drop left' (ἔμοιγε μὲν δὴ κλαυμάτων ἐπίσσυτοι | πηγαὶ κατεσβήκασιν, οὐδ' ἔνι σταγών, *Agamemnon* 887).[8] These physiological processes of the human body are imaged as a natural phenomenon participating in the ecosystem, much like when the internal organs 'whirl in eddies' (σπλάγχνα δ' οὔτοι ματᾳ- | ζει πρὸς ἐνδίκοις φρεσὶν | τελεσφόροις δίναις κυκλούμενον κέαρ, *Agamemnon* 995–7; cf. Euripides *Suppliants* 203–7).[9]

Aeschylus is particularly fond of images of dehydration and absorption. The war god Ares is famously portrayed as a short-changing banker who dries out the living, sending the ashes back to be soaked in tears of their beloved (*Agamemnon* 438–44):

ὁ χρυσαμοιβὸς δ' Ἄρης σωμάτων
καὶ ταλαντοῦχος ἐν μάχᾳ δορὸς
πυρωθὲν ἐξ Ἰλίου
φίλοισι πέμπει βαρὺ

ψῆγμα δυσδάκρυτον ἀν-
τήνορος σποδοῦ χεμί-
ζων λέβητας εὐθέτους.

Ares, the moneychanger of bodies,
holding his scales in the battle of spears,
sends back from Ilium to their dear ones
heavy dust that has been through the fire,
to be sadly wept over,
filling easily-stowed urns
with ash given in exchange for men.

Aeschylus, Agamemnon 438–44

Ares' funerary receptacles are curious, since *lebēs* (λέβης) is normally a container for liquids. By repurposing the word, Aeschylus has Ares practically drain the life fluid and return what is left in the same 'package'. A *lebēs* is allowed to contain dried remains because in the *Oresteia*'s network of imagery it will be treacherously lethal even when it contains fluid; the most ominous occurrence of *lebēs* containing liquid is in reference to Agamemnon's fatal bathtub (*Agamemnon* 1129; see Fraenkel 1950: 515–16, ad loc.). This specific semantic restriction of the term to denote lethal fluidity (and, as a consequence, dryness) seems like an ad hoc Aeschylean innovation which was not lost on Sophocles. Namely, the only other two instances where *lebēs* denotes a cinerary urn is for the vessel with Orestes' alleged ashes in both the *Libation Bearers* (686; Garvie 1986: 232, ad loc.) and Sophocles' *Electra* (1401). Here the traces of semantic intervention are still visible—the vessel *does not really* contain Orestes' ashes—but thus making it in a way all the more effective: while the *lebēs* does not materialise the death of Orestes, it announces his return alive, which spells doom for Clytemnestra and Aegisthus; as Patrick Finglass comments on the Sophoclean instance: 'To [Electra] he brings new life, to [Clytemnestra] death' (Finglass 2007: 513).[10] Sophocles continues the sinister symbolism of *lebēs* as he uses it for the container in which Deianeira received the poisonous blood of the centaur Nessus presented as love-potion (*Trachinian Women* 556; numerous parallels between this play and the *Oresteia* have long been on the record: see e.g. Garner 1990: 100ff.)

Shortly after the Ares passage, the dust cloud raised by the messenger's arrival is called 'thirsty dust, the sister and neighbor of mud' (κάσις | πηλοῦ ξύνουρος διψία κόνις, *Agamemnon*, 494–5).[11] This same dust in the *Eumenides* absorbs human blood irreversibly (ἀνδρὸς δ' ἐπειδὰν αἷμ' ἀνασπάσῃ κόνις | ἅπαξ θανόντος, οὔτις ἔστ' ἀνάστασις, *Eumenides* 646–7); once the Erinyes are appeased, they pray that 'the dust not drink up the dark blood of citizens' (μηδὲ πιοῦσα κόνις μέλαν αἷμα πολιτᾶν, *Eumenides* 980; see also Lebeck 1971: 86–8). The image of dry dust absorbing life turns out to be a very Aeschylean one, and once again it impressed Sophocles (see Cairns 2014 on *Seven against Thebes* 734–7 and *Antigone* 599–603).

The corresponding image in the *Oresteia*, especially prominent in the *Libation Bearers*, is that of compensating for bloodshed by pouring another liquid, in a sort of rehydration (cf. Zeitlin 1965). Electra outlines the Aeschylean drainage system: 'Now my father has the drink-offerings—the earth has swallowed them' (ἔχει μὲν ἤδη γαπότους χοὰς πατήρ, *Libation Bearers* 164). The heroes' ashes returned by Ares the banker, we have seen above, expect tears. But the whole point of the trilogy is that neither tears nor libations will do, but 'it is the law that when drops of gore flow to the ground, they demand other blood' (ἀλλὰ νόμος μὲν φονίας σταγόνας | χυμένας ἐς πέδον ἄλλο προσαιτεῖν | αἷμα, *Libation Bearers* 400–2). When Clytemnestra, terrified by nightmares, sends libations to Agamemnon's tomb, Orestes is equally uncompromising (*Libation Bearers* 519–21):

> τὰ δῶρα μείω δ' ἐστὶ τῆς ἁμαρτίας·
> τὰ πάντα γάρ τις ἐκχέας ἀνθ' αἵματος
> ἑνός—μάτην ὁ μόχθος. ὧδ' ἔχει λόγος.

> The gifts do not match the crime.
> Pour out all you have in atonement for one man's blood—
> and your work is wasted; so the saying goes.

> Aeschylus, *Libation Bearers* 519–21

As Anne Lebeck summarises the dynamics,

> Taken together the pouring of libations and the flow of blood form complementary halves of a single idea. The ostensible purpose of the libations is to mollify those infernal powers whose wrath is roused by bloodshed. Yet no drink-offering can effect this but an offering of blood.[12]

> Lebeck 1971: 86

Retribution in the *Oresteia*, therefore, is routinely conceived of as a cycle of losing and replenishing liquids in various forms, notably the bodily ones. But the crucial exception that proves the rule are liquids which rather *produce* desiccation. This peculiar process is delegated to the embodiments of revenge, the Erinyes, introduced as detestable creatures with gory ooze dripping from their eyes (ἐκ δ' ὀμμάτων λείβουσι δυσφιλῆ λίβα, *Eumenides* 54). Clytemnestra's ghost urges them to wither Orestes dry with bloody breath from their bellies' fire (σὺ δ' αἱματηρὸν πνεῦμ' ἐπουρίσασα τῷ, | ἀτμῷ κατισχναίνουσα, νηδύος πυρί, *Eumenides* 137–8). Their destructive potential is elaborated in the choral ode where they threaten to sterilise the earth by raining vengeful poison from somewhere within them (*Eumenides* 780–7 = 808–17; 800–4):

> ἐγὼ δ' ἄτιμος ἁ τάλαινα βαρύκοτος   (780)
> ἐν γᾷ τᾷδε, φεῦ,
> ἰὸν ἀντιπεν-
> θῆ μεθεῖσα καρδίας,
> σταλαγμὸν χθονὶ
> ἄφορον, ἐκ δὲ τοῦ
> λειχὴν ἄφυλλος ἄτεκνος, ὦ Δίκα Δίκα,   (785)
> πέδον ἐπισύμενος
> βροτοφθόρους κηλῖδας ἐν χώρᾳ βαλεῖ.
> . . .
> Αθ. ὑμεῖς δὲ μήτε τῇδε γῇ βαρὺν κότον   (800)

σκήψητε, μὴ θυμοῦσθε, μηδ' ἀκαρπίαν
τεύξητ' ἀφεῖσαι †δαιμόνων† σταλάγματα,
βρωτῆρας **αὐχμούς** σπερμάτων ἀνημέρους.

And I, wretched that I am, am dishonoured, grievously angry,
releasing poison, poison
from my heart to cause grief in revenge
in this land—ah!—a drip falling on the land,
such that it cannot bear! And from it
a canker causing leaflessness and childlessness—Justice, Justice!—
sweeping over the soil
will fill the land with miasmas fatal to humans.

. . .

Athena: So do not send down grievous wrath against this land;
do not be angry; do not create sterility
by releasing a dripping liquid from your lungs [?]
to make savage **droughts** that devour the seed.

Aeschylus, *Eumenides* 780–7 = 808–17; 800–4

Among several textual uncertainties, one is relevant for the present discussion. Sommerstein rejects Scaliger's emendation αὐχμούς ('droughts') of the MSS αἰχμάς ('spears') at 803 for the same reason for which I accept it, namely, that it 'would create a confused picture, of a poison that *drips* on the earth and yet makes it *dry*' (Sommerstein 1989: 243–4, ad loc., original emphasis). Since Erinyes represent revenge in kind, it makes sense that fluidity, which elsewhere replenishes the loss of life, assumes a lethal force when coming from them. As was the case with Agamemnon's bath, a fluid becomes deadly precisely by perverting its otherwise presumed revitalising attributes, thereby emphasising that an act of violence is committed in response to a previous crime.

Particularly telling is how exactly that works with the Erinyes. What accounts for the capacity of their excretions to cause dryness is that they result from the substance which the Erinyes had previously dried out. Apollo describes them as indigestion personified: 'Give back in agony black foam taken from human bodies, vomiting out the clots of blood that you have sucked' (ἀνῆις ὑπ' ἄλγους μέλαν' ἀπ' ἀνθρώπων ἀφρόν, | ἐμοῦσα θρόμβους οὓς ἀφείλκυσας φόνου, *Eumenides* 183–4; the 'clot' is a significant detail in Clytemnestra's dream; see later). He compares them to a 'blood-slurping lion' (λέοντος ἄντρον αἱματορρόφου, *Eumenides* 193; a comic derivation: Sommerstein 2002: 161), anticipating their self-proclaimed mission:

Xo.   αἷμα μητρῷον χαμαὶ   (261)
δυσαγκόμιστον, παπαῖ,
τὸ διερὸν πέδῳ χύμενον οἴχεται.
ἀλλ' ἀντιδοῦναι δεῖ σ' ἀπὸ ζῶντος **ῥοφεῖν**
ἐρυθρὸν ἐκ μελέων πελανόν, ἀπὸ δὲ σοῦ   (265)
βοσκὰν φεροίμαν πώματος δυσπότου·
καὶ ζῶντά σ' ἰσχνάνασ' ἀπάξομαι κάτω,
ἀντίποιν' ὡς τίνῃς ματροφόνος δύας.

Chorus: A mother's blood on the ground
is hard to bring back up—papai!—
wet blood that is shed on to the earth and disappears.
No, you must give in return a thick red liquid
from your limbs for us to **slurp** from your living body: from you
may I draw the nourishment of a draught horrid to drink!
And having drained you dry while you live, I'll haul you off below,
so that you may pay in suffering the penalty of your matricide.

Aeschylus, Eumenides 261–8

For the Erinyes, Orestes is food to be drained into a bloodless shadow (ἀναίματον βόσκημα δαιμόνων, σκιάν, *Eumenides* 302). But eventually, Apollo says,

σύ τοι τάχ' οὐκ ἔχουσα τῆς δίκης τέλος
ἐμῇ τὸν ἰὸν οὐδὲν ἐχθροῖσιν βαρύν.

You will shortly, when you fail to gain final victory in the trial,
vomit up your poison and find it does no harm to your enemies.

Aeschylus, Eumenides 729–30

The Erinyes' primary function, therefore, is conceived of as a physiological process: they feed on defiled fluids of criminals and generate murderous fluids in return. Simply put, they exact revenge by releasing bodily waste. Appropriately, they are regularly attributed with spitting, specifically with the compounds of the verb *ptuō* (πτύω), which 'expresses a strong ritual rejection' (Catenaccio 2011: 208, with fn. 17, cf. Hesiod, *Works and Days* 726). Apollo introduces them as 'the abominable maidens', literally 'to be spat upon' (αἱ κατάπτυστοι κόραι, *Eumenides* 68), who feast on human suffering, which other gods detest, 'spit out' (ἀπόπτυστου θεοῖς, *Eumenides* 191). The Erinyes also spit on the befouled marital bed of Atreus (ἀπέπτυσαν, *Agamemnon* 1192).

This grotesque image of slurping, vomiting, and spitting avengers is not only verging on comic—confirming Sommerstein's thesis that comic locutions 'heighten the blackness and bleakness of the vicious cycle of retaliatory violence'—but is also thematically integrated, as it recalls central events in the course of the Atreid curse. Aegisthus takes time to retell the myth of his father, Thyestes, who vomited a meal of his sons:

κἄπειτ' ἐπιγνοὺς ἔργον οὐ καταίσιον
ᾤμωξεν, ἀμπίπτει δ' ἀπὸ σφαγὰς **ἐρῶν**,
μόρον δ' ἄφερτον Πελοπίδαις ἐπεύχεται   (1600)
λάκτισμα δείπνου ξυνδίκως τιθεὶς ἀρᾷ,
οὕτως ὀλέσθαι πᾶν τὸ Πλεισθένους γένος.
ἐκ τῶνδέ σοι πεσόντα τόνδ' ἰδεῖν πάρα·
κἀγὼ δίκαιος τοῦδε τοῦ φόνου ῥαφεύς·
τρίτον γὰρ ὄντα μ' ἔλιπε, καθλίῳ πατρὶ   (1605)
συνεξελαύνει τυτθὸν ὄντ' ἐν σπαργάνοις,
τραφέντα δ' αὖθις ἡ Δίκη κατήγαγεν.

Then, when he recognised the unrighteous deed,
he howled aloud, fell backwards while **vomiting out** the
slaughtered remains, and called down an unendurable fate
on the house of Pelops, kicking over the table to chime
with his curse: 'So perish all the race of Pleisthenes!'
It is because of this, you see, that you now behold this man [i.e. Agamemnon]
fallen.
And I was rightfully entitled to contrive this slaying.
I was my wretched father's third child; Atreus . . ., drove me
out together with him, when I was a tiny infant in swaddling clothes.
When I grew up, Justice brought me back again.

Aeschylus, Agamemnon 1598–607

The 'digestive curse' goes further back, about as far back in the bloodline as it can go. Thyestes' father Pelops was cooked and served to the gods by his father, Tantalus, who was punished for it by eternal thirst and continuously receding water. (Alternatively, Tantalus stole nectar and ambrosia from the Olympian banquet—another nutritional violation.) The Aeschylean liquid imagery presently discussed was indeed so fundamental for the story of Atreids that it is attested operating on other media as well. Tantalus emblematised the flow of bodily fluids. An especially fascinating example of iconographic evidence is that of amulets depicting uterine jars with Tantalus invoked to drink menstrual blood—or not, since menorrhea can be, uniquely, both pathological and beneficial (discussed by Faraone 2009). Apparently, various problems with the amount of fluids in the body—both good and bad, sometimes simultaneously—run in the family, so to speak. Calling the Atreid curse recursive is more than a pun. Revenge in the *Oresteia* operates within the body, imaged as a physiological process of intake and discharge of bodily fluids. Symptomatically, the grief for Argive heroes fallen in the Trojan war—of which the initial casualty, Iphigenia, triggered this phase of the Atreid revenge cycle covered by *Oresteia*—causes pain in the liver (ἧπαρ, *Agamemnon* 432).

Unsurprisingly, Aeschylus draws strong parallels between the trilogy's agents of revenge, Orestes and the Erinyes. Orestes is likewise spitting, as he 'disregards' their threats (*apoptueis*, ἀποπτύεις, *Eumenides* 303). The nurse, troubled with the baby's 'urge to urinate' (*lipsouria*, λιψουρία, *Libation Bearers* 756), refers to the organ responsible for it as *nēdus* (νηδύς, 757), matching the *nēdus* of the Erinyes, where, as we have seen, they are to generate the fire to dehydrate Orestes (*Eumenides* 138, mentioned earlier). The term is well chosen, since it can mean 'bowels' or any cavity generally but is often associated with a collection of fluids. In the Hippocratic *On Airs, Waters, and Places* (19), it directly reflects climatic humidity, responding virtually as a hygrometer. It is the epicentre of thirst and unrestrained appetite (Euripides, *Cyclops* 243–6, 303ff., 574–5; Napolitano 2003: 145–6) and is paired with agricultural irrigation (Euripides, *Suppliants* 205–7; Harry 1912). Finally, a fluid parallel between Orestes and the Erinyes is *thrombos*, the clot of blood which Erinyes would vomit back (θρόμβος, *Eumenides* 184, mentioned earlier) and the clot that Orestes-as-snake sucks with his mother's milk in her dream (θρόμβος, *Libation Bearers* 533, 546). Evidently, Orestes and the Erinyes share a comparable anatomy and a similar diet. While Erinyes slurp the polluted blood of transgressors and excrete poisonous ooze in return, Orestes is symbolically breastfed by Clytemnestra's toxic milk and produces murderous discharge.[13]

In such an arrangement, Cilissa washing Orestes' diapers foreshadows his eventual purification and absolution from guilt in the *Eumenides*, an outcome likewise rich in language of literal washing. Right after the Erinyes remind Apollo, the 'purifier of houses', who 'cleansed' Orestes from murder (δωμάτων καθάρσιος, *Eumenides*, 63; φόνου δὲ τοῦδ' ἐγὼ καθάρσιος, *Eumenides* 578), that Orestes spilled his mother's blood—that is, his own—they imply that no community will allow him to use their lustral water (τὸ μητρὸς αἷμ' ὅμαιμον ἐκχέας πέδοι . . . ποία δὲ χέρνιψ φρατέρων προσδέξεται; *Eumenides* 653, 656).

But just as the moral and legal absolution of Orestes will take some doing, so is Cilissa at pains to keep up with her hygienic duties. The language she uses to describe the family miseries paints an image of overflowing filth which is impossible to hold back. Orestes' death is too much for her (*Libation Bearers* 747); previously, she could 'patiently put up with all the other sufferings', literally, 'drain them out' (τὰ μὲν γὰρ ἄλλα τλημόνως ἤντλουν κακά, 748). The verb *antleō* (ἀντλέω), primarily meaning bailing out bilge water from a ship, is symbolically charged. In line with the trilogy's recurrent concern with perilous excess (e.g. *Agamemnon* 376–8), the house of Atreus is spoken of as an overburdened ship that risks going off-course and sinking (*Agamemnon* 1005–13); punishment for crime is imaged as a shipwreck resulting from excessive, confusingly mixed, illegal, unjust cargo (τὸν ἀντίτολμον δέ φαμι παρβάταν | ἄγοντα πολλὰ παντόφυρτ' ἄνευ δίκας, *Eumenides* 554, 550–65). Continuing this imagery, Cilissa's arresting alliteration τλημόνως ἤντλουν, rather than being 'probably accidental' (Garvie 1988: 247, ad loc.), literally *blends* ultimate misfortune with the inability to control the inflow and outflow of liquids. By jumbling up the letters she is practically implementing her impression that woes of this family are all 'mixed together' (συγκεκραμένα, 744).[14] This is a flood.[15] Like the Atreid dynastic ship, Orestes' diapers are overflown with crime and revenge, inseparably, and they are leaking.

This close focus on Orestes' diapers raises the question of why Cilissa does not specify what exactly the discharge is. There is some debate whether she means only urine, since only the need to urinate is mentioned at 756 (so Garvie 1986: 250, on *Libation Bearers* 757), or is there, as Sommerstein understands it, a 'veiled reference to the evacuation of solid waste', because urine traces alone 'would hardly require the services of a κναφεύς ([*knapheus*] 760)', and explicit mention of faeces would be too much for tragedy where even the otherwise decent noun *kopros* (κόπρος) is avoided (Sommerstein 2002: 159). True, clothes stained with diarrhoea are cleaned by a *knapheus* (Aristophanes, *Wasps* 1126–8), and the noun *antlia* (ἀντλία, cognate of Cilissa's ἀντλέω in *Libation Bearers* 748) can mean excrements (Aristophanes, *Peace* 18). But having seen earlier all of Aeschylus' graphic descriptions unparalleled elsewhere, one wonders why he would stop short of finding a way to express the urge to defecate if he wanted to; the opening scene of Aristophanes' *Frogs* suggests such a need was well exploited in comedy, and the word Cilissa uses for the urinary pressure is a bold tragicomic compound itself.[16] Perhaps a sufficient explanation would be that Aeschylus specifically points out urine to emphasise the liquidity of the process (though in ordinary circumstances a healthy infant's faeces is somewhat more liquid than solid anyway). But once we analyse how exactly Cilissa speaks of her duties, one particular effect of urine might add a further layer of meaning to her otherwise exceptionally symbolic role.

Namely, Cilissa says that in taking care of Orestes she did everything by her-self, so 'the nurse and laundrywoman had a combined duty' (Lattimore 1953), that is, 'washerwoman and wet-nurse shared the shop' (Fagles 1984), or 'launderer and a caterer were holding the same post' (Sommerstein 2008) and, literally, 'the nurse and launderer had the same *telos*' (κναφεὺς τροφεύς τε ταὐτὸν εἰχέτην τέλος, 760). This must convey something more than simply 'I myself practiced these two crafts', which she will say in those exact words in the following line (διπλᾶς δὲ τάσδε χειρωναξίας, 761). Rather, the verse 760, I argue, epitomises the trilogy's central theme of reciprocal circularity of cause and effect: 'the same person was giving him milk and cleaning his waste' applies not only to Orestes' nurse and surrogate mother who provides him with beneficial nourishment and then has to take care of his discharge but also to Orestes' biological mother, who instead of due mother's milk feeds him only with cursed blood-clotted heritage and consequently faces his revenge.[17] The fundamental issue of the *Oresteia* is that crime is necessarily fol-lowed by counter-crime; 'the impious deed breeds more to follow, resembling their progenitors' (τὸ δυσσεβὲς γὰρ ἔργον | μετὰ μὲν πλείονα τίκτει, | σφετέρᾳ δ᾽ εἰκότα γέννᾳ, *Agamemnon* 758–60). The 'doer must suffer' might as well be the unofficial subtitle of the trilogy:

> ὄνειδος ἥκει τόδ᾽ ἀντ᾽ ὀνείδους,
> δύσμαχα δ᾽ ἔστι κρῖναι.
> φέρει φέροντ᾽, ἐκτίνει δ᾽ ὁ καίνων·
> μίμνει δὲ μίμνοντος ἐν θρόνῳ Διὸς
> παθεῖν τὸν ἔρξαντα· θέσμιον γάρ.

> Insult comes in return for insult,
> and it is a hard struggle to judge.
> The ravager is ravaged, the killer pays;
> it remains firm while Zeus remains on his throne
> that he who does shall suffer, for that is his ordinance.
>
> Aeschylus, *Agamemnon* 1560–4

The phrase receives the aura of ancient wisdom: 'and for a bloody stroke let the pay-ment be a bloody stroke.' For him who does, suffering—that is what the old, old saying states' (ἀντὶ δὲ πληγῆς φονίας φονίαν | πληγὴν τινέτω. δράσαντι παθεῖν, | τριγέρων μῦθος τάδε φωνεῖ, *Libation Bearers* 312–14). From this angle, therefore, the remark that 'the nurse and launderer had the same *telos*' acquires additional force: not only do 'doing' and 'suffering' head towards the same goal (τέλος)—that is, to each other—but the very words κναφεύς and τροφεύς have the same *ending*. This homoeoteleuton practically binds the 'doer' to the 'sufferer'.[18] This is where the exclusive reference to urine in Orestes' diapers, with faeces left unmentioned, may come into consideration. Unlike faeces, urine was not only waste to be *cleaned* by the launderer but actually an ingredient *used* by the launderer as a detergent (Olson and Biles 2015: 416–17, on Aristophanes, *Wasps* 1127–8[19]). In other words, the *Oresteia*'s frustrating circularity would find its ultimate expression if the liquid contents of Orestes' swaddling-clothes are in fact both the filth and the purifier, simultaneously the problem and the solution.

Thus, the central premise of the trilogy is imaged as this circulation of crime-contaminated fluids through the body of baby Orestes, which, symptomatically, no one can control. Neither can Cilissa tell if Orestes is hungry, thirsty, or needs to urinate, nor can he himself speak to say it. The uninhibited and uninhibitable neonatal metabolism serves as a fit allegory for the inevitability of cyclical wrongdoing in the *Oresteia*. The rotation of inflicting and suffering injustice is a *perpetuum mobile* beyond external control, like an incontinent infant's urinary tract, run only by its own internal reflexes: in the emended text it is *autarchēs* (αὐτάρχης), 'self-governing' (757), though the manuscript reading *autarkēs* (αὐτάρκης), 'self-sufficient',[20] is very tempting, as it would convey the idea of a closed, self-sustainable loop in which urine is treated with urine.

Aeschylus' imagery belongs to broader ancient tradition. On the one hand, metabolic and hydraulic metaphors will be in circulation, as it were, in various contexts. In Plato's *Symposium* Socrates compares intellectual and moral influence to water flowing from a fuller vessel into an emptier one; the *Timaeus* allegorises cosmogenesis via a peculiar irrigation system of the body (Plato, *Symposium* 175d-e; *Timaeus* 47e–84c.). The body generally, on the other hand, is an especially potent metaphor when something goes wrong. Thucydides' graphic description of the ravaging Athenian plague is followed by what Jeffrey Rusten called the 'general breakdown of moral and social restrains' (1989: 189, on Thucydides 2.52–54.1); the gruesome account of bodies falling apart may also be seen as the figurative manifestation of the imploding social order (cf. now Serafim 2019). Comedy, expectedly, prefers the 'rear entrance' for sending political messages. The memorable scene of painful constipation in Aristophanes' *Assemblywomen* resulting in the birth of faeces (317–71) is unanimously interpreted as symbolising political defeat of the Athenian male.[21] But it is Aeschylus who cleared the ground for affirming the connection between physiological urges and external pressures and tensions—political, social, moral, and religious. He may well have been the first, and certainly for a long time the only one outside comedy and satire, to pursue so systematically this channel (so to speak) for illustrating the most pressing demands placed on humans with the basest function of the human body.

The body, in terms of consumption, digestion, and excretion, would become a widespread metaphor for social disorders, moral declines, and political crises in Latin literature, as has been surveyed by Emily Gowers (1993: 12–16), who notes, for example, that '[t]he individual body could be seen as the small-scale incarnation of national *luxuria*' (1993: 13).[22] Well-established in Latin is the socio-political metaphorical use of the adjective *intestinus*, 'internal', especially in Sallust and Cicero for referring to civil war, *bellum intestinum*, and internal conflicts generally.[23] A generation later, Livy would elaborate on internal discord as an illness requiring *remedium* (2.45.4), and on civil war burning inside the entrails (*intestino et haerente in ipsis visceribus uramur bello*, 32.27; for the politics of 'body horror' in Livy, see now Hay 2018).[24]

Especially interesting for our purposes is Livy's aetiologising of Latin intestinal metaphor with an old Greek parable of body as society. A plebeian insider delegated by patricians to pacify the seditious plebs in 494 BCE tells them a story: once upon a time, the body's limbs revolted against the stomach for doing nothing but enjoying the food the limbs provided; thus they decided to starve the belly into submission, only to end up starving themselves, and so 'with this parable, he showed the similarity between the internal revolt of the body and the anger of the plebs toward the senators, and so won over men's minds' (*comparando hinc quam intestina corporis seditio similis esset irae plebis in patres, flexisse mentes hominum*, 2.32.8–12; trans. Warrior 2006; the parable is attested at Xenophon, Aesop, and others, and attributed to various speakers: Ogilvie 1965: 312–13, ad loc.). It may be a coincidence that two metabolic metaphors found their way into two foundational legends: Livy deploys this parable to dramatise the first plebeian secession and the institution of the magistracy of *tribunus plebis*, while Aeschylus' *Oresteia* ultimately prepares the ground for introduction of the council of Areopagus. Whatever may be the case, there is something about intestinal urges that can turn them into effective means of persuasion when major measures in the public sphere need to be carried out: body–society cannot function unless bodily urges are addressed first. They are non-negotiable. To quote a delicious truth-bomb in praise of farting thrown by the freedman Trimalchio at his dinner-guests, 'That's the one thing that not even Jupiter can prohibit' (*hoc solum vetare ne Iovis potest*, Petronius, *Satyricon* 47.4).

Notes

* Arguments from this chapter have been presented on various occasions; for useful suggestions, I thank Julia Laskaris, Tom Hawkins, Catalina Popescu, Darko Todorović, Jeffrey Rusten and the volume editors.

1 Text and occasionally modified translations of Aeschylus are from Sommerstein 2008; all other translations are mine unless noted otherwise.

2 Sidgwick 1892: xvii: '[T]he Nurse, whose rustic homeliness and grotesque but natural inconsequence of speech forms, like the talk of the Herald in the *Agamemnon*, an effective contrast to the fearful drama that impends. It relieves the tension of feeling just at the crisis: and the pithy illiterate babble of the old woman about Orestes' babyhood, adds the touch of nature to the dark tragic figure of the Avenger.' For a 'comic relief' interpretation, cf. more recently Pypłacz 2009. Apparently likewise puzzled by some awkward passages, ancient sources speculated that Aeschylus wrote while inebriated: Chameleon, according to Athenaeus, *Deipnosophists* 1.21d; Plutarch, *Convivial Questions* 7.10.

3 The other two instances are *Agamemnon* 12 (Watchman's nocturnal restlessness in anticipation of Agamemnon's return and the change of ruler), and *Agamemnon* 330 (night patrol of the Greeks after the capture of Troy); cf. perhaps 'day-wandering dream vision' (ὄναρ ἡμερόφαντον ἀλαίνει, *Agamemnon* 82).

4 And prophecy: the best the nurse could do with the infant's attempts to communicate was to be a 'diviner' of his needs (πρόμαντις, 758), and the Chorus soon warns her not to be a 'bad interpreter' of the news of Orestes' death (οὔπω· κακός γε μάντις ἂν γνοίη τάδε, 777). The grown Orestes himself is an interpreter of the snake in the dream (τερασκόπον, 551; on prophecy in the *Oresteia*, see Roberts 1985). Cilissa typically mistook the baby Orestes' inarticulate signals (ψευσθεῖσα, 759), perhaps just as in the immediately preceding scene Clytemnestra mistook Orestes for a Phocaean because he spoke with a different accent (*Libation Bearers* 563).

5 cf. also Cassandra's 'brightening' prophecy at *Agamemnon* 1120.

6 Lebeck 1971: 1, 3: 'The images of the *Oresteia* are not isolated units which can be examined separately. Each one is part of a larger whole: a system of kindred imagery. . . . When related to each other and to the ideas which they illustrate or the dramatic action which translates them into visual terms, the images cease to be discrete and arbitrary pictures and emerge as important components of the play's significance.' Aeschylus' imagery has been studied extensively: see Goheen 1955; Van Nes 1963; Peradotto 1964; Smith 1965; Scott 1966; Fowler 1967; Garson 1983; Saayman 1993; Catenaccio 2011.

7 This exact ambivalence of water is nicely captured by the comedian Antiphanes (Athenaeus 1.23, fr. 228 Kassel and Austin), who parodies Sophocles' simile of stubborn trees felled by

flood (*Antigone* 710–14) by reconfiguring them into those that perish by stubbornly keeping their 'thirst and dryness' (δίψαν, ξηρασίαν); elsewhere, Antiphanes also speaks of sailing as virtually suicidal (fr. 100 Kassel and Austin).

8   cf. Statius's description of Hypsipyle (*Thebaid* 5.593–4), discussed by Krebs, present volume, Chapter 21, p. 340.

9   Compare the neat transition from physiology to meteorology in Sallust's *Jugurthine War* (75.7): after the Roman army struggled to secure enough water supply for a difficult campaign, 'it is said that such an amount of water fell suddenly from the sky that for the army it was enough and indeed too much' (*tanta repente caelo missa vis aquae dicitur, ut ea modo exercitui satis superque foret*). See also the landscape reflecting Dido's state of mind in Virgil's *Aeneid* 4.532–665, discussed by Krebs, present volume, Chapter 21, p. 344.

10 The effect is still there even if Finglass is right that there are two different vessels mentioned, and that this *lebēs* is 'not *the* urn, which was never taken inside. Clytemnestra is preparing a vessel in anticipation of the return of her son's ashes to the house' (Finglass 2007: 512).

11 Apparently a very Aeschylean locution: cf. 'smoke, the sister of fire' at *Seven against Thebes* 493–4; Clarke 1995.

12 For the idea, cf. the wine libations poured by Dido becoming '*obscenum . . . cruorem*' (Virgil, *Aeneid* 4.455): Krebs, present volume, Chapter 21, pp. 344–5.

13 On 'good breast' and 'bad breast' in the *Oresteia*, see DeForest 1993: 137–8.

14 See also the political allegory of polluting a clear spring of water with mud (βορβόρῳ, comic word) at *Eumenides* 694–5, with Sommerstein 2002: 163.

15 See Sommerstein's own (unintentional?) phrasing: 'The effect in the *Oresteia* is as though the αἰσχρότης were breaking through in spite of all efforts to contain it' (2002: 164–5).

16 Sommerstein 2002: 159: 'λιψουρία (756)—a compound that strikingly wraps together in one word a highly untragic reference to urination with a verbal root (that of λέλιμμαι) so elevated that it is hardly known otherwise except from Hellenistic epic [. . .] and two passages of *Seven against Thebes* (355, 380).'

17 The fluid connection is foreshadowed in the choral parable of the lion cub in the *Agamemnon* (717–36): the infant lion, representing Orestes, is 'fond of the nipple but deprived of its milk' (ἀγάλακτον . . . φιλόμαστον, 718–19), tame when pressed by 'intestinal urges' (γαστρὸς ἀνάγκαις, 726), eventually showing the character inherited from his parents as he returns grown up to take vengeance on the house 'soaked in blood' (αἵματι δ᾽ οἶκος ἐφύρθη, 732). The 'imagery drawn from the lion parable is used to describe every figure in the *Oresteia* who acts as an instrument of the Erinys' (Lebeck (1971: 50, with references); for the parable, see also Knox 1952; Saayman 1993: 13–16; Nappa 1994.

18 One might also hear ritual overtones, since homoeoteleuton is characteristic of such formulae; cf. Clytemnestra's rhyming prayer: Ζεῦ, Ζεῦ τέλειε, τὰς ἐμὰς εὐχὰς τέλει· | μέλοι δέ τοί σοι τῶνπερ ἂν μέλλῃς τελεῖν (*Agamemnon* 973–4); see Fraenkel 1950, Volume II: 440, ad loc. for a general discussion and Hogan 1984: 9 on the rhyme. Goldhill 1984 discusses some thematically significant semantic aspects of τέλος and its compounds in the *Oresteia*.

19 I thank Julia Laskaris for reminding me of this.

20 cf. Thucydides 2.41.1 for political and medical connotations of σῶμα αὔταρκες, with Rusten 1989: 159.

21 'While his wife has risen to the highest position possible in the city, Blepyrus has sunk to the lowest' (Henderson 1991: 102; cf. 189, §401); Sommerstein (1998: 173, commenting on Aristophanes, *Assemblywomen* 369) observes that 'during the same time that Blepyrus has been struggling with his bowels, the Assembly meeting has begun and ended on the Pnyx; thus while Blepyrus after much labour has "given birth" to a quantity of excrement, the Assembly under his wife's guidance has been giving birth (cf. 549–50) to a new Athens.' For parallels between the *Assemblywomen* and the *Oresteia*, see Vidović 2017, especially 41–2 for the constipation scene and Agamemnon's death in the tub.

22 Gowers cites examples of Seneca who 'pictures himself as an island of integrity in the swelling flood of luxury (*circumfudit me ex largo frugalitatis situ venientem multo splendore luxuria et undique circumsonuit* [*On the Tranquility of the Soul*] 1.4.10)' (14, fn. 52), and Cicero's metaphors of dregs and sewage (15); for political metaphor of bodily pollution, see Bradley 2012: 36–9.

23 *Bellum intestinum*: Sallust, *Catiline* 5.2; Cicero, *Against Catiline* 1.5.5, 2.28, etc. Cicero frequently pairs *intestinus* with *domesticus*, sometimes suggesting also insidiousness, as in *occultum intestinum ac domesticum malum* (*Against Verres* 2.1.39), or with an extended corporeal imagery of wounds to the state caused from within (*multa sunt occulta rei publicae volnera . . . nullum externum periculum est, . . . inclusum malum, intestinum ac domesticum est*; *On the Agrarian Law* 1.26.7); curiously, once when using *intestinus* it in its literal, biological sense, Cicero quasi-apologetically calls attention to the metaphor 'the liver's door' (*ad portas iecoris—sic enim appellantur*; *On the Nature of Gods*, 137.1).

24 For accumulated metaphors, cf. also *intestino bello totae gentes consumuntur*, Columella 9.9.6.8.

Bibliography

Primary sources

Fagles, R. (1984) *Aeschylus. The Oresteia*. New York: Penguin Books.

Finglass, P. (2007) *Sophocles. Electra*. Cambridge: Cambridge University Press.

Fraenkel, E. (1950) *Agamemnon, vols I–III*. Oxford: Clarendon Press.

Garvie, A. F. (1986) *Aeschylus. Choephori*. Oxford: Clarendon Press.

Lattimore, R. (1953) *Aeschylus I. Oresteia*. Chicago: University of Chicago Press.

Olson, S. D. and Z. P. Biles (2015) *Aristophanes. Wasps*. Oxford: Oxford University Press.

Rusten, J. S. (1989) *Thucydides. The Peloponnesian War: Book 2*. Cambridge: Cambridge University Press.

Sidgwick, A. (1892) *Aeschylus. Choephoroi*. Oxford: Clarendon Press.

Sommerstein, A. H. (1989) *Aeschylus. Eumenides*. Cambridge: Cambridge University Press.

———. (1998) *Aristophanes. Ecclesiazusae*. Warminster: Aris & Phillips.

———. (2008) *Aeschylus, Vol. II: Agamemnon, Libation Bearers, Eumenides*. Cambridge, MA: Harvard University Press.

Warrior, V. M. (2006) *Livy. The History of Rome, Books 1–5*. Indianapolis: Hackett.

Secondary literature

Bradley, M. (2012) 'Approaches to pollution and propriety', in M. Bradley and K. R. Stow (eds) *Rome, Pollution, and Propriety: Dirt, Disease, and Hygiene in the Eternal City from Antiquity to Modernity*. Cambridge: Cambridge University Press, 11–40.

Bremmer, J. (1986) 'Agamemnon's death in the bath: Some parallels', *Mnemosyne* 39(3/4), 418.

Cairns, D. (2014) 'The bloody dust of the Nether Gods: Sophocles, *Antigone* 599–603', in E. K. Emilsson, A. Maravela, and M. Skoie (eds) *Paradeigmata: Studies in Honour of Øivind Andersen*. Athens: Norwegian Institute at Athens, 39–51.

Catenaccio, C. (2011) 'Dream as image and action in Aeschylus' *Oresteia*', *Greek, Roman, and Byzantine Studies* 51, 202–23.

Clarke, M. (1995) 'Aeschylus on mud and dust', *Hermathena* 158, 7–26.

DeForest, M. (1993) 'Clytemnestra's breast and the evil eye', in M. DeForest (ed.) *Woman's Power, Man's Game: Essays on Classical Antiquity in Honour of Joy King*. Wauconda: Bolchazy-Carducci, 129–48.

Edwards, A. (1991) 'Aristophanes' comic poetics: Τρύξ, scatology, σκῶμμα', *Transactions of the American Philological Association* 121, 157–79.

Faraone, C. (2009) 'Does Tantalus drink the blood, or not? An enigmatic series of inscribed Hematite gemstones', in U. Deli and C. Walde (eds) *Antike Mythen: Medien, Transformationen und Konstruktionen*. Berlin: Walter de Gruyter, 248–73.

Fowler, B. H. (1967) 'Aeschylus' imagery', *Classica et Mediaevalia* 28, 1–74.

Garner, R. (1990) *From Homer to Tragedy: The Art of Allusion in Greek Poetry*. London: Routledge.

Garson, R. W. (1983) 'Observations on some recurrent metaphors in Aeschylus' *Oresteia*', *Acta Classica* 26, 33–9.

Goheen, R. F. (1955) 'Aspects of dramatic symbolism: Three studies in the *Oresteia*', *The American Journal of Philology* 76, 113–37.

Goldhill, S. (1984) 'Two notes on τέλος and related words in the *Oresteia*', *The Journal of Hellenic Studies* 104, 169–76.

Gowers, E. (1993) *The Loaded Table: Representations of Food in Roman Literature*. Oxford: Clarendon Press.

Gregory, J. (2009) 'Introduction', in P. Meineck, C. E. Luschnig, and P. Woodruff (eds) *The Electra Plays*. Indianapolis: Hackett, vi–xxxii.

Harry, J. (1912) 'ΩΣ ΑΡΔΗΙ ΝΗΔΥΝ (Eur. *Suppl.* 207)', *The Classical Review* 26(1), 8–9.

Hay, P. (2018) 'Body horror and biopolitics in Livy's third decade', *New England Classical Journal* 45(1), 2–20.

Heath, J. (1999) 'Disentangling the beast: Humans and other animals in Aeschylus' *Oresteia*', *The Journal of Hellenic Studies* 119, 17–47.

Henderson, J. (1991) *The Maculate Muse: Obscene Language in Attic Comedy*. New York: Oxford University Press.

Hogan, J. C. (1984) *A Commentary on the Complete Greek Tragedies: Aeschylus*. Chicago: University of Chicago Press.

Hooley, D. M. (2007) *Roman Satire*. Malden: Wiley-Blackwell.

Kassel, R. and C. Austin (1983–2001) *Poetae comici Graeci*. Berlin: Walter de Gruyter.

Knox, B. M. W. (1952) 'The lion in the house (*Agamemnon* 717–36 [Murray])', *Classical Philology* 47(1), 17–25.

Lebeck, A. (1971) *The Oresteia: A Study in Language and Structure*. Washington, DC: Center for Hellenic Studies, Harvard University Press.

Napolitano, M. (2003) *Euripide. Ciclope*. Venice: Marsilio.

Nappa, C. (1994) '*Agamemnon* 717–36: The parable of the lion cub', *Mnemosyne* 47(1), 82–7.

Ogilvie, R. M. (1965) *A Commentary on Livy, Books 1–5*. Oxford: Clarendon Press.

Peradotto, J. J. (1964) 'Some patterns of nature imagery in the *Oresteia*', *American Journal of Philology* 85, 378–93.

Pypłacz, J. (2009) 'Los elementos cómicos en la *Orestía* de Esquilo', *Cuadernos De Filologia Clasica; Estudios Griegos e Indoeuropeos* 19, 103–14.

Roberts, D. H. (1985) 'Orestes as fulfillment, *Teraskopos* and *Teras* in the *Oresteia*', *The American Journal of Philology* 106, 283–97.

Rousseau, G. (1963) 'Dream and vision in Aeschylus' *Oresteia*', *Arion* 2(3), 101–36.

Saayman, F. (1993) 'Dogs and lions in the *Oresteia*', *Akroterion* 38(1), 11–18.

Scott, W. C. (1966) 'Wind imagery in the *Oresteia*', *Transactions and Proceedings of the American Philological Association* 97, 459–71.

Seidensticker, B. (1982) *Palintonos Harmonia: Studien zu komischen Elementen in der griechischen Tragödie*. Göttingen: Vandenhoeck & Ruprecht.

Serafim, A. (2019) 'Sicking bodies: Stasis as disease in the human body and the body politic', in H. Gasti (ed.) *ΔΟΣΙΣ ΑΜΦΙΛΑΦΗΣ: A Volume in Honour of Professor Katerina Synodinou*. Ioannina: Carpe Diem, 673–95.

Silk, M. S. (2002) *Aristophanes and the Definition of Comedy*. Oxford: Oxford University Press.

Smith, O. (1965) 'Some observations on the structure of imagery in Aeschylus', *Classica et Mediaevalia* 26, 10–72.

Sommerstein, A. H. (2002) 'Comic elements in tragic language: The case of Aeschylus' *Oresteia*', in A. Willi (ed.) *The Language of Greek Comedy*. Oxford: Oxford University Press, 151–67.

Van Nes, D. (1963) *Die maritime Bildersprache des Aischylos*. Groningen: J. B. Wolters.

Vidović, G. (2017) 'Hijacking Sophocles, burying Euripides: Clytemnestra, Erinyes, and Oedipus in Aristophanes' *Assemblywomen*', *Lucida Intervalla* 46, 34–67.

Zeitlin, F. I. (1965) 'The motif of the corrupted sacrifice in Aeschylus' *Oresteia*', *Transactions and Proceedings of the American Philological Association* 96, 463–508.

21

OPEN WOUNDS, LIQUID BODIES, AND MELTING SELVES IN EARLY IMPERIAL LATIN LITERATURE

Assaf Krebs

A wound is a break in the continuum of any corporeal tissue, inflicted by external agency. This definition entails various subdivisions, most significantly the one between open and closed wounds:

> open wounds are those in which the protective body surface (the skin or mucous membrane) has been broken, permitting the entry of foreign material into the tissues. In closed wounds, by contrast, the damaged tissues are not exposed to the exterior.
>
> *Encyclopaedia Britannica*

This short definition is similar to that of Cornelius Celsus in a much earlier encyclopaedia, composed in the first century CE and known as *On Medicine*. In its fifth book, Celsus specifies five classes of lesions harmful to the body (*noxa corpori*): injuries that occur when a new thing is formed within the body (such as stones in the bladder); injuries which occur when something grows bigger or swells (such as swelling veins); injuries related to loss of bodily matter (mutilation, for example); injuries derived from internal harms; and injuries resulting from external factors, such as wounds (*vulnera*, 5.26.1a). Although not using the terms 'open' and 'closed', Celsus distinguishes wounds on the external surface of the body, which are '*oculis subiecta*'—exposed to sight—from internal wounds, which are hidden inside the body.[1] He also enumerates three kinds of fluid discharge from wounds: blood, which 'everybody knows' flows when the wound is fresh or when it starts healing; pus, which is thicker and brighter than blood; and *sanies*, a fluid substance thinner than blood that has various colours and textures, and is discharged from ulcers between the two stages. Secretion of *sanies* usually indicates that the wound has begun to heal.[2]

Open wounds and bodily fluids constitute the main themes of this chapter. It focuses on material and symbolic representations, especially blood and tears, the latter being closely related to mental wounds. Like the previous chapter by Goran Vidović, it explores the dual function of bodily liquids as both physical and symbolic phenomena. However, it uses literary and cultural theories, particularly psychoanalytical and sociological approaches, to analyse Roman literary representations of wounds and bodily fluids and their metaphoric use in de-constructing the human physical and mental world. The chapter draws mostly on Ovid's *Metamorphoses*, which 'uses the body as its focus for its view of the human condition' (Segal 1998: 9). Its main argument is that, upon entering the metaphorical domain, wounds confer liquidity on the subject's

DOI: 10.4324/9780429438974-28

structure: just as corporeal wounds pierce the skin and allow the exit of secretions, which may endanger the body, so metaphoric wounds, particularly wounds of the soul, perforate the self, liquefy the ego's structure, and dissolve the language and social order. These wounds thereby challenge the solid, coherent, and steady position of the subject, turning the subject into a fluid process of becoming, replacing ontology with metaphysics. The chapter will thus progress from fluid bodies to melting language and dissolving communication and from the liquefied ego and fluid melancholic soul to the dissolution of the entire social order.

The phenomenon of fluidity is that quality which differentiates liquids and gases from solids. The contrast between liquids and solids is related to the nature of their structure: in solids the molecular connections are consolidated and therefore resist the separation of atoms more strongly than fluids, whose connections are weaker. Hence fluids continue to deform as long as 'shear force is applied, and this continuing deformation under stress is characteristic of all fluids. Fluids can thus be defined as any material that is unable to prevent the deformation caused by a shear stress' (Shaughnessy et al. 2005: 15). In this respect we may say that solids have distinct spatial dimensions and are able to retain their form over time much more easily than fluids, which are inconsistent and have no steady spatial position. Liquids can hardly fix space or time, and for them it is the 'flow of time that counts, more than the space they happen to occupy' (Bauman 2006: 2). For solids, however, time is less substantial than space, and in a way we might say that time is a less relevant category for them. Through these qualities, fluidity can be examined not only as a physical phenomenon but also as a metaphor for the human condition.

Open wounds are a cutaneous phenomenon: they lacerate and perforate the skin in which they are located. As such, they draw symbolic attributions from the skin, which in itself has both material and symbolic meanings. The skin is a vulnerable material surface in constant contact with inner bodily fluids and parts and with the external world; it is a private organ which is also a visible public site; it is connected to one's identity (e.g., through colour, texture, and flaws) and serves as part of the subjectivity. In modern psychoanalytic theories the skin is a site through which the self develops, by experiences of surface and contiguity, and the ego is constructed.[3] The skin's psychic dimension is related to three major cutaneous functions: its quality as a container; its nature as an interface between outside and inside; and its feature as a primary means of communication and relationships (Anzieu 1989: 40). When a wound lacerates the skin, this 'first of all first places' (Connor 2004: 36) loses the functions of holding, supporting, and communication, thereupon perforating the boundaries of the ego. This is evident in metaphors of emotional wounds (such as wounds of love and melancholy, which will be discussed later in the chapter), as well as in mental pain. Open wounds are thus much more than a mere physical phenomenon: they damage the psychic envelope, melting the steadiness of the ego, and dissolve the sense of the self.

Liquefied bodies and melting language

Battles in antiquity were a messy (and wet) business. Mixtures of blood, sweat, vomit, and other fluids erupted from injured bodies, spurting all over the men and their surroundings. Streams of warm blood flow from wounds in Lucan's battle description in the fourth book of *On the Civil War*;[4] blood streams everywhere in Virgil's description

339

of Nisus and Euryalus butchering the enemy's soldiers (*Aeneid* 9.333–456); in Ovid's depiction of the fight between Phineus and Perseus the floor of the palace is so drenched in bodily liquids that the men slither and fall (*Metamorphoses* 5.75–6.). Open wounds and flowing liquids go together not only in battle wounds: in the ninth book of *On the Civil War*,[5] Lucan uses a stream of fluid attributes to describe a man wounded by a snakebite: his limbs float (*natare*) in *sanies*, his calves flow (*fluxere*), his knees and thigh muscles melt (*liqui*), his groin leaks (*destillare*) a black discharge; and his entrails flow (*fluere*).[6] The shoulder, he continues, and the strong arms turn into water (*manare*), the neck and head flow (*fluere*), and the flesh drips away (*stillare*) faster than snow that melts in the south wind or than wax that melts in the sun (9.781–2). Eventually the solid corporeal frame dissolves in a process of melting, leaving 'only wound and no body'.[7] Another snakebite appears in the story Hypsipyle and Opheltes in the fifth book of the *Thebaid*.[8] Statius describes the frantic wanderings of the horrified woman looking for the boy while the grass around her is drenched with his wet blood. When she finally catches sight of the wounded body the view is of ravaged skin, exposed bones, and sinews soaked (*madere*) in a shower (*imber*) of fresh blood.[9] 'The whole body', concludes Statius, 'is in the wound'.[10] This total transformation of the body into a wound signifies the destruction of the corporeal order and the melting of the boy's solid frame: the inner liquids drain out, the hidden organs are revealed, the chest disappears, and the face, previously conveying his identity, is gone.[11] The body loses its meaning as it changes from a coherent bounded object into a fragmented agglomeration of organs, fluids, and substances. The multiple liquid metaphors Statius chooses to describe the scene—sprayed blood (5.590), dripping sinews (5.597), and a shower of blood (5.598)—all intensify its fluid atmosphere.

'Fluids exuded by the body', write the editors of this volume in the Introduction, 'seeping through various orifices and beyond the boundary of the skin, usually signal change within the body and point to its fundamental instability and permeability' (p. 4). But, as they further state, there is more than merely a corporeal issue at stake. When Hypsipyle realises that the child is injured her pain is such that her tears as well as her words are gone.[12] In contrast to the open and fluid wounded body, her own body is dry, sealed, and tearless. Nevertheless, her trauma (a word for 'wound' in Greek) is accompanied by immediate destruction of the language, and this wounds the texture of the story itself, severing its succession and allowing a short simile to erupt onto the literary surface.[13] When the simile congeals and the narrative continues. Hypsipyle gathers the child's torn limbs with her hair and, creating a sort of 'second skin', she reconstructs a sense of container to the boy. Only then, with this symbolic sealing of the wounded body, can language regain its communicative function: 'her voice released, found a way for her pain, and her groans dissolved into words.'[14]

One of the most remarkable examples of a body whose transformation into 'nothing but a wound'[15] liquefies it completely, appears in Ovid's version of the myth of Marsyas (*Metamorphoses* 6.382–400). The contest between Apollo and Marsyas features a series of oppositions between Greeks and barbarians, Apollonian and Dionysiac rituals, solid rational order and fluid uncontrollable nature. The Ovidian version of the myth focuses on the very moment of flaying, which turns the satyr into a wound:

'Why do you tear me from myself?' he cried,
'Ah! I repent, Ah! it is not worth it', he shouted, 'the flute.'

The skin of the screaming satyr torn from the surface of his limbs
And there was nothing which was not a wound; blood pours from every side,
The sinews lie open with no cover, the vibrating veins foaming
Without the skin; you could count the leaping entrails
And the shining vessels in the chest.[16]

Ovid, *Metamorphoses* 6.385–91

The wound caused by the destruction of the skin dissolves the hierarchy of the satyr's body: internal becomes external, depth becomes surface, hidden becomes exposed. Blood covers everything, replacing the stable cutaneous envelope. The double-negation structure of the wound—the becoming nothing but the annihilation of the skin—liquefies the corporeal boundaries but also the borders between Marsyas, the world, and those who watch the scene. The spectators' tears penetrate earth and re-erupt from its viscera in the form of the River Marsyas. The end of the satyr's solid existence is the beginning of liquid flow; the steady subject turns into a process of becoming.

Marsyas' wound extends beyond his material existence: his cry, 'Why do you tear me from myself?' implies an identification of the self and the skin, alluding to the formation of a mental wound as well. Indeed, modern psychoanalytical theory from Freud to the present argues for a close relation between the body's surface and the formation of the self.[17] The French psychoanalyst Didier Anzieu ties the structure and function of the skin to the ego through his concept of 'Skin-Ego', which he uses as a metaphor and metonym.[18] According to Anzieu, the function of the Skin-Ego is parallel to and supported by three major skin characteristics: its nature as a sac that contains, retains, and nourishes; its function as an interface between the world and the subject; and its disposition as a site of primal communication together with the mouth (Anzieu 1989: 40). The myth of Marsyas, argues Anzieu, represents the intuition that a personal soul and a psychic self subsist only as long as a bodily envelope guarantees their individuality. The gushing river, in his interpretation, represents life instincts and power, but this energy is available only to those who preserve the wholeness of the skin-ego, which is supported by the skin's intact surface (Anzieu 1989: 52). Indeed Marsyas' wound, the loss of the cutaneous envelope, transforms his body from solid to fluid and dissolves his ego as well. This process is accompanied by loss of language, albeit different from Hypsipyle's: twice his words break into syllables, once the narrator cuts his direct speech: ' "Ah! I repent, Ah! it is not worth it", he shouted, "the flute" ';[19] after that his voice is rendered mute, leaving the rest of the story to be told by the narrator. When Marsyas' wound melts the boundaries between his 'self' and the world, no external reference is possible (as nothing can be external when the boundary is gone), and his language dissolves entirely, together with the corporeal structure. As mentioned earlier, Anzieu pointed to the centrality of the skin as a significant site of communication and acquisition of language. Indeed, as seen in the previous example, the skin's wounding can be followed by a loss of language and communication. Breaching the lingual structure by moaning, meaningless syllables and sounds, or total silence, dissolves the solid and fixed sense that enables language to be shared by people through its common-sense. Once this common-sense melts, the wounded language loses its ability to communicate, turning sense into non-sense.[20]

The story of Narcissus and Echo (*Metamorphoses* 3.339–510) is another example of intersection across wound, fluids, and language. When Narcissus rejects Echo, who has fallen in love with him, she is deeply hurt. Vulnerable and exposed, she hides under leaves and cages as if they were 'second skin' (since her own skin provides neither cover nor protection), but her strong emotions do not abate: she 'clings to her love and her pain increases'.[21] The emotional wound has corporeal implications:

> Her sleepless concerns weaken her poor body,
> Leanness shrinks her skin (*adducit cutem macies*), and into the air
> All the moisture of her body evaporates; only voice and bones are left.
> The voice remains, and they say that her bones received the figure of stones.[22]
>
> Ovid, *Metamorphoses* 3.396–9

The phrase *adducit cutem macies* is a metaphoric expression for wound, originating from the gladiatorial world, probably because once the blood of the injured gladiator congealed, it shrank the flesh and pulled the skin with it (Asso 2009: 107).[23] Echo's wound affects her bodily fluids as well as her lingual communication: it dries her inner moisture (*sucus corporis*), turning her solid bones into firm rocks, and leaves her voice shapeless. Without her body and the ability to use corporeal gestures, her language—whose communicative function Juno had already destroyed—is empty, and only meaningless sound (*sonus*) is left (3.401). As for Narcissus, he addresses his own reflection in the water but receives no answer. The barren dialogue of the emotionally wounded boy with his liquid reflection drains the communicative function of language and undermines his steady position, in which he 'sticks to the place [of his fluid reflection, A.K.] with motionless expression'.[24] In vain he tries to catch the image in the water, in vain he sees the lips of his fluid simulacra move; no words (*verba*) reach his ears (3.462). In his agony he weeps, and his own tears mix with the lake water,[25] destroying his reflection and dissolving the firm boundary between in and out, subject and object, man and nature. Unable to stand the pain, Narcissus melts away (*intabescere*), like pieces of wax under heat, or like hoar frost (*pruina*) in the sun (3.487–9). Just as fluids are unable to retain their form under physical stress, so Narcissus and Echo cannot retain either their corporeal frame or the stable emotional structure under the burden of their wounds: their language turns into non-sense, their ego collapses, their bodies dissolve, and they become a process of metamorphosis.

In the tale of Procne and Philomela (*Metamorphoses* 6.412–74) the dissolution of bodies, language, and souls melts the entire social order. Procne begs her husband Tereus to bring her sister Philomela for a visit. Tereus agrees and sails to Procne's home country, where, after obtaining permission from her father (whose intense sorrow and fear breaks his words into sobbing, 6.510), he takes the girl, and they set out for home. However, Tereus, inflamed by his unbridled love and passion for the girl, takes her to a remote place, rapes her, and cuts out her tongue to prevent her from communicating the deed; he then tells Procne that her sister died on the journey. After several months Philomela resolves to overcome her wounded body and muted tongue and to convey her story on a piece of cloth. She weaves purple signs on the white surface as if it were congealed blood on fair skin.[26] Like crimson scars (such as those of slaves after being punished[27]), her signs fix and consolidate the past events and make them public; like scars they function as a reminder of the former wound, engraving death on the corporeal surface. When Procne is given the scarred cloth a wound is created in her own soul, and the pain is such that her tears immediately dry up. She cannot find words to express her agony, and her speech is repressed.[28] She then punishes Tereus by having him unknowingly eat his own son, inflicting a mental wound on his soul as well: like a wound, which is created by penetration of an external object into the body, so Tereus' viscera are penetrated by his son's flesh;[29] like the eruption of fluids from the wound, so Tereus, on discovering the deed, wishes to open up (*reserare*) his chest and vomit the boy out (6.663–4). Tereus himself becomes a wound.

The whole tale of Procne, Philomela, and Tereus is a story of wounds and fluids, injuries and traumata. It is a story of melting solid conventions and norms: the husband rapes his sister-in-law, the mother kills her son, the father consumes his own offspring, and the wife prefers her sister to her husband and son. Corporeal, ethical, and moral boundaries are breached in this story as one wound leads to another in a chain of fluid becomings, and the wounded becomes the one that wounds the other. The language transforms into signs, and in a certain reversal it is actually the symbolic blood that survives all the transformations as if it were solid, as it reappears as a red mark on the sisters' feathers, after they have been turned into birds. This blood marks the former bodies, fluids, and order, and like scars on skin it serves as a constant reminder of the past.[30] That which cannot be uttered is told by blood and pain.

Liquefied egos, melancholic souls, and dissolving social order

Whereas Procne and Philomela were transformed into other bodies, Cyane the nymph transforms into a liquid substance. When she realises Proserpina's unfortunate faith, a deep grief creates an inconsolable wound in her soul:

> But Cyane, mourning for the ravished goddess and the spurned rights
> Of her spring, carries quietly an inconsolable wound in her soul
> And she is all consumed by her tears;
> And into these waters, of which she was once a great divinity,
> She is now diminished: you could see her members softening,
> Her bones become bent, her nails lose their rigour.
> First of all the softest parts liquefy,
> Her dark hair, her fingers, legs, and feet;
> For it is a small transformation from thin limbs to
> Cold waves. After these her shoulders, back, sides,
> And chest all vanished into thin streams.
> And then, instead of living blood in the vitiated veins,
> Water surges, and nothing you could catch is left.[31]
>
> Ovid, *Metamorphoses* 5.425–37

Cyane mourns (*maerere*) her loss. She endures her mental wound, her tears keep flowing, and her deep sorrow shatters her psychic integrity.[32] According to Freud, both mourning and its pathological form melancholy are mental states that react to a real or symbolic loss, and both are involved in 'profoundly painful dejection' (Freud 2001a: 244). In these mental states, Freud maintains, the ego's energy ceases to be directed outwards to the lost object but is directed inwards to the self. However, whereas in mourning the ego focuses on detaching from the lost object and replacing it with a new one, in melancholy the attempt to preserve the lost object fails, and in a complex process it eventually erupts in acts of self-violence. In a striking sentence in the closing paragraph of *Mourning and Melancholia*, Freud contends that the whole complex of melancholia 'behaves like an *open wound*, drawing to itself cathectic energies (. . .) from all directions, and emptying the ego until it is totally impoverished' (Freud 2001a: 252, emphasis mine). Unable to separate from her lost object (Proserpina), Cyane is consumed by her own tears in a symbolic act of self-punishment, and her melancholic wound reduces her completely, melting her body and soul, turning her into liquid.

Recognition of the existence of a melancholic state of mind (or character) is of course not a Freudian innovation: the Greeks and their Roman successors referred to melancholy among other things as a mental state like agitation, sorrow, restlessness, or anger.[33] The ancients did not connect melancholia directly to wounds as Freud did; however, they related it (as well as other mental states) to the balance of bodily fluids and moisture, as part of humoral theory.[34] In *On Memory and Recollection* (453a14ff.) Aristotle asserts that in melancholic people the process of recollection creates constant movement, which does not cease until the sought object is found. This disorder, Philip van der Ejik stresses, 'manifests itself particularly in people whose region of sensory perception is surrounded by moisture, "for once moisture is set in motion, it does not readily stop moving until the sought object is found and the movement has taken a straight course".'[35] Cyane's wound sets her melancholic soul in motion; not being able to find Proserpina, her lost object, she is trapped in constant mental motion, as if she herself were a fluid. Indeed fluids are characterised by their mobile nature: they ' "flow", "spill", "run out", "splash", "pour over", "leak", "flood", "spray", "drip", "seep", "ooze"; unlike solids, they are not easily stopped—they pass around some obstacles, dissolve some others and bore or soak their way through others still'.[36] Thus, whether we refer to the Freudian melancholic wound or the ancient fluid imbalance and moisture's movement, Cyane's melancholic wound transforms her solid shape into a process of transformation and finally melts her completely into a liquid substance.

One of the most famous cases of melancholic wound in Roman literature is that of Dido as it appears in Book Four of Virgil's *Aeneid*. This banal story of a broken heart leads to a state of continuous melancholy whose consequences are dramatic. Dido's mental wound makes its first appearance in the opening lines of Book Four, once the image of the virtuous Aeneas 'comes back' (*recursare*) to her memory and wounds her.[37] This memory sets her soul in restless motion (in accordance with Aristotle's theory mentioned earlier), whose waves will hit the corporeal shores seven hundred lines later, as the emotional injury will transform into a corporeal wound that will terminate her life. Dido's wound releases a burst of body liquids and fluid metaphors that appear throughout the book: her tears burst out when she realises she has fallen in love with Aeneas (*lacrimis . . . obortis*, 4.30); when she obliges Aeneas to take an oath (4.314); when she cries (*ire in lacrimis*) on realising that he has left her (4.413); and just before killing herself, when she sees Aeneas' garments (4.649). Her union with Aeneas, the very first cause of her troubles, happens during a stormy tempest, as gushing flows of water cascade from the mountains.[38] She constantly fluctuates (*fluctuare*, 4.532) between anger and love as if she were a turbulent sea; her stormy soul keeps moving, frantic and restless;[39] blood blinds her eyes (4.643); and the wine she pours to placate the gods turns into foul blood (*obscenum . . . cruorem*, 4.455), widening the metaphor of the bloody wound to the gods' response. In the final scene of the story she decides to put an end to her melancholy and wounds herself: 'and in the middle of her words her friends watch her collapse on the sword, and her blood foams on the weapon and sprayed over her hands.'[40] This blood continues to flow onto her sister's lap as she tries to wash the wound, and the sister's tears mingle with the cleaning water and the pouring blood, creating a mixture of fluids that dissolves the boundaries between the two. By the end of the book neither the streams of blood nor the flow of events can be stopped. The psychic wound becomes visible at the intersection between the internal injury and the external wound on the surface of her body, and the queen dies. It turns out that the only stable thing that has remained steady throughout the story is Dido's wound itself, which even in her death is still 'infixed under her chest'.[41]

Dido's pain feeds her wound through the veins, drawing her cathectic energy,[42] making her restless and disrupting her rational judgement. She herself is like a fluid, which cannot sustain any shape for long and is constantly ready and prone to change.[43] Besides Dido there are other solids that metaphorically melt in the story: sacred loyalties, traditional alliances, and conventions of hospitality. To melt solids, argues the sociologist Zygmunt Bauman, is 'by definition dissolving whatever persists over time and is negligent of its passage or immune to its flow' (Bauman 2006: 2). Indeed, Dido's disloyalty to her dead husband, Aeneas' treachery, Anna's deception—all melt the social structure, while the lover's alliance is turned into solid and relentless hostility between Rome and Carthage.[44]

One more tragic story that dissolves social conventions is that of Pyramus and Thisbe (*Metamorphoses* 4.55–167). The two lovers dwell in the city of Queen Semiramis, surrounded by high walls (4.57–8), confined by its rigid social norms and separated by a wall between their neighbouring houses. A small fissure (*rima*, 4.65) in this wall allows the passage of their voices and emotions. They resolve, through this crack, to escape social constraints and run away. They set a meeting point in nature outside the city walls, whither Thisbe is first to arrive. Near a stream she encounters a lioness, her mouth foaming with blood after she has slaughtered her prey. The frightened girl escapes, and in her haste she drops her garment, which—once detected by the lioness—is torn to pieces in the animal's bloody mouth.[45] When Pyramus arrives and sees the bloodstained garment, he is convinced that Thisbe has been eaten by a beast. Lamenting his loss he sheds tears over the vestment, entreats earth to receive his blood as well,[46] and then stabs himself. His withdrawing the sword from the wound causes the blood to spurt high up—'no different from when a decayed lead water pipe is cut, and through the thin fissure, hissing, the water spouts forth far away, slicing the air with its jets'.[47] As in previous examples, here too the act of wounding is followed by the brief pipe simile that cuts the narrative itself. This simile alludes to the flow of blood as a hydraulic system, which is characterised by a tight network of tubes containing a finite amount of liquids in a state of balance and stasis. Bodily fluids, however, do not obey this principle: instead of relying on the principle of scarcity, they are characterised by abundance and plenitude, and by endless emanation; furthermore, as argued by Naomi Segal, they not only 'replenish themselves by a logic of plenty' but also prove the ego's inability to act as a container (Segal 2009: 94). Indeed, in his loss, Pyramus' ego collapses and is unable to contain his sorrow; his mental structure breached, his tears and his blood erupt. The endless flow of the blood—a metaphor for his endless loss—dissolves the boundary between in and out, man and nature: the internal blood exits the body, absorbed by the tree under which he lies, and colours its fruits red.

When, soon after, Thisbe reaches the place, she espies human body parts lying on the ground, still beating and covered with blood (4.133–4). The traumatic scene stuns her, and she 'shivers like the surface of the sea that trembles as the breeze grazes its surface'.[48] As soon as she realises that she is looking at her lover's corpse she starts beating herself, wailing and weeping; she then embraces the wounded body, 'filled the wound with tears and mixed them with his blood, giving him kisses on his frozen lips'.[49] Pyramus' wound, the place whence the blood spurts out, becomes a receptive container for Thisbe's tears. The bodily fluids of the couple mix, and the corporeal boundary between them follows the emotional one and melts. Then, in a further act of union, Thisbe chooses to take her own life as well by falling on Pyramus' sword, still warm from his own blood (4.163). Two wounded souls, two wounded bodies, two perforated melancholic egos that could not contain the loss. The entire event occurs in a liminal space, outside the stagnant society and tamed culture, in wild and moist surroundings, near a stream of water, and on the dewy grass. The wounds of the two create a mechanism of anti-structure that not only negates their bodies and souls but also symbolises the liquefaction of tradition, norms, and conventions. Also, it challenges linear time by duplication and contingency: the whole scene repeats itself in a loop: Thisbe reaches the place twice, the lamentations and mourning of loss occur twice, the stabbing and the suicide are double, and so is the effect on the fruits of the tree.

Conclusions: between solidity and fluidity

The primal particles of our world, writes Lucretius, are made of solid materials (*On the Nature of Things* 1.951). They differ from liquid materials in their structure, as the atoms of liquids are round and smooth, so their motion is easy and rapid (2.452–61). This structural difference, as Lucretius himself notes, have profound implications: that which keeps moving and changing constantly exits its own boundaries, and this process of change means the death of its former essence; therefore a world without steady things would mean nothing to men.[50] Modern psychoanalytic theories suggest a similar argument from a different point of view, maintaining that the human soul comprises two principles: emergence and continuity. Emergence is related to the constant flux of change of the self and the world, which is perceived subjectively and cannot be explained through shared rules and common sense; continuity, on the other hand, is related to the perception of the world as continuous and predictable, therefore logical and sensible (Amir 2016: xiii). Fluidity is also associated with a process of change that melts solid social norms and structures. In this respect it is a metaphor for states characterised by a movement that breaches boundaries and challenges unities and common orders. These ancient and modern perceptions all point to a substantial contrast between fluids and solids, the latter of which symbolises steady structures that enable perception and understanding of the world as a steady phenomenon.

The bodily fluids discussed in this chapter are the result of corporeal and mental wounds. Besides rupturing the body's boundary, wounds also tell the story of the changing boundaries of language, literature, and human experience. Once the corporeal or mental surface is torn, the possibility of different types of existence opens up: the sense of continuity is damaged, and instead a chaotic and uncontrollable movement appears. The skin becomes a strainer through which the human interior (body and soul) pours out (physically or metaphorically) and falls into endless and shapeless space, which drains the surface's previous definitions as well as the definition of the former structure. Indeed, as Steven Connor writes,

> when . . . the frail containing envelope of the skin is torn, dissolved, melted and lacerated, this is perhaps an apprehension in a violent mode of the growing fluidity of relations between the self and its contexts and secondary instruments, a condition in which the skin is no longer primarily a membrane of separation, but a medium of connection or greatly intensified semiotic permeability, of codes, signs, images, forms, desires.
>
> Connor 2004: 65–6

With the wound, the fluids, and the loss of surface, both the principle of continuity of the subject and the text fall apart and dissolve: the body is inverted, the language loses its meaning, and common-sense becomes non-sense. If writing about the wounded body represents a fragile encounter between the corporeal and the lingual, writing about bodily fluids generates a process, an occurrence, a becoming. This process is always involved in a physical or emotional trauma, which is connected to the potential end of former existence and of leakage into a state of annihilation or chaos. Identification of the ego with the body collapses, and the detachment from the stable positions melts the historicity of the body and the subject, leaving them in a constant flow of becoming. Once the whole body becomes a wound there is no turning back. The cases of Ophletes and Marsyas represent two sides of this becoming: whereas the first dissolves into death (the end of existence and of life), the second transforms into a source of other lives (the ending of Marsyas the satyr and the becoming of the River Marsyas).

Fluid subjects change from a functioning system into a system of functions with no super-structure; they lose their ontological status and become fusion of subject–object. Echo becomes sound and rocks, and Procne and Philomela are metamorphosed into birds. They undergo a process of separation from their bodies and subjectivity, and the one becomes the multiple—many rocks, many birds, located in many places at any time. Time becomes contingent and relative or mingled with the past (e.g. through similes—as shown in the examples of Hypsipyle and Pyramus; or through scars—as suggested in the cases of Procne and Philomela, Pyramus, and Thisbe). Law and order are lost, since in a fluid and contingent world there is no need to prefer one form of existence over another, there is no 'single' or 'many', nor private and public. Conventional laws and traditions are destroyed in Dido's story, as well as in others—such as that of Procne and Philomela—melting the boundaries of society and its organised structure. Indeed, the real threat seems not to lie in the lacerated skin or in the wound itself, nor in the breach of the boundary or its relation to disorder and 'abject'. It is the metamorphoses between solid to liquid or, more specifically, the state of fluidity that is alarming. Fluidity, as I hope to have shown in this chapter, produces transformations, processes, contingency, lack of control, and a potential threat to the human order. It replaces ontology by metaphysics and represents anti-structure that eliminates the fixed meaning of the world in favour of alternative and unpredictable paths. In this sense, bodily fluids, unlike wounds, have not only a subversive potential but a productive one, by making room for new solids by the melting of former ones.

Notes

1 'Interestque vulnus in summa parte sit an penitus penetraverit, necessarium est notas sub-
icere, per quas, quid intus actum sit, scire possimus, et ex quibus vel spes vel desperatio
oriatur', Celsus, *On Medicine* 5.26.7.
All translations from the Latin are my own.

2 Celsus, *On Medicine* 5.26.20a–c. Majno argues that for Celsus the difference between
wound and ulcer lies in its ability to be cured (Majno 1975: 360). As regards this chapter, this division is of no importance, and I therefore do not refer to it throughout my
discussion.

3 E.g. Bick 1968; Tustin 1981; Anzieu 1989; Ogden 1989; Freud 2001b.

4 See also Lucretius, *On the Nature of Things* 4.210ff.; 4.240ff.; 4.354.

5 Lucan, *On the Civil War* 9.767–81.

6 Lucan, *On the Civil War* 9.770–76: 'Membra natant sanie, surae fluxere, sine ullo/ tegmine
poples erat, femorum quoque musculus omnis/ liquitur, et nigra destillant inguina tabe./ Dis-
siluit stringens uterum membrana, fluuntque/ uiscera; nec, quantus toto de corpore debet,/
effluit in terras, saeuum sed membra uenenum/ decoquit, in minimum mors contrahit omnia
uirus.'

7 'Sine corpore vulnus', Lucan, *On the Civil War* 9.769. cf.: 'sine vulnere corpus', Ovid,
Metamorphoses 12.99; 13.267.

8 Statius, *Thebaid* 5.505 ff.

9 'Rapta cutis, tenuia ossa patent nexusque madentes/ sanguinis imbre novi', Statius, *Thebaid*
5.597–8.

10 'Totumque in vulnere corpus', Statius, *Thebaid* 5.598.

11 'Non ora loco, non pectora restant', Statius, *Thebaid* 5.596.

12 'Non uerba in fulmine primo/ non lacrimas habet', Statius, *Thebaid* 5.593–4.

13 Statius, *Thebaid* 599–604. Cf. Aurbach's analysis of the Homeric scene of Eurycleia, who
identifies Odysseus by his scar. When the old woman touches the scar, at the moment of
crisis, the story breaks and a long simile cuts through the narrative: Auerbach 2003 (1946):
chap. 1.

14 'Tandem laxata dolori/ uox inuenit iter, gemitusque in uerba soluti', Statius, *Thebaid*
606–7.

15 'Nec quicquam nisi vulnus erat', Ovid, *Metamorphoses* 6.388.

16 Ovid, *Metamorphoses* 6.385–91: 'Quid me mihi detrahis?' inquit/ 'A! piget, a! non est'
clamabat 'tibia tanti.'/ clamanti cutis est summos direpta per artus,/ nec quicquam nisi vulnus erat; cruor undique manat,/ detectique patent nervi, trepidaeque sine ulla/ pelle micant
venae; salientia viscera possis/ et perlucentes numerare in pectore fibras.'

17 Freud himself wrote that 'the ego is first and foremost a bodily ego; it is not merely a surface
entity, but is itself the projection of a surface' (Freud 2001b: 26).

18 Anzieu 1989. For Anzieu 'every psychical activity is analytically dependent upon a biological function' (40).

19 ' "A! piget, a! non est" clamabat "tibia tanti" '; cf. Ovid, *Metamorphoses* 6.386–91. See
Newlands's discussion on this line and its comparison with the myth in the *Fasti* and *Ars
Amatoria* 3.505 (Newlands 2018: 172).

20 This is conspicuous when pain is involved (whether physical or emotional), and—as Elaine
Scarry shows—the language and its representational function are shattered (Scarry 1985: 5).

21 'Haeret amor crescitque dolore', Ovid, *Metamorphoses* 3.395.

22 'Extenuant vigiles corpus miserabile curae/ adducitque cutem macies et in aera sucus/ corporis omnis abit; vox tantum atque ossa supersunt:/ vox manet, ossa ferunt lapidis traxisse
figuram', Ovid, *Metamorphoses* 3.396–9.

23 Cf.: 'Ossaque nondum adduxere cutem', Lucan, *On the Civil War* 4.288.

24 'Sibi vultuque immotus eodem/ haeret', Ovid, *Metamorphoses* 3.418–19.

25 'Lacrimis turbavit aquas', Ovid, *Metamoprhoses* 3.475.

26 'Purpureasque notas filis intexuit albis', Ovid, *Metamorphoses* 6.577. For purple blood, see: 'purpureus lunae sanguine vultus erat', Ovid, *Amores* 1.8.12; 'purpureus venit in ora pudor', Ovid, *Metamorphoses* 2.5.34.

27 See for example: 'cras Phoenicium poeniceo corio invises pergulam', Plautus, *Pseudolus* 228; 'ita ego vestra latera loris faciam ut valide varia sint,/ ut ne peristromata quidem aeque picta sint Campanica/ neque Alexandrina beluata tonsilia tappetia', Ibid. 145–7. See also: 'fiet tibi puniceum corium', Plautus, *Rudens* 1000. Segal even compares the purple skin of slaves to the purple stripes on the senators' togas, claiming that just as the purple stripes signified the difference between the senators to the common people, so did the purple scars symbolised the gap between slaves and Roman citizens (Segal 1968: 139).

28 'Dolor ora repressit,/ verbaque quaerenti satis indignantia linguae/ defuerunt, nec flere vacat', Ovid, *Metamorphoses* 6.585. Henry is right to acknowledge that *ora* is not to be taken literally as 'face' but as a figurative metaphor for speech (Henry 2013: 3 ff.).

29 'Vescitur inque suam sua viscera congerit alvum', Ovid, *Metamorphoses* 6.651.

30 'Neque adhuc de pectore caedis/excessere notae, signataque sanguine pluma est', Ovid, *Metamorphoses* 6.669–70.

31 Ovid, *Metamorphoses* 5.425–37: 'At Cyane, raptamque deam contemptaque fontis/ iura sui maerens, inconsolabile vulnus/ mente gerit tacita lacrimisque absumitur omnis/ et, quarum fuerat magnum modo numen, in illas/ extenuatur aquas: molliri membra videres,/ ossa pati flexus, ungues posuisse rigorem;/ primaque de tota tenuissima quaeque liques-cunt,/ caerulei crines digitique et crura pedesque/ nam brevis in gelidas membris exilibus undas/ transitus est; post haec umeri tergusque latusque/ pectoraque in tenues abeunt eva-nida rivos;/ denique pro vivo vitiatas sanguine venas/ lympha subit, restatque nihil, quod prendere possis.'

32 For a discussion on the social function of tears in the process of mourning, see Corbeill 2004: chapter 3. See also: Erker 2009. Erker mentions Wagner-Hael's claim that 'a public display of emotion did not serve as much to express the individual grief, but was rather a medium for values and standards, a display of loyalty and of affiliation' (137). However, the very individual result of Cyane's tears (as well as the other examples discussed in this chapter) suggests, as I will try to demonstrate in the following, that her lament is not merely ritual but rather emotional, hence its nullifying effect.

33 Nevertheless, we should keep in mind that the complex ancient notion of melancholy was quite different from the Freudian conceptualisation of the term. For a discussion on ancient melancholy, see for example: Toohey 1990, 2004; van der Eijk 2005.

34 For example, in the opening chapter of Book 30 of *Problemata*, Aristotle asserts that when the humour (*chumoi*) of black bile is in a state of excess, one is prone to develop melancholia in accordance with the mixtures created in one's body. Later views (such as that of Galen) maintained that the character is determined by a mixture (*krasis*) of the four humours in accordance with external factors such as diet and lifestyle. For more on humours and humoral theory, see Wilkins' chapter in this volume.

35 Van der Eijk 2005: 142. Van der Eijk cites Aristotle's *On Memory*.

36 Bauman 2006: 2

37 'At regina gravi iamdudum saucia cura/ Vulnus alit venis', Virgil, *Aeneid* 4.1–2.

38 'Ruunt de montibus amnes', Virgil, *Aeneid* 4.663–5.

39 'Partis animum versabat in omnis', Virgil, *Aeneid* 4.630.

40 'Atque illam media inter talia ferro/ conlapsam aspiciunt comites, ensemque cruore/ spu-mantem sparsasque manus', Virgil, *Aeneid* 4.663–5.

41 'Infixum . . . sub pectore vulnus', Virgil, *Aeneid* 4.689. In the suicide scene the narra-tor's voice interrupts Dido's speech, and the tempestuous flux of events is intensified by proliferation of connective particles and phrases as if the words themselves join the flood: *atque . . . media inter . . . ensemque . . . sparsasque*, Ibid. 4.663–5. Her second appear-ance in the Aeneid is in the land of the dead, where she is still described as wounded (Ibid. 6.450).

42 'Vulnus alit venis', Vergil, *Aeneid* 4.663–5. Page's words that 'the wound drains her life-blood' (Page 1957: 345) bring to mind the Freudian cathectic draw of energy as part of a melancholic wound.

43 cf. Bauman 2006: 2

44 'Let us remember', Bauman reminds us, 'that all this was to be done not in order to do away with the solids once and for all and make the brave new world free of them for ever, but to clear the site for new and improved solids' (Bauman 2006: 3).

45 'Ore cruentato', Ovid, *Metamorphoses* 4.104.

46 ' "Accipe nunc" inquit "nostri quoque sanguinis haustus!" ' Ovid, *Metamorphoses* 4.118.

47 'Non aliter quam cum vitiato fistula plumbo/ scinditur et tenui stridente foramine longas/ eiaculatur aquas atque ictibus aera rumpit', Ovid, *Metamorphoses* 4.122–4.

48 'Exhorruit aequoris instar,/ quod tremit, exigua cum summum stringitur aura', Ovid, *Metamorphoses* 4.135–6.

49 'Vulnera supplevit lacrimis fletumque cruori/ miscuit et gelidis in vultibus oscula figens', Ovid, *Metamorphoses* 4.140–1.

50 'Nam quod cumque suis mutatum finibus exit,/ continuo hoc mors est illius quod fuit ante./ proinde aliquid superare necesse est incolume ollis,/ ne tibi res redeant ad nilum funditus omnes', Lucretius, *On the Nature of Things* 2.753–6

Bibliography

Primary sources

Asso, P. (2009) *A Commentary on Lucan, De Bello Civili IV: Introduction, Edition, and Translation*. Berlin: Walter de Gruyter.

Encyclopaedia Britannica [Online]. Available at: https://www.britannica.com/science/wound (Accessed February 2021).

Page, T. E. (1957) *The Aeneid of Virgil, Books I-IV*. London: Palgrave Macmillan.

Secondary literature

Amir, D. (2016) *On the Lyricism of the Mind: Psychoanalysis and Literature*. London: Routledge.

Anzieu, D. (1989) *The Skin Ego* (translated by C. Turner). New Haven: Yale University Press.

Auerbach, E. ([1946] 2003) *Mimesis: The Representation of Reality in Western Literature*. Princeton, NJ: Princeton University Press.

Bauman, Z. (2006) *Liquid Modernity*. Malden: Polity Press.

Bick, E. (1968) 'The experience of the skin in early object-relations', *The International Journal of Psychoanalysis* 49(2–3), 484–6.

Connor, S. (2004) *The Book of Skin*. Ithaca: Cornell University Press.

Corbeill, A. (2004) *Nature Embodied: Gesture in Ancient Rome*. Princeton, NJ: Princeton University Press.

Erker, D. Š. (2009) 'Women's tears in ancient Roman ritual', in T. Fögen (ed.) *Tears in the Graeco-Roman World*. Berlin: Walter de Gruyter, 135–60.

Freud, S. ([1917] 2001a) 'Mourning and melancholia', in J. Strachey (ed.) *The Standard Edition of the Complete Psychological Works of Sigmund Freud*, vol. XIV. London: Vintage and Hogarth Press, 237–58.

———. ([1923] 2001b) 'The ego and the id', in J. Strachey (ed.) *The Standard Edition of the Complete Works of Sigmund Freud*, vol. XIX. London: Vintage and Hogarth Press, 1–66.

Henry, J. ([1889] 2013) *Aeneidea or Critical, Exegetical, and Aesthetical Remarks on the Aeneis*, vol. II. Cambridge: Cambridge University Press.

Majno, G. (1975) *The Healing Hand : Man and Wound in the Ancient World*. Cambridge, MA: Harvard University Press.

Newlands, C. E. (2018) 'Violence and resistence in Ovid's *Metamorphoses*', in M. Gale and J. H. D. Scourfield (eds) *Texts and Violence in the Roman World*. Cambridge: Cambridge University Press, 140–78.

Ogden, T. H. (1989) *The Primitive Edge of Experience*. Northvale: Jason Aronson Inc.

Scarry, E. (1985) *The Body in Pain: The Making and Unmaking of the World*. New York: Oxford University Press.

Segal, C. (1998) 'Ovid's metamorphic bodies: Art, gender, and violence in the *Metamorphoses*', *Arion* 5(3), 9–41.

Segal, E. (1968) *Roman Laughter : The Comedy of Plautus*. Cambridge, MA: Harvard University Press.

Segal, N. (2009) *Consensuality: Didier Anzieu, Gender and the Sense of Touch*. Amsterdam: Rodopi.

Shaughnessy, E. J., I. M. Katz, and J. P. Schaffer (2005) *Introduction to Fluid Mechanics*. Oxford: Oxford University Press.

Toohey, P. (1990) 'Some ancient histories of literary Melancholia', *Illinois Classical Studies* 15(1), 143–61.

———. (2004) *Melancholy, Love, and Time: Boundaries of the Self in Ancient Literature*. Ann Arbor: University of Michigan Press.

Tustin, F. (1981) *Autistic States in Children*. London: Routledge.

van der Eijk, P. (2005) *Medicine and Philosophy in Classical Antiquity*. Cambridge: Cambridge University Press.

Part VII

ANCIENT FLUIDS
Afterlife and reception

22

THE RECEPTION OF CLASSICAL CONSTRUCTIONS OF BLOOD IN MEDIEVAL AND EARLY MODERN MARTYROLOGIES

Anastasia Stylianou

'The blood of his dear saints (like good seed) never falleth in vain to the ground' wrote John Foxe (1583), England's most famous Early Modern martyrologist. Why was Foxe so preoccupied with martyrs' *blood*, and where did the imagery of blood as seed come from? The answer can be found primarily in the classical heritage upon which Early Modern Christianity was built. This chapter will first examine how the Bible and early Church constructed blood, especially martyrs' blood. It will then consider the reception of these constructions in the Medieval and Early Modern periods, exploring the striking variations in how elements of this heritage were accepted, rejected, adapted, and deployed by different models of Western Christianity. I shall also analyse how Early Modern confessions drew upon both classical and early Church conceptions of bodily fluids as each sought to present itself as the true heir of apostolic and patristic Christianity.

Through focusing on the specific case study of martyrs' blood, this chapter will also contribute to the wider argument that, from the classical to Early Modern periods (and beyond), bodily fluids lay at the very heart both of religion and of believers' perceptions of the body. I shall suggest that later eras shared with the classical period a prevalent assumption that bodily fluids, especially blood, expressed and carried key facets of the character of the individual who embodied them (namely their holiness, in regard to martyrs' blood). It will be seen that blood was viewed as a unique medium between the inner body and outer world, particularly during and after death, and could thus extend the boundaries of an individual's body and its characteristics far beyond the limits of their skin.

While there has been very fruitful interest among Medieval scholars (above all Bynum 2007) in the perceptions and treatments of blood in Medieval Western Christianity, Early Modern scholars have largely neglected the question of what role constructions of blood played in the Reformations, particularly in relation to both the English Reformations and martyrdom. There is a need for the findings of the rich scholarship on the roles of blood in Early Modern medicine (e.g. Horstmanshoff et al. 2012), literature, imagery, and perceptions of gender, sexuality, and race (e.g. Lander Johnson and Decamp 2018), to be brought into dialogue with Early Modern English confessional depictions of blood, and also for analysis of the changes and continuities between Medieval and Early Modern religious constructions of blood. This chapter

DOI: 10.4324/9780429438974-30

and my other works to date (Stylianou 2017, 2018) are intended as a starting point for such a dialogue.

Blood as life and sacrifice: the Bible and early Church

Early Christian communities inherited a multifaceted and theologically rich conception of blood from Judaism. These remained present in the Old Testament of the Christian Bible. In the Old Testament, blood plays a number of critical roles. First, it carries life. For this reason, animal sacrifice redeems human sin, since blood sacrifice gives one life up to God in order to redeem another (Leviticus 17:11). The life-carrying quality of blood is so important that God forbids the Israelites to eat blood on pain of exile; they, and even every foreigner residing among them, must drain the blood from animal flesh before eating it. Secondly, innocent (animal) blood functions as the medium of atonement for sin. Judaism was the first religion in which blood restored the sacrificial shrine and the whole world to its original purity, allowing mankind to commune with the divine (Geller 1992: 97–124). Thirdly, blood seals covenants. Every biblical covenant is sealed in blood, starting with the blood of circumcision and of sacrifice sealing the Abrahamic covenant, which founded the relationship between God and his chosen nation. The indispensable role blood plays in sealing covenants highlights its critical theological importance, not only above every other part of the body but above every other natural substance; it alone can create a binding agreement between God and his people. Finally, unrighteous shedding of innocent human blood has dire consequences. It stains and pollutes; it also calls out for retributive justice and can bring down God's vengeance (e.g. Deuteronomy 19:13). The image of earth stained by innocent blood which cries out to God recurs throughout the Old Testament, beginning with pious Abel, murdered by his jealous brother Cain (Genesis 4:10). Bloodguilt extends beyond the guilty individuals, to their descendants, their entire city, or beyond (e.g. I Kings 2:32–3). Blood shed unrighteously pollutes the land, which then requires cleansing, usually through shedding the blood of the blood-shedder (e.g. Numbers 35:33, II Samuel 21; see Sachs 2008: 261–2), perhaps drawing on the idea that blood equates to life, and so blood for blood equals a life for a life. Innocent blood is a potent force, bringing about divine retribution if humanity has failed to avenge it (e.g. Genesis 4, II Maccabees 8:3; see Jordaan 2012: 4).

The New Testament reaffirmed and extended these concepts of blood. The blood most commonly referred to in the New Testament is Christ's, and in this there is both continuity with and divergence from Old Testament constructions of animal blood sacrifices and innocent bloodshed. Christ's blood is the fulfilment of earlier sacrifices, bringing about the eternal atonement, salvation, and sanctification that they could not. It is also different from earlier martyrdoms and innocent bloodshed. Whereas Abel's blood and the blood of other innocent and pious people called for God's vengeance, Christ's blood calls primarily for God's mercy and is an instrument of peace, saving people from God's wrath (Hebrews 12:23–4; Colossians 1:20).

In both the Old and New Testaments, there is not a particularly developed theology of martyr's blood (in striking contrast to later Christian texts). While blood is a major theme in the Old Testament, martyrdom is not (Spronk 2004: 993). In general, the Old Testament affords little place to the notion of martyrdom, and where it does occur it is not usually presented as a cause for celebration or emulation.[1] The New Testament stresses the importance of martyrdom, based on the notion of sharing in Jesus' sufferings and death (e.g. cf. Matthew 16:24–5; Romans 8:17). For St Paul, the martyrs' sufferings also complete Christ's sufferings; like Christ's, they are efficacious for the Church (Colossians 1:24). At times, the New Testament depicts martyrdom as sacrificial. In II Timothy 4:6 and Philippians 2:17, Paul uses the Greek verb *spendo-mai* to describe his sufferings and the probability that he will be executed for his faith, denoting that he is being poured out like the libation which the Law prescribed for the conclusion of animal sacrifices (e.g. Exodus 29:40–1; see Zamfir 2017: 75–94). Sacrificial imagery is also seen in Revelation, where the martyrs cry out to God from beneath the altar for their blood to be avenged (Revelation 6:10). These passages sometimes connect martyrdom with blood; for example, Paul writes of the first Christian martyr's death, 'And when the blood of your martyr Stephen was shed, I stood there' (see also, Matthew 23:29–36; Luke 11:50–1; Hebrews 12:4).[2] However, the nascent ideology of martyrdom as a sacrifice is not overtly connected to blood (unlike in later Christian texts). While the belief is already present that believers are united with Christ's body and blood through participation in the Eucharist, this is not extended into a theology of martyrs' sacrificial blood imitating Christ's sacrificial blood.

As the Christian community experienced waves of mass martyrdom during the first three centuries after Jesus' death, its theology of martyrs' blood developed. In post-biblical, early Church texts, martyrs' bloodshed was sometimes presented as a sacrifice which imitated Christ's (Klawiter 2015). For example, Origen argued that, just as Christ 'has wiped out our sins by his death', so the 'martyrs, take away the sins of the saints' by offering themselves as expiatory sacrifices; 'as we are bought "by the precious blood of Jesus" . . . so by the precious blood of the martyrs certain have been bought' (1984: 222–3; trans. Bettenson). By partaking in the Eucharist of Christ's sacrificed body, Christians were prepared to be sacrificial victims with him in martyrdom (Mayes 2010: 322). 'But how can we shed our blood for Christ, who blush to drink the blood of Christ?' wrote St Cyprian (1886: 362–3; trans. Donaldson and Coxe). This theology came full circle in some texts where martyrdom itself was depicted as a Eucharistic offering.[3]

Because martyrs' blood was believed to imitate and be united with Christ's blood, it was depicted as echoing the functions of his blood: it purified, expiated, and sanctified. For martyrs themselves, it did so through 'baptism in blood'. Origen presented martyrdom as either a first or a second baptism (depending on whether the individual had previously been baptised): '[The martyrs] being baptized in their own blood and washing away every stain at the altar in heaven' (cited in, and trans. by, Fergusson 2014: 145, 86; see also Fergusson 2009: 417–19). Similarly, St Cyril of Jerusalem wrote: '[The wound in the crucified Christ's side] shed forth blood and water; that men, living in times of peace, might be baptized in water, and, in times of persecution, in their own blood' (1894: n.p.; trans. Gifford). Martyrs' blood was also presented as having a communal impact, functioning as a sacrifice 'offered for the benefit of the community of the faithful' (Salisbury 2004: 138).

In many early Church texts, martyrs' blood was represented as a vessel and conduit of grace, both bringing about conversions and strengthening pre-existing faith. It also performed miracles, transforming bodies, places, and objects. Since it was 'precious blood', a thriving cult of martyrs' blood relics developed (Salisbury 2004: 71); for example, at the beheading of St Cyprian in 258 and at the torture of St Vincent in 304, Christians in the crowd collected their blood on linen cloths (Salisbury 2004: 59–62).

As well as being connected with God's mercy, martyrs' blood was also connected with God's just vengeance: it would be avenged, and unrepentant blood shedders punished, including first and foremost the demonic powers, who were seen as bloodthirsty and blood-guilty (Leyerle 2001: 37). For example, in St Victorinus of Pettau's commentary on Revelation, the devil was presented as, above all, not a tempter but a murderer, and associated with bloodshed (1886: 144–364; trans. Wallis). Likewise, Victorinus wrote that the beast, which is the right hand of the Antichrist and an agent of the devil, has a 'mouth armed for blood . . . and a tongue which will proceed to nothing else than to the shedding of blood' (1886: 17:3).

When we consider that martyrs' blood was believed to witness to the true faith, imitate Christ's blood sacrifice, and be a vessel of grace, we can see why Tertullian described martyrs' blood as the seed of the Church (Tertullian, edition Glover 1931: 50.13). In accordance with biblical and (classical) medical ideas of blood relating to life and fertility, the early Christians saw martyrs' blood as a fecund substance, bringing fresh life to the Church despite the death of some of her members (Leyerle 2001: 45, 47). The fact that martyrs' blood was depicted as the seed of the Church both reflects the multitude of important functions it was believed to play and alerts us to the thriving cult which arose around it.

A world awash in Christ's blood: c. 1100–1500

During the latter half of the first millennium in the West, martyrs' blood gradually declined in importance, as martyrdom became less common. Meanwhile, Christ's blood increased still further in importance. By the high and late Medieval periods, 'the bloodbath from an exsanguinated Christ moved to the centre of European piety. . . [and] northern European devotional art and poetry seem awash in [his] blood' (Bynum 2007: 1–2). While martyrs' blood was less important in this period than in the early Church, it was still mentioned, albeit less frequently, in popular hagiographical compilations, and there were still popular cults of martyrs' blood. Moreover, it retained its importance in some textual genres, such as crusading martyrologies and anti-clerical and heretical writings.

The most popular saints and martyrs were those of the early Church, their lives featuring prominently in art, literature, and material culture. In some cases, relics of their blood worked spectacular miracles and drew pilgrims. From the late fourteenth century, there were many reports of St Januarius' blood in Naples miraculously liquefying; the same was reported of St Panteleimon's blood in Ravello (de Ceglia 2014). Early Church martyrs' blood relics were listed in the relic collections of churches, abbeys, and cathedrals; for example, the Collegiate Church of St Mary in Warwick possessed the stone upon which St George shed his blood during his martyrdom (Davidson 1997: 438).

In both popular piety and popular religious texts across Medieval Europe, early Church beliefs about martyrs' blood endured, as illustrated by three popular Medieval hagiographical compilations: Jacobus de Voragine's *Golden Legend* (1260s), the anonymous *Speculum Sacerdotale* (fifteenth century) (edition Weatherly 1936), and

John Mirk's *Festial* (c. 1380s). Many of the saints whose lives appeared in these collections were early Church martyrs, and the compilations drew on and referenced early Church texts for their source material. When these texts mentioned martyrs' blood, it often appeared as a vessel of grace, whether described as a holy object, depicted as working miracles, bringing unbelievers to faith, or connected with baptism in blood. For example, Mirk's *Festial*, in relating the tale of the children slain by King Herod, painstakingly explained the doctrine of baptism in blood (1905: 36). The martyrdom of St Christopher in the *Golden Legend* exemplifies enduring belief in the supernatural powers of martyrs' blood to heal the sick and bestow faith upon unbelievers. Christopher's persecutor, a pagan king, has the saint tied to a pillar and arrows shot at him, but the saint remains unharmed. The king, thinking that Christopher must now be dying, comes to mock him, and is struck through the eye by one of the arrows and blinded. Christopher says, 'Tyrant, I will be dead by tomorrow. Then make a paste with my blood and rub it on your eyes, and you will recover your sight.' The king orders the saint to be beheaded, and his blood is brought to the king, who rubs it on his eyes, and his sight is restored. He then converts to Christianity and is baptised. The *Golden Legend* attributed this tale to the Church Father Ambrose (de Voragine 1993, Volume I: 14).

Some early Church ideas about martyrs' blood, in contrast, received very little attention in Medieval texts, particularly the concept of martyrs' deaths being sacrifices which imitate Christ's blood sacrifice, with power to purify, expiate, and sanctify the community. Martyrs' blood expiates and cleanses the sins of the martyrs themselves (baptism in blood) at points in the *Golden Legend*, *Festial*, and *Speculum Sacerdotale*, but almost never the sins of others.

In early Church writings, as we have seen, explicit parallels were sometimes drawn between Christ's blood and martyrs' blood, and the functions of both were very similar. In Medieval texts, such direct comparisons were rare, and martyrs' blood was not often presented as so similar to Christ's blood. The closest case is the cult of Thomas Becket (Archbishop of Canterbury c. 1119–70). After his murder by followers of King Henry II, during the protracted power struggle between the English Church and State, he immediately began to be venerated as a martyr. Local lay people and monks collected the blood flowing from his corpse, and the monks began offering it mixed with water at his shrine (Koopmans 2016: 537, 543–4). People from across northern Europe came on pilgrimage to Canterbury Cathedral to venerate Becket's shrine and drink his diluted blood; many healing miracles were attributed to it (Koopmans 2016). Becket's cult was depicted in twelve stained-glass windows in Canterbury Cathedral and related in Benedict of Peterborough's hagiography. Both Benedict and the windows explicitly drew a parallel between Becket's blood mixed with water and the Eucharistic mixture of Christ's blood and water (Koopmans 2016: 537–8, 541, 543). Benedict also aligned Becket's death with Christ's role as the high priest who, by his holy blood, entered the sanctuary (Koopmans 2016: 541).

While rhetoric of martyrs' blood was less common in most Medieval religious texts than in early Church ones, in some crusading martyrologies and heretical texts a rhetoric of martyrs' blood features prominently. It always involved conflict between a church of martyrs and their bloody enemies and often placed this within an apocalyptical context drawn from Revelation. These texts echoed the apocalyptical mind set of early Christianity, from which mainstream Medieval Catholicism had largely moved away. In the fifteenth-century Hussite Crusades, for example, the Hussites wrote of the 'bloody hands' of the crusaders, and the crusaders accused the Hussites of 'shedding the blood of Christians' (cited in Fudge 2004: 128–9). Both sides saw themselves as bleeding martyrs, the Catholic poet Hans Rosenblüt describing the crusaders' defeat at the battle of Domažlice as 'the bloodshed of . . . the martyrs' (cited in Fudge 2004: 122). In these texts, the backdrop to this war between bleeding martyrs and bloody enemies was the apocalypse—the 'night of antichrist' (cited in Fudge 2004: 131).

The Hussites were not alone in seeing the Catholic Church as bloody. Heretics across the Medieval period depicted it thus. This was often situated within an apocalyptic framework. Monta of Cremona's *Summa* (c. 1241) against the Cathars claimed that they believed that the beast and the woman drunk with the blood of the saints in Revelation referred to the Catholic Church, particularly the pope (Wakefield and Evans 1969: 328). According to the Catholic inquisitor Bernard Gui, the (heretical) Beguins concurred (Wakefield and Evans 1969: 423–5, 432). This Medieval bloody, anti-Catholic rhetoric was part of a widespread, enduring, and increasingly apocalyptical construction of the Catholic Church as a blood-shedding and bloodthirsty Other. It culminated in the rhetoric of English Protestant martyrological polemic.

Bleeding martyrs and bloody enemies: c. 1500–c. 1700

While the early Church's preoccupation with martyrs' blood had subsided during the Medieval period, it re-emerged during the European Reformations. Confessional polemic became saturated with a rhetoric of martyrs' blood, and, once again, martyrology became the dominant genre of hagiography [Gregory 1999]. From the first English Protestant works, a rhetoric of martyrs' blood started to appear. As Protestant theology developed, so did this rhetoric, encapsulating many key Protestant beliefs and polemic strategies. The Protestant and Catholic battle over 'right' belief was also a battle over the meaning and ownership of words and verbal imagery [Cummings 2002: 15]. Additionally, it was a war over the meaning and ownership of Christian history, as both sides attempted to present themselves as the theological and rhetorical heirs of early Christianity. The two confessions fundamentally disagreed over what the early Church had believed and how it had worshipped, including over the early Church's theologies of martyrdom and of martyrs' blood.

Early Church authors often tended to view martyrs' blood primarily as drawing down God's mercy and redemption, and only secondarily his vengeance, in imitation of Christ's blood. Most English Protestant apologists moved away from this stance, depicting martyrs' blood crying exclusively for vengeance, not for mercy and redemption. This was closely connected to their Eucharistic theology, most of them being Sacramentarians (Christians who did not believe that Christ's body and blood were physically present in the Eucharist).

In early Church thought, the lineage of martyrdom began again from the death of Christ. Indeed, the whole sacrificial and salvatory system began afresh, and as part of this, so did the nature and meaning of martyrdom. The blood of the Christian martyrs was not so much like the blood of Abel as like the blood of Christ, the new Adam and the new model for martyrdom. However, Early Modern Sacramentarian Christians rejected this proposition. Their martyrs imitated Christ's death in outward form, but they rejected the notion that their deaths could imitate Christ's in terms of being sacrificial, sanctifying, and salvatory. This created a dichotomy between Christ's blood and martyrs' blood: Christ's blood functioned like an Old Testament blood sacrifice, albeit with vastly amplified effects, while martyrs' blood functioned primarily in terms of the Old Testament notion of bloodguilt, demanding vengeance. In short, for most English Protestant apologists, martyrs' blood was not a vessel of grace and had no salvific value.[4]

Traditional cults of martyrs' blood were thus attacked and suppressed. Over the course of the sixteenth century, John Bale and John Foxe, the leading English Protestant martyrologists, particularly criticised the cult of Thomas Becket's blood. Given that this was the most popular English cult of martyrs' blood and that it included pronounced comparisons between Christ's blood and Becket's blood, it was an obvious target, and one through which the cult of martyrs' more generally could also be attacked. Bale wrote mockingly:

> [Becket] was so gloryouse a martyr and precyouse advocate of theirs, that they made hys bloude equall with Christes bloude and desyred to clyme to heauen therby. Many wonderfull myracles coulde that mytred patrone of theirs do in those dayes, whan the monkes had fryre Bakons bokes and knewe the bestowynge of fryre Bongayes mystes but now he can do non at all.
>
> Askew and Bale 1996: 80

This second sentence refers to the belief that Becket's blood worked miracles, which Bale attributes to demonic powers, in an allusion to Roger Bacon, a thirteenth-century Franciscan suspected of sorcery (Beilin, edition Askew and Bale 1996: note 80). Thus, Bale reduces the cult of Becket's blood to no more than a delusion and a hoax: not only is Becket's blood not salvific, unlike Christ's, but it does not even possess any miraculous powers; rather, the 'wonders' it has performed are due to the devil.

It is important to remember that most Early Modern Protestant theologians and apologists were cessationists—they believed that miracles (supernatural events, where God broke the laws of nature) had ceased with the end of the apostolic age (e.g. Calvin 1996: 13). Thus, for Protestant writers, the apostolic miracles recounted in early Church authors such as Eusebius were perfectly plausible, but the subsequent continuation of miracles was not. Foxe wrote scathingly when discussing Becket: 'If God in these latter dayes geveth no myracles to glorify the name of his owne sonne: much lesse wil he geve miracles to glorify, T. Becket' (1563, edition Foxe 2013: 100).[5] This meant that Protestant apologists often criticised the assumption that martyrs' blood could work miracles: their writings sometimes presented it as working wonders (spectacular natural events with a divine message), but not miracles.

For Early Modern Protestants, martyrs' blood was most often associated with vengeance. John Foxe's *Acts and Monuments* has a consistent narrative thread of 'blood for blood'. Innocent martyrs' blood is providentially avenged with the blood of their persecutors, whether more metaphorically, through their deaths, since blood equated to life in Judeo-Christian thought, or more literally, in gruesome tales of persecutors bleeding to death. This drew upon both Old Testament notions of bloodguilt only being avenged with the blood of the murderer and upon the narrative of Revelation, where God avenges the blood of the martyrs by sending plagues of blood and destroying their persecutors. For example, Foxe relates that Raphe Ellerker, who sentenced a

Protestant martyr to death and would not even let the martyr give a testimony of faith at his execution, was brutally killed and mutilated in a battle shortly after, his body being a terrible example to 'al bloudy and merciles men' (1563, edition Foxe 2013: 722). Foxe, likewise, strongly underlined that God avenged martyrs' blood with their persecutors' blood in tales where the martyrs' persecutors bled to death, as in the case of the French Catholic king, Charles IX:

> his bloud gushing our by divers partes of his body, he . . . layed upon pillowes with his heeles upwards, and head downeward, voyded so much blood at his mouth, that in a few hours he dyed. Which story . . . may be a spectacle and example to all persecuting kinges and Princes polluted with the blood of Christian Martyrs.
>
> 1583, edition Foxe 2013: 1231

For Foxe, these persecutors' gory deaths were a mirror of the unspeakable punishment they would face on the impending day of judgement. Protestant authors, like their forebears in the early Church, often expected an imminent apocalypse. Indeed, Foxe had confidently dated the end of the world to 1594, or earlier if God shortened the days for the sake of the elect (Fudge 2014: 155). This apocalyptical framework had a strong influence on their constructions of martyrdom, which were largely drawn from Revelation. From the 1520s, Protestant apologists and polemicists often painted the Catholic Church as the bloodthirsty and blood-guilty Antichrist. Bale, writing from exile in the 1540s, penned England's first Protestant martyrologies and a lengthy commentary on Revelation. In both, he depicted the blood of the martyrs (including early Christians, Medieval heretics, and Early Modern Protestants) as crying out to God for vengeance, and argued that God would thus soon wreak upon the Catholic Church 'the great vengeaunce . . . for shedynge of innocentes bloude' (Askew and Bale 1996: 94). In Foxe's commentary on Revelation, as in his *Acts and Monuments*, he depicted the Catholic Church, and especially the papacy, as the Antichrist and Whore of Babylon; and in both texts he elided the classical pagan persecutors of the early Christians with the Catholic persecutors of the Protestants, presenting both as blood-guilty idolaters (Foxe 1587: q5^r, C1^r). Foxe saw the Early Modern Protestant Churches as the heirs of the early Church, and himself as a second Eusebius.

English Catholics, however, laid claim to exactly the same heritage. They saw themselves, through unbroken apostolic succession and tradition, as the very same one, holy, Catholic, and apostolic Church of the first centuries, and the Protestants as like the early heretics. Thomas More (1485–1535) began the Early Modern Catholic tradition of labelling the Protestant martyrs as mere pseudo-martyrs, drawing on the theology the early Church had developed against 'heretical' groups' claims to martyrdom (Foxe 1587, Q6^r). Again and again, Catholic apologists quoted St Augustine's 'martyrem non facit poena sed causa' (cf. Ployd 2017). Protestant apologists quickly adopted the same language of pseudo-martyrdom and deployed the same quotation (Dillon 2002).

While Early Modern Protestants rejected the notion that martyrs' blood functioned like Christ's blood, Early Modern Catholics placed renewed emphasis upon it. Their martyrological texts, from the 1580s onwards, frequently depicted martyrs' blood similarly to Christ's blood. Martyrs' blood was 'holy blood' (Allen 1581: F7[v]), 'sacred blood' (Mush 1877: 98), 'sacred streams', and 'gracious moisture' (Mush 1877: 363). It possessed expiatory powers, functioning as a sacrifice, similar to Old Testament sacrifices and to Christ's expiatory, sacrificial death; it called for God's mercy and forgiveness of sins (see Dillon 2002: 137). The Jesuit priest and martyr Edmund Campion wrote,

> Very many even at this present being restored to the Church, new soldiers give up their names [ordination], while the old offer up their blood [martyrdom]. By which holy hosts and oblations, God will be pleased: and we shall no question, by him overcome.
>
> Allen 1582: E7[r]

Echoing the writings of early Church authors like Origen, Early Modern Catholics saw their martyrs' blood as an expiatory sacrifice on behalf of the community. Robert Parsons, a leading English Catholic apologist, wrote:

> I beseeche God to accept the Innocent bloude of his vertuouse preests, for some part of pacification of his wrathe toward us, and towards oure persecutors, that they having the miste of errour taken from their eyes, may see the truthe of Christs Catholique religion.
>
> Parsons 1582: M4[r]

Martyrs' blood, in Catholic theology, was a material vessel of grace, comparable to the Eucharist. Martyrs became sanctified through baptism in blood,[6] and their blood, in turn, became a holy object, which could transmit grace to people, objects, and places. While Protestant texts rarely presented martyrs' blood as miraculous, Catholic texts demonstrated an enduring belief in the miraculous nature of martyrs' blood. One of the most notable examples concerns the blood of Henry Garnett, the Jesuit Superior in England, martyred in 1606. At Garnett's execution, a bloodstained husk of straw 'did leap into the hand of a Catholic who stood by with great desire to get some part of the martyr's blood' (Gerard 1871: 297). It was given to a devout Catholic woman, who observed that after six weeks Garnett's face appeared on the straw, including 'his beard bespotted with bloude, and a bloudy circle compassing the necke where yet was . . . cutt of' (Anon Bod. MS, date unknown: 135[r]).[7] The bloody straw rapidly became famous, and was taken to 'many of the chiefest Catholics about London', as well as members of the King's Council (Protestants) (Gerard 1871:, 303). It worked healing miracles in some of those who touched it, curing the gentlewoman who owned it of a severe sickness and another 'gentlewoman in great peril of her life by danger of childbirth' (Gerard 1871: 304).

Martyrs' blood was also believed to have the power to confer or strengthen faith. The future missionary priest and martyr Henry Walpole was converted after a drop of Campion's blood splashed on him (Gerard 1988: 130; Alfield 1582: F1[r]). This echoes tales in early Church and Medieval martyrologies, such as St Christopher's blood healing and converting the king who sentenced him to death, and St Longinus, the centurion who pierced Christ with a spear, being healed of blindness by Christ's blood, and so converted (de Voragine 1993: 14). Martyrs' blood could also sanctify inanimate matter, such as objects and places. Allen wrote that Campion

> was hanged on the new galloes, which is now called among catholikes the Gibbet of Martyrs, because it was first set up and dedicated in the bloud of an innocent Catholike Confessor, and afterwards by the mans, and divers Priests and others Martyrdoms, made sacred.
>
> Allen 1582: C2ᵛ

Catholic texts go so far as to portray blood re-sanctifying whole countries. They viewed the Elizabethan Catholic martyrs' blood as essential to the reconsecration of England (Allen 1564): 'it is not force nor might / . . . that must convert the land, / It is the blood by martirs shed' (Alfield 1582: G1ʳ).

While Protestant writers commonly connected contemporary martyrs' blood with the apocalypse, Catholic writers almost never did so. Whereas most English Protestant writers believed the apocalypse to be imminent, most English Catholic writers did not. Rather, they awaited a future time when England would be Catholic again, and some saw martyrs' blood as an agent of this change (Alfield 1582: G1ʳ). Allen wrote confidently: 'God never suffereth it [the true faith] to cease or fail in any Country: though it stand with travail and blood.' He saw martyrdom as 'a joyful sign of mercy' that God 'willl not forsake the place nor people' and will send 'a calm, or the conversion of the whole'. Allen states unequivocally that the Catholics await this calm (the lifting of persecution) or the reconversion of the whole of England, and see present tribulations as a period of God's chastisement for their sins (Allen 1581: 112).

Both confessions drew heavily upon early Church heritage; however, they disagreed over what the early Church had actually believed. For most Protestants, early Christians had been Sacramentarians, whose martyrs' blood did not function like Christ's and did not feature in intercessory cults or possess expiatory powers; moreover, early Christians had witnessed the cessation of miracles as the apostolic age passed. For Catholics, early Christians had believed in transubstantiation, and their martyrs' blood functioned like Christ's and had attracted cults centred on its intercessory, salvific, and supernatural powers. Moreover, for most Protestants, the long-awaited age of the apocalypse, heralded in Biblical and early Church writings, had come; martyrs' blood called for and signalled God's impending vengeance. For most Catholics, the apocalypse was not imminent, any more than it had been in the early Church, but rather England would be reconverted, partly through the power of martyrs' blood.

Blood and other fluids: c. 1000–c. 1700

In the relationship between martyrs' blood and other bodily fluids in Medieval and Early Modern texts, we can see the enduring influence of both the early Church and of classical medical ideas discussed earlier in this volume (see especially Chapters 3, 9, 20, 13, and 15, by Salvo, Fallas, Mulder, Flemming, Wilkins, and Totelin, respectively). For Medieval and Early Modern authors, there was a close relationship between blood and other bodily fluids, although there was no consensus on how far other bodily fluids actually stemmed from blood, as opposed to simply being similar or comparable to it. Most Medieval and Early Modern hagiographers and apologists had little detailed knowledge of medicine, so their depictions, rather than reflecting contemporary, evolving medical thought, derived from a fusion of popular and theological understandings.

Martyrs' blood was repeatedly connected to two fluids associated with generation: milk and semen. Martyrs bleeding milk are found in five accounts in the *Golden Legend*, one in the *Speculum Sacerdotale* and two in the *Festial*. This association continued in Early Modern Protestant and Catholic writings, Foxe portraying St Paul bleeding milk when beheaded (Foxe 1583, edition Foxe 2013: 2133) and the Catholic martyr Margaret Clitherow stating, 'I mind by God's assistance to spend my blood in this faith, as willingly as I ever put my paps to my children's mouths' (Mush 1877: 427). Martyrs bleeding milk first emerged in apocryphal early Church writings, like *The Acts of Paul and Thecla*, before moving into orthodox Christian martyrologies and being widely accepted by the Medieval period. Classical medicine allows us to make sense of these tableaux, since milk was understood in some classical writings to be a form or product of blood. Aristotle depicted mother's milk as a concoction of menstrual blood in *Generation of Animals*, and Hippocrates implicitly did so in *On the Nature of the Child* (see Lawrence and Totelin, Chapters 14 and 15, inter alia, in this volume for further information). In Christian texts, this notion was fused with biblical ideas of martyrdom and spiritual conversion being connected to motherhood (II Maccabees 7; Matthew 23:37; Galatians 4:19), as evident in tropes of martyrs bleeding milk.

When we consider that martyrdom was understood as a fertile act, that the Church was gendered as female, and that male seed was believed to be produced from or closely related to blood, we can better understand why Tertullian depicted martyrs' blood as the *seed* of the Church. His image was repeated in many Protestant and Catholic apologetical and martyrological texts. Foxe wrote 'Tertullian hath well sayde, that the bloud of the Martyr is the seed of the Gospell', and the Catholic martyrologist John Mush stated: 'as she (the early Church) then cast her seed of blood to the generation of many, so now she fighteth with blood to save those that she hath borne' (Mush 1877: 363).

From the early Church, martyrs' blood had also been paralleled with tears. For example, St Gregory of Nazianzus had compared baptism in blood and baptism in tears, presenting the former as more august and the latter as more laborious (Gregory of Nazianzus 1894: 39). Similarly, the Jesuit John Gerard wrote: 'I . . . was left to . . . wash with many tears a soul which I was not counted fit to wash—once and quickly— with my blood' (Gerard 1988: 143; trans. Caraman). Allen paralleled tears and blood as modes of intercession: '[Into] misery our country to us most dear, being fallen, and having no other human help to recover it . . . we will not fear nor fail to pray and ask it of God with tears and blood' (edition Kingdon 1965: 77). Moreover, we see a close relationship between blood, tears, and even sexual fluids, in Mush's depiction of how Clitherow's husband reacted to her martyrdom: '[When he] heard that they condemned her, he fared like a man out of his wits, and wept so vehemently that the blood gushed out of his nose in great quantity' (Mush 1877: 418). Clitherow and her husband, in a sense, bleed together here, as one body joined by the conjugal bond of the exchange of fluids.

It was, therefore, upon classical foundations that the Medieval and Early Modern worlds built their constructions of martyrs' blood. Subsequent writers were eager to depict themselves as the heirs of Patristic thought. In the Reformations, both sides competed to depict themselves standing in the place of the bleeding early Church, facing bloodthirsty enemies and emerging victorious. The Catholic apologist Francisco Suarez taunted the English Protestants, 'elegantly does Tertullian . . . say to the tyrants: "Inflict torment, torture, for . . . [w]e become more as often as we are by you cut down; the blood of Christians is seed"' (2011: 19:3; trans. Simpson).

Notes

1 The main exceptions appear in some of the deuterocanonical/apocryphal books, where martyrdom is an expiatory sacrifice for sin, which pleases God, pacifies his wrath, and leads him to be merciful, e.g. 'The Prayer of Azariah' in Daniel and the books of Maccabees (see Heard 1987; Klawiter 2015: 553–73).
2 Quotations from the Bible throughout, unless otherwise stated, are from *The Holy Bible, New International Version* (2011), Biblica, Inc., accessed at: www.biblegateway.com/ (Accessed February 2021).
3 This is found in Ignatius of Antioch, Irenaeus, and in the anonymous martyrology of Polycarp (see Klawiter 2015: 567–8).
4 For further discussion of these points, see Stylianou 2018.
5 See, in contrast, Foxe's acceptance of the possibility that at St Paul's execution he bled milk instead of blood, *A&M* 1570: 68–9.
6 Early Modern Catholic apologists' awareness of the patristic basis of such language is evident when Allen explains it in his *Apologie* (1581), with references to Cyprian and Augustine (Allen 1581: P2ʳ).
7 My very grateful thanks to Katie McKeog for bringing this text to my attention.

Bibliography

Primary sources

Alfield, T. (1582) *A True Report of the Death & Martyrdome of M. Campion Jesuite and Prieste, & M. Sherwin, & M. Bryan preistes.* London: R. Rowlands or Verstegan.
Allen, W. (1564) 'A true, sincere and modest defense of English Catholics that suffer for their faith both at home and abroad', in R. Kingdon (ed.) *The Execution of Justice in England by William Cecil and a True Sincere and Modest Defense of English Catholics by William Allen.* Ithaca, NY: Oxford University Press.
Allen, W. (1581) *An Apologie and True Declaration of the Institution and Endevours of the Two English Colleges.* Rheims: J. De Foiguy.
———. (1582) *A Briefe Historie of the Glorious Martyrdom of XII. Reverend Priests.* Rheims: J. De Foiguy.
Anon. Bodleian Manuscript English Th. b. 2. Date unknown.
Askew, A. and J. Bale (1996) *The Examinations of Anne Askew.* Edited by Elaine V. Beilin. Oxford: Oxford University Press.
Calvin, J. (1996) *The Bondage and Liberation of the Will: A Defence of the Orthodox Doctrine of Human Choice Against Pighius.* Edited by A. N. S. Lane and translated by G. I. Davies. Carlisle: Paternoster.
Cyprian (1886) 'Epistle 62, to Caecilius, on the sacrament of the cup of the Lord', in A. Roberts, J. Donaldson, and A. Cleveland Coxe (eds) *Ante-Nicene Fathers,* vol. V. Buffalo: Christian Literature Publishing Co., 258–64.
Cyril of Jerusalem (1894) 'Catechetical Lecture 3L "On baptism"', in P. Schaff and H. Wace (eds) *Nicene and Post-Nicene Fathers,* vol. VII. Buffalo: Christian Literature Publishing, 14–18.

De Voragine, J. (1993) *The Golden Legend: Readings on the Saints*. Edited and translated by W. Granger Ryan. Princeton, NJ: Princeton University Press.

Foxe, J. (1587) *Eicasmi seu meditationes in sacrum apocalypsin*. London: Thomas Dawson.

———. (2013) *The Unabridged Acts and Monuments Online* [Online]. The Acts and Monuments Online. Available at: www.johnfoxe.org (Accessed February 2021). (This modern compilation includes the 1563, 1570, 1576, and 1583 editions.)

Gerard, J. (1871) *The Condition of Catholics Under James I: Father Gerard's Narrative of the Gunpowder Plot*. Edited with His Life, by John Morris. London: Longmans, Green & Co.

———. (1988) *The Autobiography of a Hunted Priest. John Gerard S.J.* San Francisco: Ignatius Press.

Glover, T. R. (1931) *Tertullian. Apologeticum*. Cambridge, MA: Harvard University Press.

Gregory of Nazianzus (1894) 'Oration 39, "Oration on the holy lights"', in P. Schaff and H. Wace (eds) *Nicene and Post-Nicene Fathers,* vol. VII. Buffalo and New York: Christian Literature Publishing, 352–60.

Kingdon, R. (1965) *The Execution of Justice in England by William Cecil and a True Sincere and Modest Defense of English Catholics by William Allen*. Ithaca: Oxford University Press.

Mirk, J. (1905) *Festial: A Collection of Homilies*. London: Published for the Early English Text Society.

Mush, J. (1877) 'A true report of the life and martyrdom of Mrs Margaret Clitherow', in J. Morris (ed.) *The Troubles of our Catholic Forefathers Related by Themselves,* vol. III. London: Burnes and Oates, 331–40.

Origen (1984) '*Hom. in numeros,* 10.2, 222, and *Exhortatio ad martyrium,* 20, 223', in H. Bettenson (ed.) *Christian Fathers: A Selection from the Writings of the Fathers from St Clement of Rome to St Athanasius*. Oxford: Oxford University Press, 220–2.

Parsons, R. (1582) *An Epistle of the Persecution of Catholickes in England Translated out of the Frenche into Englishe and Conferred withe the Latyne Copie*. Douai: Fr Parson's Press.

Suarez, F. (2011) *Defense of the Catholic and Apostolic Faith Against the Errors of Anglicanism*. Translated from the Latin by Peter L. P. Simpson [Online]. Aristotelophile. Available at: www.aristotelophile.com/Books/Translations/Suarez%20Defense%20Whole.pdf (Accessed February 2021).

Victorinus of Pettau (1886) 'Commentary on the apocalypse', trans. R. Ernest Wallis, in A. Roberts, J. Donaldson, and A. Cleveland Coxe (eds) *Ante-Nicene Fathers,* vol. VII. Buffalo: Christian Literature Publishing Co, 341–60.

Weatherly, H. (1936) *Speculum Sacerdotale*. London: Oxford University Press.

Secondary literature

Bynum, C. (2007) *Wonderful Blood: Theology and Practice in Late Medieval Northern Germany and Beyond*. Philadelphia, PA: University of Pennsylvania Press.

Cummings, B. (2002) *The Literary Culture of the Reformation: Grammar and Grace*. Oxford: Oxford University Press.

Davidson, C. (1997) 'Sacred blood and the late medieval stage', *Comparative Drama* 31(3), 436–58.

De Ceglia, F. P. (2014) 'Thinking with the saint: The miracle of saint Januarius of Naples and science in early modern Europe', *Early Science and Medicine* 19(2), 133–73.

Dillon, A. (2002) *The Construction of Martyrdom in the English Catholic Community, 1535–1603*. Aldershot: Ashgate.

Fergusson, E. (2009) *Baptism in the Early Church: History, Theology, and Liturgy in the First Five Centuries*. Grand Rapids: William B. Eerdmans.

————. (2014) *The Early Church at Work and Worship, Vol II: Catechesis, Baptism, Eschatology, and Martyrdom*. Cambridge: James Clarke & Co.

Fudge, T. (2004) ' "More glory than blood": Murder and martyrdom in the Hussite crusades', in D. R. Holeton and Z. V. David (eds) *The Bohemian Reformation and Religious Practice*, vol. V. Prague: Academy of Sciences of the Czech Republic.

————. (2014) 'Jan Hus as the apocalyptic witness in John Foxe's *History*', *Communio Viatorum* 56(2), 136–68.

Geller, S. (1992) 'Blood cult: Towards a literary theology of the priestly works of the Pentateuch', *Prooftexts* 12(3), 97–124.

Gregory, B. S. (1999) *Salvation at Stake: Christian Martyrdom in Early Modern Europe*. Cambridge, MA: Harvard University Press.

Heard, W. J. (1987) 'Maccabean martyr theology: Its genesis, antecedents and significance for the earliest soteriological interpretation of the death of Jesus', PhD thesis, University of Aberdeen, Aberdeen.

Horstmanshoff, M., H. King, and C. Zittel (eds) (2012) *Blood, Sweat and Tears: The Changing Concepts of Physiology from Antiquity into Early Modern Europe*. Leiden: Brill.

Jordaan, P. J. (2012) 'Ritual, rage and revenge in II Maccabees 6 and 7', *HTS Teologiese Studies* 68(1).

Klawiter, F. (2015) ' "Living water" and the sanguinary witness: John 19:34 and martyrs of the second and early third century', *The Journal of Theological Studies* 66(2), 553–73.

Koopmans, K. (2016) ' "Water mixed with the blood of Thomas": Contact relic manufacture pictured in Canterbury Cathedral's stained glass', *Journal of Medieval History* 42(5), 535–58.

Lander Johnson, B. and E. Decamp (2018) *Blood Matters: Studies in European Literature & Thought, 1400–1700*. Philadelphia, PA: University of Pennsylvania Press.

Leyerle, B. (2001) 'Blood is seed', *The Journal of Religion* 81(1), 26–48.

Mayes, R. (2010) 'The Lord's supper in the theology of Cyprian of Carthage', *Concordia Theological Quarterly* 74, 307–24.

Ployd, A. (2018) '*Non poena sed causa*: Augustine's anti-Donatist rhetoric of martyrdom', *Augustinian Studies* 49(1), 25–44.

Sachs, G. (2008) 'Blood feud', *Jewish Bible Quarterly* 36(4), 261–2.

Salisbury, J. (2004) *The Blood of Martyrs: Unintended Consequences of Ancient Violence*. London: Routledge.

Spronk, K. (2004) 'Good death and bad death in ancient Israel according to biblical lore', *Social Science & Medicine* 58, 987–95.

Stylianou, A. (2017) 'Martyrs' blood in the English reformations', *British Catholic History* 33(4), 534–60.

————. (2018) 'Martyrs' blood in reformation England', PhD thesis, University of Warwick, Warwick.

Wakefield, W. and A. Evans (1969) *Heresies of the High Middle Ages*. New York: Columbia University Press.

Zamfir, K. (2017) 'The departing Paul: Some reflections on the meaning of *spendomai* and its early Christian reception', *Ephemerides Theologicae Lovanienses* 93(1), 75–94.

23

'EXPELLING THE PURPLE TYRANT FROM THE CITADEL'

The menstruation debate in book 2 of Abraham Cowley's *Plantarum Libri Sex* (1662)

Caroline Spearing

Anastasia Stylianou's chapter in this volume (Chapter 22) shows that the bloodshed in Early Modern martyrologies needs to be read in terms of its classical underpinnings, in a contested and sectarian discourse in which blood can act as a cry for vengeance or, conversely, as a 'vessel of grace' (p. 358). In England a century later, blood again became an important literary trope, this time in the form of the bloodshed of civil war. Charles I was denounced by his enemies as 'that man of blood', in a formula recalling the persecutors of Christian martyrs (Stylianou, Chapter 22, pp. 361–2); Royalists read the regicide as a martyrdom and the king's blood as a miracle-working conduit of divine grace (Purkiss 2005: 123–4). The close identification between the physical body of the sovereign, the concept of the 'body politic', and the nation as geographical entity encouraged readings of the effects of the English civil war in terms of a bleeding and mutilated human body (Purkiss 2005: 109–13).[1] In a 7,000-line Latin poem ostensibly on the subject of plants, Abraham Cowley (1618–67) sharpens this imagery to focus on the porous and leaky female body, specifically on the flow of menstrual blood that originates in that body itself. Like the martyrologists before him, he developed this metaphor using themes and tropes from classical literature. In Cowley's case, however, reference to classical medical theory is combined, to startling effect, with allusion to literary authors, and above all to Ovid.

The *Plantarum Libri Sex* is an odd work in many ways. One of the longest extant neo-Latin poems, its stated aim is to address the properties and uses of herbs, flowers, and trees. Along the way, however, the work incorporates lively debate between animated plants, witty stories of Ovidian metamorphosis, a battle between the gods of Europe and the gods of America, and an extended account of the English Civil War and Restoration. Book 2, the focus of this chapter, is set in the Oxford Botanic Garden, where the plants are magically animated by the full moon and meet for vigorous debate as to the ethics of abortion, the treatment of hysteria, and the purpose of menstruation: whether it is ridding the body of a toxic substance or rather represents the natural draining of the over-accumulation of a harmless or even benign fluid.

Literary discussions of menstruation before the twentieth century are very rare indeed. Information from the ancient world is overwhelmingly confined to medical texts and to the archaeological record, as in Rosalind Janssen's contribution to this volume, Chapter 2; studies of Medieval and Early Modern menstruation are

DOI: 10.4324/9780429438974-31

increasingly able to draw on women's own experience in the form of female-authored letters and diaries, while the explosion of cheap print in the mid-seventeenth century saw the publication of satirical pamphlets which (as we shall see) pull no punches in their graphic depiction of bodily functions (Read 2013: 2–3, 31). Cowley's *Plantarum Libri Sex* is thus highly unusual: a male-authored work, it addresses menstruation and menstrual blood not in the form of a medical treatise but rather in the elevated mode of Latin verse, engaging overtly with the classical literary texts—notably Ovid, Horace, and Virgil—to underpin and inform this discussion of a subject 'both taboo and mundane' (Read 2013: 105).

What makes this male poet address a topic so rigorously excluded from general public discourse? In this chapter, I argue that the menstruation debate belongs not to medical literature but rather to contemporary political controversy. In debating whether menstrual blood is harmful or benign, Cowley engages with seventeenth-century misogyny, notably that directed at the queen, Henrietta Maria. The debate represents an attempt here to depict the female as an innocent and subservient helpmeet in the face of an oppositional rhetoric which instead represented the gender as toxic, incontinent, and emasculating. In political terms, Cowley's rehabilitation of the female represents an expression of support for traditional dynastic monarchy, with its focus on the domestic and familial, over the aggressively masculine and public government of the Protectorate.

Abraham Cowley (1618–67) was one of the most celebrated English writers of the seventeenth century, and his funeral was more spectacular than any previously afforded to a poet (Lindsay 2004). Most of what is known of his life comes from his literary executor Thomas Sprat, supplemented by the account in Aubrey's *Brief Lives*; the most recent book-length biographies were both published in 1931, by Arthur Nethercot and Jean Loiseau, respectively.[2] Alexander Lindsay's article in the *Oxford Dictionary of National Biography* is the most recent treatment (Lindsay 2004). Born the seventh and posthumous child of a London stationer, Cowley was educated first at Westminster School, where at the age of 15 he published a collection of poems, *Poetical Blossoms* (Cowley 1633). He then proceeded to Trinity College, Cambridge, in 1636 and was elected a minor fellow in 1640. When civil war broke out in 1642, he declared for the King, fleeing Cambridge for Oxford before he could be formally ejected from his fellowship.

While in Oxford, he came into the orbit of the Queen, the French Catholic Henrietta Maria, and by 1644 was part of her staff in Paris, ciphering and deciphering the correspondence between her and her husband, the increasingly beleaguered Charles I. After Charles' execution in 1649, Cowley took on a more active role in international diplomacy and espionage. In 1653 he returned to England on a secret mission, was arrested, interrogated, imprisoned, and finally released only when his old college friend Dr Thomas Scarborough paid the then-enormous bail of £1,000. It may have been at Scarborough's suggestion that Cowley proceeded to study medicine at Oxford, apparently as a cover for continued intelligence work (Sprat 1668).

In 1656, Cowley published a volume of his collected poems, prepared for the press during his time in the Tower (Cowley 1656). The Preface to this volume contained a statement which, it seems, was to haunt him for the rest of his life:

> Now though in all *Civil* Dissentions, when they break into open hostilities, the War of the *Pen* is allowed to accompany that of the Sword, and every one

is in a maner obliged with his Tongue, as well as Hand, to serve and assist the side which he engages in; yet when the event of battel, and the unaccountable *Will* of *God* has determined the controversie, and that we have submitted to the conditions of the *Conqueror*, we must lay down our *Pens* as well as *Arms*, we must *march* out of our *Cause* it self, and *dismantle* that, as well as our *Towns* and *Castles*, of all the *Works* and *Fortifications* of *Wit* and *Reason* by which we defended it.

Poems (1656), sig. a4^r

Modern scholars continue to debate the extent to which these words represented a genuine change of sides rather than a pragmatic accommodation.[3] It is, however, clear that, for whatever reason, Cowley never again enjoyed the full confidence of Charles II's inner circle; equally, he was mistrusted by the Cromwellian regime. At the Restoration in 1660, he failed to secure the court appointment promised at various times by both Charles I and his son, and in 1663 he retired to the country, first to Barnelmes and later to Chertsey, where he died in 1667.

The first two books of the *Plantarum* were published as a stand-alone volume in 1662. While their publication dates from the early years of the Restoration, when Cowley was still trying to establish himself at court, the author's own preface to the volume, the *Praefatio*, states clearly that they were composed before that, in the final period of the Interregnum: *haec omnia scripta sunt paulo ante foelicissimum regis reditum*—'all this was written shortly before the most happy return of the king' (Cowley 1662: sig. A7v.). In drawing attention to the gap between composition and publication, Cowley attempts to manage the reception of the text, compelling the reader to consider it in the context of the 1650s, when the expression of Royalist loyalty could be dangerous, rather than in that of its actual publication after the Restoration. Given the poet's experience with codes and secret communication, the reader might well wonder whether this work, like much Royalist poetry of the Interregnum, may not be all that it seems, and whether this botanical verse treatise might contain a political message.[4] It would not, after all, be surprising if Cowley were to attempt to invoke his Interregnum writings as evidence of his unswerving Royalist allegiance.

For all its interest, the *Plantarum* is a particularly inaccessible work. There is no modern edition, and the most recent English translation dates from 1689 (Cowley 1689). Professor J. Daniel Kinney is in the course of preparing a complete edition, with notes and facing translation, and has edited a website, *The Abraham Cowley Text and Image Archive*, which contains a wealth of related material (Kinney 2007). The full text of the 1662 *Plantarum Libri Duo* and the 1668 *Poemata Latina*, in which all six books appeared, can be accessed at Early English Books Online; and a transcribed electronic text is available via The Philological Museum.[5] Discussions of the work as a whole can be found in Leicester Bradner's *Musae Anglicanae*; chapter 8 of Robert Hinman's monograph on Cowley; Victoria Moul's introduction for EEBO; and Ruth Monreal's *Flora Neolatina*.[6] A recent article by Victoria Moul addresses Ovidian intertextuality in books 1–2 (Moul 2015).

Cowley opens *Plantarum* 2 by dismissing his male readers: he is celebrating the women-only Roman rite of the Bona Dea. He invokes the moon and the Roman deities associated with the female body and asks them to assist in the birth of his poem (1–20). He then proceeds to set the scene: it is the April full moon in the

Botanic Garden in Oxford; the plants assemble to discuss how to be most useful to humankind, dividing into groups according to their particular pharmacological applications. The poet explains that, as a man, he was unable to be present: his source of information is the Laurel (*Laurus*), who attended the gynaecological subcommittee (21–50). Then comes a catalogue of the Laurel's fellow attendees, beginning with the president, the appropriately named Artemisia (Mugwort), and culminating in Laurel herself (51–132).

Artemisia suggests that the plants discuss menstruation (133–70). The herbs Pennyroyal (*Pulegium*) and Dittany (*Dictamnus*), who are both emmenagogues, argue that menstrual blood is a toxic substance which needs to be expelled from the body. Its harmful effects include maddening dogs, deforming the foetus, turning leaves yellow, souring wine, and clouding mirrors (171–304). But this lurid exposition is mocked by Plantain (*Plantago*) and Bramble (*Rubus*), to universal acclaim (317–72). The Rose (*Rosa*) then replies, arguing instead that menstrual blood is a benign substance which is used to feed the foetus as it grows; moreover, after childbirth this substance is transformed into breast milk. Menstruation is merely the natural excretion of an accumulated surplus (373–474).

Finally, Laurel is summoned to give an authoritative view. She argues that menstruation is essential to maintaining difference between the sexes. Humans are the only species to menstruate, though all species reproduce. Differences in appearance between the sexes are less marked in other species; but men are attracted by female beauty, so that sex is not merely a physical act. Hence menstruation flushes away the blood which would otherwise cause women to become more masculine in appearance, as in the Hippocratic case of Phaethousa (479–604; for Phaethousa, see [Hippocrates] *Epidemics* 6.8.32). Laurel's words are received enthusiastically, and the meeting moves on: Birthwort (*Aristolochia*) describes her use in childbirth (613–726); Mastic-Tree (*Lentiscus*) and Savin (*Sabina*) argue about the ethical responsibilities of plants used as abortifacients (727–918); Myrrh (*Myrrha*) gives a dramatic account of the effects of hysteria (989–1181). But her speech is interrupted by the arrival of the gardener, whose wife is in labour and who is looking for cyclamen to ease her pains. The meeting ends abruptly as the plants scurry back to their places (1181–204).

The speeches of Pennyroyal, Dittany, and Rose reflect the two alternative explanations for menstruation which were current in Early Modern medicine (Crawford 1981; Read 2013: 16–21). The first, deriving ultimately from Hippocrates, was that menstruation represented a purgation of 'superfluous humours' which the warmer bodies of men enabled them to exude through sweat. This is the view expounded by Pennyroyal, who emphasises the dangerous consequences of the failure to expel menstrual blood. The second was the Galenic view that menstrual blood was accumulated in the female body in order to nurture a growing foetus (and was later converted into breast milk); menstruation was the natural expulsion of this 'plethora' if conception did not occur. It is this account that we find in Rose's speech. Dittany's lurid account of the malign properties of menstrual blood represents beliefs found in Pliny the Elder which were still widely held, though falling out of scientific favour, in the seventeenth century. Cowley dutifully notes these sources in his extensive footnotes, which refer to Galen, Aristotle, Pliny, and Hippocrates as well as to more recent authorities such as the French physician Jean Fernel (1497–1558).

Laurel's argument departs from both the Hippocratic and the Galenic model. The masculinisation which accompanies amenorrhea and which was observed in the case of Phaethousa was noted by both Aristotle and the Early Imperial physician Soranus (Aristotle, *Generation of Animals* 2.7, 747a; Soranus, *Gynaecology* 1.23).[7] Neither authority, however, suggests that the maintenance of physiological difference is the *purpose* of menstruation; and, as far as I am aware, this is Cowley's own invention.[8] Moreover, the evidence Laurel adduces in support of her argument is far from compelling: only humans, she says, display marked differences between the sexes. As if to render the weakness of the argument inescapable, Laurel explicitly includes the examples of cattle and lions as species whose male and female are identical. Footnote references not to medical texts but to Ovid's *Metamorphoses* further highlight the departure from conventional scientific discourse.

Laurel's argument, puzzling in the extreme in a physiological content, can be read politically. As we saw earlier, by pinpointing so exactly the work's date of composition, Cowley encourages the reader to locate the *Plantarum* in a political moment, specifically that of the late Interregnum. Furthermore, throughout book 2 Cowley uses metaphors of cities, battles, and sieges to describe the female body, the processes within it, and its vulnerability to external forces. This imagery is a natural fit with the argument for the toxicity of menstrual blood: so in Plantain's speech, for example, the antagonistic relationship between the body and this blood is fully developed:

> *Purpureum* meritò depellitis arce *Tyrannum*,
>   Ejicitur patria jure *Homicida Cruor*.
> *Formosam* Imperio *vitam* premit ille superbo
>   Cogit & insultans multa nefanda pati.
> Lurida lethalis circumfert arma Veneni,
>   Et Comites *Morbos*, agmina longa, trahit.
>   [italics in original throughout]

> You rightly drive out the purple tyrant from the citadel, and the murderous blood is lawfully expelled from its country. It constrains the beautiful spirit with its arrogant sway, and insultingly forces it to endure many unspeakable horrors. It surrounds it with the ghastly weapons of lethal poison, and brings in its train the long columns of sickness.
>
>   *Plantarum* 2.319–24[9]

However, the imagery recurs in Rose's account of menstrual blood as nurturing and benign, when she compares menstruation to a city dealing with overpopulation:

> Sic si forte olim numerosae prolis abundat,
>   Nec populis *Regio* sufficit ipsa suis,
> Haud illi indignum est alias dimittere partem
>   Quaesitum terras, *Exiliúmque* pium est.

> Thus if by chance a region should teem with abundant offspring, and not have room for its own peoples, there is no disgrace whatever in sending some of them to find other lands, and it is a pious exile.
>
>   *Plantarum* 2.391–4

Cowley uses civic and military imagery to describe the processes of the female body; conversely, mid-seventeenth-century discourse often figured political tensions in highly gendered terms. Diane Purkiss and Laura Knoppers have shown how opposition to Charles I in the 1630s and 1640s frequently centred on the queen: as a woman, a foreigner, and a Catholic, Henrietta Maria was regularly labelled a malign influence and generated a raft of misogynistic rhetoric focusing on the inappropriate nature of female political involvement (Purkiss 2005: 71–9; Knoppers 2011: 33–41). Oppositional discourse insists upon the natural separation of the domestic and female from the political and male; Cowley's imagery, on the other hand, denies this separation.

More generally, the state was regularly represented as a female body in both Royalist and Parliamentarian discourse. When Charles I wore white satin to his coronation in 1625, he underlined the mystic marriage taking place between monarch and kingdom (Sandstroem 1990: 96–8). In a very different social register, in 1648 one Elizabeth Poole reported to the Army Council her vision of 'a woman which should signifie the weake and imperfect distressed state of this land' and a man who 'should improve his faithfulnesse to the Kingdome, by his diligence in the cure of this person' (Clarke 1891–1901). Nearly forty years later, marriage imagery was to feature heavily in Restoration panegyric. John Dryden's *Astraea Redux* (1660; see Clarke 1891–1901), for example, figures the ship carrying Charles to England as a bride:

> The *Naseby* now no longer *Englands* shame,
> But better to be lost in *Charles* his name
> (Like some unequal Bride in nobler sheets)
> Receives her Lord.
>
> John Dryden, *Astraea Redux* 230–3

In the hands of Restoration satirists, the notorious sexual incontinence of Charles II became politicised. Paul Hammond has shown how the traditional discourse of 'the king's two bodies' lay beneath anxieties that long-established distinctions between public and private were becoming blurred by the increasing domination of the body politic by the king's private body (Hammond 2006: 117–27). In Andrew Marvell's 'Last Instructions to a Painter' (1667), for example, Charles' immediate reaction to the vision of a naked female England, bound and gagged, is to grope her (889–904). Although Marvell's poem post-dates the *Plantarum Libri Duo*, the trope can also be found much earlier, as here in John Cleveland's lines on the Earl of Essex:

> Impotent Essex! Is it not a shame
> Our Commonwealth, like to a Turkish Dame,
> Should have an Eunuch-Guardian? May she bee
> Ravish'd by Charles, rather then sav'd by thee.
>
> John Cleveland, 'To P Rupert' (1642), 45–9

Diane Purkiss has commented on this startling image of Charles as 'the successful rapist of the feminine nation' (Purkiss 2005: 125–6).

The blood shed in the course of the Civil War was perceived as polluted and corrupt, lending itself to analogy with ancient beliefs in the toxicity of menstrual blood. When Parliament described Charles I as a 'man of blood', they alluded not only to his alleged shedding of innocent blood but also to the polluting nature of blood itself (Crawford 1977: 42). In calling the conflict a 'monstrous flux of blood', Richard Fanshawe makes explicit the link between Civil War and menstruation ('Presented to His Highnesse in the West, Ann. Dom. 1646' 1–2, Fanshawe 1997). Diane Purkiss's psychoanalytic account of masculinity in the English Civil War goes further in reading any flow of blood as intrinsically female, associated with both menstrual bleeding and the gore of childbirth. The blood that flows when the male body is wounded in battle risks collapsing the distinction between a hard and impermeable maleness and the chaos, liquidity, and flux of the female body (Purkiss 2005: 35–45, 111–13). This anxiety as to the potential erosion of masculinity in warfare further strengthens a reading of the female body as filthy, leaky, and corrupt. Underlying the focus in *Plantarum* 2 on menstrual blood is the blood of civil war itself.

Finally, Cowley's meeting of female plants recalls hostile contemporary portrayals of all-female assemblies. For Parliament, the private and domestic character of the emasculating influence of Henrietta Maria became indicative of a fundamental weakness in the institution of monarchy, revealing its vulnerability to governance by the closeted and feminine rather than by the public and masculine Parliament (Purkiss 2005: 75–9). This particular anxiety was brought to a head in *The King's Cabinet Opened*, the publication by Parliament of captured royal correspondence (1645; see Purkiss 2005: 71–97; Knoppers 2011: 42–67).

The absurdity of female political involvement was fully exploited in the contemporary satirical trope of the Parliament of Women, Parliament of Ladies, or Mistress Parliament, which could—and did—directly target the ladies of the queen's circle, playing on stereotypes of female incontinence and sexual voracity. Transgressive pregnancy and monstrous birth became standard metaphors for political change. One Royalist pamphlet, for example, depicts Parliament as the promiscuous lover of Cromwell and several other generals, who after a pregnancy of seven years gives birth to a headless monster with the feet of a bear (1648a). In another pamphlet of 1648, Captain Army steps in to forbid the banns of marriage between Mr. King and Mrs. Parliament, and calls in Doctor Period to purge the would-be bride of her retained menses (1648b). Published just a month before 'Pride's Purge', the removal by the military of Members of Parliament likely to oppose the execution of Charles I, the pamphlet allegorises relations between King, Parliament, and army in terms of the reproductive functions of the female body.

The female body as metaphor for the state; its processes read as analogues of political change; the nation as bride of the monarch; the topicality of female political involvement; the 'flux of blood' caused by civil war and the emasculating nature of that blood—all these tropes combine with Cowley's own signposting in the *Praefatio*, and with the originality and implausibility of Laurel's argument, to suggest that something other than gynaecology is under discussion here. Cowley is using his discourse of the female body, written allegedly in the late 1650s, to express a Royalist allegiance which, he hopes, will demonstrate his unbroken loyalty.

Contemporary rhetoric depicts the female body as unclean and liquescent; women are grotesque, irrational, and sexually incontinent; female political involvement is at best absurd and at worst toxic. By contrast, Cowley's plants are decorative, restrained, and subservient. Engaging directly with the 'Parliament of Women' tradition, he calls his meeting of plants a 'green senate' (*viridis senatus*, 2.129) and has his president address them as 'conscript mothers' (*matres conscriptae*, 2.133) in a humorous allusion to the conventional designation of the Roman Senate as *patres conscripti*. He proceeds to show how orderly, sober, and learned they are—they have even read William Harvey on the circulation of the blood (2.242–6)! Their overriding concern is to be useful, meeting for the explicit purpose of sharing their pharmacological expertise (2.157–62). And of course this concern is put into action at the end of the book, when the gardener comes in search of herbs to assist his wife in childbirth. In contrast to the intemperate harridans of Parliamentary rhetoric, the Royalist Cowley presents female plants who are acutely conscious of their subordinate role and ready to abandon their intellectual pursuits for service at a moment's notice. It is not difficult to read this as a spirited defence of Henrietta Maria's role in government from a long-standing and loyal servant.

The menstruation debate reinforces this reading of the female as subservient and benevolent. Plantain pushes back against Dittany's insistence on the polluted nature of menstrual blood, dismissing her lurid account of its toxic effects.

> Sed quid opus justo superaddere vana timori
>   *Terricula* & magnum magnificare malum?
> Dum nimiùm tragicè saevum depingitis hostem,
>   *Horribilis* magìs est, *Credibilísque* minús.

> But what need is there to add empty terrors on top of a legitimate fear, and to make a great trouble greater? While you depict the fierce enemy in over-tragic vein, it is more horrible, but less believable.
>
> *Plantarum* 2.325–8

It is not the blood, she continues, which drives men mad, but its source (meaning the female sexual organs); bees may or may not avoid menstruating women, but women should certainly avoid bees; if menstrual blood actually does blunt swords, that should surely be considered a blessing (2.343–54).

Laurel's speech further reinforces Plantain's sanitising approach. First she argues against the Galenic view put forward by Rose, that menstrual blood nurtures the foetus: Nature, she claims, is too gentle to make children into cannibals by feeding them on their mothers' blood:

> An pascet prolem *Natura cruore* tenellam?
>   Proh scelus! Haec morum *saeva elementa* dabit?
> Tale *rudimentum* quid mirum si genus ipsum
>   Efferat humanum, *Cannibalésque* facit?

> Or will Nature feed a tender little child on blood? What villainy! Will she provide these first seeds of its upbringing? What surprise would there be if such a beginning brutalised the human race itself, and made them cannibals?
>
> *Plantarum* 2.545–8

Laurel here decouples gestation from blood and helps to erode the association of the liquescent female body with the bloodshed of civil war.

Moreover, Laurel's account of menstruation is as a civilising process, which by maintaining difference in male and female appearance distinguishes humans from animals and elevates sexual relations to a higher, aesthetic, plane.

> At gentem humanam formosior ignis adurit,
>   Nec rationis egens, nec sine *Luce Calor.*
> Nec facit hunc *Coïtûs* brevis & coenosa voluptas:
>   Ah! Pereant, quibus haec sola vocatur *Amor.*
> Larga *venustatem* tribuit Natura *puellis,*
>   Et nivei blandas corporis illecebras.
> Aurataque viros voluit retinere catena,
>   Et fraenare vagos *dulcibus imperiis.*
> Nec solùm incestas turpissima pascere membra,
>   Sed spectantum *oculis* gaudia *casta* dare.

But a finer flame burns up the human race, one not lacking in reason, nor a heat without light. The brief and murky pleasure of the sexual act does not generate this: ah! may they perish, who call only this by the name of Love. Bountiful Nature endowed girls with charm, and the alluring enticements of a snow-white body. She also wanted to hold back wandering men with a golden chain, and restrain them with sweet commands. She did not merely wish for women of low morals to provide sustenance for the most depraved members, but to give chaste delight to the eyes of onlookers.

Plantarum 2.579–88

According to Laurel, sexual pleasure belongs to the eyes as much to the body: its joys are not the 'murky pleasure' of physical release but rather the 'chaste delight' of aesthetic enjoyment. Instead of being an emanation of dark and elemental forces, menstrual blood becomes a sign of a bountiful nature, *larga natura*, who by this means maintains the differences between men and women which give sexual attraction its aesthetic and rational dimension. As such, Laurel's brand of sexual love has much in common with the Neoplatonic ideal of the chaste love engendered by contemplation of female beauty, which was in turn strongly associated with the court of Henrietta Maria.[10]

Significantly, female sexuality is wholly occluded: the gaze satisfied by the contemplation of the female is an exclusively male one. Hence menstruation becomes, not the expulsion of a toxic substance, nor a manifestation of the mysterious and occult workings of gestation and pregnancy, but rather a process designed for the ultimate benefit of the male. Menstruation provides an object of aesthetic pleasure and elevates the sexual act above the merely physical. In this way it acts as guardian and guarantor of a refined and chaste male heterosexuality.

The focus in *Plantarum* 2 on menstruation, gestation, and childbirth thus becomes more intelligible in the light of contemporary discourse, which regularly used these female bodily processes in the rhetoric of political change. Whereas Parliamentary discourse emphasised the threat posed by the dominance of a sex depicted as unstable, corrupting, irrational, and sexually rapacious, Cowley chooses instead to portray the female as innocuous and subservient. The potentially corrupting and toxic menses are sanitised and rationalised as essential to an elevated male sexuality. The dominant, subversive, foreign, and Catholic queen of Parliamentary rhetoric becomes instead the submissive and decorous helpmeet of the Botanic Garden. The purple tyrant is expelled indeed.

Notes

1 Helen King's chapter in this volume (Chapter 24) shows how Elizabeth I aligned her bodily continence with her good judgement as a ruler (pp. 391–3).
2 Sprat 1668; Loiseau 1931; Nethercot 1931; Aubrey 2015.
3 Nethercot argues that Cowley genuinely switched allegiance at this point (Nethercot 1931: 149–57). For the contrary view, see e.g. Corns 1992: 256–9; Revard 1993: 394.
4 The seminal works on the polyvalent nature of Interregnum writing are those of Annabel Patterson and Lois Potter (Patterson 1984; Potter 1989). Subsequent scholarship which has adopted or adapted Potter's methodology includes work by Thomas Corns, Stephen Zwicker, James Loxley, Robert Wilcher, and Syrithe Pugh (Corns 1992; Zwicker 1996; Loxley 1997; Wilcher 2001; Pugh 2010).
5 Cowley 1668, edition Sutton 2006/7.
6 Bradner 1940; Hinman 1960; Monreal 2010; Moul, *Introduction*.
7 For a detailed recent discussion of the Phaethousa story and its reception, see King 2013.
8 I am very grateful to Professor Helen King and Dr Sara Read for their help on this question.
9 All text is cited from Cowley 1662a. Translations are my own.
10 The scholarship on Neoplatonism and the Caroline court is considerable. For Graham Parry, it represented a facile mode of ornamentation (Parry 1981: 184–203). More sympathetic and nuanced accounts can be found in Britland and Veevers (Britland 2006: 6–13; Veevers 1989: 19–39).

Bibliography

Primary sources

(1645) *The King's Cabinet Opened, or, Certain Packets of Secret Letters & Papers, Written with the King's own Hand, and Taken in his Cabinet at Naseby Field June 18 1645*. London.

(1648a) *Mistris Parliament Presented in her Bed After the Sore Travaile and Hard Labour Which She Endured Last Week, in the Birth of Her Monstrous Offspring, the Childe of Deformation, the Hopefull Fruit of Her Seven Yeers Teeming, and a Most Precious Babe of Grace. By Mercurius Melancholicus*. London.

(1648b) *A New Marriage, between Mr. King and Mrs. Parliament. The Banes Forbidden by Captaine Army, with the Grounds and Reasons He Gives for the Same*. London.

Aubrey, J. (2015) *Brief Lives: With an Apparatus for the Lives of Our English Mathematical Writers*. Edited by Kate Bennett. Oxford: Oxford University Press.

Clarke, W. S. (1891–1901) *The Clarke Papers. Selections from the Papers of William Clarke, Secretary to the Council of the Army, 1647–1649, and to General Monck and the Commanders of the Army in Scotland, 1651–1660*. Edited by C. H. Firth. London: Printed for the Camden Society.

Cowley, A. (1633) *Poetical Blossomes. By A. C. (Abraham Cowley.) [With a Portrait Engraved by Robert Vaughan.]*. London: B. A. & T. F. for Henry Seile.

———. (1656) *Poems: Viz. I. Miscellanies. II. The Mistress, or, Love Verses. III. Pindarique Odes. and IV. Davideis, or, a Sacred Poem of the Troubles of David*. London: Humphrey Moseley.

———. (1662) *A. Couleii Plantarum: libri duo.* London: Nathaniel Brooks.

———. (1668) *Abrahami Couleij Angli Poemata latina. In quibus continentur, sex libri plantarum, viz. duo herbarum, florum, sylvarum. Et unus miscellaneorum.* London: John Martyn.

———. (1689) *The Second and Third Parts of the Works of Mr Abraham Cowley, the Second Containing What was Written and Published by Himself in His Younger Years: Now Reprinted Together. The Sixth Edition. The Third Containing his Six Books of Plants . . . Now Made English by Several Hands (J. O., C. Cleve, N. Tate, Mrs. A. Behn), etc.* London: Charles Harper.

Fanshawe, R. (1997) *The Poems and Translations of Sir Richard Fanshawe.* Edited by Peter Davidson. Oxford: Oxford University Press.

Marvell, A. (1667) 'Last instructions to a painter', in N. Smith (ed.) *The Poems of Andrew Marvell* (rev. edn). London: Longman, 362–96.

Sprat, T. (1668) *An Account of the Life and Writings of Mr. Abraham Cowley. The Works of Mr Abraham Cowley. Consisting of Those Which were Formerly Printed: And Those Which He Design'd for the Press, Now Published Out of the Authors Original Copies.* London: Henry Herringman.

Sutton, D. (2006/7) *Abraham Cowley, De Plantis Libri VI (1668) A Hypertext Critical Edition* [Online]. The Philological Museum. Available at: www.philological.bham.ac.uk/plants/ (Accessed February 2021).

Secondary literature

Bradner, L. (1940) *Musae Anglicanae: A History of Anglo-Latin Poetry 1500–1925.* New York: Modern Language Association of America.

Britland, K. (2006) *Drama at the Courts of Queen Henrietta Maria.* Cambridge: Cambridge University Press.

Corns, T. N. (1992) *Uncloistered Virtue: English Political Literature, 1640–1660.* Oxford: Clarendon Press.

Crawford, P. (1977) 'Charles Stuart, that man of blood', *The Journal of British Studies* 16, 41–61.

———. (1981) 'Attitudes to menstruation in seventeenth-century England', *Past & Present* 91, 47–73.

Hammond, P. (2006) *The Making of Restoration Poetry.* Cambridge: DS Brewer.

Hinman, R. B. (1960) *Abraham Cowley's World of Order.* Cambridge, MA: Harvard University Press.

King, H. (2013) 'Sex and gender: The Hippocratic case of Phaethousa and her beard', *EuGeStA: Journal on Gender Studies in Antiquity* 3, 124–42.

Kinney, D. (2007) *The Abraham Cowley Text and Image Archive* [Online]. The Abraham Cowley text and image archive. Available at: http://cowley.lib.virginia.edu/small/bk4country.htm (Accessed February 2021).

Knoppers, L. L. (2011) *Politicizing Domesticity from Henrietta Maria to Milton's Eve.* Cambridge: Cambridge University Press.

Lindsay, A. (2004) *Cowley, Abraham (1618–1667)* [Online]. Oxford dictionary of national biography. Available at: www.oxforddnb.com/view/article/6499 (Accessed February 2021).

Loiseau, J. (1931) *Abraham Cowley. Sa vie, son œuvre.* Paris: Didier.

Loxley, J. (1997) *Royalism and Poetry in the English Civil Wars: The Drawn Sword.* Basingstoke: Palgrave Macmillan.

Monreal, R. (2010) *Flora Neolatina: Die Hortorum libri IV von René Rapin SJ und die Plantarum libri VI von Abraham Cowley. Zwei lateinische Dichtungen des 17. Jahrhunderts.* Berlin: Walter de Gruyter.

Moul, V. Introduction to Cowley's Plantarum Libri Sex [Online]. Available at: https://proquest. libguides.com/ld.php?content_id=49656325 (Accessed February 2021). Date unknown.

————. (2015) 'The transformation of Ovid in Cowley's herb garden', in P. Mack and J. North (eds) *The Afterlife of Ovid*. London: University of London, Institute of Classical Studies, School of Advanced Study, 221–34.

Nethercot, A. H. (1931) *Abraham Cowley, the Muse's Hannibal*. London: Oxford University Press.

Parry, G. (1981) *The Golden Age Restor'd: The Culture of the Stuart Court, 1603–42*. Manchester: Manchester University Press.

Patterson, A. M. (1984) *Censorship and Interpretation: The Conditions of Writing and Reading in Early Modern England*. Madison: University of Wisconsin Press.

Potter, L. (1989) *Secret Rites and Secret Writing: Royalist Literature, 1641–1660*. Cambridge: Cambridge University Press.

Pugh, S. (2010) *Herrick, Fanshawe and the Politics of Intertextuality: Classical Literature and Seventeenth-Century Royalism*. Farnham: Ashgate.

Purkiss, D. (2005) *Literature, Gender and Politics during the English Civil War*. Cambridge: Cambridge University Press.

Read, S. (2013) *Menstruation and the Female Body in Early Modern England*. Basingstoke: Palgrave Macmillan.

Sandstroem, Y. L. (1990) 'Marvell's "Nymph Complaining" as historical allegory', *Studies in English Literature, 1500–1900* 30, 93–114.

Veevers, E. (1989) *Images of Love and Religion: Queen Henrietta Maria and Court Entertainments*. Cambridge: Cambridge University Press.

Wilcher, R. (2001) *The Writing of Royalism 1628–1660*. Cambridge: Cambridge University Press.

Zwicker, S. N. (1996) *Lines of Authority: Politics and English Literary Culture, 1649–1689*: Ithaca: Cornell University Press.

24

OPENING THE BODY OF FLUIDS

Taking in and pouring out in Renaissance readings of classical women

Helen King

In this chapter I would like to offer a concluding perspective on the transmission and appropriation of classical ideas about the female body in Renaissance and Early Modern Europe, drawing together a number of the key themes which have emerged in this volume: gender and fluidity; the permeation of boundaries; classification of fluids and fluid classification; the interface between literary and visual representations; medical theories and lay interpretations; age and decay; excretion and regeneration. My theme here is the nature of virginity and of the virgin body in classical and Renaissance Europe, as represented in the story of how the Vestal Virgin Tuccia carried water in a sieve. Tuccia's sieve is most commonly thought simply to offer a paradigm for the integrity of the 'whole unimpaired body' of the virgin, in contrast to the openness and unpredictable flows of the mature woman (Warner 1985: 242). By looking at later appropriations of the image of this sieve, including a pair of sixteenth-century images—Mantegna's 'Two Exemplary Women of Antiquity'—as well as the better-known 'sieve portraits' of Elizabeth I, the Virgin Queen, I shall suggest that Tuccia's receptions return us to the ancient versions of the story with fresh questions about the role of this piece of agricultural or cooking technology in representing the body. Is it intended to retain the good while letting out the bad or to refine the contents to improve them? Is this, in our terms, a sieve or a strainer?

Many contributors to this volume have already considered the extent to which fluids have their own identities, and their capacity for transformation within the body. In the models taken from ancient medicine, the bodies of both men and women use heat to 'cook' their fluids. In looking at Tuccia, I shall be investigating not only the movement of bodily fluids around the body but also the means of control when they move out into the external world. What is it that Tuccia's sieve, and the virgin body itself, does not let out? Bearing in mind the wider issues of the control of fluids in the female body, in particular the role of urination in virginity tests, I shall suggest that, while the virginal sieve may relate to the organs of reproduction, it also strongly references the bladder.

In the process of exploring sixteenth-century images through classical texts, I shall also raise the question of the different methodologies which scholars of the classical worlds use to address their questions, methodologies which are revealed very clearly in various chapters in this volume. Depending on their backgrounds in literature, history, art or archaeology, two responses are made and need always to be held in

DOI: 10.4324/9780429438974-32

tension: the close reading of a single text or image with little attention to social context, and the synthetic method of creating a single picture from a few scattered sentences featuring in texts far apart in time or in genre. The nature of the evidence may lead to us focusing more on one of these responses, but at the very least, we should be aware of our methods. Here, while concentrating on detailed study of just a few examples, I shall also be assuming some level of continuity in the representation of the virginal body, continuity which persists despite different religious and social contexts. Those in more recent historical periods consciously drew on the classics to define what it is to be a woman and, in the case of Elizabeth I, to create and manipulate their own image. By the sixteenth century, allusions to Tuccia's sieve played with a range of meanings, and those who made these allusions were aware not only of the classical texts but also of later ideas about the workings of the body. A strong constant in all this is a pre-modern interest in fluids as more important than organs in understanding the body, as discussed in the Introduction to this volume (pp. 2–3).

Fluid bodies in history

Characterisation of the pre-modern body as what I have elsewhere called a 'body of fluids' took off after Gail Kern Paster's 1993 book *The Body Embarrassed: Drama and the Disciplines of Shame in Early Modern England*, which started from the insight that our apparently fixed, biological bodies can only be understood 'in terms of culturally available discourses' (Paster 1993: 4; King 2011). People in Early Modern Europe approached their bodies through humoral theory; because this understanding of the body in terms of fluids was not merely 'metaphor', but represented how the body was actually experienced, as scholars we should try to appreciate 'the formative effects of physiological theory on the subject' rather than regarding the four humours as an intellectual curiosity and relegating them to a few footnotes in our work (Paster 1993: 4, 7). While Paster (1993: 39, 24) never suggested that this humoral 'body of fluids' should be gendered as female—men, too, were composed of fluids—she noted that the belief that women were wetter or more leaky than men was 'a given of contemporary scientific theory'. An important element of this was control; women were presented as less able to control their fluids, and thus as more like children, and this formed part of the argument for patriarchy (Paster 1993: 25).

The body of fluids was, however, not only a humoral body. Its contents included fluids which we no longer recognise, or to which we attribute very different roles, as well as parts of the body in which we no longer believe. For example, Michael Stolberg (2012: 513) has drawn attention to the pre-modern idea of a space between the flesh and the skin, linked to sweat being given a more important role in the body in that period, and has pointed out just how many different kinds of sweat there were in Early Modern medicine (for more on sweat, see the Envoi, this volume). Fluids were not only spoken about with overlapping language but were also presented as transforming into each other; Paster (1993: 40) mentioned the proverb 'Let her cry, she'll piss the less', repeated by Ambroise Paré.

A particularly powerful example of transformation is the production of breast milk from menstrual blood (Lawrence, Stylianou, Chapters 14 and 22, this volume). In the famous image from Leonardo da Vinci, created before he dissected any complete human cadavers, an imagined male and female body meet in generation (Royal

Collection Trust, RL 19097v). Noble et al. (2014) argue that in 1508 Leonardo corrected his views and removed the connection seen in this image between the spinal cord and the channels carrying the seed.[1] Leonardo's hemisected man, his spine clearly visible and linked to his seed production, recalls the Hippocratic *On Airs, Waters, and Places* (22) description of the Scythians, who cut the channels behind the ears and thus prevent their semen from moving properly down the body; in the male body, the head and the spine are thus also considered relevant to generation. In this treatise, the explanation for some Scythians dressing as women and taking women's roles was that they cut behind the ears to cure varicose veins: 'such treatment is destructive of the semen owing to the existence of the vessels behind the ears, which, if cut, cause impotence and, it seems to me that these are the vessels they divide.'

Returning to Leonardo's image, in the much sketchier figure on the left—the woman—there is an imaginary channel from womb to nipple to carry the blood-which-becomes-milk. In 1993 Paster (1993: 40) had noted the Early Modern view that breast milk was 'blood made white'. Barbara Orland (2012) has since discussed how, while milk could be seen as 'white blood', blood could be regarded as 'slightly coloured' milk: each fluid uses the language of the other. In Leonardo's drawing, we can therefore see a flow between gendered bodies, from the male head, down the spine into the penis, from the penis to the vagina, and from the womb to the woman's breasts, as fluids transform into each other and mingle to create a new life.

Alongside blood and milk, urine also featured prominently. Michael McVaugh (2012: 117) has drawn attention to the sixteenth-century Berengario da Carpi combining sweat, urine and milk when trying to understand the body, so that urine 'is sweated out like milk from the breast'. Less prosaically, in 1518 Lorenz Fries (Fries 1518: lxiii[v] cited by Stolberg 2015: 49) wrote that urine is 'water that is strained from blood and other humours'. The image of the strainer is important here, and I shall return to it later. As Karine van't Land (2012: 379–80) notes, Aristotle's description of the formation of the embryo is that it is like turning milk into cheese, another straining/sieving process. Within blood, serum is named for the Latin word for 'whey', the liquid by-product of cheese manufacture. Urine, Greek *ouron*, may also be related to a Greek word for whey, *oros* (Stolberg 2015: 49).

Testing Tuccia

The imagery of the fluids used in the household pervades the history of women's bodies, but it exists alongside another set of images in which those bodies are fields to be ploughed and planted. Straining/sieving unites the two, being associated with the transformation of fluids and the refinement of solids. Tuccia, famously, carries water in a sieve in order to prove that she is still a virgin. Nor does she simply carry the water; when she reaches the forum, she pours it out at the feet of the priests. Kathleen Coyne Kelly opened her 2000 book on *Performing Virginity and Testing Chastity in the Middle Ages* with an extract from Robertson Davies's novel *The Rebel Angels*, in which the character Maria Theotoky describes how, when teaching first-year engineering students the history of science and technology, she told the story of how Vestal Virgins could carry water in a sieve and challenged the few girls in her class to repeat this miracle. After they all failed, Maria succeeded, because 'mine was greased, which proved that the Vestal Virgins had a practical understanding of colloid chemistry'

(Kelly 2000: 1, quoting Davies 1983: 52). For Maria Theotoky, any Vestal could achieve this result, but in the ancient myths Tuccia stands alone.

Many ways to detect a virgin were proposed in the pre-modern world—by her demeanour, her sensitivity to touch or smell, the size of her breasts, her modesty, posture, or voice—but some could very easily be counterfeited. As Kathleen Coyne Kelly (2000: 66) observes for the Middle Ages in particular, 'It seems that virginity is so overdetermined that apparently just about anything can signify its presence or absence.' Challenges to the existence of the hymen by medical writers from antiquity onwards and comments about how bleeding at intercourse could, for example, be menstrual rather than hymeneal took one potential physical marker out of the debate; for example, Ambroise Paré described what Marie Loughlin (1997: 31–2, 195–6, n.78) characterised as the 'fiction of the hymen as the guarantee of virginity'. Kelly (2000: 10) has recently described the hymen as a 'notoriously unstable and ambiguous concept, with an anxious and uncertain history'. Another possibility, mentioned by Helkiah Crooke in 1615, was to measure the distance from the tip of the nose to the sagittal suture and check whether the diameter of the woman's neck was greater than this. This test was not, however, separate from ideas of the hymen, because it draws on the classical idea that the loss of virginity widens the neck and deepens the voice (Hanson and Armstrong 1986).

Tuccia's action is not some established procedure or ordeal applied to women accused of unchastity, but as presented in the sources, it is her own idea to clear her name. In attributing agency to Tuccia in this way, the story contrasts with the long tradition of 'virginity tests' administered in Western medicine, literature, and popular tradition, even though these too involved fluids. As the body itself could not prove virginity, the actions of the virgin could be used instead. Some later virginity tests involved walking over fire. As a Vestal, Tuccia's role was to prevent the fire of the goddess Vesta—virgin goddess of the hearth—from going out. Later images of Tuccia sometimes included the fire, for example, Louis-Joseph Le Lorrain's *La Vestale Tuxia*, from the mid-eighteenth century.[2]

More commonly, tests for virginity involved fluids. While these seem far less dangerous than using fire, their aim too is to go beyond the appearance of the body in order to avoid any suspicion that the virgin is merely acting the part. In the tests found in literature, two fluids in particular feature, and one of these is water. For example, the thirteenth-century romance *Floris and Blauncheflur* involves a magic fountain in which, if an unchaste woman should wash her hands, the water screams out and becomes 'red as blood' (Kelly 2000: 8–9).

The other fluid is urine. Some tests assume that you can detect a virgin by the way in which she passes urine, or simply by giving her a lot of water to drink and seeing if she can hold it. In a Medieval commentary on pseudo-Albertus Magnus, sniffing 'the fruit of a lettuce' would make a corrupted virgin pass urine immediately (Lemay 1992: 127). In his *Popular Errors*, Laurent Joubert—chancellor of the faculty of medicine at Montpellier from 1556—challenged other writers' enthusiasms for virginity tests involving drinking lignum aloe or smelling the smoke of dock leaves to see if the woman immediately passed urine (de Rocher 1989: 46; Paster 1993: 43–4). He criticised invasive virginity tests which could damage a virgin's 'womb pipe' by widening it (de Rocher 1989: 210–11). The fifteenth-century Niccolò Falcucci copied out some tests from Gilbertus Anglicus which suggested that non-virgins will urinate as soon as they drink anything containing coal or cockles. According to Falcucci, virgins also become pale if fumigated with dock flowers (Lemay 1992: 127; Kelly 2000: 29–30). Guilielmus de Saliceto (*Summa conservationis*, f. i3ra, cited in Kelly 2000: 29) described what Laurent Joubert later called a woman's 'manner of pissing' as potential evidence of virginity; a virgin 'urinates with a subtle hiss'. Joubert himself regarded the virgin's pissing as more like that of a man than of a mature woman: before she is deflowered a virgin pisses 'straight and far' (Paster 1993: 43–4). Alternatively, the urine of virgins may itself be seen as distinctive in some way; according to a medical text of 1580, *De secretis mulierum*, it is 'clear and lucid, sometimes white, sometimes sparkling' (Lemay 1992: 128, cited in Kelly 2000: 29).

'Holding your water', then, was evidence of virginity in Medieval and Early Modern virginity tests and, in a different way, for Tuccia. In the absence of any faith in anatomical structures to provide the answers, there is something interesting here about the reality, the materiality, of the fluids as providing true evidence. The ability to hold one's water suggests being in control, and this is linked to the bodily integrity of the virgin. This should direct our interpretation of another holder of Tuccia's quite unusual, but not unique, name: in Juvenal's sixth satire (line 64), where, watching the performance of the *pantomimus* Bathyllus, '*Tuccia vesicae non imperat*', Tuccia can't control her bladder.[3] Jim Adams (1982: 92) suggested that these fluids are sexual. Perhaps; certainly, Tuccia is excited by watching this man play a woman being raped by Jupiter (Gunderson 2005: 235). But perhaps not; we could be moving here towards what Paster (1993: 23) called 'not the relatively comfortable subject of women but its much less comfortable analogue—bladder incontinence'. What is important here is the moisture; this Tuccia is wet, not dry, leaking, not contained. Unlike her Vestal namesake, she is not in control of the moment when she pours out her fluids at the feet of the priests.

Following Tuccia

In the original ancient context, as Amy Richlin noted in 'Carrying Water in a Sieve: Class and the Body in Roman Women's Religion', Tuccia demonstrates that 'Roman women's cults show a preoccupation with the female body . . . and with the class divisions relevant to the female body: *matrona*, slave, prostitute' (Richlin 1997: 331; 2014: 197, 202). Richlin argued that the Tuccia story shows that women's bodies are permeable, but must become impermeable: as they live their lives, all Roman women, therefore, 'must carry water in a sieve' (1997: 357; 2014: 232). Is that the case? Where does this position the virgin body?

The story of Tuccia is told in a number of sources; the first extant account appears to be that of Livy, who composed the first books of his history in the early 20s BCE (Champion 2015: 201). It is in the epitome of Livy's Book 20 that we find the simple statement: 'Tuccia, a Vestal Virgin, was convicted of unchastity' (*incesti damnata est*; *incestum* is 'an older form of *in-castum*, the negation and antonym of *castum*' or purity, Wildfang 2006: 54). This is the only surviving version in which her trial has proceeded as far as the point where she could be found guilty, which Robin Lorsch Wildfang has suggested supports the idea that, as the outlier here, Livy is 'reworking . . . Vestal history to suit his own purposes' (Wildfang 2006: 90 n. 37).

The fullest version is that of Dionysius of Halicarnassus, 2.69.1, perhaps written in 9 BCE; his account and that of Livy are independent. Here, Tuccia is never found guilty. The story comes immediately after that of another Vestal miracle, Aemilia, who throws a band from her clothing on to the 'cold ashes' to make the sacred fire burn. Tuccia's story is introduced with the rider that it is 'even more like a myth' than this previous miracle.

> They say that somebody unjustly accused one of the holy virgins, whose name was Tuccia, and although he was unable to point to the extinction of the fire as evidence, he advanced false arguments based on plausible proofs and depositions; and that the virgin, being ordered to make her defence, said only this, that she would clear herself from the accusation by her deeds (τοῖς ἔργοις). Having said this and called upon the goddess to be her guide, she led the way to the Tiber, with the consent of the pontiffs and escorted by the whole population of the city; and when she came to the river, she was so hardy as to undertake the task which, according to the proverb, is among the most impossible of achievement: she drew up water from the river in a sieve (ἀρυσαμένην ἐκ τοῦ ποταμοῦ κοσκίνῳ), and carrying it as far as the Forum, poured it out at the feet of the pontiffs. After which, they say (φασι), her accuser, though great search was made for him, could never be found either alive or dead.
>
> Dionysius of Halicarnassus 2.69.1; trans. Earnest Cary[4]

The reference 'according to the proverb' covers a number of sayings in which either the impossibility or the pointlessness of using a sieve to carry water is picked up; for example 'It's no more use than pouring water into a sieve', used in Plautus, *Pseudolus* 102 (*non pluris refert quam si imbrim in cribrum geras*; Hansen 2002: 71).

In contrast to Aemilia, who made a formal prayer to Vesta and asked her to confirm that she had served chastely, here Tuccia says very little: 'only this, that she would clear herself from the accusation *by her deeds*' (my italics) followed by a request to the goddess to 'be her guide'. The story is about action and materiality, not words and arguments. It is also unusual in other ways. It explicitly lacks the classic sign of Vestal unchastity: the fire at the temple going out (Parker 2004: 581 n.72). Recalling the later tests for virginity by using urination, Tuccia can most definitely 'hold her water', and she is in complete control of when and where it comes out.

Tuccia on display

At the beginning of the sixteenth century, Mantegna painted 'Two Exemplary Women of Antiquity' on two panels, made to look like gilt bronze against marble; one of these, now in London's National Gallery, is *The Vestal Virgin Tuccia with a sieve*.[5] The lighting is interesting and suggests they may have been positioned in a room in such a way that each of them appeared to be correctly lit by the natural light. Were these designed as a pair or part of a larger set for a bedroom? We know that Mantegna painted a total of four *donne illustri* in all: *Tuccia, A Woman Drinking, Dido,* and *Judith*. The surviving *Dido* and *Judith* are done in a different way to the other pair of 'exemplary women'; there is no painted marble background, no paint made to look like bronze in the figures themselves. So we could perhaps see this as two sets of two, not one set of four. The original location of this image of Tuccia is not known, but Renaissance representations of her were found in a range of settings: from public spaces to bedrooms, including on wedding chests (Baskins 1998; Edwards 2008: 314; Baert 2018: 60).

Mantegna was the court artist at the time when Isabella d'Este, born in Ferrara in 1474, had moved to Mantua on her marriage at 16 to Francesco Gonzaga. Mantegna painted *Minerva Expelling the Vices from the Garden of Virtue* (c. 1499–1502) for Isabella's *studiolo*, the room where she read and where she received guests. While the contents of the *studiolo* are scattered across the museums of Europe, this painting survives (Verheyen 1971). He also painted for her *studiolo* an image in 'feigned bronze'. It is not clear to what this refers, but it was perhaps one of the pair of images from which we have Tuccia, suggesting that this was not a bedroom image. Isabella was one of the most educated women of her day, able to read both ancient Greek and Latin (although there is scholarly disagreement as to how well she knew these languages; see Furlotti and Rebecchini 2008: 92). She used the ancient world as part of constructing herself as exceptional and innovative and, unlike similar women of her period and status, she collected antiquities rather than commissioning religious works (San Juan 1991; Franklin 2006: 149 ff.; Furlotti and Rebecchini 2008: 92–115). What would she have made of Tuccia? The contents of her *studiolo* were intended to 'display refinement of taste and learning'; what one displayed there could be used to defend oneself against any accusation that humanist education in a woman brought her virtue into question (San Juan 1991: 69–70). Subtle uses of classical mythology demonstrated that learning and chastity could go together.

Isabella had access to several of the Renaissance collections of lives or stories of famous women; some of these were in the Biblioteca Estense in Ferrara, and another, Jacopo Foresti's *De claris selectisque mulieribus*—a 1497 Latin continuation of Boccaccio's *De mulieribus*—was in her personal library (Benson 1992: 33). She is on record as writing that one of these collections, Sabadino degli Arienti's 1483 *Gynevera de le clare donne*, which the author himself sent to her, was useful for those who would 'attempt to follow in the footsteps of those illustrious ladies'; this is interesting because Sabadino specified that his work was intended to entertain rather than to offer exemplars (Franklin 2006: 152 citing Benson 1992: 41 n.16).[6] The modern women he describes are characterised by their 'chastity, charity, and devotion to family' and include women rulers, soldiers—and scholars (Benson 1992: 41–2). In contrast to some of the other Italian defences of women, Sabadino does not see illustrious women as 'exceptional' women, somehow going beyond normal women, but rather as capable of doing what men do, using 'talents that are natural to all women' (Benson 1992: 33 and 41). The source for Tuccia which Isabella would be most likely to have known well is Petrarch, who in *Triumph of Chastity* 1.148–51 (1355) wrote 'Amongst others the pious Vestal Virgin who ran boldly to the Tiber and to exonerate herself from wicked blame brought water from the river to the temple in a sieve' (trans. Anne Dunhill); this was quoted for example in 1601 by Lucrezia Marinella in her *The Nobility and Excellence of Women, and the Defects and Vices of Men* (Dunhill 1999: 103; see also Adler 1978: 3; Warner 1985: 242–3). Although Petrarch did not name Tuccia, this is probably because she was so famous that he had no need to do so. Tuccia's place as exemplar of chastity reached its peak at the end of the sixteenth century, when she was even compared to the Virgin Mary in the 1591 *Se pur sù ne li stellanti chiostri* by Leonora Bernardi (Cox 2011: 67–8). Sarah Eycken (2018: 74 and 83) notes that Mantegna's *The Vestal Virgin Tuccia with a Sieve* includes a vase containing two lilies above Tuccia's head and, bearing in mind that both the vase and the lilies symbolise Mary, even suggests that 'Tuccia can be seen as a pagan version' of her.

Returning to the question of how Isabella d'Este and other sixteenth-century women would have read Tuccia's story if it was represented in their bedroom or *studiolo*, the pairing with *A Woman Drinking* is also relevant in this case. While Tuccia has 'a steady gaze fixed on something or someone in the spectator's space' (Franklin 2006: 153), *A Woman Drinking* is looking up. She has been identified as Sophonisba (Livy 30.15), who also features in Petrarch's *Triumph of Love*, drinking poison sent, or actually administered, to her by her husband as her escape route from being enslaved; unlike Tuccia, then, she is not a solo active subject. Tuccia carries (and in Dionysius then pours out) fluid: Sophonisba takes in. But there are some questions about the identification of the second Mantegna figure as Sophonisba.

In art, Sophonisba can be shown about to drink, drinking, or even as having drunk, as for example in Giambattista Pittoni's 1730 *La morte di Sofonisba*. As *A Woman Drinking* is specifically shown in the process of drinking, a better pairing for Tuccia could be to see this as Artemisia, who drank the ashes of her husband Mausolus, thus becoming the perfect widow who is also her husband's tomb; the link is made explicit by Aulus Gellius, who describes how she drank his ashes with spices and water (10.18: *ossa cineremque eius mixta odoribus contusaque in faciem pulveris aquae indidit ebibitque*) in what Valerius Maximus calls a 'potion' 'to become a living and breathing tomb (4.6.1: *vivum ac spirans sepulcrum fieri*). Drinking ashes also provides a suggestive contrast to Tuccia who, in Valerius Maximus' version of her story—particularly popular in the Middle Ages (Kelly 2000: 63)—carries water in a *cribrum*, a sieve, which she 'snatched up'. A conveniently available sieve for a Vestal would be not an agricultural implement but the bronze sieve in which the Vestals carried fire to the temple when the flame was renewed each year on March 1 (Valerius Maximus 8.1.5; Richlin 1997: 357; 2014: 232; Stamper 2005: 35). Ash and fire, death and life, thus unite the two stories. The fire tended by the Vestals has been seen as a symbol of male fertility—perhaps the greater internal heat which, in medical understandings of the body, enables men to convert blood into semen. The presence of fire in the temple may be saying that masculinity depends on women to maintain it, echoing Richlin's point that Roman religion is 'an organic system that involved men and women together' (Richlin 2014: 199). But women also need heat; theirs, less powerful than men's, enables menstrual blood to be transformed into breast milk.

In structuralist terms, if Tuccia's sieve full of water veers too far towards the wet, Artemisia's consumption of ash is located too far towards the dry. This is replicated in the contrast between the purity of Tuccia and the excessive passion of Artemisia— in Aulus Gellius she is 'inflamed with grief and with longing for her spouse', and he writes of the 'violence of her passion'. Artemisia is unduly consumed by desire; it becomes ash. Like the virgin Tuccia, but in a negative way, the widow Artemisia is 'exemplary'. Neither of them should be emulated; their juxtaposition may concern chastity versus lust, as framed by bodily fluids. The message is that the good wife (recalling the bedroom setting of many such images), no longer a virgin but not yet a widow, is positioned between the poles of chastity and lust, feeling just the right amount of chaste desire for her husband. Before considering another dimension of these stories—the ease with which Tuccia moves from model of chastity to dangerous witch—let us explore further the different roles of the sieve and its later uses.

The place of the sieve

What was the sieve in which Tuccia carried water, and why does she use it? If it is not a part of the Vestals' ritual, what is it? There is no reference to the sieve in the summary of Livy (where there is also no mention of her going on to prove her innocence), but as I have already noted, it features in Valerius Maximus' version of the story, from the time of Tiberius. Here, Tuccia's story is told in a list of reasons why 'ill-famed defendants were acquitted or condemned':

> the chastity of the Vestal Virgin Tuccia, charged with impurity, emerged from an obscuring cloud of ill fame. In the certain knowledge of her innocence she dared to seek hope of salvation with an argument of doubtful issue. Seizing hold of a sieve (*arrepto enim cribro*), 'Vesta,' she said, 'if I have always brought pure hands to your sacred service, make it so that with this I draw water from the Tiber and bring it to your temple.' To the priestess' prayer thrown out boldly and rashly the Nature of Things gave way.
>
> Livy 8.1.5; trans. D. R. Shackleton Bailey

The *cribrum*, which is also the word used in Latin sieve proverbs, has as its core meaning 'divider' or 'separator', as does Dionysius's *koskinon* (Harrison 1903: 310); the root of both the Latin *cribrum* and the Greek *koskinon* is the proto-Indo-European *kreidhrom* (Baert 2018: 62–3). Its normal use is certainly not to carry fire, as happens in the context of Vestal ritual, but to assist in various agricultural processes. In the Renaissance visual images, Tuccia carries a large flat sieve like those still used in improving soil quality by retaining stones and releasing fine earth, although Marina Warner's description of her 'panning' in the Tiber suggests the variant in which the sieve retains what is valuable and lets out what is not (Warner 1985: 242). The *koskinon* can come in other sizes; it features in Galen, *Method of Medicine* 5.10 (10.355 Kühn), in preparing a remedy which 'has been pounded and sifted thoroughly with a very fine sieve and then ground down until it becomes a fine powder'. Here, the idea is again that passing something through a sieve improves its quality. Athenaeus, always a good source for anything related to cookery, refers twice to a bronze *koskinon* through which mashed-up cheese is pushed; here again, the purpose of the sieve is to refine what passes through it.[7] In Plutarch (*Table Talk* 7.3, 699b), the lung is made like a sieve (*ēthmoeidēs*) 'for the sake of the liquids and the solids that slip in with them'.

An alternative focus is on losing what one should retain. In Plato's *Gorgias* (493c), the soul of the thoughtless person is like a sieve, *koskinon*, as it is unable to hold anything; again, letting out the 'good stuff', but not to a good end. In an earlier book of *Method of Medicine*, Galen uses *koskinon* in this way: 'It is a statement filled with so many errors that for me it stands comparison with the dullard when he said to the sieve that he could not discover what would plug it up and what would not' (*Method of Medicine* 1.9, 10.68 Kühn; trans. Ian Johnston and Greg Horsley). The word *koskinon* is also used for the sieve in which the Danaids carry water; a typical punishment in Hades is to 'fetch water in a *koskinon*', and in the *Republic* (2, 363d) Plato mentions a punishment in Hades in which the impious and unjust are forced to carry water in a sieve. While the husband-murdering Danaids have to carry water to put in a jar full of holes, at some point the vessels in which they carry that water are themselves described as having holes in them (Costa and Whittle 1973: 289–91; Keuls 1974; Paster 1993: 47).

When *cribrum* refers to an agricultural sieve, it explicitly lets through the fine earth and retains the stones and larger chunks of earth, as in Cato, *On Farming* 151 (quoted by Aulus Gellius, 3.14.17), describing sowing the seeds of the cypress tree: 'Over it sift earth from a sieve (*cribrum*), to the depth of a half-finger' (trans. W. D. Hooper and Harrison Boyd Ash). *Cribrum* is also used for a sieve in Columella (5.6.6), on planting elm seeds: 'we shall thickly cover the beds all over with the seed and scatter crumbling earth over them with a sieve to the depth of two inches' (trans. E. S. Forster and E. H. Heffner). This contrasts with another device, the *liknon*, a winnowing tool which is open at one side. In winnowing, the woman may shake the grain in a sieve so that small seeds and dust pass through the holes, while stalks, husks, and lumps of earth collect at the top and can be picked off or, in a particular kind of sieve, are allowed to slide over one, lower, part of the edge of the sieve (Harrison 1903: 309). The *liknon* retains the good and lets out the bad, while the *cribrum* retains the bad and lets out the good.

The *liknon* is used in harvesting—the end of the process—and the *cribrum* in planting—the beginning of the process. This may be significant in terms of the *cribrum* as virginal body; remembering that Vestal Virgins could, and did, go on to marry, so their bodies would in time need to retain the male and female fluids of generation rather than letting out this 'good stuff'. Recalling Aristotle on generation as being like making cheese (see p. 383), the womb as a sieve can solidify the fluids that make a baby.

In ancient Greek medical texts, there is also a *krēserēs* in the Hippocratic text *On Barren Women* 10, translated in the Loeb as 'flour sieve'.[8] Its use here is to strain some mare's milk which is going to be used to clean out pus from the womb of a woman who is unable to become pregnant because the pus is melting the male seed: 'She should then remove the covering and insert the syringe in the direction of her uterus, knowing herself where to put it' (trans. Paul Potter). Like the *cribrum* used in soil preparation—and of course, in a dominant image of the female body, women are the earth in which men plant seed (DuBois 1988: 39–64)—the Hippocratic use of sieving here is about retaining the bad, to let through the good, sometimes retaining the solids so the fluid is purer. The same is true of the Greek *kinachura* in Aristophanes (*Assemblywomen* 730) translated as 'bran sieve'—the sieve retains the bran so the flour is finer.

The *cribrum* lets out the good; Tuccia retains her water. Bearing in mind the Western interest in virgins and urine, relevant here is the work of Michael McVaugh (2012). Writing about the history of the kidney, he extended the image of sieves to body parts, asking whether the kidney was seen historically as a strainer (separating the solid from the liquid), or a sieve (here, removing smaller solid particles and not the larger ones). In Galen, the kidneys allow only the thinnest part of the blood and bile to pass through; the density of the kidney holds the thicker part. Galen discusses why this thicker part does not act as a barrier preventing more blood coming forward for filtration. Blood moves through the kidneys 'as if through an *ēthmos*'; the Greek term *ēthmos* designates a woven, usually reed, basket used to filter juice from grapes or olives, or a metal, pierced tray used to filter wine as it is served. The purpose here is to separate liquid from solid, and this distinguishes the *ēthmos* from the *kostikon*, a woven tray used to separate different sized solids, such as grades of flour. So here, the kidneys are more like a strainer than a sieve.

Returning to Tuccia's *cribrum*, this is the sort of sieve that is used in planting, not harvesting, but it remains unclear whether it holds back or releases the 'good'. In *Greek Virginity* (1990: 139–40), Giulia Sissa drew a structuralist parallel between the jar and the sieve:

> There are rewards to be gleaned from delving into the technical details of agricultural literature: jar and sieve are part of the indispensable equipment of civilized life, forever used for storing what is necessary and for getting rid of what is unnecessary. The twin expressions *tetrêmenos pithos* (leaky jar) and *koskinôi antlein* (to draw [liquid] with a sieve) literally invert these fundamental actions of storing and separating. A jar that allows its contents to leak out, a sieve used as a container: what better way to suggest improvidence in the head of a family?
>
> Sissa 1990: 139–40

While I am drawn to the opposition jar/sieve, this does not seem to me to represent what the *cribrum* tends to signify. It is not about 'getting rid of what is unnecessary' so much as improving the material; retaining the bad to let out the good. Furthermore, in Tuccia's story, using a sieve as a container to carry water is about providence rather than improvidence.

I have not yet mentioned the womb, but Sissa's comments evoke its representation as a jar in the Hippocratic *On Diseases of Women* (King 1988: 34–5). If all are women 'sieves', this must work both ways: they not only let out fluid but also allow in men's sexual fluids. Releasing the fluids in menarche, defloration, and the lochia transformed the ancient Greek *parthenos* into a mature woman (King 1988: 72–3). The virgin who can hold her water will in time need to let out fluids in order to remain healthy, and then to retain her menstrual blood and the seed of her partner in order to produce a foetus.

Elizabeth and Tuccia

In the sixteenth century, Tuccia's story could be evoked by the mere presence of a sieve. But does this mean that all sieves are Tuccia's sieve? This question brings us to the central example of a woman using the imagery of Tuccia for her own purposes: this woman is Elizabeth I. Portraits of Elizabeth are recognised as highly complex, with 'a distinctively "linguistic" orientation'; they use metaphor and allegory, as well as including text within the image. But, as Mary Hazard (1990: 61) has shown, they also use images to 'convey attitudes that were too bold to be expressed in words'. In her portraits, particularly during and immediately after the ultimately unsuccessful negotiations to marry her to François, Elizabeth I was often depicted with a sieve; the portraits of this period were seen by Roy Strong (1987: 97–9) as 'symbolic arguments against the Anjou match' (Montrose 2006: 121–2; McBurney and Kimbriel 2014).

She also had sieve motifs on her clothing and jewellery (McBurney and Kimbriel 2014: 642 nn.10, 11).

Were the 'sieve portraits' an assertion of Elizabeth's virginity and thus an objection to the idea of this marriage, or could they support the union? François, Duke of Alençon, was Elizabeth's 'Frog Prince'; this was how he signed his letters to her and, while he remained a possible husband, she was given presents with the frog theme (Adler 1978: 1–2). Doris Adler (1978: 2) has linked this to a folk story circulating at the time, 'The well at the world's end', in which a girl who has been sent by her wicked stepmother to collect water in a sieve is told by a frog the secret she needs: she must 'Stop it with moss, daub it with clay.' The frog turns out to be a prince, so the presence of the sieve in portraits could be because Elizabeth had found her prince.[9]

Even if we accept this connection, this does not exhaust the symbolism. The 'sieve portraits' themselves contain many clues as to their interpretation. In the Siena sieve portrait, now dated to 1583, two years after the marriage negotiations had ended, the Queen holds the sieve, on the rim of which is inscribed A TERRA IL BEN / MAL DIMORA IN SELLA, 'The good falls to the ground while the bad remains in the saddle' (Eycken 2018: fig. 54). We could read this as a 'good *cribrum*' message, suggesting not just that the Queen can tell the good from the bad, but that she acts on her knowledge. The sieve itself is empty, not in use.

While the sieve references Tuccia through the powerful image of the Virgin Queen, it also connects with another series of images, associated with the virtue of Prudence.[10] For example, in Pieter Bruegel the Elder's *Prudentia* (1559), everyone is preparing for the future, salting meat, putting out a fire, having both a priest and a physician at the sickbed; and the cardinal virtue of Prudence herself is shown with a mirror (for self-knowledge) and carrying a sieve on her head to separate good from bad.[11]

Paster suggested that the sieve is powerful as an image precisely because of its indeterminacy, as

> a paradoxical symbol of Elizabeth's virginity. Full of holes, the sieve refers unmistakably to the symbolically leaky, hence unreliable nature of ordinary women's bodies even as it asserts, through its link to Tuccia, the queen's transcendence as virgin monarch of ordinary women. For Elizabeth in her capacity as ruler, the sieve is an emblem not of leakiness but of discernment, of the good judgment requisite in rulers.
>
> Paster 1993: 50

So is Elizabeth's sieve referencing both her virginity and her prudence, and are these two connected? Is it that by maintaining her virginity Elizabeth is also a better ruler? Her motto, seen in the Plimpton sieve portrait on the coat of arms, is 'semper eadem', always the same (Eycken 2018: fig. 51). That is a challenge to the traditional view of female maturation and the transition from virgin to wife and mother. This portrait, which also includes a quotation from Petrarch, dates to 1579. What is exceptional in the Plimpton sieve representation is that the sieve is tied to the Queen's dress by a ribbon, like 'an extension of her body' (Baert 2018: 58). In another of Livy's Vestal Miracle stories, Claudia pulled a ship up the Tiber by tying it to her girdle, so this could be a double Vestal reference. Furthermore, the ribbon is coming out at a point which suggests the sexual organs, in what Louis Montrose (2006: 124; Eycken 2018: 101) called 'a displacement of the Queen's sexuality'. She appears to be showing us, very precisely, that she has a non-leaky sieve. Or is it simply a sieve, with the inherent capacity both to hold and to release? To me this recalls a 1522 image of the womb turned inside out, in Berengario da Carpi.[12] In this inside-out womb image, the many black dots show the cotyledons, which end in the orifices through which the menses flow into the womb.

Does the sieve relate not to her virginity but to the rumours of her sexual appetites (particularly in relation to Dudley), which had scandalised Europe (Montrose 2006: 126–7)? Constance Jordan (1990) proposed that the presentation of the sieve as a 'saddle' in the Siena portrait adds a further level of complexity here; being in the saddle suggests riding, and thus sexual passion. Is the rider in control of the passions or giving them rein? Is this 'being in the saddle' about a dominance that is more masculine than feminine? 'To ride on a saddle-sieve is to rule with and through passions that the ruler/lover has the power both to express and to deny' (Jordan 1990: 167). The fluid in the sieve 'falls to the ground' in a way that Tuccia's Tiber water did not, and water to the earth suggests fertilisation, not virginity.

In the Siena sieve portrait, the roundels on the column at the top left show scenes from the story of Dido and Aeneas; from Aeneas fleeing from Troy to his departure from Dido. Scholars have interpreted this as showing Elizabeth as a superior version of Aeneas, abandoning entirely any possible sexual relationships in order to meet her destiny as founder of a new empire. However, this is complicated too. While Elizabeth was known as 'Dido' in her own time, this was a reference to Petrarch's version in the *Triumph of Chastity*, where the emphasis was not on her lack of chastity by surrendering to Aeneas but on her death in order to remain true to her husband—not for Aeneas (11.10–11). By keeping her body pure, Dido kept Carthage independent (168).

So is this about celebrating Elizabeth's chastity while at the same time acknowledging the rumours of her unchastity (Montrose 2006: 127)? For Barbara Baert (2018: 70), referencing the imagery of the two bodies of the ruler, for Elizabeth the sieve is 'a filter between the I as woman and the I as queen' as she herself 'flows back and forth between both bodies'. For her pupil Sarah Eycken (2018: 103), what is important is that while Tuccia's sieve is impermeable, that of Elizabeth is permeable, because it is 'the sieve of discernment'.

Tuccia as witch

Elizabeth retained control of her own imagery; Tuccia's story was open to interpretations in which her exemplary nature was denied. How did she achieve her reversal of forces of Nature? In one sixteenth-century tradition, Tuccia could demonstrate the 'vertue of unstain'd virginity', as in the list of exemplars given in Thomas Cranley's 1635 poem, 'Amanda: or, The reformed whore'.[13] However, in a contrary interpretation, she was a witch. In Thomas Twyne's chapter 'On virgins', in a work written in 1576 and thus within Elizabeth's reign:

> Valerius in the seventh book and third chapter, writeth of a certain vestal virgin in Rome called Tuccia, whose chastity being obscured with a sinister report of incontinency, and she her self being privy of her own innocency, took into her hands a sieve, and thus prayed unto the goddess Vesta, saying:

Vesta, if I have always touched thy sacrifices with chaste and undefiled hands, command that I may take water out of the river Tyber with this Sieve, and carry it into thy Temple. Which indeed she accomplished, the common course of Nature giving place to her bold attempts.

Twyne 1576: 36[14]

Why should Nature 'give place' to her? One possibility in the sixteenth century was witchcraft; Elizabeth passed a witchcraft law in 1563, the 'Act against Conjurations, Inchantments and Witch Craft', and the 1590s were the period of the largest number of accusations of witchcraft during her reign (Levin 2002: 108). Quoting Cotta's *Discovery of Witchcraft*, published 13 years after Elizabeth's death in 1603, Richard Bovet in 1684 linked Tuccia to Livy's account of that other Vestal,

Claudia, who (unassisted by any humane help) did (only with a small string fastened thereunto) draw a mighty Ship along the River of Tyber; which by reason of its vast weight and greatness, could not be moved by the force of many strong Men, assisted by Cattle that were used to draw heavy burthens, which with good reason he concludes she could not have performed without the co-operation of some evil Spirit. He likewise mentions Tuccia, another of that Sister-hood, who by muttering some Invocation, or Inchantment, could take up water in a Sieve, and carry it at a good distance from the River Tyber, without spilling a drop.

Bovet 1684: 110

In the passage to which Bovet refers here, Cotta (1625: 32–3 makes it clear that pulling a heavy object is 'supernatural and above her power and nature'; he describes Tuccia as 'reported by mumbling of a certain prayer, to keep water within a sieve, or a riddle, as witnesseth not only Pliny, but even Tertullian'.[15] Yet in Roman sources there is no connection between these two women and witchcraft or magic.

In the trials of witches, one of the grounds for conviction was confession. Another was to produce witnesses who would swear to the accused person speaking or acting in a way that only those familiar with the Devil could do. In Increase Mather's account of the 1692 Salem witch trials, which states this principle, we read that 'Claudia was *seen* by witnesses enough, to draw a ship which no human strength could move. Tuccia a Vestal Virgin was *seen* to carry water in a sieve: the Devil never assists men to do supernatural things undesired' (Mather 1693: 38; my italics). Tuccia is included in a discussion of the power of words to affect matter. This is also the context of the reference to Tuccia in Pliny (*Natural History* 28.12). It comes in a section on 'the remedies derived from man', and specifically on whether words and chants work; people think they do but, Pliny writes, the wisest men say no. After giving examples of how very formal patterns of prayer have been used, for example to stop something being omitted, Pliny gives the example, 'extant too is the plea of innocence uttered by the Vestal Tuccia when, accused of unchastity, she carried water in a sieve in the year of the City six hundred and nine' (trans. W. H. S. Jones).[16] So for Pliny, this is not Cotta's 'mumble' but a properly patterned prayer. Philemon Holland's translation of Pliny (1601: 295) has it as a 'charm'.

The Tuccia stories, then, are far more complex than they may at first appear. The various brief versions in ancient literature are incomplete but rich in possible signification. Mantegna's alignment of Tuccia with what may be Artemisia brings out further meanings: fire and ash, chastity and lust, purity and passion, agency and lack of agency. Yet the presence of fire is already hinted at in Tuccia's 'snatching up' an implement that may have been used in ritual to carry fire. This implement, the sieve, has taken us into the complexities of both agricultural tools and the workings of bodily organs. While it concerns separation of good and bad, and thus discernment, the sieve may be more like a strainer, letting through what is good, or used more to retain the good things. In Early Modern literature, while Tuccia can be a model of Christian womanhood, analogous to the Virgin Mary, and thus suitable for display to respectable women, her invocation of Vesta can become evidence of her witchcraft, so that this otherwise exemplary woman becomes a villain.

The body's fluids are presented here as key to sexual and social identity. As a way of thinking about the female body, the *cribrum*-sieve normally refines materials, improving the quality of what passes through it. Virgins retain rather than release, unlike the uncontrolled body of the leaking Tuccia of Juvenal. The Vestal Tuccia, however, does not let anything pass through until she is at the feet of the priests; if the sieve is a sort of womb, she is perhaps showing that she will suppress and then use her fertility in the service of the state, but she also makes one think of later virginity tests which depend on the reluctance of the virgin bladder to pour out its contents. She is, however, in control; she knows what must be retained and when it must be retained, whether that is urine, menstrual blood or male semen. Elizabeth's sieve, which references both chastity and the potential for reproduction, is also about knowing when to keep bodily fluids in and when to let them out; only at the proper time, and only in the interests of the greater good.

Notes

1 They attribute the erroneous view to one of the books probably owned by Leonardo, Alessandro Benedetti, *Anatomice sive historia corporis libri V* (Venice, 1498), chapter 18 'De semine'.

2 This is one of the many Vestal images usefully collected in Eycken 2018; here, fig. 46.

3 There is also Tuccia daughter of Caeso in Valerius Maximus 4.10.

4 κατηγορῆσαί τινά φασιν ἀδίκως μιᾶς τῶν παρθένων τῶν ἱερῶν Τυκκίας ὄνομα, ἀφανισμὸν μὲν πυρὸς οὐκ ἔχοντα προφέρειν, ἄλλας δέ τινας ἐξ εἰκότων τεκμηρίων καὶ μαρτυριῶν ἀποδείξεις φέροντα οὐκ ἀληθεῖς· κελευσθεῖσαν δ'ἀπολογεῖσθαι τὴν παρθένον τοῦτο μόνον εἰπεῖν, ὅτι τοῖς ἔργοις ἀπολύσεται τὰς διαβολάς.

5 www.nationalgallery.org.uk/paintings/andrea-mantegna-the-vestal-virgin-tuccia-with-a-sieve (Accessed February 2021).

6 Sabadino's work is online at https://it.wikisource.org/wiki/Gynevera_de_le_clare_donne (Accessed February 2021).

7 Athenaeus 647e, quoting from Chrysippus, *The Art of Baking*. Athenaeus also mentions sieve-sellers at the market (126f.).

8 This is chapter 222 in the edition of Émile Littré (Volume VIII: 430) and chapter 10 in the Loeb edition of Paul Potter (Volume X: 356).

9 In some variants of this folk tale, the virgin is sent by her mother to fetch water in a sieve because she has broken the only pitcher they own. In return for the frog teaching her how to make the sieve watertight, she has to sleep with the frog for two nights, and he then becomes a prince. See www.pitt.edu/~dash/frog.html (Accessed February 2021) from Halliwell-Phillipps 1849: 43–7.

10 Although the statement that this portrait shows Elizabeth 'as Tuccia' seems to be going too far; e.g. https://commons.wikimedia.org/wiki/File:Metsys_Elizabeth_I_The_Sieve_Portrait_ c1583.jpg (Accessed February 2021).

11 www.pieterbruegel.net/object/prudence-prudentia (Accessed February 2021).

12 Jacopo Berengario da Carpi, *Isagogae breves* (Bologna: Benedictus Hectoris, 1522): www. nlm.nih.gov/exhibition/historicalanatomies/Images/1200_pixels/Berengario_p24r.jpg (Accessed February 2021).

13 *Amanda* is a 'key text in the history of pornography' couched as a description of converting a prostitute to virtue; Craik 2007: 126.

14 The structure of Tuccia's powerful and answered prayer—invocation of a named deity, claim for divine favour, specific request—is identified by Mueller 2002: 52.

15 This is taken from the second edition of Cotta, *The Trial of Witch-Craft*. The juxtaposition of Claudia and Tuccia is also found in Augustine, *City of God*, 10.16.2, where Tuccia is not named but is clearly the person meant in his 'a Vestal virgin suspected of unchastity removed all doubt when she filled a sieve with water from the Tiber and it did not run out through the holes.' See also *City of God*, 22.11.3.

16 This would be 145 BCE, making this one possible date for Tuccia; others, such as Parker 2004: 593, give 230 BCE, based on the location of her story in Livy.

Bibliography

Primary sources

Bovet, R. (1684) *Pandaemonium, or, The Devil's Cloyster*. London: J. Walthoe.

Cotta, J. (1625) *The Trial of Witch-Craft*. London: John Legat for Richard Higgenbotham.

Cranley, T. (1635) *Amanda: Or, The Reformed Whore*. London: John Norton.

Crooke, H. (1615) *Microcosmographia: A Description of the Body of Man*. London: William Jaggard.

De Rocher, G. D. (1989) *Laurent Joubert, Popular Errors (1578)*. Tuscaloosa: University of Alabama Press.

Dunhill, A. (1999) *Lucrezia Marinella, The Nobility and Excellence of Women, and the Defects and Vices of Men*. Chicago: University of Chicago Press.

Fries, L. (1518) *Spiegel der Artzny*. Strasbourg: Grüninger.

Halliwell-Phillipps, J. O. (1849) *Popular Rhymes and Nursery Tales: A Sequel to the Nursery Rhymes of England*. London: John Russell Smith.

Holland, P. (1601) *The Historie of the World, Commonly Called, The Naturall History of C. Plinius Secundus*. London: Adam Islip.

Lemay, H. R. (1992) *Women's Secrets: A Translation of Pseudo-Albertus Magnus' De Secretis with Commentaries*. New York: SUNY Press.

Mather, I. (1693) *A Further Account of the Trials of the New England Witches*. London: J. Dunton.

Twyne, T. (1576) *The Schoolmaster, or Teacher of Table Philosophy*. London: Richard Jones.

Secondary literature

Adams, J. (1982) *The Latin Sexual Vocabulary*. London: Duckworth.

Adler, D. (1978) 'The riddle of the sieve: The Siena Sieve Portrait of Queen Elizabeth'. *Renaissance Papers* 1–10.

Baert, B. (2018) 'Around the sieve: Motif, symbol, hermeneutic', *TEXTILE. Journal of Cloth and Culture* 16(1), 1–23.

Baskins, C. L. (1998) *Cassone Painting, Humanism and Gender in Early Modern Italy*. Cambridge: Cambridge University Press.

Benson, P. J. (1992) *The Invention of the Renaissance Woman: The Challenge of Female Independence in the Literature and Thought of Italy and England*. University Park: Pennsylvania State University Press.

Champion, C. B. (2015) 'Livy and the Greek historians from Herodotus to Dionysius', in B. Mineo (ed.) *A Companion to Livy*. Chichester: Wiley-Blackwell, 190–204.

Costa, C. C. N. and E. W. Whittle (1973) 'Holed pitchers for the Danaids: A first allusion in Seneca?', *Mnemosyne* 26(3), 289–91.

Cox, V. (2011) *The Prodigious Muse: Women's Writing in Counter-Reformation Italy*. Baltimore, MD: Johns Hopkins University Press.

Craik, K. A. (2007) *Reading Sensations in Early Modern England*. Basingstoke: Palgrave Macmillan.

Davies, R. (1983) *The Rebel Angels*. Harmondsworth: Penguin Books.

DuBois, P. (1988) *Sowing the Body: Psychoanalysis and Ancient Representations of Women*. Chicago: University of Chicago Press.

Edwards, N. (2008) 'The vestal virgin Tuccia', in A. Bayer (ed.) *Art and Love in Renaissance Italy*. New York: Metropolitan Museum of Art, 313–14.

Eycken, S. (2018) 'Tuccia and her sieve: The Nachleben of the vestal in art', MA thesis, KU Leuven, Leuven.

Franklin, M. A. (2006) *Boccaccio's Heroines: Power and Virtue in Renaissance Society*. Aldershot: Ashgate.

Furlotti, B. and G. Rebecchini (2008) 'Isabella d'Este and the Culture of the Studiolo', in B. Furlotti and G. Rebecchini (eds) *The Art of Mantua: Power and Patronage in the Renaissance*. Los Angeles: The J. Paul Getty Museum, 92–115.

Gunderson, E. (2005) 'The libidinal rhetoric of satire', in K. Freudenberg (ed.) *The Cambridge Companion to Roman Satire*. Cambridge: Cambridge University Press, 224–40.

Hansen, W. F. (2002) *Ariadne's Thread: A Guide to International Tales Found in Classical Literature*. Ithaca: Cornell University Press.

Hanson, A. and D. Armstrong (1986) 'Two notes on Greek tragedy: The virgin's voice and neck: Aeschylus, *Agamemnon* 245 and other texts', *Bulletin of the Institute of Classical Studies* 33, 97–102.

Harrison, J. E. (1903) 'Mystica Vannus Iacchi', *Journal of Hellenic Studies* 23, 292–324.

Hazard, M. A. (1990) 'The case for "case" in reading Elizabethan portraits', *Mosaic* 23(2), 61–88.

Jordan, C. (1990) 'Representing political androgyny: More on the Siena portrait of Queen Elizabeth I', in A. M. Haselkorn and B. S. Travitsky (eds) *The Renaissance Englishwoman in Print: Counterbalancing the Canon*. Amherst: University of Massachusetts Press, 157–76.

Kelly, K. C. (2000) *Performing Virginity and Testing Chastity in the Middle Ages*. London: Routledge.

Keuls, E. (1974) *The Water Carriers in Hades: A Study of Catharsis Through Toil in Classical Antiquity*. Amsterdam: Hakkert.

King, H. (1988) *Hippocrates' Woman: Reading the Female Body in Ancient Greece*. London: Routledge.

———. (2011) 'Inside and outside, cavities and containers: The organs of generation in seventeenth-century English medicine', in P. A. Baker, H. Nijdam, and K. van 't Land (eds) *Medicine and Space: Body, Surroundings and Borders in Antiquity and the Middle Ages; Visualising the Middle Ages*. Leiden: Brill, 37–60.

Levin, C. (2002) *The Reign of Elizabeth I*. Basingstoke: Palgrave Macmillan.

Loughlin, M. H. (1997) *Hymeneutics: Interpreting Virginity on the Early Modern Stage*. Lewisburg: Bucknell University Press.

McBurney, H. and C. S. Kimbriel (2014) 'A newly discovered variant at Eton College of the Queen Elizabeth I Sieve portrait', *The Burlington Magazine* 156, 640–9.

McVaugh, M. (2012) 'Losing ground: The disappearance of attraction from the kidneys', in M. Horstmanshoff, H. King, and C. Zittel (eds) *Blood, Sweat and Tears: The Changing Concepts of Physiology from Antiquity into Early Modern Europe, Intersections*. Leiden: Brill, 103–37.

Montrose, L. (2006) *The Subject of Elizabeth: Authority, Gender, and Representation*. Chicago: The University of Chicago Press.

Mueller, H.-F. (2002) *Roman Religion in Valerius Maximus*. London: Routledge.

Noble, D., D. DiFrancesco, and D. Zancani (2014) 'Leonardo da Vinci and the origin of semen', *Notes and Records (Royal Society Journal of the History of Science)* 68, 391–402.

Orland, B. (2012) 'White blood and red milk: Analogical reasoning in medical practice and experimental physiology (1560–1730)', in M. Horstmanshoff, H. King, and C. Zittel (eds) *Blood, Sweat and Tears: The Changing Concepts of Physiology from Antiquity into Early Modern Europe, Intersections*. Leiden: Brill, 443–78.

Parker, H. N. (2004) 'Why were the Vestals virgins? Or the chastity of women and the safety of the Roman state', *American Journal of Philology* 125(4), 563–601.

Paster, G. K. (1993) *The Body Embarrassed: Drama and the Disciplines of Shame in Early Modern England*. Ithaca: Cornell University Press.

Richlin, A. (1997) 'Carrying water in a sieve: Class and the body in Roman women's religion', in K. L. King (ed.) *Women and Goddess Traditions in Antiquity and Today*. Minneapolis: Fortress Press, 330–74.

———. (2014) 'Carrying water in a sieve: Class and the body in Roman women's religion', in A. Richlin (ed.) *Arguments with Silence. Writing the History of Roman Women*. Ann Arbor, MI: Michigan University Press, 197–240.

San Juan, R. M. (1991) 'The court lady's dilemma: Isabella D'Este and art collecting in the Renaissance', *Oxford Art Journal* 14(1), 67–78.

Sissa, G. (1990) *Greek Virginity*. Cambridge, MA: Harvard University Press.

Stamper, J. W. (2005) *The Architecture of Roman Temples: The Republic to the Middle Empire*. Cambridge: Cambridge University Press.

Stolberg, M. (2012) 'Sweat: Learned concepts and popular perceptions, 1500–1800', in M. Horstmanshoff, H. King, and C. Zittel (eds) *Blood, Sweat and Tears: The Changing Concepts of Physiology from Antiquity into Early Modern Europe, Intersections*. Leiden: Brill, 503–22.

———. (2015) *Uroscopy in Early Modern Europe*. Aldershot: Ashgate.

Strong, R. (1987) *Gloriana: The Portraits of Queen Elizabeth*. London: Thames and Hudson.

van't Land, K. (2012) 'Sperm and blood, form and food: Late Medieval medical notions of male and female in the embryology of *membra*', in M. Horstmanshoff, H. King, and C. Zittel (eds) *Blood, Sweat and Tears: The Changing Concepts of Physiology from Antiquity into Early Modern Europe, Intersections*. Leiden: Brill, 363–92.

Verheyen, E. (1971) *The Paintings in the Studiolo of Isabella d'Este at Mantua*. New York: New York University Press.

Warner, M. (1985) *Monuments and Maidens: The Allegory of the Female Form*. London: Weidenfeld and Nicholson.

Wildfang, R. L. (2006) *Rome's Vestal Virgins: A Study of Rome's Vestal Priestesses in the Late Republic and Early Empire*. London: Routledge.

ENVOI

Mark Bradley and Victoria Leonard

Alexander the Great's sweaty countenance caught the eyes—and noses—of several ancient observers. Plutarch's *Life of Alexander* (4.18) recalls Alexander's contemporary Aristoxenus, who—presumably picking up on one of the king's traits that was widely recognised—observed that 'a most agreeable odour exhaled from his skin, and that his breath and body all over was so fragrant as to perfume even his clothes.'[1] This fragrant excretion was interpreted by some as unambiguous evidence of Alexander's divinity: like the gods, he ingested and exuded scented vapour. But others had keener noses and wits. Elsewhere, in one of his *Table Talk* dialogues, Plutarch imagines a debate about Alexander's drinking habits: the fragrant sweat that stained his clothes was evidence of a hot constitution from his excessive drinking. Just as hot, dry climates produce frankincense and cassia, Alexander's heavy binges caused an imbalance in the humours, driving out noxious moisture in the form of fragrant perspiration.[2] Plutarch's source for this wisdom was the short treatise *On Sweat*, which was written by the Peripatetic philosopher Theophrastus, a contemporary of Alexander who set out to understand the role and function of this bodily fluid and what it told us about the bodies of those who exuded it. Plutarch's smart analysis, in which the worlds of biographical eulogy and medical diagnosis clash, underscores the complexity and ambiguity of bodily fluids as markers of bodily health, character, and behaviour that has been a major theme throughout this volume.

Bodily fluids loom large in stories about Alexander, both as a sign of his mortality and as a provocative talking point. In *Life of Alexander* 28, Plutarch discusses contemporary myths about Alexander's divinity and recounts an episode where the king was struck by an arrow and points out 'This, my friends, is flowing blood (*to rheon haima*), and not [as Homer, *Iliad* 5.340] "Ichor, such as flows through the blessed gods".' Elsewhere (14.5), at the outset of one of Alexander's expeditions, a wooden statue of Orpheus was seen to sweat profusely (*hidrōta polun . . . aphēke*), interpreted by many as a fearful omen but by Alexander's soothsayer as a sign that he was going to achieve such remarkable things that would cause the poets and musicians themselves to profusely sweat (*polun hidrōta*) in celebration. Along with blood and sweat, tears are also telling in stories of Alexander: his mortal grief over the death of Cleitus sees him lying on the floor weeping like a slave (52.3); 'tears flow from his eyes' when he witnesses his suffering army (Arrian, *Anabasis* 7.11.3); and later (7.14.1) he lays over the body of Hephaestion in tears all day long. In Plutarch's parallel life (*Life of Caesar* 11.3), Julius Caesar in turn weeps on reading a biography of Alexander, tears that he explains are due to how little he himself has

DOI: 10.4324/9780429438974-33

achieved in comparison to the Macedonian king.[3] Later, upon Caesar's assassination (66.7), the pedestal of Pompey's statue is drenched with the dictator's blood (*polu kathēimaxen . . . ho phonos*), the ultimate fluid release that demonstrates Caesar's mortality and Pompey's revenge from beyond the grave. Fluids in Plutarch's parallel lives link Greeks and Romans in a common bond of mortality.

As a Peripatetic treatise, Theophrastus' On Sweat—the source of Plutarch's wisdom in the anecdote with which this Envoi started—was principally concerned with what we can learn from observation, and so was very much in touch with the common, everyday experiences of bodily function and behaviour that have pervaded this volume.[4] For Theophrastus, as for Aristotle, sweat could be interpreted as breath (*pneuma*) that penetrated out through the skin and condensed on its surface. It was also an excretion through which foreign matter was expelled from the body and could be either superficial sweat (which was watery and thin) or sweat that came from deeper in the body, heavier and more odorous. Theophrastus connected salty sweat and bad odour to *kachexia* ('bad condition of the body').[5] Like many of the philosophical and literary texts examined throughout this volume, Theophrastus' treatise was keenly in tune with the Hippocratic Corpus, especially the treatise *On Internal Affections*, which associated certain types of disease with both bad breath and ill-smelling sweat.[6] For Theophrastus, sweat varied according to age: children and old people sweat least and sweat cleanest; adolescent sweat smells because of moisture, heat, and sexual drive. As well as age, bodily condition, the ingestion of food, and external temperature all affected sweat levels—as did fear and anxiety. And different parts of the body sweat differently:

> Foul odour characterizes a region and what is in it, if it does not breathe well—for there is rotting, and bad odour is a kind of rot. Therefore sweat from the armpits and generally that from cavities have the worst odour.
> Theophrastus, *On Sweat* 9, trans. Fortenbaugh et al. 2003

Sweat, then, allows us to close this volume by teasing out several overarching themes that cut across the fluids studied within it. Most of the fluids discussed in this volume are localised, but sweat can seep out anywhere. The human body has between two and four million sweat glands, most concentrated on the feet and least on the back. Most of these are 'eccrine glands', which are situated all over the body and produce the clean, healthy sweat that is exuded during a good workout and comprises water, salt, and potassium. The 'apocrine glands', often positioned in hairy places like the armpits and the groin, are the glands that are activated when individuals are nervous or under stress: this sweat contains fatty acids and proteins and is broken down by bacteria on the surface to produce the body odour and yellow patches that we associate with this kind of sweat, which often appears on the palms of the hands, soles of the feet, and forehead.[7] On these grounds, Theophrastus was correct: even in modern medical terms there is 'good sweat' and there is 'bad sweat' (what we sometimes call 'regular sweat' and 'stress sweat'), and Helen King in Chapter 24 shows that Early Modern medicine also diagnosed many different types of sweat (see p. 382). And one of the most interesting things about sweat as part of the human condition is its connection, like the fluids discussed by Irene Salvo, Catalina Popescu, and Michael Goyette in this volume (Chapters 3, 5, and 17), to particular types of emotion: anger, shock, fear, anxiety, stress, nightmares, and traumatic memories all trigger sweats. As we have seen, Theophrastus and other ancient observers were acutely aware of these connections, and this left the opportunity ripe for physiognomic observation, moral commentary, and the association of fluids with social identities, character, and behaviour.

Like many of the fluids examined within this volume, and as Plutarch's account of Alexander and Theophrastus' *On Sweat* have demonstrated, sweat was an inherently ambiguous exusion. It alluded to honest hard toil and extreme physical exertion (think of Winston Churchill's famous proclamation 'I have nothing to offer but blood, toil, tears and sweat', itself based on a set of classical proverbs).[8] As Plutarch's *Lives of Alexander and Julius Caesar*, mentioned earlier, have shown, sweat could sit alongside blood and tears in signalling a degree of humanity, mortality, and humility among the otherwise godlike: this very theme is explored by Emily Kearns in Chapter 12 on blood in early Greek literature. Moreover, sweat could even be configured positively as a fragrant and erotic fluid connected metaphorically to spices or perfumes, as Jane Burkowski in Chapter 4 in this volume has explored, linking to the erotic and generative fluids examined by Julie Laskaris (Chapter 6), Rebecca Fallas (Chapter 7), and other contributors.

But sweat also functioned as a diagnostic feature of deviance, a fluid produced on the surface of bodies as a result of inappropriate, uncivilised behaviour, spiritual weakness, or inner corruption. Like other bodily fluids, once expelled, sweat was pollutive and threatening. For this reason, Galen, scornfully undermining the practice of using products of the human body as medical remedies, lists earwax, menstrual blood, urine, excrement, and sweat as substances that were both useless and vile for reingestion into the body (*Simple Medicines* 10.1, 12.249 Kühn). Sweat, disease, and death were closely linked together, alongside other fluids associated with dissolving and putrefying bodies (see Chapters 19, 20, and 21 by Tasha Dobbin-Bennett, Goran Vidović, and Assaf Krebs): just like with Alexander's body exuding the alcohol it had consumed, ancient physicians recognised sweat as a diagnostic feature of a wide range of maladies, associated in particular with different kinds of fever.[9] As well as disease, sweat was also linked to fear, embarrassment, and lovesickness, betraying mental trauma and bodily weakness. As Burkowski's chapter demonstrates, excessive sweating was sometimes considered an animalistic trait, with 'goatiness' appearing as an enduring and pervasive metaphor for sweaty body odours (see pp. 80–1), and when emperors such as Nero or Maximinus Thrax sweated, it was a telltale sign of their corrupt regimes.[10]

Sweat was inherently ambiguous, and this ambiguity manifested itself in some of the categories used to describe it, as Amy Coker's opening chapter argues about the full repertoire of linguistic categories, including sweat (both *hidrōs* and its variants, and the suggestive *kikkasos*, 'thigh sweat', see p. 19); and it is unsurprising that we find some of these same ambiguities playing out in Latin satire (see the chapters by Claude-Emmanuelle Centlivres Challet, Chapter 8, and Andreas Gavrielatos, Chapter 18). This ambiguity between the natural, nutritive or erotic effusions of the female body, their interstitial and contaminating properties, and their function in procreation and as signals of female fluidity has been one of the overarching themes of this volume: Chapters 14 and 15 by Thea Lawrence and Laurence Totelin, for example, explore some of these anxieties and uncertainties about breastmilk across a range of Roman literary contexts.

Like blood, milk, and tears, and as John Wilkins in Chapter 13 on Galen has shown us, sweat was a multi-sensory phenomenon intimately connected to all the body's other fluids: it made itself visible to the eyes and olescent to the nose; but it was also something felt and tasted. It was a silent witness to fear and embarrassment and a characteristic of the brutish and the beastly. In order to become such a powerful diagnostic tool, sweat, like other bodily fluids, permeated the boundaries of the body, bringing the inside of the body to the outside. But unlike blood, semen, milk, phlegm, tears, and so on, sweat did not exude from visible orifices in the body over which there was any straightforward form of control. Sweat seeps out of the skin, a corporeal barrier with invisible holes (as Peter Kelly explores in Chapter 16 on dissolving bodies in Lucretius and Ovid, esp. pp. 265–8): for this reason, and because it is spontaneous and cannot be controlled, sweat functioned for some as a reliable diagnostic feature of behaviour, character and emotion—although actors and orators were sometimes able to feign even perspiration, as Gavrielatos' chapter shows us. At the same time, excessive sweating could demonstrate that the barriers between the inner body and the outside world were not functioning as they should—a corporeal state underpinning the theories of Mary Douglas and Mikhail Bakhtin about the foul, the dangerous, and the 'grotesque', and representing the same anxieties about fluid bodies that are examined in Chapters 2 and 3 by Rosalind Janssen and Irene Salvo on menstruation in this volume.[11] 'By simply issuing forth', Douglas argues, these fluids 'have traversed the boundary of the body' at its most vulnerable points (1966: 122). These fluids are especially pollutive, argued Valerio Valeri in *The Forest of Taboos* (1999: 102), when they threaten to gain entry back into the body and cause it to decay and die along with them.

These ideas about the pollutive capacity of the body were developed further by William Miller in his influential monograph *The Anatomy of Disgust*:

> When our inside is understood as soul the orifices of the body become highly vulnerable areas that risk admitting the defiling from the outside. But when our inside is understood as vile jelly, viscous ooze, or a storage area for excrement, the orifices become dangerous as points of emission of polluting matter, dangerous both to us and to others. Not all of the orifices are equally dangerous in their power to pollute or vulnerable in their capacity to admit pollution, nor are they the only sources of secretions that are polluting: the skin is home, after all, to glands producing sweat and oil, which have varying capacities to elicit disgust.
>
> Miller 1997: 89

The tension between the spiritual body as a repository for the soul and the profane body as an unreliable container for 'vile jelly', 'viscous ooze', and excrement did not diminish in later ancient periods. Indeed, anxieties around the porous, unbounded body intensified following the shift towards ascetic self-discipline and counter-eroticism that dominated Christian thought from the fourth century onwards. Susan Ashbrook Harvey has emphasised the increasing prominence of bodily experience and sensory engagement in the behaviour and discourse of late antique Christians over their earlier counterparts (2001: 90). The Christian reconceptualisation of what the body was, what it was for, and how it could be reconciled with an existence that extended far beyond corporeal birth and death into eternity, was not a straightforward process.

In the early fifth century CE, Augustine argued that the virtue of martyred saints who were violently damaged or dismembered would be preserved; they would rise physically restored but marked by seams, with the glory of suffering inscribed on their skin (*The City of God*, edition Dombart 1905–8, 22.19). But what about toe- and finger-nail clippings, and hair cut over a lifetime? Would this be reunited with the resurrected body? What happens when parts of the body are separated and are irretrievable? Which body is favoured for resurrection, the young, middle-aged, or old? And are imperfect bodies resurrected, such as those marked by deformities, or fat or thin bodies? Augustine attempts to give conclusive answers, but these invariably raise more questions about the awkward body that now has to transcend two physical realms rather than one.

By the mid-seventh century, these intransigent questions were still being asked, revealing the inability of early Christian thinkers to uncomplicate the body. Braulio of Saragossa warned against inquiring too far into the spiritual properties of fluids once emitted and separated from their body of origin to avoid transgressing into superstition, like those who ask about the eternal status of aborted foetuses, menstrual blood, and semen. Braulio understood that although blood would once again course through resurrected veins, not all superfluous fluids emitted over the course of an earthly life would resurge, especially those humours 'by which corruptions are born or vices generated' (*Letter* 42, edition Migne 1863b).[12]

Reconciling the fluid-producing body, which muddled gender and enflamed desire, with the spiritually perfect body preoccupied late ancient and early Medieval somatic thought. Flesh, whose acquaintance with filth begins in the mother's dirty innards, is the occasion and condition of redemption, as Jennifer A. Glancy observes (2010: 127). As in earlier periods, bodily fluids like semen, vomit, tears, and menstrual blood were perceived as messy and troublesome, provoking horrified revulsion to fascinated voyeurism. Excessive attempts to control and subdue the physical self paradoxically only emphasised bodily function and its resistance to regulation, making the obtrusion of the body inevitable.

In her theory of the material turn in Late Antique Christianity, Patricia Cox Miller stresses the diminishing separation between body and spirit, where physical bodies were invested with the holy (2009: 3–9). Especially in hagiographical narratives, the irrepressible, fluid-producing body becomes a locus of spirituality. Moments where the borders of the body dissolve and fluids become conspicuous function as pressure-points or indicators of particular sanctity, as we have seen in Laurence Totelin's chapter (pp. 247–9). Felicity's maternal martyred body is in a transitional state from occupation by the prenatal foetus to the autonomy of the postnatal infant, facilitated by breastfeeding. Totelin highlights how the awkward placing of Perpetua and Felicity's bodies with their offspring, not quite occupying the same body but recently separated and still co-dependent, generates tension and anxiety. Thea Lawrence's contribution examines the paternalistic regulation of the female body through breastfeeding in earlier Roman periods (pp. 224–39), and we see this pattern here in the removal of the infant from Perpetua by her father, and divine intervention to ensure that Perpetua did not feel pain through breast engorgement or fever.

The miraculous regulation of the ebb and flow of women's bodies, often involving some form of correction, is a trope that extends far beyond early martyr narratives. In the Life of Symeon Stylites the Younger, which dates from the late sixth or early seventh centuries, Symeon's mother Martha has visions of John the Baptist instructing her on how to conceive and feed a baby, including only feeding him from her right breast (*Life of Symeon Stylites the Younger* 3). Martha's resistance results in bodily dysfunction and deformity; when she tries to feed Symeon from her left breast, it shrivels up (*Life of Symeon Stylites the Younger* 4). The control of female physicality through miraculous maleness extends even to Symeon as a baby, who refuses to drink her milk whenever she eats sacrificial meat (*Life of Symeon Stylites the Younger* 6). Childbirth, menstruation, and lactation can make female bodily integrity appear more fragile and easily compromised, and Helen King's chapter (Chapter 24) shows how women were represented with a lesser capacity to control their fluids (pp. 381–98).

Yet women were not only subject to regulation through bodily fluids and indeed mobilised them in bodily resistance. Attempts to rebaptise a consecrated virgin into Arianism are repulsed when she spontaneously begins to menstruate, staining the baptismal waters with blood (Gregory of Tours, *History of the Franks* 2.2, edition Latouche 1963–5). Similarly, a widow dissuades a monk against raping her by describing the bad odours exuding from her body while menstruating (John Moschus, *Spiritual Meadow* 87.3, edition Migne 1863a, *caput* 205). Gorgonia, the sister of Gregory of Nazianzus, shrunk from the treatment and gaze of male physicians during illness and instead cured her own disease by mixing her tears with the Eucharistic elements at the altar and smearing the 'ointment' on her body (Gregory of Nazianzus, *Oration* 8.35, edition Migne 1857: 789–818). Especially in erotic, reproductive, and violent contexts, exceptionally holy women could use their fluids exceptionally, subverting shame and expending fluids to oppose attempts to reduce their agency. The body, especially the generative female body, can therefore both disrupt and affirm holiness, and the ambivalence of the body remains difficult to resolve with the ideal of Christian piety. More broadly, discomfort with a body lacking clearly defined boundaries or conspicuous with fluids transcends Classical, Late Ancient, and Medieval receptions.

It is with Miler's 'varying capacities' of bodily fluids, from the repulsive to the miraculous, that we would like to close this volume. We might like to think that fluids expelled from the body are biologically predetermined as sources of pollution, but—as this volume has demonstrated—it is not as simple as that. Fluids served as powerful signs that could be used in diverse ways in medical, political, moral, and biographical dialogues, a set of diagnostics with pollution and disgust at one end, and the miraculous and divine at the other. The interpreter carried significant agency in determining the value of these signs and using them to explore cultural anxieties about bodies and the identity and status of those who inhabited them. Through these interpretations, ancient bodies could be regulated and hierarchies established. This volume, then, has revealed how the fluid-producing body calls attention to itself through fantastic, gory, and transcendental fluids pushing to the textual surface and put to spectacular use, not to be ignored.

Notes

1 Unless otherwise stated, all translations are our own.
2 Plutarch, *Table Talk* 1.6.6. On sweat and illness, see the Hippocratic Corpus, e.g. *Aphorisms* 4; *Epidemics* 1.2–4; Celsus, *On Medicine* 2.5; Virgil, *Aeneid* 9.778. Cf. Plutarch, *Aemilius Paullus* 25, on horses reeking of sweat.

3 A story also recounted by Suetonius, *Life of Caesar* 7, and Dio Cassius 37.52.2.

4 William W. Fortenbaugh et al. (2003) published a critical text, translation, and commentary on this treatise, in a volume that also examined Theophrastus' treatises *On Dizziness* and *On Fatigue*, which offers a comprehensive interpretation of this text.

5 Theophrastus, *On Sweat* 3: 'It is clear from many things that [bad odour is caused] by the bad condition [of the body]: from those who are sick and from those who, being rather frequently engaged in sexual activity, are already in a [bad] condition, and generally those who [as convalescents] are already attending to [the body].'

6 Hippocratic Corpus, *On Internal Affections* 49.

7 A wide-ranging recent overview of the physiology of sweat, with a focus on 'hyperhidrosis' (excessive sweating) is Loureiro et al. (2018), especially the chapter by S. Neves.

8 Churchill's words and their classical origins are examined by Crum (1947).

9 Celsus (*On Medicine* 2.4–5) provides much detail about the relationship of different types of sweat to particular ailments and diseases.

10 Suetonius, *Nero* 24.1; *Historia Augusta, The Two Maximini* 4.3.

11 Douglas 1966; Bakhtin 1968.

12 For a discussion of Medieval and Early Modern blood, holiness, and martyrdom, see Anastasia Stylianou's chapter (Chapter 22) in this volume.

Bibliography

Primary sources

Dombart, B. (1905–8) *Sancti Aurelii Augustini Episcopi De Civitate Dei Libri XXII, Bibliotheca Scriptorum Graecorum et Romanorum Teubneriana*. 2 vols. Leipzig: Teubner.

Fortenbaugh, W. W., R. W. Sharples and M. G. Sollenberger (2003) *Theophrastus. On Sweat, On Dizziness and On Fatigue*. Leiden: Brill.

Latouche, R. (1963–5) *Histoire des Francs*. 2 vols. Paris: Belles Lettres.

Migne, J.-P. (1857) Gregory Nazianzus, *Oratio 8. Patrologia Graeca*, vol. 35, 789–818.

———. (1863a) John Moschus, *Pratum spirituale, Patrologia Graeca*, vol. 87.3, 2847–3116.

———. (1863b) Braulio of Saragossa, *Epistula 42 to Taius, Patrologia Latina*, vol. 80, 687–90.

van den Ven, P. (1962) *La vie ancienne de S. Syméon le jeune (521–592)*, vol. 1. Brussels: Société des Bollandistes.

Secondary literature

Ashbrook Harvey, S. (2001) 'On holy stench: When the odor of sanctity sickens', *Papers Presented at the Thirteenth International Conference on Patristic Studies Held in Oxford, 1999*. Vol. 2. Leuven: Peeters, 90–101.

Bakhtin, M. M. (1968) *Rabelais and His World*. Translated by H. Isvolsky. Cambridge, MA: MIT Press.

Cox Miller, P. (2009) *The Corporeal Imagination: Signifying the Holy in Late Ancient Christianity*. Philadelphia, PA: University of Pennsylvania Press.

Crum, R. H. (1947) 'Blood, sweat and tears', *The Classical Journal* 42(5), 299–300.

Douglas, M. (1966) *Purity and Danger: An Analysis of Concepts of Pollution and Taboo* (reprinted 2002 with additional preface). London: Routledge.

Glancy, J. A. (2010) *Corporal Knowledge: Early Christian Bodies*. Oxford: Oxford University Press.

Loureiro, M. de P., J. R. M. de Campos, N. Wolosker, and P. Kauffman (eds) (2018) *Hyperhidrosis: A Complete Guide to Diagnosis and Management*. New York: Springer.

Miller, W. I. (1997) *The Anatomy of Disgust*. London: Harvard University Press.

Valeri, V. (1999) *The Forest of Taboos: Morality, Hunting, and Identity among the Huaulu of the Moluccas*. Madison: University of Wisconsin Press.

INDEX